Lecture Notes in Engineering

The Springer-Verlag Lecture Notes provide rapid (approximately six months), refereed publication of topical items, longer than ordinary journal articles but shorter and less formal than most monographs and textbooks. They are published in an attractive yet economical format; authors or editors provide manuscripts typed to specifications, ready for photo-reproduction.

Lecture Notes in Engineering

Edited by C. A. Brebbia and S. A. Orszag

19

Finite Rotations in Structural Mechanics

Proceedings of the Euromech Colloquium 197, Jabłonna, Poland, 1985

Edited by
W. Pietraszkiewicz

Springer-Verlag Berlin
Heidelberg GmbH

Series Editors
C. A. Brebbia · S. A. Orszag

Consulting Editors
J. Argyris · K.-J. Bathe · A. S. Cakmak · J. Connor · R. McCrory
C. S. Desai · K.-P. Holz · F. A. Leckie · G. Pinder · A. R. S. Pont
J. H. Seinfeld · P. Silvester · P. Spanos · W. Wunderlich · S. Yip

Editor
Wojciech Pietraszkiewicz
Institute of Fluid-Flow Machinery
of the Polish Academy of Sciences
ul. Fiszera 14
80-952 Gdańsk, Poland

ISBN 978-3-540-16737-2 **ISBN 978-3-642-82838-6 (eBook)**
DOI 10.1007/978-3-642-82838-6

2161/3020-543210

PREFACE

The deformation near a material particle of the classical continuum is produced by successive superposition of a rigid-body translation, a pure stretch along principal directions of strain and a rigid-body rotation of those directions. The rotational part of deformation is particularly important in the non-linear analysis of thin-walled solid structures such as beams, thin-walled bars, plates and shells, since in this case finite rotations may appear even if the strains are infinitesimal.

It seems that the research concerning the application of finite rotations is carried out independently in different fields of structural mechanics. Theoretical and numerical methods developed and the results obtained for a particular type of the structure or for a particular material behaviour not always are used to analyse similar problems for other types of structures or for another material behaviour.

Since the research in this field had been growing rapidly, it was decided to organize an informal international meeting, under the auspices of the European Mechanics Committee, entitled: Euromech Colloquium 197 "Finite Rotations in Structural Mechanics". The meeting was held on 17 - 20 September 1985 in Jabłonna, a small suburbian area of Warsaw. The aim of this Colloquium was to bring together scientists, mainly from European countries, and provide them with an opportunity to discuss their current research activities concerning the role of finite rotations in the non-linear analysis of thin-walled structures. The goal of this research was to provide more accurate theoretical models for the behaviour and to develop more efficient methods for computerized analysis of flexible beams, thin-walled bars, plates and shells in the non-linear range of deformation. The Colloquium was the first international meeting devoted entirely to this field of research.

Participants from ten countries attended the Colloquium and 33 lectures were presented and discussed during the meeting. 12 lectures were delivered by participants from the Federal Republic of Germany, 6 from Poland, 4 from the USSR, 3 from France, 2 from the USA, 2 from Denmark and participants from Canada, Japan, Hungary and West Berlin presented one lecture each.

This volume contains 26 selected and revised papers which (except one) were presented and discussed during the meeting. The paper by K.F. Chernykh was sent to the organizers already before the meeting, but was not presented and discussed, because the author was not able to come. Since two other prospective contributors to the volume were not able to finish their revised manuscripts before the last deadline (mid-February 1986) it seemed reasonable to include the report by K.F.Chernykh into the free place.

The themes of the original papers published in this volume contain several non-classical formulations of the non-linear theory of shells and bars as well as various associated variational principles, expressed explicitly or implicitly in terms of rotations, the results of non-linear numerical analysis of elastic and inelastic thin-walled structures with the use of rotations as well as some problems of solid mechanics associated with the notion of a finite rotation. The volume should provide the reader with a unique source of information about the current status and trends of the research in the field of application of finite rotations in the non-linear structural analysis.

I would like to express gratitudes to the European Mechanics Committee for approving this original Colloquium and for providing me with important guidelines, based on the vast experience with organization of almost 200 meetings of this kind. I am indebted to the Polish Academy of Sciences for some financial support and for allowing the organizers to use the House of Meetings and Conferences in Jabłonna, where convenient and inexpensive accomodation with full board was available and where the scientific sessions of the Colloquium took place. I am also grateful to Springer-Verlag for their courteous and effective production of these Proceedings.

W. Pietraszkiewicz

EUROMECH COLLOQUIUM NR 197

FINITE ROTATIONS IN STRUCTURAL MECHANICS

17-20 September 1985, Jabłonna, Poland

LIST OF PARTICIPANTS

J.ARBOCZ, Prof. Dr.
T.H. Delft
Department of Aerospace Engineering
Kluyverweg 1
2629 HS Delft, Holland

S.N.ATLURI,Prof.
Georgia Institute of Technology
Center for Advancement of Computational Mechanics
School of Civil Engineering
Atlanta, Georgia 30332,USA

E. AXELRAD,Prof. Dr.
Probst.-Heinrich-Str. 4
8000 München 40, FRG

J.BADUR, Dr.inż.
Instytut Maszyn Przepływowych PAN
ul.Fiszera 14
80-952 Gdańsk, Poland

W.BARAŃSKI, Doc.Dr.hab.inż.
Politechnika Łódzka
Instytut Inżynierii Budowlanej
Al.Politechniki 6
93-590 Łódź, Poland

M.BERNADOU, Prof.
Institute National de Recherche en Informatique et en Automatique
Domaine de Voluceau-Rocquencourt
B.P. 105-78153 Le Chesnay Cedex
France

J.BIELSKI, mgr inż.
Politechnika Krakowska
Instytut Mechaniki i Podstaw Konstrukcji Maszyn
ul.Warszawska 24
31-155 Kraków, Poland

W.R. BIELSKI, Dr
Instytut Geofizyki PAN
ul.Pasteura 3
02-093 Warszawa, Poland

R. de BOER, Prof.Dr.-Ing.
Universität Essen
FB Bauwesen
Postfach 6843
Universitätstr. 15
D-4300 Essen 1, FRG

H.BUFLER, Prof.Dr.-Ing.
Universität Stuttgart
Institut für Mechanik /Bauwesen/
Lehrstuhl II
Pfaffenwaldring 7
7000 Stuttgart 80, FRG

Cheng YUAN-SHENG, Prof.
Institute of Applied Mathematics and Mechanics
Shanghai University of Technology
149 Yan Chang Lu
Shanghai 20 00 79, China

A.COMBESCURE, Ing.
Centre d'Etudes Nucleaires de Saclay
DEMT/SHTS/LAMS
F-91191 GIF-SUR-YVETTE Cedex
France

N.DAHAN, Dr
Laboratoire de Mechanique et Technologie
ENSET/UNIVERSITE PARIS 6/CNRS
61, Avenue du President-Wilson
94230 Cachan,France

M.C.DÜKMECI, Prof.Dr.-Ing.
Instanbul Teknik Universitesi P.K.9
Instanbul, Turkey

F. A. EMMERLING
Hochschule der Bundeswehr München,
FB LRT
Institut für Mechnik
Werner-Heisenberg-Weg 39
D 8014 Neubiberg, FRG

M.EPSTEIN, Prof.
The University of Calgary
Faculty of Engineering
Department of Mechanical Engineering
2500 University Drive N.W.
Calgary, Alberta, Canada T2N 1NA

O.A.GAJL, Dr.inż.
Politechnika Łódzka
Instytut Inżynierii Budowlanej
Al.Politechniki 6
93-590 Łódź,Poland

T. HINKELMANN, Dipl.-Ing.
Institut für Statik
Universität Hannover
Callinstr. 32
3000 Hannover 1, FRG

M. IURA, Dr.
Dept.of Civil and Structural Engg.
Tokyo Denki University
Hatoyama, Hiki-Gun
Saitama 350-03, Japan

S. KONIECZNY, doc. dr hab.inż.
Politechnika Łódzka
Instytut Inżynierii Budowlanej
Al.Politechniki 6
93-590 Łódź, Poland

S.KRENK, Prof.Dr.
Department of Structural Engineering/ABK
Building 118, Lundtoftevej 100
DK 2800 Lyngby, Denmark

V.N.KUKUDŻANOV, Prof. D.Ph.-M.N.
Institute of Problems of Mechnics
Academy of Science USSR
Vernadski Avenue 101
Moscow 117526, USSR

G.LA'MER,Prof.
Technical University
Department of Civil Engineering Mechanics
H-1521, Budapest XI
Müeqyetem RKP. 3.
Hungary

K.-H. LAMBERTZ, Dipl.-Ing.
Universität Hannover
Institut für Baumechanik
und Numerische Mechanik
Callinstr. 32
D-3000 Hannover, FRG

J.MAKOWSKI, Dr inż.
ul.Chabrów 10 m 25
45-221 Opole, Poland

A. MATZENMILLER,Dipl.-Ing.
Universität Stuttgart
Institut für Baustatik
Pfaffenwaldring 7
D-7000 Stuttgart, Postfach 1140,FRG

B.MICHALAK, dr inż.
Politechnika Łódzka
Instytut Inżynierii Budowlanej
Al.Politechniki 6
93-590 Łódź, Poland

H. MØLLMANN, Dr
Department of Structural Engineering
Building 118
Technical University of Denmark
2800 Lyngby, Denmark

L.-P. NOLTE, Dr.-Ing.
Ruhr-Universität Bochum
Institut für Mechanik
Postfach 102148
Universitätsstr. 150 IA 3/125
4630 Bochum 1, FRG

W. PIETRASZKIEWICZ,prof.dr hab.inż.
Instytut Maszyn Przepływowych PAN
ul.Fiszera 14
80-952 Gdańsk, Poland

L. RECKE, Dipl.-Ing.
Ruhr-Universität Bochum
Institut für Konstruktiven Ingenieurbau
Lehrstuhl IV, Postfach 102148
Universitätsstr. 150 1A-6/151
4630 Bochum 1, FRG

E.REISSNER, Prof. Dr
B-D10
University of California
San Diego
La Jolla, CA 92093, USA

M. SAJE, Doc.Dr.Ing.
Edward Kardelj University
Faculty for Architecture, Civil Engineering and Survey
Jamova 2
61000 Ljubljana, Yugoslavia

R. SCHMIDT, Dr.-Ing.
Institute of Civil Engineering
University of Wuppertal
D-5600 Wuppertal 2
FRG

H.K. SCHOOP, Prof. Dr.-Ing.
Technische Universität Berlin
2 Institut für Mechanik - FB 9
1 Berlin 12 - Jebensstr. 1
West Berlin

M. SEKULOVIČ, Prof. Dr.
Gradjevinski Fakultet
Bul.Revolucije 73
11000 Beograd
Yugoslavia

L.I. SHKUTIN, Dr.
Computer Center SB of the USSR
Academy of Sciences
660036 Krasnoyarsk, USSR

J. SKRZYPEK, doc. dr hab. inż.
Politechnika Krakowska
Instytut Mechaniki i Podstaw
Konstrukcji Maszyn
ul.Warszawska 24
31-155 Kraków, Poland

S. SRPČIČ, Ing.
Edward Kardelj University
Faculty for Architecture
Civil Engineering and Survey
Jamova 2
61000 Ljubljana
Yugoslavia

E. STEIN, Prof.Dr.-Ing.
Universität Hannover
Institut für Baumechanik und
Numerische Mechanik
Callinstr. 32
D-3000 Hannover,FRG

H. STUMPF, o.Prof.Dr.-Ing.
Ruhr-Universität Bochum
Lehrstuhl für Mechanik II
Universitätsstr. 150
4630 Bochum 1, FRG

M.L. SZWABOWICZ, Dr inż.
Instytut Maszyn Przepływowych PAN
ul.Fiszera 14
80-952 Gdańsk, Poland

J.J. TELEGA, Dr
Instytut Podstawowych Problemów
Techniki PAN
ul.Świętokrzyska 21
00-049 Warszawa, Poland

I.T. TEREGULOV, Prof.
Kazań Engineering Construction Institute
1 Zelyonaya Str.
Kazań, 420043 Tartar Autonomous Republic
USSR

A. TESÁR, Dr.-Ing.
Institute of Structures and Architecture
Dúbravská Cesta
842 20 Bratislava
Czechoslovakia

R.VALID, Prof.
ONERA - BP 72
92322 Chatillon Cedex, France

N.V. VALISHVILI, Prof. Dr.
Av. Rustaveli 104, 30
Kutaisi, USSR

A. VLAHINOS, Dr
Department of Civil Engineering
University of Colorado at Denver
1100 14th St., Denver
CO. 80202 USA

W. WALTHER, Dr.-Ing.
Universität Essen
Fachbereich 10-Bauwesen
4300 Essen 1, FRG

F. Y.-M. WAN
University of Washington
Seattle, WA 98195, USA

D. WEICHERT, Dr.-Ing.
Ruhr-Universität Bochum
Lehrstuhl für Mechanik II
Universitätsstr. 150
4630 Bochum 1, FRG

G. WEMPNER, Prof.
School of Engineering,
Sciences and Technology
Georgia Institute of Technology
Atlanta, GA 30340, USA

Cz. WOŹNIAK, Prof. dr. hab.inż.
Uniwersytet Warszawski
Instytut Mechaniki
PKiN, p. 935
00-901 Warszawa, Poland

W.WUNDERLICH, Prof. Dr.-Ing.
Ruhr-Universität Bochum
Institut für Konstruktiven Ingenieurbau
Lehrstuhl IV
Postfach 102148
Universitätsstr. 150
4630 Bochum 1, FRG

EUROMECH COLLOQUIUM Nr 197

FINITE ROTATIONS IN STRUCTURAL MECHANICS

17-20 September 1985, Jabłonna, Poland

P R O G R A M

Tuesday, 17 September

9.00 - 9.10	Opening of the Colloquium
	THEORY OF SHELLS
9.10 - 9.40	*R. Valid,* Finite rotations, variational principles and buckling in shell theory
9.40 - 10.10	*E. Axelrad, F.A. Emmerling,* Intrinsic shell theory formulation effective for large rotations
10.10 - 10.40	*R. de Boer, W. Walther,* Fundamental equations and extremum principles in the theory of thin shells
10.40 - 11.10	*J. Badur, W. Pietraszkiewicz,* On the non-linear theory of shells derived from Cosserat continuum with constrained micro-rotations
11.10 - 11.30	Coffee break
	CONTINUUM MECHANICS
11.30 - 12.00	*M. Epstein,* Inhomogeneity and rotation
12.00 - 12.30	*H. Bufler,* Finite rotations and complementary extremum principles
12.30 - 13.00	*W.R. Bielski, J.J. Telega,* On the really complementary energy principle in the case of finite deformations
13.00 - 14.30	Lunch break
	ONE-DIMENSIONAL PROBLEMS
14.30 - 15.00	*M. Iura,* Finite displacement theory of naturally curved and twisted beams with finite rotations
15.00 - 15.30	*L.-P. Nolte,* One-dimensional finite rotation shell problems in displacement formulation
15.30 - 16.00	*T. Hinkelmann, G. Lumpe, H. Rothert,* On a general theory of large rotations and small strains with applications to complex beam-in-space structures
16.00 - 16.30	*S. Krenk, O. Gunneskov,* Finite deformations of rotating blades with pretwist
16.30 - 17.00	Coffee break

PLATES

17.00 - 17.30	*E. Stein, K.-H. Lambertz, L. Plank,* Ultimate load analysis of thin-walled steel structures with elasto-plastic deformation properties using FEM
17.30 - 18.00	*L. Librescu, R. Schmidt,* Higher-order moderate rotation theories for elastic anisotropic plates
18.00 - 18.30	*H. Møllmann,* Theory of thin-walled elastic beams with finite displacements

Wednesday, 18 September

ANALYSIS OF SHELLS

9.00 - 9.30	*W. Wunderlich, L. Recke,* Rotations as primary unknowns in the non-linear theory of shells and corresponding finite element models
9.30 - 10.00	*N.V. Valishvili,* On non-linearity problems of revolution shells
10.00 - 10.30	*L.-P. Nolte,* On the derivation and efficient computation of large rotation shell models
10.30 - 11.00	*E. Ramm, A. Matzenmiller,* Large rotation analysis of shells by isoparametric, degenerated finite elements
11.00 - 11.30	Coffee break

THEORY OF SHELLS

11.30 - 12.00	*J. Makowski, H. Stumpf,* Finite strains and rotations in shells
12.00 - 12.30	*G. Wempner,* Mechanics of shells with finite rotations and transverse shear strains
12.30 - 13.00	*M.L. Szwabowicz,* Admissibility of finite rotations in the nonlinear deformation of shells undergoing small strains
13.00 - 14.30	Lunch break

CONTINUUM MECHANICS

14.30 - 15.00	*C. Woźniak,* On the linear elasticity with finite displacements and rotations
15.00 - 15.30	*L.I. Shkutin,* Separation of the finite rotation field when constructing non-linear deformation models of thin bodies
15.30 - 16.00	*D. Weichert,* Elasto-plastic structures under variable loads at small strains and moderate rotations
16.00 - 16.30	*N. Dahan, M. Predeleanu,* Study of finite rotations in large torsional deformation
16.30 - 17.00	*G. Lamer,* Equations of motion of thin-walled structures undergoing large displacements
17.00 - 17.30	Coffee break

MEMBRANES

17.30 - 18.00 *F.Y.M. Wan*, New boundary layer phenomena in nonlinear membrane problems with finite rotations

18.00 - 18.30 *H. K. Schoop*, Membranes and shells from the viewpoint of plane reference

Thursday, 19 September

9.00 - 14.00 Touristic trip to Warsaw

14.00 - 16.00 Lunch break

16.00 - 18.00 Round-table discussion

18.30 - 20.30 Conference dinner

Friday, 20 September

SHELLS

9.00 - 9.30 *I. Teregulov*, Constitutive equations for physically non-linear anisotropic shells in case of finite strains and rotations

9.30 - 10.00 *M. Bernadou*, Variational formulations of some nonlinear thin shell problems

10.00 - 10.30 *R. Schmidt*, On the nonlinear theory of thin elastic shells with finite strains and rotations

10.30 - 11.00 Coffe break

11.00 - 11.30 *V.N. Kukudżanov*, On the finite deformation of the elastoplastic media with explicit use of a finite rotation

11.30 - 12.00 *J. Bielski, J. Skrzypek*, Optimal design of a geometrically non-linear elastic toroidal shell subject to external pressure and bending

12.00 - 12.10 Closing of the Colloquium

12.10 - 13.00 Lunch

CONTENTS

INTRINSIC SHELL-THEORY FORMULATION EFFECTIVE FOR LARGE ROTATIONS AND AN APPLICATION *)

E.L. AXELRAD and F.A. EMMERLING
Institut für Mechanik
Universität der Bundeswehr
D-8014 Neubiberg, F.R. Germany

Specialization of the general thin-shell theory for the class of large deformations realizable by small strain is considered. This leads in three different ways to the intrinsic nonlinear theory of flexible shells. As an illustration, large displacements and rotations culminating in collapse of finite-length tubes are investigated.

1. Introduction

The thin-shell theory is or must be (as an approximate theory) occupied with endeavours to obtain results important in applications.

Since two decades, in particular, since the rise of the finite element method, the nonlinear problems have become the prime concern of the shell-theory.

However, though it went without saying, a fundamental restriction prevailed. The nonlinear problems considered in the theory of (linearly elastic) shells were mostly those of prediction of stability and analysis of postbuckling. This is a consequence of the preoccupation with only a class of shells - with those designed for strength and stiffness. Such shells must bar out substantial displacements and nonlinearity before failure.

The other end of the scale of deformability is occupied by shells, which are designed for maximum displacements by small strain.

We witness a further shift of emphasis of the shell-theory research. Not any thinkable large displacements and rotations but preferably those actually realizable by small strain attract growing interest. The origin of this approach is clearly recognizable in the renowned pioneering work of REISSNER [1], MUSHTARI and GALIMOV [2] and others. Recent investigations of many a distinguished mechanicians such as KOITER [3,4,5], LIBAI and SIMMONDS [6,7] go further. The problems of shells, which can put into effect large deformations by small strain, develop the profile of the essential mission of the nonlinear theory.

*) This work is dedicated to the recent 70th birthday of Professor W.T. Koiter.

Such shells are employed in industry in highly diverse shapes for over a century (a review of their types and of the relevant literature can be found in [8]). The experience gained has facilitated the perception of the unity of the basic features of both geometry and stress state of all these shells. Despite the diversity of their shapes they constitute a clearly recognizable class. The term *flexible shells* (FS) proposed for this class (in 1976) appears to be accepted.

The kind of contribution of the shell theory desirable for the FS applications is manifested by the experience of the analysis of "stiff" shells. Most of its practical results are due not to the general theory but to the specialized branches. The membrane theory is applied for prebuckling, the Donnell-Koiter theory for stability and postbuckling. The indication is unequivocal. The FS can be effectively treated by means of specialized theoretical tools. It is in this sense we value as still up to date and highly precise the statement of SIMMONDS [7] "To mount a frontal assault on the shell problem - to solve the equations of motion for arbitrary geometry and arbitrary initial conditions - seems patently absurd".

The stability and the (bifurcational) postbuckling of FS shells is adequately covered by the Donnell-Koiter theory.

The situation is different for the working state of FS shells. Any strain-state assuring flexibility must be as far as possible from that of a stiff shell. - The membrane theory is irrelevant here.

Formulation of a theory appropriate for FS has been made easier by the analysis of the various individual problems of FS. It disclosed the immanent feature of any small-strain state rendering large displacements: such deformation varies much less strongly with one of the surface coordinate than with the other.

Thus, the FS problems occupy the ground between the domains of the membrane theory and the Donnell-Koiter theory. These classic theories are known to encompass strain states which vary, respectively, slowly or strongly with both coordinates.

In what follows we start (Sect. 2) with a brief discussion of the three mutually complementary approaches to the formulation of the FS theory.

Is a purely mathematical, as general as possible, statement of an approximate theory always optimal? The pure mathematics was seen by its authors (K. Gauss, for one) differently. Their view was formulated by HALMOS [9]: "The heart of mathematics consists of concrete problems". This judgement appears hardly less relevant for the shell theory.

In the following the FS theory is discussed (Sect. 3) in connection with the nonlinear flexure of thin-walled, both cylindrical and curved, tubes (Sect. 4). The flexure of finite tubes is one of a very few known to the authors investigated cases of two-dimensionally variable strain involving large rotations.

The formulation of the theory is strictly intrinsic. It does not require any explicit reference to displacements. The evaluation of displacements is also not needed for the boundary conditions.(The kinematic conditions are no exception).

Of course, the intrinsic formulation leans on the fundamental works [1-7]. This tradition goes back to the founders of shell theory. (The first break-through towards the intrinsic theory has been achieved in the memorable H. Reissner (1912) work. It has in fact introduced compatibility conditions, discovered in their generality by A.L. Goldenveizer in 1939. Further references can be found in [2,8,10]).

The intrinsic approach makes the problem involving finite, and, in particular large, rotations immensely more tractable. Moreover, as pointed out by SIMMONDS [11] the displacment form of the field equations can be illconditioned. And this just for the nearly inextensional bending which is immanent for small-strain large displacements.

For the stability analysis the flexibility causes serious complications. Both the prebuckling shape and stress state are substantially perturbed. On the other hand this perturbation leads to localization of the buckling modes. - The resistance to buckling is weakened in certain parts of the shell. Specialization of the analysis for the (initially) local buckling saves a great deal in difficulties. The local stability analysis is applied (in Sect. 4) to the evaluation of collapse of bent tubes.

2. On Small Strain Leading to Large Displacements

Obviously not any shape and-edge-constraints enable a shell to realize large displacements by small strain. This is possible only for certain type of situations. Specializing the general shell theory (consistently with its inherent accuracy) for the relevant strain-states leads to a virtually self-contained simplified theory of flexible shells (FS).

The inherent features of the FS deformation constitute the *basic hypothesis* of the relevant theory. The hypothesis can be formulated either as a set of physical assumptions or as a purely mathematical restriction on the variation of strain and the shell shape. Such formulations of the hypothesis and of the corresponding mutually equivalent versions of the theory have been presented previously [8,12-14]. In what follows we discuss the different approaches and formulations of the FS-theory in so far as they complement each other in exposing features of the flexible shells and their theory. Apparently the first set of physical assumptions sufficient to characterize small-strain large displacements can be discerned in the Th. v. Kármán - L. Beskin (1911-1945) [15,16](other references in [8,12]) linear analysis of curved tubes. These are three assumptions (notation of Fig. 1 and [14] is used):

A. The extensional strain along the coordinate lines x^2 and the wall-bending moment M_1 can be neglected in the analysis of strain and of equilibrium, respectively.

B. The shear-strain resultant γ and the torsional moment $H_1 = H_2 = H$ can be neglected, respectively, in the analysis of strain and equilibrium.

C. Those elasticity relations of the thin-shell theory required in the FS theory, can be used in the form simplified by neglecting the Poisson-factor terms.

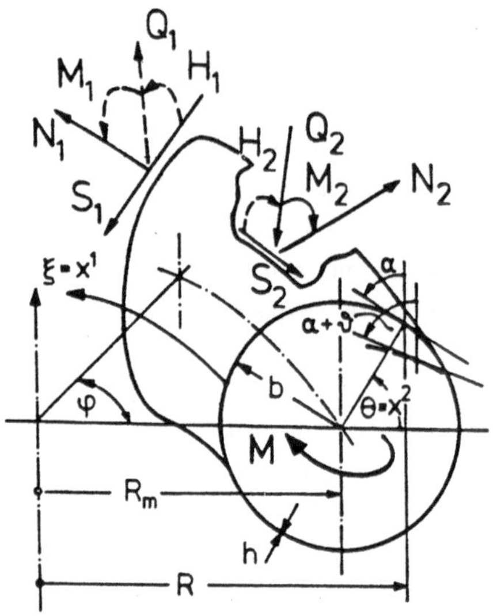

Fig. 1 Thin-walled tube. Coordinates and stress resultants.

For isotropic homogeneous shells

$$\varepsilon_1 Eh = N_1 - \nu N_2 \cong N_1, \quad M_2/D = \varkappa_2 + \nu\varkappa_1 \cong \varkappa_2, \quad D = Eh^3/[12(1-\nu^2)] \qquad (1).$$

The state of elastic deformation possessing the features A,B,C is extensionless, that is, purely flexural in the x^2 direction. At the same time it is nearly momentless or membrane in the x^1-direction. These features are described by several synonimous terms. Such a stress state may be properly termed semi-membrane, semi-extensionless or *semi-momentless*. The last named term sounds less ambiguously. It will be employed in what follows.

Assumptions identical to A,B,C have lead to the "semi-momentless" linear theory of cylindrical shells proposed by V.S. Vlassov and several other authors in 1932-1936.

GOLDENVEIZER [10] has shown for circular shells that the semi-momentless strain constitutes the main part of the deformation. It is the fraction determined by the small roots of the characteristic equation and thus varying slowly along the cylinder (with x^1). The complementary fraction was shown to vary with x^1 much (b/h times) stronger; it constitutes the edge effect.

Extended to shells of double curvature and to large displacements [17] the assumptions A,B,C had constituted the first basis for the FS theory.

NOVOZHILOV [18] pointed out in 1951 that the semi-momentless theory of cylindrical shells follows from the general shell theory for a stress state which varies much less strongly with x^1 than with x^2. It is, namely, the simplification consistently due for any stress state, in which all substantial strain and stress resultant (here represented by $F(x^1, x^2)$) satisfy the conditions:

$$\frac{|F,_{11}|}{(a_1)^2} \sim \frac{|F,_{22}|}{(a_2)^2} \quad \varepsilon, \quad \varepsilon << 1, \ ((\ldots),_\alpha = \frac{\partial(\ldots)}{\partial x^\alpha}), \tag{2}$$

where a_α, $\alpha = 1,2$ denotes the Lamé parameters $a_\alpha = |\underset{\sim}{r},_\alpha|$; $\underset{\sim}{r}$ is the radius vector of a point (x^1, x^2) on the middle surface.
The derivation of the FS-theory (from the general shell-theory) on the basis of a minimum number of purely mathematical restrictions is very desirable. It assures the consistency of the theory.

Consider briefly three different approaches to such a specialization of the shell theory based on (2). The first two differ in the kind of additional restriction imposed on the shape of the shells considered. (Just as (2) was originally proposed and is sufficient only for cylindrical shapes). The third approach involves besides (2) a restriction on the stress state.

(a) Initially axisymmetrical shells and those having the form of such a shell locally (as in Fig. 1). The FS equations for these shells can be obtained [8, 13] in a way which extends the consideration leading to the Reissner-type nonlinear equations (those of axisymmetrical deformation of shells of revolution and of curved beams [8,19]). The basis of this derivation is constituted by two integrals of the compatibility equations and two integrals of the equilibrium equations.
It renders *nonlinear* equations of Reissner type for *nonaxisymmetrical deformation*. Simplification consistent with (2) produces equations, which, significantly, are identical to those following from the general shell equations when the semi-momentless assumptions A,B,C are introduced. This shows equivalence of the semi-momentless shell model to the restriction (2). In other words, this model is adequate for any strain which satisfies the condition (2).

(b) For shells with a nonaxisymmetrical local shape the semi-momentless theory follows from general shell theory when the assumption (2) is introduced in conjunction with an appropriate restriction of the shell shape. Rather realistic appears the requirement that the local shape vary along the shell surface not more strongly than the stress state, being investigated (one satisfying the condition (2)). Consider briefly the relevant argumentation (its detailed presentation is in print elsewhere).
The mentioned, additional to (2), restriction is imposed by conditions

$$(\underset{\sim}{r},_\alpha F),_\beta \sim \underset{\sim}{r},_\alpha F,_\beta \qquad (\alpha,\beta = 1,2) \tag{3}$$

being applied for estimating the small terms (those with $F = \varepsilon_2, M_1, \gamma, H$), which are to be neglected in the specialized theory.

We start with the compatibility and equilibrium equations of the nonlinear shell theory in the vector form ([20]*), (1.33), (1.57)):

$$(a_1\underset{\sim}{\varkappa}_1)_{,2} - (a_2\underset{\sim}{\varkappa}_2)_{,1} + a_1a_2\underset{\sim}{\varkappa}_1 \times \underset{\sim}{\varkappa}_2 = \underset{\sim}{0}, \tag{4}$$

$$(a_1\underset{\sim}{\varepsilon}_1)_{,2} - (a_2\underset{\sim}{\varepsilon}_2)_{,1} - a_1a_2(\underset{\sim}{t}'_1 \times \underset{\sim}{\varkappa}_2 - \underset{\sim}{t}'_2 \times \underset{\sim}{\varkappa}_1) = \underset{\sim}{0}, \tag{5}$$

$$(a_1\underset{\sim}{N}_2)_{,2} + (a_2\underset{\sim}{N}_1)_{,1} + |\underset{\sim}{r}_{,1} \times \underset{\sim}{r}_{,2}|\, \underset{\sim}{q} = \underset{\sim}{0}, \tag{6}$$

$$(a_1\underset{\sim}{M}_2)_{,2} + (a_2\underset{\sim}{M}_1)_{,1} + a_2\underset{\sim}{r}^*_{,1} \times \underset{\sim}{N}_1 + a_1\underset{\sim}{r}^*_{,2} \times \underset{\sim}{N}_2 = \underset{\sim}{0}. \tag{7}$$

The vector parameters $\underset{\sim}{\varkappa}_\alpha$, $\underset{\sim}{\varepsilon}_\alpha$ and $\underset{\sim}{N}_\alpha$, $\underset{\sim}{M}_\alpha$ are composed of the strain and stress resultants. With the convenient basis $\underset{\sim}{t}'_\alpha \approx \underset{\sim}{t}^*_\alpha = \underset{\sim}{r}^*_{,\alpha}/a^*_\alpha$, $\underset{\sim}{n}^* = \underset{\sim}{t}'_1 \times \underset{\sim}{t}'_2/|\underset{\sim}{t}'_1 \times \underset{\sim}{t}'_2|$ and $\underset{\sim}{t}'^1 = -\underset{\sim}{n}^* \times \underset{\sim}{t}'_2$, $\underset{\sim}{t}'^2 = \underset{\sim}{n}^* \times \underset{\sim}{t}'_1$ we have, in particular:

$$\underset{\sim}{\varkappa}_\alpha = \underset{\sim}{n}^* \times (\varkappa_\alpha \underset{\sim}{t}'_\alpha + \tau_\alpha \underset{\sim}{t}'^\beta) + \underset{\sim}{n}^*\lambda_\alpha, \quad \underset{\sim}{\varepsilon}_\alpha = \varepsilon_\alpha \underset{\sim}{t}'_\alpha + \frac{\gamma}{2} \underset{\sim}{t}'^\beta,$$
$$\underset{\sim}{M}_\alpha = \underset{\sim}{n}^* \times (M_\alpha \underset{\sim}{t}'_\alpha + H_\alpha \underset{\sim}{t}'^\beta), \quad \underset{\sim}{N}_\alpha = N_\alpha \underset{\sim}{t}'_\alpha + S_\alpha \underset{\sim}{t}'_\beta + Q_\alpha \underset{\sim}{n}^*. \qquad (\alpha \neq \beta) \tag{8}$$

We estimate the relative order of magnitude of the terms with ε_2 and those with γ, which appear in equation (4) after elimination of the $\underset{\sim}{n}^*\lambda_\alpha$ components of $\underset{\sim}{\varkappa}_\alpha$ with the help of (5). Using the relations (3) it is not difficult to find that this order is $\nu\varepsilon$ for ε_2 and $(E/G)\varepsilon = 2(1+\nu)\varepsilon$ for γ(with the quantity ε defined in (2)). Similarly, the order of the M_1 and H_α-terms in (6) is $\nu\varepsilon$ and $2(1-\nu)\varepsilon$, respectively.

The error of the semi-momentless assumptions A and B is thus disclosed to be indeed of the order of that in the basic hypothesis of the FS theory (2).

Further analysis shows (as was actually done in a different way in the preceding subsection (a)) that relations (1) assumed under C are a consequent outcome of those in A. And this without any relevance to the assumptions A being justified or not.

This conclusion is arrived at in the course of a variational derivation of the elasticity relations on the lines of [20] and of a statically-geometrically analogous deliberations. Simultaneously, this renders the simplified elastic-energy expressions of the FS-theory and the boundary conditions at x^1 = const , reduced for this theory.

---

*) Equations analogous to (4), (5) for the "bending" $\underset{\sim}{k}_\alpha$ as well as the corresponding tensor-component scalar equations can be found in the monograph [6], Libai and Simmonds (eqs. (3.54),(3.58) and (3.63), (3.64) in [6]). However, the meaning of $\underset{\sim}{k}_\alpha$ is different from $\underset{\sim}{\varkappa}_a$. Specifically, so with respect to the $\underset{\sim}{n}^*$-component.

The boundary conditions at the edges x^2 = const are those (four) of the general theory.

In contrast to the edges x^2 = const there are only two conditions to be fulfilled on the edges x^1 = const. The variational analysis indicates that these conditions can be imposed on the following resultants of the edge stress and strain, respectively:

$$N_1,\ S_1,\ \varkappa_2,\ \tau_2. \tag{9}$$

(c) A simple, purely mathematical, step leads to the FS-theory from the vector complex shell-equation. This equation has been obtained [14] as Novozhilov complex equations [18], extended to include nonlinearity and presented in a vector form:

$$(a_2\tilde{\underset{\sim}{N}}_1)_{,1} + (a_1\tilde{\underset{\sim}{N}}_2)_{,2} + ih'\ \Big(\underline{\frac{\partial}{\partial x^1} \frac{a_2}{a_1} \underset{\sim}{n}^* \frac{\partial}{\partial x^1}} + \frac{\partial}{\partial x^2} \frac{a_1}{a_2} \underset{\sim}{n}^* \frac{\partial}{\partial x^2} \Big)(\tilde{N}_1 + \tilde{N}_2) =$$

$$= (C\, \underset{\sim}{\varkappa}_1 \times \underset{\sim}{\varkappa}_2 - \underset{\sim}{q})\, a_1 a_2, \tag{10}$$

where the coordinates x^α are orthogonal and

$$\tilde{\underset{\sim}{N}}_1 = \underset{\sim}{N}_1 + C\underset{\sim}{\varkappa}_2,\ \tilde{\underset{\sim}{N}}_2 = \underset{\sim}{N}_2 - C\underset{\sim}{\varkappa}_1,\ \tilde{N}_\alpha = N_\alpha - C\varkappa_\beta,\ (\alpha \neq \beta),$$

$$C = \sqrt{-EhD},\ h' = h/\sqrt{12(1-\nu^2)}.$$

The equation (10) results from (4) - (7) and the (full) elasticity relations of the shell theory under the following assumptions (which express those leading in the linear theory to the Novozhilov complex equations):

$$(1 + i\,\frac{h'}{R})\,[\underset{\sim}{N}_\alpha\ \underset{\sim}{M}_\alpha] \cong [\underset{\sim}{N}_\alpha\ \underset{\sim}{M}_\alpha]\,. \tag{11}$$

These relations can be decoded as a requirement that flexural and torsional stresses be not too different in their order of magnitude from the respective membrane stress

$$\frac{h'}{R_{min}} \ll \left|\frac{N_\alpha/h}{E\varkappa_\beta h/2}\right|,\ \left|\frac{S_\alpha/h}{E\tau_\alpha h/2}\right| \ll \frac{R_{min}}{h'}\ ,\ \beta \neq \alpha \tag{12}$$

with $1/R_{min}$ denoting the minimum of the normal-section curvature. The eq. (9) is reducible to a set of six scalar equations for six real dependent variables N_1, N_2, $S = S_\alpha$, $\varkappa_1$, $\varkappa_2$, $\tau = \tau_\alpha$.

To obtain the FS equations - those of the semi-momentless stress state - one needs only to drop the underlined term in (10). The affinity of such a simplification to that in (3) is telling.

Thus, we have at our disposal a consequent simple nonlinear theory specialized for the clearly delimited class of problems, those encountered in the analysis of large shell-deformations realizable by small strain. Whether a specific problem can be adequately investigated with the aid of the FS theory must be usually recognizable without calculations. It is always possible to check a solution obtained by applying to it the criterium (2).

3. Nonlinear Flexure of Elastic Tubes

Consider circular tubes bent by forces, applied on the edges, which are stiffened by flanges (Fig. 2). For the sake of simplicity the two flanges will be identical.

Besides the straight tubes, i.e. cylindrical shells, the initially curved tubes will be investigated. However, as the nonlinearity of flexure is more pronounced for straight tubes and those with small initial curvature, we concentrate on tubes which are "slender", satisfying the condition *)

$$r = \frac{R}{R_m} = 1 + \frac{b}{R_m} \cos\theta \cong 1. \qquad (13)$$

We choose the surface coordinates $x^1 = \xi$, $x^2 = \theta$ shown in Fig. 1. The condition (13) makes $a_1 \cong a_2 = b$. Denoting the tube length by L ($L = \varphi R_m$) we have the coordinates of the tube edges $\xi = 0, L/b$.

Constraints at the edges generate edge-effect deformation. It disappears at a distance (from the respective edge) which is short. The well-known assessment of the distance is $3.5\sqrt{hb}$ [20]. Provided the tube is substantially longer, its behaviour and final collapse are fully determined by the deformation of the main part of the tube, away from the edges. It is this deformation we must investigate. Such deformation varies along the tube weakly and must satisfy the condition (2). It can be adequately described by the nonlinear semi-momentless theory. This will be done in what follows.

Two limiting-cases of flanges on the tube edges will be considered (Fig. 2). The corresponding edge conditions of the semi-momentless theory are indicated by the set of resultants (9).

---

*) The simplification leads to no substantial loss of accuracy even for $b/R_m \sim 1$ [13]. A solution not using (13) demands no essential changes in the following

(a) "Thin" flanges resist deformation of the edge in its plane but allow free warping and transmit to the edge given longitudinal forces. With the edge forces distributed linearly and statically equivalent to a moment M (Fig. 2), the relevant conditions are

$$\xi = 0,\ L/b : \varkappa_2 = 0, \qquad N_1 = \frac{M}{\pi b^2} \cos\theta \quad . \tag{14}$$

(b) "Stiff" flanges prevent any deformation of a tube edge. The relevant boundary conditions are

$$\xi = 0,\ L/b : \varkappa_2 = 0,\ \tau = 0 \ . \tag{15}$$

Turning to the field equations we start with those under (4) - (7). Omitting in the corresponding scalar equations the terms with M_1, ε_2, H, γ (to be neglected according to A and B) and specializing the equations for the shape being considered, renders ([20], p. 181)

$$\begin{aligned}
&\varkappa_{2,1} - \tau_{,2} = 0 \ , && N_{1,1} + S_{,2} = - q_1 b \ , \\
&\tau_{,1} - \varkappa_{1,2} + \frac{\varepsilon_{1,2}}{R_2'} - \varkappa_2 \frac{b}{\rho_1} = 0, && S_{,1} + N_{2,2} + N_1 \frac{b}{\rho_1'} - \frac{M_{2,2}}{R_2'} = - q_2 b \ , \\
&\frac{\varkappa_1}{R_2'} + \frac{\varkappa_2}{R_1} + \frac{\varepsilon_{1,22}}{b^2} = \tau^2, && \frac{N_2}{R_2'} + \frac{N_1}{R_1'} + 2\, S\tau - \frac{M_{2,22}}{b^2} = q,
\end{aligned} \tag{16}$$

where $1/R_2'$, $1/\rho_\alpha'$ denote the normal-section and geodetic curvature parameters of the deformed shell,

$$\frac{1}{R_1} = \frac{\cos\theta}{R_m}\ , \ \frac{1}{\rho_1} = \frac{\sin\theta}{R_m}, \ \frac{1}{R_2'} = \frac{1}{b} + \varkappa_2, \ \frac{1}{R_1'} = \frac{1}{R_1} + \varkappa_1, \ \frac{1}{\rho_1'} = \frac{1}{\rho_1} - \frac{\varepsilon_{1,2}}{b} \ . \tag{17}$$

Together with the elasticity relations (1) the eqs. (16) constitute a closed system. The solution of the problem will be sought with the aid of numeric integration. The system of ordinary differential equations appropriate for numeric integration will be derived by applying to the system (1), (16), (17) the Fourier-series method. The equations will be integrated in the form:

$$\frac{d}{d\xi} \underset{\sim}{X} = \underset{\approx}{A}\underset{\sim}{X} + \underset{\sim}{B} \ . \tag{18}$$

The first step of the derivation of (18) consists in eliminating from eqs. (1), (16), (17) most of dependent variables. We retain in (18) only the variables N_1, $\varkappa_2$, S, τ, which determine the stresses and deformations of sections and tube edges ξ = const. The resulting system of four equations is reduced to the following nondimensional form (18).

$$\frac{1}{\sqrt{h^o}}\frac{\partial}{\partial\xi}\begin{bmatrix} T \\ \varkappa \\ S^o \\ \tau^o \end{bmatrix} = \begin{bmatrix} 0 & 0 & -\frac{1}{r}\frac{\partial}{\partial\theta}r^2 & 0 \\ 0 & 0 & 0 & \frac{1}{r}\frac{\partial}{\partial\theta}r^2 \\ -U' & -V & 2\frac{\partial}{\partial\theta}r\rho\tau^o & 0 \\ -V & U & 0 & \frac{\partial}{\partial\theta}r\rho\tau^o \end{bmatrix}\begin{bmatrix} T \\ \varkappa \\ S^o \\ \tau^o \end{bmatrix} . \quad (19)$$

The distributed-load terms - $\underset{\sim}{B}$ in (18) - are not retained for the sake of brevity, they can be easily reintroduced.

We use in (19)the dimensionless variables

$$T = \frac{N_1}{Ehh^o}, \; \varkappa = \varkappa_2 b, \; S^o = \frac{S}{Eh\sqrt{h^{o3}}}, \; \tau^o = \frac{\tau b}{\sqrt{h^o}}, \; \rho = \frac{R_2'}{b} = \frac{1}{1+\varkappa}, \; h^o = \frac{h'}{b} \quad (20)$$

and the operators:

$$U' = \frac{br}{\rho_1' h^o} - \frac{\partial}{\partial\theta}\rho\frac{br}{R_1' h^o} = U - \frac{\partial}{\partial\theta}(rT) + \frac{\partial}{\partial\theta}\rho^2\left[\frac{\partial^2}{\partial\theta^2}(rT) + \mu c\varkappa - r\tau^{o2}\right],$$
$$U = \mu s - \mu\frac{\partial}{\partial\theta}\rho c, \; V = \frac{\partial}{\partial\theta}\rho\frac{\partial}{\partial\theta^2} r + \frac{1}{\rho}\frac{\partial}{\partial\theta}r, \; c = \cos\eta, \; s = \sin\eta, \; \mu = \frac{b}{R_m h^o}. \quad (21)$$

For the deformation symmetric with respect to the plane $\theta = 0, \pi$, we seek the solution in the series form:

$$\begin{bmatrix} T \\ \varkappa \end{bmatrix} = \begin{bmatrix} g_o \\ 0 \end{bmatrix} + \sum_1^m \begin{bmatrix} g_n \\ f_n \end{bmatrix} \cos n\theta, \quad \begin{bmatrix} S^o \\ \tau^o \end{bmatrix} = \sum_1^m \begin{bmatrix} S_n^o \\ \tau_n^o \end{bmatrix} \sin n\theta \; . \quad (22)$$

We insert these expansions into (19) and equate the respective Fourier coefficients on both sides of each equation. This renders for the coefficients $g_o(\xi)$, $g_1(\xi)$,..., $\tau_m^o(\xi)$ a set of ordinary differential equations. The equations can be readily written out, using the matrix procedure described in [20], directly in the form (18). The column matrix $\underset{\sim}{X} = \{\underset{\sim}{T}, \underset{\sim}{\varkappa}, \underset{\sim}{S}^o, \underset{\sim}{\tau}^o\}$ is composed of the four column matrices encompassing the Fourier coefficients of the respective dependent variables:

$\underset{\sim}{T} = \{g_o \; g_1 \; g_2 \; \dots \; g_m\}$, $\underset{\sim}{\varkappa} = \{0 \; f_1 \; f_2 \; \dots \; f_m\}$, ...

The system (18) contains 4m + 2 equations, with m being the mumber of the last coefficient retained in each series.

The equations (18) are nonlinear: the system matrix $\underset{\sim}{A}$ depends as defined by (19) on the actual shape and stressed state. It must be "up-dated" for a current level of loading.

The integration of the system (19) was carried out for each step of loading with the aid of multiple-shooting method (described, for instance, by R. Bulirsch and J. Stoer, Einführung in the Numerische Mathematik II, Springer, Berlin-New York, 1973). Details of the programming can be found in the Report 03/84 of the Institut für Mechanik, HSBw München (1984).

The longitudinal coordinate is in the equations (19) represented only with the quantity $\xi\sqrt{h^{\sigma}}$. Correspondingly the length of the tube L is represented by the dimensionless parameter l

$$l = \frac{L}{\pi b} \sqrt{h^{\sigma}} \quad . \tag{23}$$

For a relatively unfavourable dimensions and loading - a cylinder of length l = 0.65 under nearly critical bending moment - it is sufficient to retain m = 5 terms of each Fourier series. The Pascal-written program requires on the Burroughs 7800 the machine time 500 - 700 sec.

The second-order-theory calculation of the solution described in [19,20] lasts 10^3-times less. A slower personal computer HP-85 does it in several minutes.

For the flexure of tubes with the two cases of end constraints, described by the boundary conditions stated in (14), (15) the results of large-deformation analysis are presented in Fig. 2-6.

4. Failure of Bent Tubes

Experiments invariably show elastic tubes to collapse by way of bifurcation-buckling inside a comparatively small zone on the compressed side (Fig. 2). This was reported already in the famous work [21] by L.G. Brazier (1927). However later, at least until [17] (1965), the opinion absolutely prevailed in the literature, that a long tube collapses when a maximum bending moment is reached, that is - by way of limit-point instability. A comprehensive investigation of the tube-failure with the STAGS computer program is due to W.B. Stephens, J.H. Starnes and B.U. Almroth [22].

After this work and those in [24] and [25] the role of the prebuckling deformation in the collapse of tubes should be clear. Namely, large precritical elastic deformation produces new shape of the tube (Fig. 2) and new stress distribution. This causes in an area on the compressed side of the tube a special situation. With the increasing load (bending moment) not only the stresses concentrate here but also the local geometry becomes more susceptible to local bifurcation buckling (Fig. 3).

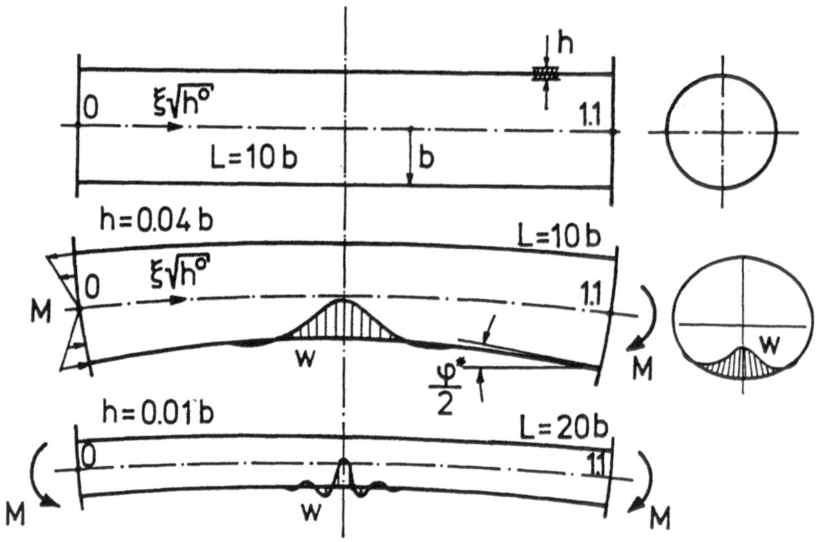

Fig. 2 *Finite length tubes with "thin" end flanges. Bifurcation buckling starts locally.*

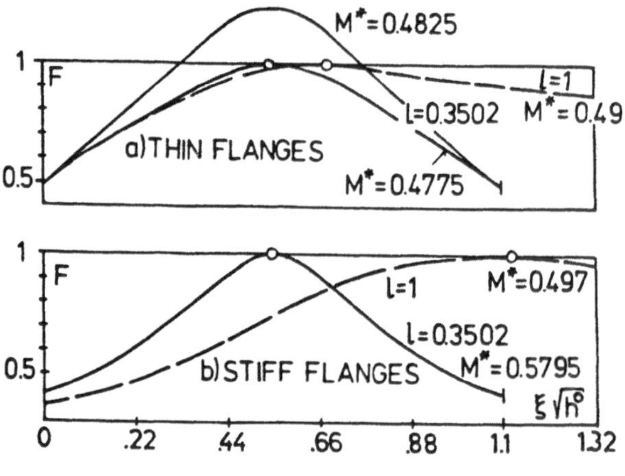

Fig. 3 *In the deformed tube there appears a zone where with the increase of loading the resistance to local bifurcation buckling nonlinearly falls and stresses grow.* $M^* = \sigma_B / \sigma_{cl}$, $\sigma_B = M/W$.

It should be emphasized: all this has in principle nothing in common with imperfections and postbuckling analysis.

Imperfections can, of course, weaken the tube additionally. Analysis of postbuckling is of particular importance in the case of <u>local</u> buckling: besides the imperfection sensitivity the fate of the shell <u>after</u> the buckling may pose serious questions. But first we have to determine the prebuckling deformation.

It is here that the nonlinear theory must be applied. The geometry and the stress distribution depend on the level of loading. The variety of possible situations is immense. The use of nondimensional parameters helps in reducing it. Nevertheless we must concentrate on analysis of examples and a presentation of graphs for only the parameters determining the collapse load.

Two initially cylindrical tubes presented in their deformed shape in Fig. 2 have the same nondimensional length l = .3502. The shell with h/b = 0.01, L/b = 20 is identical in its dimensions to the longest of the cylinders investigated in [22]. (We note, that the longer a tube the easier, obviously, the limit-point collapse can occur compared to the local buckling). The other tube has a four times larger relative wall thickness and two times smaller relative length. Both these shells belong to a variety which is described in the prebuckling state by a quantitatively identical dimensionless solution (18) - (22). The shapes shown in Fig. 2 correspond to a bend moment M defined by the same dimensionless quantity (σ_B/σ_{cl})

$$\sigma_B = \frac{M}{W} = 0.4925\ \sigma_{cl},\ \sigma_{cl} = Eh/[b\sqrt{3(1-\nu^2)}],\ W = \pi b^2 h. \tag{24}$$

The quantities σ_B and σ_{cl} denoted in (24) have physical sense. Of course, σ_B is the stress caused by the moment M in a(straight) tube with undeformed circular cross-section; σ_{cl} is the classical critical stress of such a tube under uniform axial compression.

As mentioned, the local nondimensional stress and strain resultants are for the two shells of Fig. 2 identical. But a glance at Fig. 2 shows that the displacements and rotations are substantially different. In the first case the angle of rotation φ^*, between the tangent planes at the points $\theta = \pi$ (Fig. 1) of the tube edges, is 0.412 radian. For the other tube, one with smaller wall-thickness, φ^* = 0.206 radian. It is still a <u>large rotation</u> (according to the now usual classification of [2]). The rotations in the cross-section, denoted by ϑ in Fig. 1 are also large.

The problem of tube flexure can illustrate the complications involved in using the displacements and rotations as main dependent variables (cf. the citation of [11] at the end of Sect. 1). Indeed, for a short cylindrical tube, say with L = 2b, which deforms fully in accordance the linear theory of elasticity, we have $\varphi = L/R^*$ with R^* - the radius of elastic curvature of the bent tube ($1/R^* = M/E\pi b^3 h$). The longitudinal extension is $\varepsilon = (b/R^*)\cos\theta$. The theory of shells with moderate rotations assumes $\varphi < |\varepsilon_1|$. Even in this case (of a stiff shell) the condition is clearly violated.

The large displacements and the resulting local geometry vary with respect to both surface coordinates. This and the resultant "weakening" in the midlength part of the tube is evident in Fig. 2. The extent of this (nonlinear) change of the situation with the increasing load should be exposed also by Fig. 3.

The graphs of Fig. 3 illustrate variation along the tube of the factor (F) indicating a certain susceptibility to local buckling. We use here a *local-stability* approach [8,20,26] based on the following *hypothesis:*

The bifurcation instability is determined by the stress state and the shape of shell inside the zone of the initial buckle(s).

The hypothesis directly leads to a corollary: In so far as in the domain of the initial buckle(s) the stress resultants and the curvature parameters of the shell are approximately constant, they may be assumed constant in the stability analysis. The local buckling zone diminishes with the wall-thickness. Therefore the corollary renders a way of checking the stability, which is asymptotically exact for the limiting case of very thin shells. The stability condition in this approximation contains only the stress and strain resultants at a point of the shell. It has the form $S(N_i^*, R_j') \geqq 0$. If this condition is broken at some point it means bifurcation instability of the shell.

Of course the parts of shell surrounding the "instability point" are in a better position with respect to buckling. These better conditions are in fact replaced in the asymptotic analysis by conditions identical to those at a point. We neglect thereby some reserves of the shells' ability to resist buckling. The asymptotic stability condition $S(N_i^*, R_j') \geqq 0$ determines the N_i^* as characteristics of *resistance to buckling* at a point of the shell.
It is almost evident that for a uniaxial compression N_1 the asymptotic stability condition is

$$|N_1| \leq N_1^*,\ N_1^* = E\,h^2/[R_2'\sqrt{3(1-\nu^2)}] \quad . \tag{25}$$

The quantity N_1^* characterizes the local *resistance to buckling* at a point (where the curvature parameter of the deformed normal section is R_2'). The N_1^*/h is equal to the critical stress σ_{cl} of uniform compression for a circular cylinder with the radius equal to R_2'. The characteristic F used in Fig. 3 is defined by (cf. (25)):

$$F = |N_1(\xi,\pi)|/N_1^* . \tag{26}$$

The Fig. 3 shows the prebuckling deformation to lead to a serious deterioration of the local resistance to buckling at some zone of the tube. This zone is usually at midlength of the tube, where $1/R_2' = 1/b + \varkappa_2$ is minimal. But for longer tubes, those with $l = 1$ the examples of Fig. 3 show the zone susceptible to buckling to be not at midlength. The zone moves to where the compressive stress (N_1) is larger.

So far the approximation, exact for the limiting case of very thin shells, was discussed. Is, however, the variation of the stress and curvature inside the buckling zone substantial, so must the asymptotic condition be corrected. For the buckling zone experiencing uniaxial compression, being the case in our problem, the appropriate stability condition (26) has the form similar to (25):

$$|N_1| \leq \lambda N_1^* = \lambda Eh^2/[R_2'\sqrt{3(1-\nu^2)}] \quad . \tag{27}$$

The factor $\lambda = N_{1cr}/N_1^* \geq 1$ depends on h/R_2' - on the wall thickness in the buckling zone. (Actually λ depends on the intensity of variation at R_i', N_i with respect to $x = \xi b/c$, $y = \theta b/c$, $c = \sqrt{hR_2'(\xi_0,\theta_0)}$ with ξ_0, θ_0 - the coordinates of the center of the buckling zone).

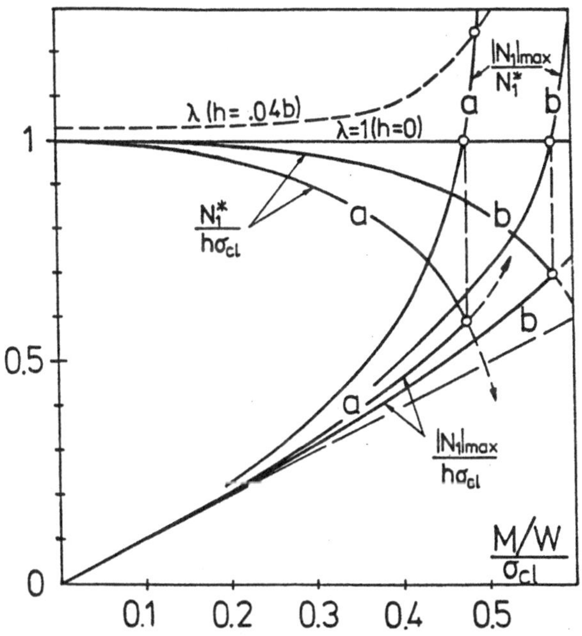

Fig. 4 *At the center of the zone susceptible to buckling the compressive stress (N_1/h) increases nonlinearly with the load (M). Still more intensively falls the resistance to buckling. N_1^* determined by (25) - the R_2' grows. All curves for tubes with l = 0.3502. The λ curve for h/b = 0.04.*

The variation of the parameters of stress and strain in the center of eventual buckling zone, caused by the increase of loading is shown in Fig. 4. Curves denoted "a" and "b" concern tubes with "thin" and "stiff" flanges respectively. The bifurcation-buckling security measured by the relation (27) falls catastrophically when

σ_B/σ_{cl} increases. Even for the tubes - with h = 0.04 b (very thick for this problem), when the dimensions of a buckle (shown in Fig. 2) are large and therefore λ can be substantial (Fig. 4), the dimensionless buckling moment σ_B/σ_{cl} is merely some 3% over that of $h/b \to 0$.

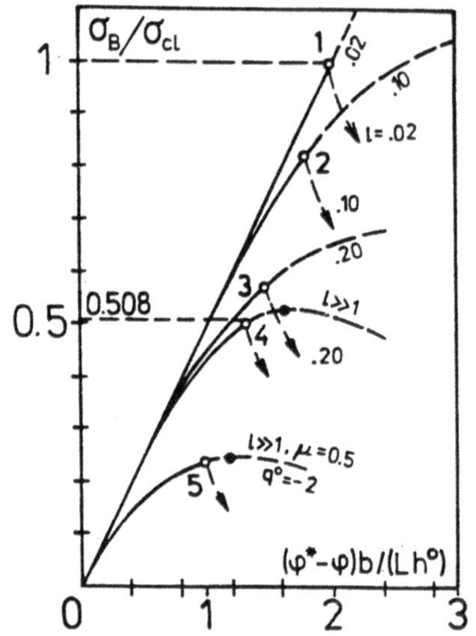

Fig. 5 *Nonlinear flexure depends on the tube-length. Bifurcation buckling occurs before the maximum bending moment could be reached.* $\sigma_B = M/W$.

The nonlinear characteristics of several tubes are displayed in Fig. 5. In all cases the bifurcation buckling occurs before the maximum bending moment can be reached, this means the limit-case instability is not possible. The maximum moment is the more in excess of the critical one the shorter the tube.

For the infinitely long tubes this is amply demonstrated in [27]. A review of the critical bending moments for tubes of different length and initial curvature is given in Fig. 6.

The critical bending moments are the lowest not for the longest tubes. Moreover, one notices that with the increase of the tube length the value of the critical moment (σ_B/σ_{cl}) remains under that calculated for the infinitely long tubes. This phenomenon is explained by the buckling occuring by longer tubes not in the midlength part of the tube. As has been discussed in connection with Fig. 3, the influence of edge loading may cause by sufficiently long but finite tubes a certain maximum of compressive stress, which is not in the midlength of the tube.

We remark, finally that the initial curvatures considered, those corresponding to μ = 0.25 or 0.5, are small. They are possible in the market tubes manufactured according to standards as straight tubes.

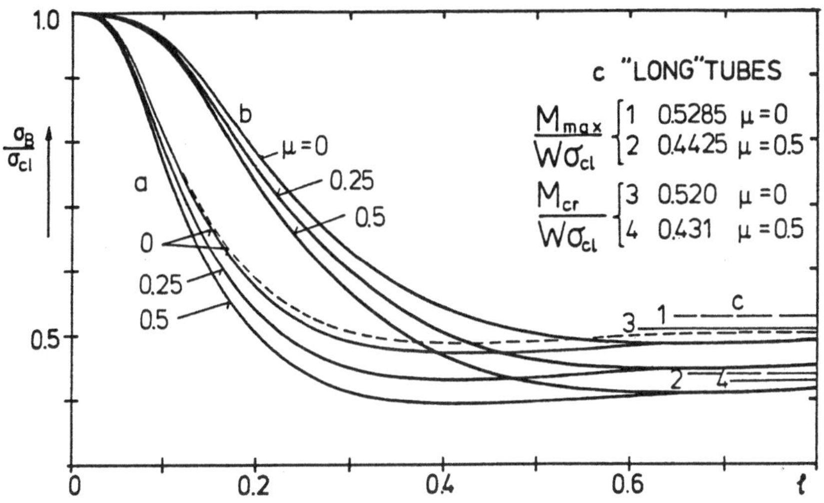

Fig. 6 *Dependence of the critical bending moment on the tube length (l), on the kind of edge flanges (a or b) and on the initial curvature.* $\sigma_B = M/W$.

References

1. REISSNER E., On axisymmetrical deformations of thin shells of revolution. Proc. Symposia in Appl. Math. 3 (1950), 27-52.

2. MUSHTARY KH.M.,GALIMOV K.Z., Nonlinear theory of thin elastic shells. Kasan 1957 (In Russian). Translation NASA-TT-F62 (1961).

3. KOITER W.T., On the nonlinear theory of thin elastic shells. Proc. Koninkl. Nederl. Akad. van Wet.; Ser. B 69 (1966), 1-54.

4. KOITER W.T., Foundations and basic equations of shell theory. Proc. 2nd IUTAM Symp. on the Theory of Thin Shells. Springer 1969, 93-105.

5. KOITER W.T., The intrinsic equations of shell theory with some applications. Mechanics Today, Ed. S. Nemat-Nasser, Pergamon Press, 1980, 139-154.

6. LIBAI A., SIMMONDS J.G., Nonlinear elastic shell theory. Adv. in Appl. Mech. 23 (1983) 271-371.

7. SIMMONDS J.G., Special cases of nonlinear shell equations. Trends in Solid Mechanics. (Edited by J.F. Besseling and A.M.A. van der Heijden). (Proc. of the Symp. dedicated to the 65th birthday of W.T. Koiter). Sijthoff & Nordhoff (1979), 211-224.

8. AXELRAD E.L., Flexible shells. Theoretical and Applied Mechanics.Proc. of the 15th Intern. Congr., Toronto 1980; ed. by F.P.J. Rimrott, B. Tabarrok. North-Holland 1980.

9. HALMOS P.R., How to write mathematics. L'Enseignement Mathematique. 16 (1970), 123-152.

10. GOLDENVEIZER A.L., Theory of Elastic Thin Shells. Translation from the Russian Edition (1953); ed. by G. Herrmann. Oxford Pergamon Press (1961).

11. SIMMONDS J.G., A set of simple, accurate equations for circular cylindrical elastic shells. Int. J. Solids & Structures, 2 (1966), 525-541.

12. AXELRAD E.L., EMMERLING F.A., Eine geometrische nichtlineare Halbmembrantheorie elastischer Schalen. Forschung im Ing.Wes. 49 (1983), 31-36.

13. AXELRAD E.L., Elastic tubes-assumptions, equations, edge conditions. Thin-Walled Structures 3 (1985), 193-215.

14. AXELRAD E.L., Flexible shells. In "Flexible shells, Theory and Applications", eds. E.L. Axelrad, F.A. Emmerling, Springer, Berlin, 1984, 44-63.

15. KARMAN v. TH., Über die Formänderung dünnwandiger Rohre, insbesondere federnder Ausgleichsrohre. VDI-Z. 55 (1911), 1889-1895.

16. BESKIN L., Bending of curved thin tubes. J. Appl. Mech., Tr. ASME, 65 (1943), A 105 - A 120.

17. AXELRAD E.L., Refinement of critical-load analysis for tube flexure by way of considering precritical deformation. Izv. AN SSSR OTN, Mekh. i. Mash. (1965), n4, 133-139 (In Russian).

18. NOVOZHILOV V.V., Thin Shell Theory. Groningen, P. Noordhoff 1970.

19. AXELRAD E.L., Nonlinear equations of axisymmetric shells and bending of thin-walled beams. Izv. AN SSSR, OTN, Mekh. i.Mash. (1960), n4, 84-92 (in Russian). Translation in: Amer. Rocket Soc. J. Supplement 32 (1962) 1147-1151.

20. AXELRAD E.L., Schalentheorie. B.G. Teubner, Stuttgart, 1983.

21. BRAZIER L.G., On the flexure of thin cylindrical shells and other "thin" sections. Proc. Roy. Soc. London, Ser. A. 116 (1927) 104-114.

22. STEPHENS W.B., STARNES J.H., ALMROTH B.O., Collapse of long cylindrical shells under combined bending and pressure loads. AIAA J., 13 (1975), 20-25.

23. ATLURI S.N., Computational analysis of finitely deformed solids with application to plates and shells - I. Computer & Structures, 18 (1984), 93-116.

24. EMMERLING F.A., Nonlinear bending of curved tubes. In "Flexible Shells, Theory and Applications", eds. E.L. Axelrad, F.A. Emmerling. Springer, Berlin 1984, 175-192.

25. AXELRAD E.L., EMMERLING F.A., Finite bending and collapse of elastic pressurized tubes. Ing.-Arch. 53, (1983), 41-52.

26. AXELRAD E.L., On local buckling of thin shells. Int. J. Non-Linear Mechanics, 20 (1985) 249-259.

27. EMMERLING F.A., Nichtlineare Biegung und Beulen von Zylindern und krummen Rohren bei Normaldruck. Ing.-Arch. 52, (1982), 1-16.

ON GEOMETRICALLY NON-LINEAR THEORY OF ELASTIC SHELLS DERIVED FROM PSEUDO-COSSERAT CONTINUUM WITH CONSTRAINED MICRO-ROTATIONS

J. BADUR and W. PIETRASZKIEWICZ
Institute of Fluid-flow Machinery
of the Polish Academy of Sciences
ul. J.Fiszera 14, 80-952 Gdańsk, Poland

1. Introduction

The set of equations for the geometrically non-linear theory of thin elastic shells is usually expressed in terms of displacements as basic independent variables of the shell deformation. Various general and reduced displacemental forms of bending shell equations are summarized, for example, by MUSHTARI and GALIMOV [1], KOITER [2], PIETRASZKIEWICZ [3,4], SCHMIDT [5] and BAŞAR and KRÄTZIG [6], where further references may be found. When displacement field is determined from the shell equations, strains, rotations and stresses may be obtained by prescribed algebraic or differential procedures.

The displacemental form of non-linear shell equations is very complex even in the tensor notation. When strains are small, components of the strain tensor are quadratic and components of the tensor of the change of curvature are cubic polynomials in displacements and their surface derivatives [4]. Additionally, the tensor of change of curvature, even in the simplest linearized case, depends upon the second derivatives of the displacements. It means that the strain energy density for the geometrically non-linear theory of shells is the polynomial up to the sixth order in displacements as well as their first and second surface derivatives. In modern computerized structural analysis, based on finite elements or finite differences, the need for discretization of the second derivatives causes many problems associated with inter-element continuity as well as discrete formulation of the boundary conditions. As a result, higher-order shape functions and difference schemes are required, which lead to additional degrees of freedom, complex schemes of numerical integration, reformulated boundary conditions etc.

The complexities associated with the displacemental shell equations can make it more attractive an alternative approach to shell theory, based on the polar decomposition of the shell deformation gradient into the rigid-body rotation and the pure stretch along the principal directions of strain. Then the finite rotation field is used as an independent or intermediate variable of the shell theory. This approach, originated by ALUMÄE[7] for the general shell geometry and by REISSNER [8] for axisymmetric deformation of shells of revolution, has been developed by WEMPNER [9], SIMMONDS and DANIELSON [10,11], PIETRASZKIEWICZ [3,12-14], SHKUTIN [15], LIBAI and SIMMONDS [16], ATLURI [17], KAYUK and SAKHATSKIY[18] and MAKOWSKI and STUMPF [19]. In this approach the structure of the non-linear shell equations becomes similar to the structure of the Cosserat surface theory [20,21,22]. However, in the latter theory displacements and rotations are, by definition, two independent kinematic field variables and the surface strain energy density is postulated from two-dimensional considerations,without any reference to the three-dimensional continuum mechanics.

In this paper we propose a new procedure for the derivation of the non-linear shell equations directly from the three dimensional constrained elastic Cosserat continuum. It is known [20,23-26] that within the Cosserat continuum each material particle can translate and independently rotate. This micro-rotation field does not coincide, in general, with the macro-rotation of the particle's neighbourhood as calculated from the displacement field in the classical continuum mechanics. The stress state is described by two, generally non-symmetric, stress and couple stress tensors.

In this paper we assume that the couple-stress tensor vanishes everywhere, what leads to the so-called pseudo-Cosserat continuum [24,25]. Additionally, micro-rotations are constrained to coincide everywhere with the macro-rotations. As a result, the Cosserat elastic continuum with the two constraints becomes entirely equivalent to the classical non-polar non-linear elasticity, but written here in different form, in terms of the Cosserat field variables. In particular, with the help of Lagrangian multipliers the second constraint is explicitly introduced into the strain energy function of the elastic Cosserat body.

In order to describe such constrained thin Cosserat body by a two-dimensional shell theory, we introduce the Kirchhoff-Love kinematic constrains and take the strain energy density in the form used in the classical first-approximation theory of thin isotropic elastic shells [27]. As a result, in the case of dead external loads, the functional of the total potential energy is constructed. It depends upon three displacements, three rotations and three Lagrangian multipliers (the scew-symmetric part of the internal surface stress resultant tensor and two shearing forces) as independent variables subject to variation. The functional is linear in the Lagrangian multipliers and is rational function containing at most fourth-order polynomials in displacements, rotations and their first derivatives. The latter feature seams to be very attractive for the numerical applications. It allows to use the simplest shape functions or difference schemes which assure the high efficiency of the numerical analysis. The stationarity conditions of the functional lead tosix equilibrium equations, three constraint conditions and appropriate static boundary and corner conditions for the nine unknowns to be determined in the solution process. All the relations are given through components with respect to the rotated basis. Some advantages, similarities and differences of such a formulation of the non-linear shell theory in comparison with the ones proposed in [7,11,15, 17] are discussed.

2. Some relations of the Cosserat continuum

Here we briefly discuss a deformation of the body $\mathcal{B}$, consisting of material particles X , Y , ..., in the three-dimensional Euclidean point space $\mathcal{E}$. Let $P=\kappa(\mathcal{B})$ and $\bar{P}=\bar{\kappa}(\mathcal{B})$ be regions of $\mathcal{E}$ occupied by the body $\mathcal{B}$ in the reference (undeformed) and in the actual (deformed) configuration, respectively. The places P and $\bar{P}$ occupied by the particle $X \in \mathcal{B}$ in both configurations are given by the respective position vectors

$$\underset{\sim}{p} = x^k(\theta^i)\underset{\sim}{i}_k \quad , \qquad \bar{\underset{\sim}{p}} = y^k(\theta^i)\underset{\sim}{i}_k = \chi(\underset{\sim}{p}) = \underset{\sim}{p} + \underset{\sim}{w} \quad , \tag{2.1}$$

where θ^i , i = 1,2,3, are the curvilinear convected coordinates, $\underset{\sim}{i}_k$ is the common orthonormal basis attached to an origin $\mathcal{O} \in \mathcal{E}$, χ is the macro-deformation function and $\underset{\sim}{w}$ is the displacement vector.

In P we introduce the base vectors $\underset{\sim}{g}_i = \underset{\sim}{p}_{,i}$, $\underset{\sim}{g}^i \cdot \underset{\sim}{g}_j = \delta^i_j$, the metric tensor $\underset{\sim}{1} = g_{ij}\underset{\sim}{g}^i \otimes \underset{\sim}{g}^j = g^{ij}\underset{\sim}{g}_i \otimes \underset{\sim}{g}_j$ with components $g_{ij} = \underset{\sim}{g}_i \cdot \underset{\sim}{g}_j$, $g^{ij} = \underset{\sim}{g}^i \cdot \underset{\sim}{g}^j$ and the scalar $g = |g_{ij}|$. Analogously defined functions in $\bar{P}$ are marked by a dash: $\bar{\underset{\sim}{g}}_i$, $\bar{\underset{\sim}{g}}^j$, $\bar{\underset{\sim}{1}}$, $\bar{g}_{ij}$, $\bar{g}^{ij}$, $\bar{g}$ etc. All properties of the macro-deformation of $\mathcal{B}$ can be described entirely in terms of $\underset{\sim}{w}$ (see [28] for details).

Within the Cosserat theory it is assumed that during deformation each material particle $X \in \mathcal{B}$ can translate and independently rotate [20, 23,24,26]. The translation is described by the vector field $\underset{\sim}{w}$ while the rotation is described by an independent proper orthogonal tensor field $\underset{\sim}{R}$, such that $\underset{\sim}{R}^{-1} = \underset{\sim}{R}^T$, $\det \underset{\sim}{R} = +1$. The tensor field is called the micro-rotation of $X \in \mathcal{B}$. In order to make it more convenient let us introduce an aholonomic triad of vectors $\underset{\sim}{d}_i$ associated with each $X \in \mathcal{B}$, rigidly rotating with the particle during its deformation. Assuming, for convenience, that in the reference configuration those vectors coincide with the base vectors $\underset{\sim}{g}_i$, in the deformed configuration we obtain

$$\underset{\sim}{d}_j = \underset{\sim}{R}\underset{\sim}{g}_j \quad , \quad \underset{\sim}{R} = \underset{\sim}{d}_i \otimes \underset{\sim}{g}^i \quad , \quad \underset{\sim}{d}_i \cdot \underset{\sim}{d}_j = g_{ij} \ . \tag{2.2}$$

The complete information about deformation of the neighbourhood of the material particle $X \in \mathcal{B}$ contain now two fields: the deformation gradient $\underset{\sim}{F}$ and the micro-rotation gradient $\underset{\sim}{D}$, defined by

$$\begin{aligned} \underset{\sim}{F} &= \operatorname{Grad} \underset{\sim}{\bar{p}} = \underset{\sim}{\bar{g}}_j \otimes \underset{\sim}{g}^j \quad , \qquad 0 < \det \underset{\sim}{F} < \infty \quad , \\ \underset{\sim}{D} &= \operatorname{Grad} \underset{\sim}{R} = \underset{\sim}{d}_{i;j} \otimes \underset{\sim}{g}^i \otimes \underset{\sim}{g}^j \ . \end{aligned} \tag{2.3}$$

where $(\ )_{;j}$ is the covariant derivative with respect to the reference metric g_{ij} .

In classical continuum mechanics it is usual to apply the polar decomposition $\underset{\sim}{F} = \underset{\sim}{R}\underset{\sim}{U} = \underset{\sim}{V}\underset{\sim}{R}$, where $\underset{\sim}{U}$ and $\underset{\sim}{V}$ are the (symmetric) right and left stretch tensors and $\underset{\sim}{R}$ is the (proper orthogonal) rotation tensor of the macro-deformation χ . However, $\underset{\sim}{V}$, $\underset{\sim}{U}$ and $\underset{\sim}{R}$ do not describe properly the strains and rotations of the neighbourhood of the Cosserat material particle $X \in \mathcal{B}$.

The Lagrangian strain measures appropriate for the Cosserat continuum are defined by [20,23,24]

$$\underset{\sim}{\mathcal{U}} = \underset{\sim}{R}^T \underset{\sim}{F} = \mathcal{U}_{ij} \underset{\sim}{g}^i \otimes \underset{\sim}{g}^j \quad , \quad \underset{\sim}{K} = \tfrac{1}{2}\, \underset{\sim}{\epsilon} \cdot (\underset{\sim}{R}^T \underset{\sim}{D}) = K_{ij} \underset{\sim}{g}^i \otimes \underset{\sim}{g}^j \ , \tag{2.4}$$

where $\underset{\sim}{\epsilon} = -\, \underset{\sim}{1} \times \underset{\sim}{1} = \epsilon^{ijk} \underset{\sim}{g}_i \otimes \underset{\sim}{g}_j \otimes \underset{\sim}{g}_k$ and dot means double scalar product performed on the second-order tensors according to $\underset{\sim}{\epsilon} \cdot \underset{\sim}{A} = \epsilon^{ijk} A_{jk} \underset{\sim}{g}_i$.

In what follows it will be convenient to use another strain measures (in analogy to $\underset{\sim}{V}$ in the classical continuum mechanics) defined according to

$$\underset{\sim}{V} = \underset{\sim}{R}\underset{\sim}{\mathcal{U}}\underset{\sim}{R}^T = \mathcal{U}_{ij} \underset{\sim}{d}^i \otimes \underset{\sim}{d}^j \quad , \quad \underset{\sim}{L} = \underset{\sim}{R}\underset{\sim}{K}\underset{\sim}{R}^T = K_{ij} \underset{\sim}{d}^i \otimes \underset{\sim}{d}^j \ . \tag{2.5}$$

Since $\underset{\sim}{R}^T \underset{\sim}{R}_{,j}$ and $\underset{\sim}{R}_{,j} \underset{\sim}{R}^T$ are scew-symmetric let us introduce their axial vectors [28]

$$\underset{\sim}{k}_j = -\tfrac{1}{2} \underset{\sim}{\epsilon} \cdot (\underset{\sim}{R}^T \underset{\sim}{R}_{,j}) \quad , \quad \underset{\sim}{l}_j = -\tfrac{1}{2} \underset{\sim}{\epsilon} \cdot (\underset{\sim}{R}_{,j} \underset{\sim}{R}^T) = \underset{\sim}{R}\underset{\sim}{k}_j = K_{ij} \underset{\sim}{d}^i . \tag{2.6}$$

The relative Lagrangian strain measures are given by

$$\underset{\sim}{E} = \underset{\sim}{U} - \underset{\sim}{1} = E_{ij}\underset{\sim}{g}^i \otimes \underset{\sim}{g}^j \quad , \quad \underset{\sim}{H} = \underset{\sim}{V} - \underset{\sim}{1} = E_{ij}\underset{\sim}{d}^i \otimes \underset{\sim}{d}^j \ ,$$
$$\underset{\sim}{h}_j = E_{ij}\underset{\sim}{d}^i = \bar{\underset{\sim}{g}}_j - \underset{\sim}{d}_j = \underset{\sim}{g}_j' + \underset{\sim}{w}_{,j} - \underset{\sim}{R}\underset{\sim}{g}_j \ . \tag{2.7}$$

The micro-rotation $\underset{\sim}{R}$ can be performed with an equivalent finite rotation vector $\underset{\sim}{\theta} = 2\,\mathrm{tg}\,\omega/2\,\underset{\sim}{e}$, where $\underset{\sim}{e}$ is the unit vector of the rotation axis and ω is the angle of rotation about $\underset{\sim}{e}$ associated with $\underset{\sim}{R}$. Then [28]

$$\underset{\sim}{d}_j = \underset{\sim}{g}_j + \frac{1}{t}\underset{\sim}{\theta} \times (\underset{\sim}{g}_j + \frac{1}{2}\underset{\sim}{\theta} \times \underset{\sim}{g}_j) \quad , \quad t = 1 + \frac{1}{4}\underset{\sim}{\theta}\cdot\underset{\sim}{\theta} \ ,$$
$$\underset{\sim}{l}_j = \frac{1}{t}(\underset{\sim}{\theta}_{,j} - \frac{1}{2}\underset{\sim}{\theta}_{,j} \times \underset{\sim}{\theta}) \ . \tag{2.8}$$

The relations $(2.7)_2$, $(2.8)_2$ and (2.6) allow to express the strain measures E_{ij} and K_{ij} in terms of $\underset{\sim}{w}$ and $\underset{\sim}{\theta}$. Solving $(2.7)_2$ and $(2.8)_2$ for $\underset{\sim}{w}_{,j}$ and $\underset{\sim}{\theta}_{,j}$ and applying the integrability conditions $\underset{\sim}{w}_{,ji} - \underset{\sim}{w}_{,ij} = \underset{\sim}{0}$, $\underset{\sim}{\theta}_{,ji} - \underset{\sim}{\theta}_{,ij} = \underset{\sim}{0}$ we obtain the vector form of compatibility conditions of the Cosserat continuum [26]

$$\epsilon^{sij}(\underset{\sim}{h}_{i;j} + \underset{\sim}{l}_j \times \underset{\sim}{d}_i) = \underset{\sim}{0} \quad , \quad \epsilon^{sij}(\underset{\sim}{l}_{i;j} + \frac{1}{2}\underset{\sim}{l}_i \times \underset{\sim}{l}_j) = \underset{\sim}{0} \ . \tag{2.9}$$

Suppose that in the deformed configuration $\bar{P}$ there is an oriented differential area $d\bar{A}$ with a unit outward normal $\bar{\underset{\sim}{n}}$ at $\bar{\underset{\sim}{p}}$. Let the image at $\underset{\sim}{p}$ in P corresponding to $\bar{\underset{\sim}{n}}d\bar{A}$ is $\underset{\sim}{n}dA$. Then [29]

$$\bar{\underset{\sim}{n}}d\bar{A} = J\underset{\sim}{n}\underset{\sim}{F}^{-1}dA \quad , \quad J = \det\underset{\sim}{F} = \sqrt{\bar{g}/g} \ . \tag{2.10}$$

Let $\underset{\sim}{t}$ nad $\underset{\sim}{m}$ be the stress and couple-stress vectors acting on the oriented differential area $\bar{\underset{\sim}{n}}d\bar{A}$ in $\bar{P}$. Then the "true" Cauchy--type stress and couple-stress tensors $\underset{\sim}{\tau}$ and $\underset{\sim}{\mu}$ are given by

$$\underset{\sim}{t}d\bar{A} = \underset{\sim}{\tau}\bar{\underset{\sim}{n}}d\bar{A} \ , \quad \underset{\sim}{m}d\bar{A} = \underset{\sim}{\mu}\bar{\underset{\sim}{n}}d\bar{A} \ , \quad \underset{\sim}{\tau} = \tau^{ij}\bar{\underset{\sim}{g}}_i \otimes \bar{\underset{\sim}{g}}_j \ , \quad \underset{\sim}{\mu} = \mu^{ij}\bar{\underset{\sim}{g}}_i \otimes \bar{\underset{\sim}{g}}_j \tag{2.11}$$

Using (2.10) and $\underset{\sim}{R}$ we can define alternative stress and couple-stress tensors referred to the undeformed configuration

$$\underset{\sim}{t}d\bar{A} = \underset{\sim}{T}_R\underset{\sim}{n}dA = \underset{\sim}{R}\underset{\sim}{T}_J\underset{\sim}{n}dA = \underset{\sim}{T}_B\underset{\sim}{R}\underset{\sim}{n}dA \ ,$$
$$\underset{\sim}{m}d\bar{A} = \underset{\sim}{M}_R\underset{\sim}{n}dA = \underset{\sim}{R}\underset{\sim}{M}_J\underset{\sim}{n}dA = \underset{\sim}{M}_B\underset{\sim}{R}\underset{\sim}{n}dA \ , \tag{2.12}$$

$$\underset{\sim}{T}_R = J\underset{\sim}{\tau}\underset{\sim}{F}^{-T} \ , \quad \underset{\sim}{T}_J = J\underset{\sim}{R}^T\underset{\sim}{\tau}\underset{\sim}{F}^{-T} \ , \quad \underset{\sim}{T}_B = J\underset{\sim}{\tau}\underset{\sim}{F}^{-T}\underset{\sim}{R}^T \ ,$$
$$\underset{\sim}{M}_R = J\underset{\sim}{\mu}\underset{\sim}{F}^{-T} \ , \quad \underset{\sim}{M}_J = J\underset{\sim}{R}^T\underset{\sim}{\mu}\underset{\sim}{F}^{-T} \ , \quad \underset{\sim}{M}_B = J\underset{\sim}{\mu}\underset{\sim}{F}^{-T}\underset{\sim}{R}^T \ . \tag{2.13}$$

In analogy to the classical continuum, $\underset{\sim}{T}_R$ and $\underset{\sim}{M}_R$ are the first

Piola-Kirchhoff type, $\underset{\sim}{T}_J$ and $\underset{\sim}{M}_J$ are the Jaumann type and $\underset{\sim}{T}_B$ and $\underset{\sim}{M}_B$ are the Biot type stress and couple-stress tensors, respectively.

In terms of those stress and strain measures various but equivalent sets of basic balance laws for the Cosserat continuum may be given [23,24]. In particular, the stress working rate is given in the following equivalent forms

$$\begin{aligned} \dot{U} &= \int_P \dot{W}\, dV = \int_P (T^{ij}\dot{E}_{ij} + M^{ij}\dot{K}_{ij})\, dV = \\ &= \int_P (\underset{\sim}{T}_J \cdot \dot{\underset{\sim}{E}} + \underset{\sim}{M}_J \cdot \dot{\underset{\sim}{K}})\, dV = \int_P (\underset{\sim}{T}_B \cdot \overset{\nabla}{\underset{\sim}{H}} + \underset{\sim}{M}_B \cdot \overset{\nabla}{\underset{\sim}{L}})\, dV \quad , \end{aligned} \tag{2.14}$$

where the superposed dot denotes the material time derivative, the superposed triangle is the co-rotational time derivative and T^{ij} and M^{ij} are components of the Jaumann stress measures given by

$$\begin{aligned} &\overset{\nabla}{\underset{\sim}{H}} = \dot{\underset{\sim}{H}} + \underset{\sim}{H}\underset{\sim}{\Omega} - \underset{\sim}{\Omega}\underset{\sim}{H} \quad , \qquad \underset{\sim}{\Omega} = \dot{\underset{\sim}{R}}\underset{\sim}{R}^T = -\underset{\sim}{\Omega}^T \quad , \\ &T^{ij} = \underset{\sim}{g}^i \underset{\sim}{T}_J \underset{\sim}{g}^j = \underset{\sim}{d}^i \underset{\sim}{T}_B \underset{\sim}{d}^j = J(\underset{\sim}{d}^i \cdot \bar{\underset{\sim}{g}}_k)\tau^{kj} \quad , \\ &M^{ij} = \underset{\sim}{g}^i \underset{\sim}{M}_J \underset{\sim}{g}^j = \underset{\sim}{d}^i \underset{\sim}{M}_B \underset{\sim}{d}^j = J(\underset{\sim}{d}^i \cdot \bar{\underset{\sim}{g}}_k)\mu^{kj} \quad . \end{aligned} \tag{2.15}$$

The respective scalar products in (2.14) contain the work-conjugate pairs of stress and strain measures [30]. The pairs $\underset{\sim}{T}_B$, $\underset{\sim}{H}$ and $\underset{\sim}{M}_B, \underset{\sim}{L}$ are particularly interesting here, since they are defined by components in the rotated triad $\underset{\sim}{d}_i$.

It follows from (2.14) that for an elastic Cosserat body the strain energy density depends explicitly only upon E_{ij} and K_{ij} , i.e. $W = W(E_{ij}, K_{ij})$. However, in what follows we are interested in analysing the Cosserat elastic body with two additional constraints. The first constraint states that the couple-stress tensor identically vanishes during an arbitrary motion, i.e. $\underset{\sim}{\mu} = \underset{\sim}{M}_R = \underset{\sim}{M}_J = \underset{\sim}{M}_B \equiv \underset{\sim}{0}$. In such pseudo-Cosserat elastic body the strain energy density does not depend explicitly upon K_{ij} , i.e. $W = W(E_{ij})$. The second constraint requires the micro-rotation $\underset{\sim}{R}$ to coincide with the macro-rotation $\underset{\sim}{R}$ during an arbitrary motion. As a result $\underset{\sim}{U} = \underset{\sim}{R}^T\underset{\sim}{F} = \underset{\sim}{U} = \underset{\sim}{U}^T = \underset{\sim}{U}^T$ and, similarly, $\underset{\sim}{V} = \underset{\sim}{V}^T$, $\underset{\sim}{E} = \underset{\sim}{E}^T$, or $\epsilon^{ijk}E_{ij} = 0$. This constraint can be introduced into the strain energy density with the help of Lagrangian multipliers λ_k , redefining it as follows

$$E = W(E_{ij}) + \epsilon^{ijk}E_{ij}\lambda_k \quad . \tag{2.16}$$

The strain energy density (2.16) defines the constrained Cosserat elastic body which is equivalent to the classical non-polar non-linear elastic body, only written here in terms of the Cosserat strain measure $\underset{\sim}{E}$. From (2.16) follow the constitutive equations of the constrained Cosserat body

$$T^{ij} = \frac{\partial E}{\partial E_{ij}} = \frac{\partial W}{\partial E_{ij}} + \epsilon^{ijk}\lambda_k \;, \qquad M^{ij} = \frac{\partial E}{\partial K_{ij}} = 0 \;, \quad \frac{\partial E}{\partial \lambda_k} = \epsilon^{ijk} E_{ij} = 0 \tag{2.17}$$

which incorporate explicitly the two constraints.

3. Deformation of a thin shell under K-L constraints

Let in the region $P \in E$, occupied by the constrained Cosserat body B in the reference configuration, the normal coordinate system $\{\theta^\alpha, \theta^3 \equiv \zeta\}$ $(\alpha = 1,2)$ is introduced, where $-h/2 \leqslant \zeta \leqslant h/2$ is the distance from the middle surface M of P and h is the thickness of P, assumed to be small.

The geometry of M is described by the position vector $\underset{\sim}{r} = \underset{\sim}{r}(\theta^\alpha)$. At each point $M \in M$, we have the natural surface base vectors $\underset{\sim}{a}_\alpha = \underset{\sim}{r}_{,\alpha}$, $\underset{\sim}{a}^\alpha \cdot \underset{\sim}{a}_\beta = \delta^\alpha_\beta$, the components $a_{\alpha\beta} = \underset{\sim}{a}_\alpha \cdot \underset{\sim}{a}_\beta$ and $a^{\alpha\beta} = \underset{\sim}{a}^\alpha \cdot \underset{\sim}{a}^\beta$ of the surface metric tensor $\underset{\sim}{a} = a_{\alpha\beta}\underset{\sim}{a}^\alpha \otimes \underset{\sim}{a}^\beta = a^{\alpha\beta}\underset{\sim}{a}_\alpha \otimes \underset{\sim}{a}_\beta$ with the scalar $a = |a_{\alpha\beta}|$, the unit vector $\underset{\sim}{a}_3 \equiv \underset{\sim}{n} = \frac{1}{2}\epsilon^{\alpha\beta}\underset{\sim}{a}_\alpha \times \underset{\sim}{a}_\beta$ orthogonal to M and the covariant components $b_{\alpha\beta} = -\underset{\sim}{n}_{,\alpha} \cdot \underset{\sim}{a}_\beta$ of the curvature tensor $\underset{\sim}{b} = b_{\alpha\beta}\underset{\sim}{a}^\alpha \otimes \underset{\sim}{a}^\beta$. The boundary contour C of M is described by $\theta^\alpha = \theta^\alpha(s)$, where s is the length parameter along C. Along C we define the unit tangent vector $\underset{\sim}{t} = d\underset{\sim}{r}/ds = t^\alpha \underset{\sim}{a}_\alpha$ and the outward unit normal vector $\underset{\sim}{\nu} = \underset{\sim}{t} \times \underset{\sim}{n}$.

Analogously defined functions at $\bar{M} \in \bar{M}$, which is the image of the $M \in M$ in the deformed configuration, are marked by an additional dash: $\bar{\underset{\sim}{r}}$, $\bar{\underset{\sim}{a}}_\alpha$, $\bar{\underset{\sim}{a}}^\alpha$, $\bar{a}_{\alpha\beta}$, $\bar{a}^{\alpha\beta}$, $\bar{\underset{\sim}{a}}$, $\bar{a}$, $\bar{\underset{\sim}{n}}$, $\bar{b}_{\alpha\beta}$, $\bar{\underset{\sim}{b}}$ etc. The deformation of M into $\bar{M}$ and C into $\bar{C}$ is described in detail in [3,4,12] where further details may be found.

Let on the middle surface M of P the following fields are introduced

$$\underset{\sim}{a}_i = \underset{\sim}{g}_i(P)|_{\zeta=0} \;, \quad \underset{\sim}{r}_i = \underset{\sim}{d}_i(P)|_{\zeta=0} \;, \quad \bar{\underset{\sim}{a}}_i = \bar{\underset{\sim}{g}}_i(\bar{P})|_{\zeta=0} \;,$$
$$\underset{\sim}{u} = \underset{\sim}{w}(P)|_{\zeta=0} \;, \quad \underset{\sim}{Q} = \underset{\sim}{R}(P)|_{\zeta=0} \;, \quad \underset{\sim}{G} = \underset{\sim}{F}(P)|_{\zeta=0} \;, \quad \underset{\sim}{\eta} = \underset{\sim}{H}(P)|_{\zeta=0} \;. \tag{3.1}$$

Material fibres of the body B, initially normal to M, after deformation may become neither straight nor normal to $\bar{M}$. In particular, $\bar{\underset{\sim}{a}}_3 \neq \bar{\underset{\sim}{n}}$, in general. However, when discussing deformation of thin elastic shells undergoing small strains it is possible in all kinematic relations to approximate it by assuming, that material fibres which are normal to M, after deformation remain straight and normal to $\bar{M}$ and do not change their lengths. Under such Kirchhoff-Love type kinematic constraint $\bar{\mathbf{a}}_3 \equiv \bar{\mathbf{n}}$. The change of the shell thickness during deformation will, however, be taken into account in the constitutive equations, see (4.2) below.

The geometry of P in the normal system of coodinates is described by the geometry of the surface M. Under the K-L kinematic constraints the same applies to $\bar{P}$ and $\bar{M}$, respectively,

$$
\begin{aligned}
&\mathbf{p} = \mathbf{r} + \zeta\mathbf{n} \ , \qquad \bar{\mathbf{p}} = \bar{\mathbf{r}} + \zeta\bar{\mathbf{n}} \ , \\
&\mathbf{g} = \mathbf{1} - \zeta\mathbf{b} \ , \qquad \bar{\mathbf{g}} = \bar{\mathbf{1}} - \zeta\bar{\mathbf{b}} \ , \\
&\mathbf{g}_i = \mathbf{g}\mathbf{a}_i \ , \quad \mathbf{g}^i = \mathbf{g}^{-1}\mathbf{a}^i \ , \quad \bar{\mathbf{g}}_i = \bar{\mathbf{g}}\bar{\mathbf{a}}_i \ , \quad \bar{\mathbf{g}}^i = \bar{\mathbf{g}}^{-1}\bar{\mathbf{a}}^i \ .
\end{aligned}
\tag{3.2}
$$

Under the K-L kinematic constraints the kinematic parameters of the pseudo-Cosserat shell are given by

$$
\begin{aligned}
&\mathbf{W} = \mathbf{u} + \zeta(\bar{\mathbf{n}} - \mathbf{n}) \ , \qquad \mathbf{R} = \mathbf{Q} \ , \\
&\mathbf{F} = \bar{\mathbf{g}}\mathbf{G}\mathbf{g}^{-1} = (\mathbf{G} - \zeta\bar{\mathbf{b}}\mathbf{G})\mathbf{g}^{-1} \ , \\
&\mathbf{H} = (\boldsymbol{\eta} + \zeta\boldsymbol{\kappa})\mathbf{g}_r^{-1} \ , \qquad \mathbf{g}_r^{-1} = \mathbf{Q}\mathbf{g}^{-1}\mathbf{Q} \ ,
\end{aligned}
\tag{3.3}
$$

$$
\begin{aligned}
&\boldsymbol{\eta} = \eta_{ij}\mathbf{r}^i \otimes \mathbf{r}^j = \boldsymbol{\eta}_j \otimes \mathbf{r}^j \ , \\
&\boldsymbol{\eta}_\beta = \eta_{i\beta}\mathbf{r}^i = \bar{\mathbf{a}}_\beta - \mathbf{r}_\beta \ , \quad \boldsymbol{\eta}_3 = \eta_{i3}\mathbf{r}^i = \bar{\mathbf{n}} - \mathbf{r}_3 \ ,
\end{aligned}
\tag{3.4}
$$

$$
\begin{aligned}
&\boldsymbol{\kappa} = \kappa_{i\beta}\mathbf{r}^i \otimes \mathbf{r}^\beta = \boldsymbol{\kappa}_\beta \otimes \mathbf{r}^\beta \ , \\
&\boldsymbol{\kappa}_\beta = \bar{\mathbf{n}}_{,\beta} + b^\alpha_\beta\mathbf{r}_\alpha = \mathbf{l}_\beta \times \mathbf{r}_3 + \boldsymbol{\eta}_{3,\beta} \ , \\
&\mathbf{l}_\beta = -\frac{1}{2}\boldsymbol{\epsilon}\cdot(\mathbf{Q}_{,\beta}\mathbf{Q}^T) = \frac{1}{t}(\boldsymbol{\theta}_{,\beta} - \frac{1}{2}\boldsymbol{\theta}_{,\beta} \times \boldsymbol{\theta}) \ .
\end{aligned}
\tag{3.5}
$$

If, additionally, the micro-rotations $\mathbf{Q}$ are constrained to coincide with the macro-rotations $\mathbf{R}$ then

$$
\begin{aligned}
&\mathbf{r}_3 = \bar{\mathbf{n}} \ , \quad \boldsymbol{\eta}_3 = \mathbf{0} \ , \quad \eta_{3\beta} = 0 \ , \quad \kappa_{3\beta} = 0 \ , \\
&\boldsymbol{\eta}_\beta = \eta_{\alpha\beta}\mathbf{r}^\alpha = \mathbf{u}_{,\beta} - \frac{1}{t}\boldsymbol{\theta} \times (\mathbf{r}_\beta - \frac{1}{2}\boldsymbol{\theta} \times \mathbf{r}_\beta) \ , \\
&\boldsymbol{\kappa}_\beta = \kappa_{\alpha\beta}\mathbf{r}^\alpha = \mathbf{l}_\beta \times \bar{\mathbf{n}} = \frac{1}{t}(\boldsymbol{\theta}_{,\beta} - \frac{1}{2}\boldsymbol{\theta}_{,\beta} \times \boldsymbol{\theta}) \times \bar{\mathbf{n}} \ .
\end{aligned}
\tag{3.6}
$$

Solving $(3.6)_2$ and $(3.5)_3$ for $\mathbf{u}_{,\beta}$ and $\boldsymbol{\theta}_{,\beta}$ and applying the integrability conditions $\underset{\sim}{u}_{,\beta\alpha} - \underset{\sim}{u}_{,\alpha\beta} = \underset{\sim}{0}$, $\underset{\sim}{\theta}_{,\beta\alpha} - \underset{\sim}{\theta}_{,\alpha\beta} = \underset{\sim}{0}$ we obtain the vector form of compatibility conditions of the non-linear shell theory

$$\epsilon^{\alpha\beta}(\underset{\sim}{\eta}_{\alpha|\beta} + \underset{\sim}{l}_{\alpha} \times \underset{\sim}{r}_{\beta}) = \underset{\sim}{0} \quad , \quad \epsilon^{\alpha\beta}(\underset{\sim}{l}_{\alpha|\beta} + \tfrac{1}{2}\underset{\sim}{l}_{\alpha} \times \underset{\sim}{l}_{\beta}) = \underset{\sim}{0} \tag{3.7}$$

given in component form in [7] and in vector form in [15,16].

Note that the strain measures $\eta_{\alpha\beta}$ and $\kappa_{\alpha\beta}$ defined in $(3.6)_{2,3}$ are rational (quadratic, at most) functions of displacements, rotations and their first surface derivatives. In general, $\eta_{\alpha\beta} \neq \eta_{\beta\alpha}$ and $\kappa_{\alpha\beta} \neq \kappa_{\beta\alpha}$. Linearizing $\underset{\sim}{\eta}_{\beta}$ and $\underset{\sim}{\kappa}_{\beta}$ with respect to $\underset{\sim}{u}$ and $\underset{\sim}{\theta}$ and taking into account that in the linear shell theory [3,12] $\underset{\sim}{\theta} = \underset{\sim}{\theta}(\underset{\sim}{u}) \simeq \epsilon^{\beta\alpha}\phi_{\alpha}\underset{\sim}{a}_{\beta} + \phi\underset{\sim}{n}$ we obtain

$$\begin{aligned} \eta_{\alpha\beta} &\simeq \tfrac{1}{2}(u_{\alpha|\beta} + u_{\beta|\alpha}) - b_{\alpha\beta}w \quad , \quad \kappa_{\alpha\beta} \simeq -\phi_{\alpha|\beta} - b^{\lambda}_{\beta}\epsilon_{\lambda\alpha}\phi \quad , \\ \phi_{\alpha} &= w_{,\alpha} + b^{\beta}_{\alpha}u_{\beta} \quad , \quad \phi = \tfrac{1}{2}\epsilon^{\alpha\beta}u_{\beta|\alpha} \quad , \end{aligned} \tag{3.8}$$

where $(\ )_{|\alpha}$ means the covariant differentiation in the reference metric $a_{\alpha\beta}$. The linearized components of $\eta_{\alpha\beta}$ and $\kappa_{(\alpha\beta)} \equiv \frac{1}{2}(\kappa_{\alpha\beta} + \kappa_{\beta\alpha})$ are the measures used in the "best" linear shell theory [31].

4. Variationally derivable non-linear shell equations

The two-dimensional strain energy density, per unit area of the undeformed middle surface M , for the constrained Cosserat shell is given in terms of (2.16) by

$$\Sigma = \int_{-h/2}^{h/2} E\sqrt{\frac{g}{a}}\, d\zeta = \Sigma_{o} + \Sigma_{con} \ . \tag{4.1}$$

Within the consistent first-approximation geometrically non-linear theory of thin isotropic elastic shells the first term in (4.1) can be approximated by the quadratic expression [27,13]

$$\begin{aligned} \Sigma_{o} &= \frac{1}{2} H^{\alpha\beta\lambda\mu}(\eta_{\alpha\beta}\eta_{\lambda\mu} + \frac{h^2}{12}\kappa_{\alpha\beta}\kappa_{\lambda\mu}) \quad , \\ H^{\alpha\beta\lambda\mu} &= \frac{E}{2(1+\nu)}(a^{\alpha\lambda}a^{\beta\mu} + a^{\alpha\mu}a^{\beta\lambda} + \frac{2\nu}{1-\nu}a^{\alpha\beta}a^{\lambda\mu}) \quad . \end{aligned} \tag{4.2}$$

The second term in (4.1) appears as the result of constrained micro-rotations of the pseudo-Cosserat body and follows from direct integration of (2.16), with $(3.6)_1$ and the accuracy of $(4.2)_1$ taken into account,

$$\Sigma_{con} = \int_{-h/2}^{h/2} \epsilon^{ijk}E_{ij}\lambda_{k}\sqrt{\frac{g}{a}}\, d\zeta = \epsilon^{\alpha\beta}\underset{\sim}{r}_{\alpha}\cdot\underset{\sim}{\eta}_{\beta}N + \bar{\underset{\sim}{n}}\cdot\underset{\sim}{\eta}_{\beta}Q^{\beta} \quad , \tag{4.3$_1$}$$

$$N = \int_{-h/2}^{h/2} \lambda_3 \, d\zeta \quad , \qquad Q^\beta = \int_{-h/2}^{h/2} \epsilon^{\beta\alpha} \lambda_\alpha \, d\zeta \quad . \tag{4.3$_2$}$$

If we introduce the symmetric surface internal stress resultant and stress couple tensors

$$N^{\alpha\beta} = \frac{\partial \Sigma_o}{\partial \eta_{\alpha\beta}} = H^{\alpha\beta\lambda\mu} \eta_{\lambda\mu} \quad , \qquad M^{\alpha\beta} = \frac{\partial \Sigma_o}{\partial \kappa_{\alpha\beta}} = \frac{h^2}{12} H^{\alpha\beta\lambda\mu} \kappa_{\lambda\mu} \tag{4.4}$$

then the strain energy (4.1) with (4.3) generates the following constitutive equations for the internal stress resultant and stress couple vectors defined with reference to the rotated basis

$$\begin{aligned} \underset{\sim}{N}^\beta &= \frac{\partial \Sigma}{\partial \underset{\sim}{\eta}_\beta} = (N^{\alpha\beta} + \epsilon^{\alpha\beta} N) \underset{\sim}{r}_\alpha + Q^\beta \bar{\underset{\sim}{n}} \quad , \\ \underset{\sim}{K}^\beta &= \frac{\partial \Sigma}{\partial \underset{\sim}{\kappa}_\beta} = M^{\alpha\beta} \underset{\sim}{r}_\alpha \quad , \qquad \underset{\sim}{M}^\beta = \bar{\underset{\sim}{n}} \times \underset{\sim}{K}^\beta = \epsilon_{\alpha\lambda} M^{\alpha\beta} \underset{\sim}{r}^\lambda \quad . \end{aligned} \tag{4.5}$$

Note that the vector $\underset{\sim}{M}^\beta$ in $(4.5)_2$ is given only through symmetric components $M^{\alpha\beta}$. This means that within the accuracy of the first-approximation non-linear shell theory only symmetric part $\kappa_{(\alpha\beta)}$ of the tensor of change of curvature enters explicitly into the shell equations. The scew-symmetric part $\kappa_{[\alpha\beta]}$ of this tensor does not contribute to the elastic strain energy density and, therefore, may be ignored in all shell relations.

The formulae (4.4) reveal the physical meaning of the Lagrangian multipliers N and Q^β to be just the scew-symmetric part of the internal surface stress resultant tensor and the shearing forces, respectively. Moreover, the tensor fields $\underset{\sim}{N} = \underset{\sim}{N}^\beta \otimes \underset{\sim}{r}_\beta$ and $\underset{\sim}{K} = \underset{\sim}{K}^\beta \otimes \underset{\sim}{r}_\beta$ are seen to be the two-dimensional counterparts of the Biot-type stress measures of the Cosserat continuum (2.13).

Let the shell be loaded by an external surface load $\underset{\sim}{p}$, per unit area of M, and by external boundary load $\underset{\sim}{f}$, per unit area of the reference boundary surface $C \times (-\frac{h}{2}, \frac{h}{2})$. For simplicity of further results let us assume $\underset{\sim}{p}$ and $\underset{\sim}{f}$ to be conservative and dead-load type. Then the total potential energy of the shell is given by the functional [4]

$$\begin{aligned} I &= \iint_M [\Sigma(\underset{\sim}{u}, \underset{\sim}{\theta}, N, Q^\beta) - \underset{\sim}{p} \cdot \underset{\sim}{u}] \, dA - \int_C [\underset{\sim}{T} \cdot \underset{\sim}{u} + \underset{\sim}{H} \cdot (\bar{\underset{\sim}{n}} - \underset{\sim}{n})] \, ds \quad , \\ \underset{\sim}{T} &= \int_{-h/2}^{h/2} \underset{\sim}{f} \sqrt{\frac{g}{a}} \, d\zeta \quad , \qquad \underset{\sim}{H} = \int_{-h/2}^{h/2} \underset{\sim}{f} \zeta \sqrt{\frac{g}{a}} \, d\zeta \quad . \end{aligned} \tag{4.6}$$

The variational principle $\delta I = 0$ states that among all possible values of independent fields $\underset{\sim}{u}$, $\underset{\sim}{\theta}$, N and Q^β, which are subject to the geometric boundary conditions, the actual solution renders the functional stationary.

Let us find the stationarity conditions of I. Taking into account that $\delta\underset{\sim}{r}_\alpha = \delta\underset{\sim}{\theta}\times\underset{\sim}{r}_\alpha$, $\delta\bar{\underset{\sim}{n}} = \delta\underset{\sim}{\theta}\times\bar{\underset{\sim}{n}}$, $\delta\bar{\underset{\sim}{a}}_\beta = \delta\underset{\sim}{u}_{,\beta}$ and $\delta\underset{\sim}{l}_\beta = \delta\underset{\sim}{\theta}_{,\beta}$ we obtain in the rotated basis

$$\delta\eta_{\alpha\beta} = \delta(\underset{\sim}{r}_\alpha\cdot\bar{\underset{\sim}{a}}_\beta) = \underset{\sim}{r}_\alpha\cdot\delta\underset{\sim}{\eta}_\beta \ , \qquad \delta\kappa_{\alpha\beta} = \delta[\underset{\sim}{r}_\alpha\cdot(\underset{\sim}{l}_\beta\times\bar{\underset{\sim}{n}})] = \underset{\sim}{r}_\alpha\cdot\delta\underset{\sim}{\kappa}_\beta ,$$
$$\delta\underset{\sim}{\eta}_\beta = \delta\underset{\sim}{u}_{,\beta} + \bar{\underset{\sim}{a}}_\beta\times\delta\underset{\sim}{\theta} \ , \qquad \delta\underset{\sim}{\kappa}_\beta = \delta\underset{\sim}{\theta}_{,\beta}\times\bar{\underset{\sim}{n}} + \bar{\underset{\sim}{n}}\times(\delta\underset{\sim}{\theta}\times\underset{\sim}{l}_\beta) \ . \tag{4.7}$$

The variation of $(4.6)_1$, performed with the help of (4.5),(4.7) and Stokes' theorem, leads to

$$\begin{aligned}\delta I = - \iint_M [(\underset{\sim}{N}^\beta|_\beta + \underset{\sim}{p})\cdot\delta\underset{\sim}{u} + (\underset{\sim}{M}^\beta|_\beta + \bar{\underset{\sim}{a}}_\beta\times\underset{\sim}{N}^\beta)\cdot\delta\underset{\sim}{\theta} - \\ - \epsilon^{\alpha\beta}\bar{\underset{\sim}{r}}_\alpha\cdot\underset{\sim}{\eta}_\beta\delta N - \bar{\underset{\sim}{n}}\cdot\underset{\sim}{\eta}_\beta\delta Q^\beta \,]\, dA + \\ + \int_C \{ (\underset{\sim}{N}^\beta\nu_\beta - \underset{\sim}{T})\cdot\delta\underset{\sim}{u} + [\bar{\underset{\sim}{n}}\times(\underset{\sim}{K}^\beta\nu_\beta - \underset{\sim}{H})]\cdot\delta\underset{\sim}{\theta} \} \, ds \ .\end{aligned} \tag{4.8}$$

The variations $\delta\underset{\sim}{u}$ and $\delta\underset{\sim}{\theta}$ at the boundary contour C are not independent, since the micro-rotations have been assumed to coincide with the macro-rotations. This constraint condition has been explicitly taken into account in the internal part of the shell. It has not been taken into account at the shell lateral boundary yet. Let $\underset{\sim}{H} = H_\nu\bar{\underset{\sim}{\nu}} + H_t\bar{\underset{\sim}{t}} + H\bar{\underset{\sim}{n}}$. Note also that for small strains $\bar{\underset{\sim}{t}} \simeq \bar{\underset{\sim}{a}}_t = \bar{\underset{\sim}{a}}_\alpha t^\alpha = \underset{\sim}{t} + d\underset{\sim}{u}/ds$, $\bar{\underset{\sim}{\nu}} = \bar{\underset{\sim}{t}}\times\bar{\underset{\sim}{n}} \simeq \bar{\underset{\sim}{a}}_t\times\bar{\underset{\sim}{n}}$. Then $(4.8)_3$ can be transformed further into

$$\int_C [(\underset{\sim}{P} - \underset{\sim}{P}^*)\cdot\delta\underset{\sim}{u} + (M - M^*)\bar{\underset{\sim}{t}}\cdot\delta\underset{\sim}{\theta}]\, ds \ + \sum_j(\underset{\sim}{F}_j - \underset{\sim}{F}^*_j)\cdot\delta\underset{\sim}{u}_j \ , \tag{4.9}$$

where

$$\underset{\sim}{F} = M^{\alpha\beta}\nu_\alpha t_\beta\bar{\underset{\sim}{n}} \ , \quad \underset{\sim}{F}^* = H_t\bar{\underset{\sim}{n}} \ , \quad M = M^{\alpha\beta}\nu_\alpha\nu_\beta \ , \quad M^* = H_\nu \ ,$$
$$\underset{\sim}{P} = \underset{\sim}{N}^\beta\nu_\beta + \frac{d}{ds}\underset{\sim}{F} \ , \quad \underset{\sim}{P}^* = \underset{\sim}{T} + \frac{d}{ds}\underset{\sim}{F}^* , \quad \underset{\sim}{F}_j = \underset{\sim}{F}(s_j+0) - \underset{\sim}{F}(s_j-0) \ . \tag{4.10}$$

Now from $\delta I = 0$, together with (4.8),(4.9) and (4.10), we obtain the following stationarity conditions of I , which consist of equilibrium equations, constraint conditions as well as static boundary and corner conditions :

$$\left.\begin{aligned} \underset{\sim}{N}^\beta|_\beta + \underset{\sim}{p} = \underset{\sim}{0} \ , \qquad \underset{\sim}{M}^\beta|_\beta + \bar{\underset{\sim}{a}}_\beta\times\underset{\sim}{N}^\beta = \underset{\sim}{0} \\ \epsilon^{\alpha\beta}\underset{\sim}{r}_\alpha\cdot\underset{\sim}{\eta}_\beta = 0 \ , \qquad \bar{\underset{\sim}{n}}\cdot\underset{\sim}{\eta}_\beta = 0 \end{aligned}\right\} \text{ in } M \tag{4.11}$$

$$\underset{\sim}{P} = \underset{\sim}{P}^* \ , \quad M = M^* \quad \text{on } C_f \quad \text{and} \quad \underset{\sim}{F}_j = \underset{\sim}{F}^*_j \quad \text{at each corner } M_j \in C_f .$$

All field variables in (4.11) are given by components in the rotated basis $\underset{\sim}{r}_i$ and are understood to be expressed in terms of $\underset{\sim}{u}$, $\underset{\sim}{\theta}$, N and Q^β as independent variables.

5. Discussion

The non-linear shell theory based on the functional (4.6) or on its incremental form (4.8) has some interesting properties. The functional depends explicitly only upon displacements $\underset{\sim}{u}$, rotations $\underset{\sim}{\theta}$ and Lagrangian multipliers N , Q^β as nine basic independent variables to be discretized. The functional is linear in N , Q^β and is rational function containing at most fourth-order polynomials in $\underset{\sim}{u}$, $\underset{\sim}{\theta}$ and their first surface derivatives. The latter property is of great importance for the computerized numerical analysis of the flexible shell structures. It allows to apply the simplest shape functions in the finite-element analysis or the simplest difference schemes in the finite-difference analysis, which assure the high efficiency of numerical algorithms applied and better convergence to the accurate results for highly non-linear problems of flexible shells.

The equilibrium equations $(4.11)_1$ were derived already in the pioneering work of ALUMÄE [7], who suggested to solve them in the intrinsic form, together with the compatibility conditions (3.7). For shallow shells the set of equations was solved in [7] in terms of two scalar stress and displacement functions. SIMMONDS and DANIELSON [10,11] expressed the basic set of shell equations in terms of finite rotations and the internal stress resultants (or the stress function vector) as basic independent variables and constructed an appropriate variational functional [10]. Several variational principles, involving the finite rotation tensor and the stress function vector,were also discussed by ATLURI [17]. In the approach used in [10,11,17] displacement field is understood to be calculated, it necessary, by additional quadratures. SHKUTIN [15] proposed to solve the equilibrium equations $(4.11)_1$ in terms of displacements and rotations, but rotations were still expressed through displacements by additional differential relations. The primary difference between the present approach and that of [15] (apart from unusual notation applied in [15]) is that in our approach the constraint conditions put on the rotations in the internal shell space have been introduced explicitly into the variational principle. As a result three additional physically meaningful independent parameters N , Q^β have explicitly appeared in the shell equations.

Acknowledgements

Partial support for this research was provided by the Ruhr-Universität Bochum , Fed. Rep. of Germany.

References

1. MUSHTARI K.M. and GALIMOV K.Z., Non-linear theory of elastic shells (in Russian), Tatknigoizdat, Kazań 1957.

2. KOITER W.T., On the nonlinear theory of thin elastic shells, Proc. Koninkl. Ned. Ak. Wet. B69(1966), 1, 1-54.

3. PIETRASZKIEWICZ W., Finite rotations in the non-linear theory of thin shells, in: Thin Shell Theory, New Trends and Applications, 155-208, ed. by W.Olszak, CISM Course No 240, Springer-Verlag, Wien 1980.

4. PIETRASZKIEWICZ W., Lagrangian description and incremental formulation in the non-linear theory of thin shells, Int. J. Non-Linear Mech., 19(1984), 2, 115-141.

5. SCHMIDT R., A current trend in shell theory: constrained geometrically nonlinear Kirchhoff-Love type theories based on polar decomposition of strains and rotations, Comp. and Str., 20(1985), 1-3, 265-275.

6. BAŞAR Y. and KRÄTZIG W.B., Mechanik der Flächentragwerke, Vieweg, Braunschweig 1985.

7. ALUMÄE N.A., Differential equations of equilibrium states of thin--walled elastic shells in the post-critical stage (in Russian), Prikl.Mat.Mekh 13(1949), 1, 95-106.

8. REISSNER E., On axisymmetric deformations of thin shells of revolution, Proc. Symp. Appl. Math., 3(1950), 27-52.

9. WEMPNER G., Finite elements, finite rotations and small strains, Int. J. Solids and Str.,(1969),5, 117-153.

10. SIMMONDS J.G. and DANIELSON D.A., Nonlinear shell theory with a finite rotation and stress-function vectors, J. Appl. Mech., Trans. ASME E39(1972), 4, 1085-1090.

11. SIMMONDS J.G. and DANIELSON D.A., Nonlinear shell theory with a finite rotation vector, Proc. Koninkl. Ned. Ak. Wet. B73(1970), 5, 460-478.

12. PIETRASZKIEWICZ W., Introduction to the Non-linear Theory of Shells, Ruhr-Universität, Mitt. Inst. f. Mech. Nr 10, Bochum, Mai 1977, 1-154.

13. PIETRASZKIEWICZ W., Obroty skończone i opis Lagrange'a w nieliniowej teorii powłok, Biul. IMP PAN 172(880), Gdańsk 1976. English transl.: Finite Rotations and Lagrangean Description in the Non--Linear Theory of Shells, Polish Sci. Publ., Warszawa-Poznań 1979.

14. PIETRASZKIEWICZ W., Finite rotations in shells, in: Theory of Shells, 445-471, Ed. by W.T.Koiter and G.K.Mikhailov, North-Holland P.Co., Amsterdam 1980.

15. SHKUTIN L.I., An exact formulation of equations of the non-linear deformation of thin shells (in Russian), in: Applied Problems of Strength and Plasticity (in Russian), 7(1977), 3-9; 8(1978),38-43; 9(1978), 19-25.

16. LIBAI A. and SIMMONDS J.G., Nonlinear elastic shell theory, in: Advances in Applied Mechanics, vol.23, 271-371, Academic Press, New York 1983.

17. ATLURI S.N., Alternate stress and conjugate strain measures, and mixed variational formulations involving rigid rotations, for computational analyses of finitely deformed solids, with application

to plates and shells -I. Theory, Comp. and Str. 18(1984), 1, 93-116.

18. KAYUK Ya.F. and SAKHATSKIY V.G., On the non-linear theory of shells based on the notion of a finite rotation, Soviet Applied Mechanics, 21(1985), 4, 65-73.

19. MAKOWSKI J. and STUMPF H., Finite strains and rotations in shells, in: Finite Rotations in Structural Mechanics, ed. by W.Pietraszkiewicz, Springer-Verlag, Berlin 1986.

20. COSSERAT E. and COSSERAT F., Theorie des Corps deformables, Herman, Paris 1909.

21. NAGHDI P.M., The theory of shells and plates, in: Handbuch der Physik, vol. VI/2, Springer-Verlag, Berlin 1972.

22. SCHROEDER F.H., Cosserat theory of shells with large rotations and displacements, lecture presented at the Euromech Colloquium 165 "Flexible Shells", 17-20 May, München 1983.

23. ERINGEN A.C. and KAFADAR C.B., Polar field theories, in: Continuum Physics, vol IV, 1-73, ed. by A.C.Eringen, Academic Press, New York 1976.

24. TOUPIN R.A., Theories of elasticity with couple-stresses, Arch. Rat. Mech. Anal., 17 (1964), 2, 85-110.

25. KOITER W.T., Couple-stresses in the theory of elasticity, Proc. Koninkl. Ned. Ak. Wet., B67(1964), 1, 17-29; B67(1964), 1, 30-48.

26. SHKUTIN L.I., Non-linear models for deformable continuum with couple-stresses (in Russian), Zhurnal Prikl. Mekh. Tekh. Fiz., 1980, 6,111-117.

27. KOITER W.T., A consistent first approximation in the general theory of thin elastic shells, in: The Theory of Thin Elastic Shells, 12--33, Ed. by W.T.Koiter, North-Holland P.Co., Amsterdam 1960.

28. PIETRASZKIEWICZ W. and BADUR J., Finite rotations in the description of continuum deformation, Int. J. Engng Sci., 21(1983), 9, 1097-1115.

29. TRUESDELL C. and NOLL W., The Nonlinear Field Theories of Mechanics, in: Handbuch der Physik, vol. III/3, Springer-Verlag, Berlin 1965.

30. HILL R., On constitutive inequalities for simple materials, Int. J. Mech. Phys. Solids, 16(1968), 5, 229-241.

31. BUDIANSKY B. and SANDERS J.L., On the "best" first-order linear shell theory, in: Progress in Applied Mechanics (Prager Anniv. Vol) 129-140, Macmillan, New York 1963.

SOME MATHEMATICAL RESULTS RELATED TO NONLINEAR THIN SHELL PROBLEMS

M. BERNADOU
I.N.R.I.A.
Domaine de Voluceau, B.P. 105
Rocquencourt, 78153 Le Chesnay Cedex, France

1. Introduction

There exist many different nonlinear models of thin shell problems depending of the geometry of the shell, of the magnitude of the displacement or of the strains. An account of these models can be found in KOITER [15]. More recently it appears to be meaningful to decompose the motion of a shell into translation, rigid-body rotation and strain. Such decomposition is particularly interesting for shells whose deformation leads to large rotations but small strains. In this way, PIETRASZKIEWICZ [19], PIETRASZKIEWICZ and SZWABOWICZ [20] derived several models according to the magnitude of the rigid-body rotations, with and without restrictions on the magnitude of the strains.

When compared to the great number of models of shells, there are very few results concerning the mathematical studies of corresponding equations, particularly the existence of solutions and their approximations. Let us mention for instance BERNADOU and CIARLET [5] for existence of linear shell solutions, BERNADOU and ODEN [8], RUPPRECHT [22], VOROVICH and LEBEDEV [25], WEINITSCHKE [26] for existence of nonlinear shallow shell solutions, NAUMANN [18] for the bifurcation buckling of thin elastic shells. A review of some results can be found in DIKMEN [14]. Perhaps the combined use of intrinsic notations (VALID [24]) and of asymptotic methods (DESTUYNDER [12]) could improve the situation. This is partly dependent of possible progresses in the mathematical studies of three-dimensional linear and nonlinear elasticity equations (CIARLET [11]), specially the part concerning existence and regularity of solutions. Another interesting way to prove such existence results for nonlinear shell equations should be the study of semi-discrete incremental methods : such results were obtained by BERNADOU, CIARLET and HU [6] in case of nonlinear three dimensional elasticity equations with boundary conditions of clamped type.

Concerning the mathematical studies of the approximation of such solutions we refer to BERNADOU [3], BERNADOU and BOISSERIE [4], CIARLET [9,10] for conforming displacement finite element methods, to DESTUYNDER and LUTOBORSKI [13] for mixed finite element methods, and to STEPHAN and WEISGERBER [23] for hybrid finite element methods. All of these results are concerned with the linear problem ; the study of nonlinear problems seems to be very open.

In this paper, we propose

(i) a unified variational formulation of some classical nonlinear thin shell problems ;

(ii) a variational formulation of corresponding total Lagrangean incremental methods ;

(iii) an account of some mathematical existence results for thin shell problems that we have developped recently.

2. Geometrical definition of a thin shell

Let Ω be a bounded open subset in a plane $\mathcal{E}^2$, with boundary Γ. Then the middle surface $\mathcal{S}$ of the shell is the image of the set $\bar{\Omega}$ by a mapping $\vec{\phi} : \bar{\Omega} \subset \mathcal{E}^2 \to \mathcal{E}^3$, where $\mathcal{E}^3$ is the usual Euclidean space. Subsequently, we shall assume that $\vec{\phi} \in (\mathcal{C}^3(\bar{\Omega}))^3$ and that all points of $\mathcal{S} = \vec{\phi}(\bar{\Omega})$ are regular, in the sense that the two vectors $\vec{a}_\alpha = \vec{\phi}_{,\alpha}$, $\alpha = 1,2$, are linearly independent for all points $\xi = (\xi^1,\xi^2) \in \bar{\Omega}$. With the covariant basis $(\vec{a}_\alpha)$ of the tangent plane, we associate the contravariant basis $(\vec{a}^\alpha)$ which is defined through the relations $\vec{a}^\alpha . \vec{a}_\beta = \delta^\alpha_\beta$, where δ^α_β is the Kronecker's symbol. We also introduce the vector $\vec{a}_3 = \vec{a}_1 \times \vec{a}_2 / |\vec{a}_1 \times \vec{a}_2|$.

The thickness e of the shell $\mathcal{C}$ can be viewed as an application $e : (\xi^1,\xi^2) \in \bar{\Omega} \to \{x \in \mathbb{R} \ ; \ x > 0\}$.

Then, the shell $\mathcal{C}$ is the set :

$$\mathcal{C} = \{M \in \mathcal{E}^3 \ ; \ \vec{OM} = \vec{\phi}(\xi^1,\xi^2) + \xi^3 \vec{a}_3, \ (\xi^1,\xi^2) \in \bar{\Omega}, \ -\frac{1}{2} e(\xi^1,\xi^2) \leq \xi^3 \leq \frac{1}{2} e(\xi^1,\xi^2)\}.$$

3. Deformation of a thin shell (KOITER [15])

The basic idea of KOITER's thin shell theories is to reduce the study of the deformation of a thin shell to the determination of the displacement field $\vec{u} = u_i \vec{a}^i$ of the particles of the middle surface.

In this way, KOITER uses the two following basic hypotheses :

(i) the normal to the undeformed middle surface, considered as a set of points, remains normal to the deformed middle surface ;

(ii) during the deformation, the stresses are approximatively plane and parallel to the tangent plane to the middle surface.

4. Total potential energy of the shell

In what follows, we assume that the shell is

(i) loaded by a distribution of forces whose resultant is $\vec{p}$ on the middle surface $\mathcal{S}$ and whose resultant moment is $\vec{0}$ on $\mathcal{S}$;

(ii) clamped on the part Γ_o of its boundary $\Gamma = \partial\Omega$;

(iii) loaded on the complementary part $\Gamma_1 = \Gamma - \Gamma_o$ of its boundary by a distribution of dead forces whose resultant is $\vec{N}$ on Γ_1 and whose resultant boundary couple is $\vec{k}$.

Then, the total potential energy of the shell for an admissible displacement field $\vec{v}$ and an admissible total finite rotation vector $\vec{\Omega}_t$ of the boundary, is given by

$$\left.\begin{aligned} J(\vec{v},\vec{\Omega}_t) = \frac{1}{2}\int_\Omega [n^{\alpha\beta}(\vec{v})\,\gamma_{\alpha\beta}(\vec{v}) + m^{\alpha\beta}(\vec{v})\,\rho_{\alpha\beta}(\vec{v})]\,\sqrt{a}\,d\xi^1\,d\xi^2 \\ - \int_\Omega \vec{p}\,\vec{v}\,\sqrt{a}\,d\xi^1\,d\xi^2 - \int_{\Gamma_1} (\vec{N}\,\vec{v} + \vec{k}\,\vec{\Omega}_t)\,d\gamma\,, \end{aligned}\right\} \quad (1)$$

where

$(n^{\alpha\beta})$ = symmetrical tensor of stress resultants
$(m^{\alpha\beta})$ = symmetrical tensor of stress couples
$(\gamma_{\alpha\beta})$ = strain tensor on the middle surface
$(\rho_{\alpha\beta})$ = tensor of changes of curvature
$d\mathcal{S} = \sqrt{a}\,d\xi^1\,d\xi^2$: area element on $\mathcal{S}$.

Remark 4.1 : At this stage, in order to get the HU-WASHIZU variational functional of the problem which is suitable even for large and finite rotations, PIETRASKIEWICZ and SZWABOWICZ [20,21] have found more convenient :

(i) to replace $\vec{k}$ by the equivalent external boundary static moment $\vec{H}$ where $\vec{k} = \vec{\bar{a}}_3 \times \vec{H}$, where $\vec{\bar{a}}_3$ is the normal to the deformed middle surface along the boundary Γ_1 ;

(ii) to use a modified tensor of change of curvature

$$\chi_{\alpha\beta} = -\left(\sqrt{\frac{\bar{a}}{a}}\,\bar{b}_{\alpha\beta} - b_{\alpha\beta}\right) + b_{\alpha\beta}\,\gamma^\kappa_\kappa\,.$$

instead of $\rho_{\alpha\beta} = (\bar{b}_{\alpha\beta} - b_{\alpha\beta})$.

□

Then, corresponding total potential energy can be written as follows :

$$J(\vec{v},n_\nu) = \frac{1}{2}\int_\Omega [n^{\alpha\beta}(\vec{v})\,\gamma_{\alpha\beta}(\vec{v}) + m^{\alpha\beta}(\vec{v})\,\chi_{\alpha\beta}(\vec{v})]\,\sqrt{a}\,d\xi^1\,d\xi^2$$
$$- \int_\Omega \vec{p}\,\vec{v}\,\sqrt{a}\,d\xi^1\,d\xi^2 - \int_{\Gamma_1} (\vec{N}\,\vec{v} + \vec{H}(\vec{\bar{a}}_3 - \vec{a}_3))\,d\gamma$$

where n_ν is, in addition to $\vec{v}$, a fourth independent variable defined on Γ_1 ($\vec{\bar{a}}_3 = n_\nu\vec{\nu} + n_t\,\vec{t} + n\,\vec{a}_3$ where $\vec{t}$ is the unit tangent to $\vec{\phi}(\Gamma_1)$ and $\vec{\nu} = \vec{t}\times\vec{a}_3$).

□

Complementary hypotheses : In addition, we assume the shell is isotropic, homogeneous and elastic, the strains are small, the stresses are approximatively plane and, for simplicity, we add $\vec{k} = \vec{0}$. So, we have the following relations

$$n^{\alpha\beta}(\vec{v}) = e\,E^{\alpha\beta\lambda\gamma}\,\gamma_{\lambda\Gamma}(\vec{v})\ ,\ m^{\alpha\beta}(\vec{v}) = \frac{e^3}{12}\,E^{\alpha\beta\lambda\Gamma}\,\rho_{\lambda\Gamma}(\vec{v})\ ,\ \vec{k} = 0\ , \tag{2}$$

where

$$E^{\alpha\beta\lambda\Gamma} = \frac{E}{2(1+\nu)}\,[a^{\alpha\lambda}\,a^{\beta\Gamma} + a^{\alpha\Gamma}\,a^{\beta\lambda} + \frac{2\nu}{1-\nu}\,a^{\alpha\beta}\,a^{\lambda\Gamma}]\ . \tag{3}$$

When substituting (2) into (1), we find out that it just remains to define the functions

$$\vec{v} \to \gamma_{\alpha\beta}(\vec{v}) \text{ and } \vec{v} \to \rho_{\alpha\beta}(\vec{v}) \tag{4}$$

in order to completely determine the functional $J(\vec{v},n_\nu) = J(\vec{v})$. Subsequently, we will see different possible choices (there are many others !).

5. Unified variational formulation of some thin shell problems

Possible solutions $\vec{u}$ minimize, at least locally, the total potential energy $J(\vec{v})$ on an admissible displacement space $\vec{W}$. For a sufficiently smooth functional J, these solutions $\vec{u}$ can also be characterized as solutions of the variational equation

$$J'(\vec{u}).\vec{v} = 0\ ,\quad \forall\vec{v}\in\vec{W} \tag{5}$$

where $J'(\vec{u})$ is the Fréchet derivative of J at point $\vec{u}$. Thanks to equations (2) and (3), relation (1) gives

$$J'(\vec{u}).\vec{v} = \int_\Omega [n^{\alpha\beta}(\vec{u})(\gamma_{\alpha\beta})'(\vec{u})\vec{v} + m^{\alpha\beta}(\vec{u})(\rho_{\alpha\beta})'(\vec{u})\vec{v} - \vec{p}\vec{v}] \sqrt{a}\, d\xi^1\, d\xi^2 - \int_{\Gamma_1} \vec{N}\vec{v}\, d\gamma \,. \tag{6}$$

Equations (5) and (6) give a unified variational formulation of different thin shell problems as soon as the functions (4) are defined. Now, let us see some examples according to the classification of the magnitude of the angle of rotation ω suggested by PIETRASZKIEWICZ [19, page 197]. From now on we assume small strains everywhere in the shell.

5.1. Small rotations [$\omega \le O(\theta^2)$] :

This gives us the linear case :

$$\left.\begin{aligned} \gamma_{\alpha\beta}(\vec{u}) &= \theta_{\alpha\beta} \\ \rho_{\alpha\beta}(\vec{u}) &= \frac{1}{2}[\phi_{\alpha|\beta} + \phi_{\beta|\alpha} + b^\lambda_\alpha(\theta_{\lambda\beta}-\omega_{\lambda\beta}) + b^\lambda_\beta(\theta_{\lambda\alpha}-\omega_{\lambda\alpha})] \end{aligned}\right\} \tag{7}$$

where the functions $\vec{u} \to \theta_{\alpha\beta}(\vec{u})$, $\phi_\alpha(\vec{u})$, $\omega_{\lambda\beta}(\vec{u})$ are given by

$$\theta_{\alpha\beta}(\vec{u}) = \frac{1}{2}(u_{\alpha|\beta} + u_{\beta|\alpha}) - b_{\alpha\beta}\, u_3 \tag{8}$$

$$\phi_\alpha(\vec{u}) = u_{3|\alpha} + b^\lambda_\alpha\, u_\lambda \tag{9}$$

$$\omega_{\alpha\beta}(\vec{u}) = \frac{1}{2}(u_{\beta|\alpha} - u_{\alpha|\beta}) \tag{10}$$

In particular, for a shallow shell, equation (7) can be simplified as follows :

$$\left.\begin{aligned} \gamma_{\alpha\beta}(\vec{u}) &= \theta_{\alpha\beta} \\ \rho_{\alpha\beta}(\vec{u}) &= u_{3|\alpha\beta} \end{aligned}\right\} \tag{7bis}$$

Since the expression (7) and (7bis) are linear, the derivatives $(\gamma_{\alpha\beta})'$ and $(\rho_{\alpha\beta})'$ are given by

$$\left.\begin{aligned} (\gamma_{\alpha\beta})'(\vec{u})\vec{v} &= \gamma_{\alpha\beta}(\vec{v}) \\ (\rho_{\alpha\beta})'(\vec{u})\vec{v} &= \rho_{\alpha\beta}(\vec{v}) \end{aligned}\right\} \tag{11}$$

5.2. Moderate rotations [$\omega = O(\theta)$] :

In this case $\gamma_{\alpha\beta}$ becomes nonlinear meanwhile $\rho_{\alpha\beta}$ remains linear :

$$\left.\begin{aligned} \gamma_{\alpha\beta}(\vec{u}) &= \theta_{\alpha\beta} + \frac{1}{2}\phi_\alpha\phi_\beta + \frac{1}{2}a_{\alpha\beta}\phi^2 - \frac{1}{2}(\theta^\lambda_\alpha\omega_{\lambda\beta} + \theta^\lambda_\beta\omega_{\lambda\alpha}) \\ \rho_{\alpha\beta}(\vec{u}) &= \frac{1}{2}[\phi_{\alpha|\beta} + \phi_{\beta|\alpha} + b^\lambda_\alpha(\theta_{\lambda\beta}-\omega_{\lambda\beta}) + b^\lambda_\beta(\theta_{\lambda\alpha}-\omega_{\lambda\alpha})] \end{aligned}\right\} \tag{12}$$

where

$$\phi(\vec{u}) = \frac{1}{2\sqrt{a}}(u_{2|1}-u_{1|2}) = \frac{1}{\sqrt{a}}\omega_{12} \,. \tag{13}$$

Now we get

$$\left.\begin{aligned} (\gamma_{\alpha\beta})'(\vec{u})\vec{v} &= \theta_{\alpha\beta}(\vec{v}) + \frac{1}{2}\phi_\alpha(\vec{u})\phi_\beta(\vec{v}) + \frac{1}{2}\phi_\alpha(\vec{v})\phi_\beta(\vec{u}) + a_{\alpha\beta}\phi(\vec{u})\phi(\vec{v}) \\ &\quad - \frac{1}{2}[\theta^\lambda_\alpha(\vec{u})\,\omega_{\lambda\beta}(\vec{v}) + \theta^\lambda_\beta(\vec{u})\,\omega_{\lambda\alpha}(\vec{v}) + \theta^\lambda_\alpha(\vec{v})\,\omega_{\lambda\beta}(\vec{u}) + \theta^\lambda_\beta(\vec{v})\,\omega_{\lambda\alpha}(\vec{u})] \\ (\rho_{\alpha\beta})'(\vec{u})\vec{v} &= \rho_{\alpha\beta}(\vec{v}) \end{aligned}\right\} \tag{14}$$

In case of a shallow shell, equations (12) can be simplied as follows

$$\left.\begin{aligned} \gamma_{\alpha\beta}(\vec{u}) &= \theta_{\alpha\beta} + \frac{1}{2}u_{3,\alpha}\,u_{3,\beta} \\ \rho_{\alpha\beta}(\vec{u}) &= u_{3|\alpha\beta} \,. \end{aligned}\right\} \tag{12bis}$$

5.3. Large rotations $[\omega = O(\sqrt{\theta})]$:

Now, both functions $\gamma_{\alpha\beta}$ and $\rho_{\alpha\beta}$ are nonlinear :

$$\left.\begin{aligned} \gamma_{\alpha\beta} &= \theta_{\alpha\beta} + \frac{1}{2}\phi_\alpha\phi_\beta + \frac{1}{2}(\theta^\lambda_\alpha-\omega^\lambda_{\cdot\alpha})(\theta_{\lambda\beta}-\omega_{\lambda\beta}) \\ \rho_{\alpha\beta} &= \frac{1}{2}[\phi_{\alpha|\beta} + \phi_{\beta|\alpha} + b^\lambda_\alpha(\theta_{\lambda\beta}-\omega_{\lambda\beta}) + b^\lambda_\beta(\theta_{\lambda\alpha}-\omega_{\lambda\alpha})] + \frac{1}{2}b_{\alpha\beta}\phi^\lambda\phi_\lambda + \\ &\quad - \frac{1}{4}\phi^\lambda\phi_\lambda(\phi_{\alpha|\beta}+\phi_{\beta|\alpha}) - (\phi^\lambda+\phi_\kappa\omega^{\kappa\lambda}\quad\theta_{\lambda\alpha\beta} \end{aligned}\right\} \tag{15}$$

where

$$\theta_{\lambda\alpha\beta} = \theta_{\lambda\alpha|\beta} + \theta_{\lambda\beta|\alpha} - \theta_{\alpha\beta|\lambda} \tag{16}$$

By analogy with (14), the reader can compute the Fréchet derivatives of $\gamma_{\alpha\beta}$ and $\rho_{\alpha\beta}$ and so, can find the corresponding variational formulation of the problem by using relations (5) and (6).

5.4. Finite rotations [$\omega \geq O(1)$] :

The following expressions are concerned by the case of finite rotations (under small strains). The case without any restricted strains leads to a somewhat complicated expression for $\rho_{\alpha\beta}$.

$$\left.\begin{aligned} \gamma_{\alpha\beta} &= \text{same expression than in } (15)_1 \\ \rho_{\alpha\beta} &= n(\phi_{\alpha|\beta} + b^{\lambda}_{\beta}\ell_{\lambda\alpha}) + n_{\lambda}(\ell^{\lambda}_{.\alpha|\beta} - b^{\lambda}_{\beta}\phi_{\alpha}) - b_{\alpha\beta} \end{aligned}\right\} \quad (17)$$

with

$$n = [1 + \theta^{\kappa}_{\kappa} + \frac{1}{2}(\theta^{\kappa}_{\kappa})^2 - \frac{1}{2}\theta^{\kappa}_{\Gamma}\theta^{\Gamma}_{\kappa} + \phi^2]\,[1 - \gamma^{\lambda}_{\lambda}]$$

$$\ell_{\alpha\beta} = a_{\alpha\beta} + \theta_{\alpha\beta} - \omega_{\alpha\beta} \quad (18)$$

$$n_{\Gamma} = [-(1+\theta^{\kappa}_{\kappa})\,\phi_{\Gamma} + \phi^{\lambda}(\theta_{\lambda\Gamma} - \omega_{\lambda\Gamma})] \;.$$

Here also the computation of the Fréchet derivatives of $\gamma_{\alpha\beta}$ and $\rho_{\alpha\beta}$ is easy but somewhat fastidious.

Moreover, it is intructive to quote the orders of the nonlinearities which appear in these different cases.

5.5. Orders on nonlinearities (with respect to the components of the displacement and their derivatives)

For clarity, we summarize the results in the following table :

γ,ρ \ rotations	small	moderate	large	finite
$\gamma_{\alpha\beta}$	1	2	2	2
$\rho_{\alpha\beta}$	1	1	3	5 (or 3 : see remark 5.1)

Table 1 : Orders of nonlinearities for $\gamma_{\alpha\beta}$ and $\rho_{\alpha\beta}$

Remark 5.1 : According to remark 4.1, the substitution of $\chi_{\alpha\beta}$ to $\rho_{\alpha\beta}$ decreases the order of nonlinearity from 5 to 3.

6. Variational formulation of corresponding total Lagrangean incremental methods

Except the case (7), all the other equations are nonlinear. In order to solve them, it is necessary to introduce some linearization algorithms. Among them, let us mention total or updated incremental methods, Newton or modified Newton methods... In this section we propose a convenient way to formulate such a total Lagrangean incremental methods.

In a recent paper by BERNADOU, CIARLET and HU [1984], we have proved that the total Lagrangean incremental method is nothing but Euler's method for approximating an appropriate differential equation in a Sobolev space. In other words, in order to approximate the solution of the equation

$$A(u) = f \tag{19}$$

where A is a nonlinear, Fréchet-differentiable operator, the incremental method amounts to introduce the following iterative scheme :

$$\left.\begin{aligned} &u^o = 0 \,, \\ &A'(u^n)(u^{n+1} - u^n) = \frac{f}{N} \,, \; n = 0,\ldots,N-1 \end{aligned}\right\} \tag{20}$$

where N is a given integer. The associated variational formulation is given by

$$\left.\begin{aligned} &u^o = 0 \; ; \; u^n \to u \; ; \; u^{n+1} - u^n \to v \; ; \\ &\text{Find } v \text{ such that} \\ &\int_\Omega [A'(u)v]\, w \, d\Omega = \int_\Omega \frac{f}{N}\, w \, d\Omega \,, \quad \forall w \end{aligned}\right\} \tag{21}$$

Of course, all the above expressions are purely formal. When considering a real problem, we need to indicate the admissible spaces which are in general of Sobolev's type (see ADAMS [1] or LIONS-MAGENES [16]). For instance, in case of three-dimensional nonlinear elasticity, we have proved in BERNADOU-CIARLET and HU [6] the following convergence theorem :

Theorem 1 : In three dimensional elasticity, the iterative schemes (20) or (21) are convergent in the sense that

$$\|\vec{u}^N - \vec{u}\|_{W^{2,p}(\Omega)} = O(\tfrac{1}{N}) \,. \tag{22}$$

□

In addition, as a by-product, let us quote that this convergence gives an existence result for the nonlinear elasticity problem.

Application to thin shell problems :

For simplicity, let us assume that $\Gamma_o = \Gamma$ (not essential at this stage). Then, variational equations (5) (6) can be written as follows

$$J'(\vec{u})\vec{v} = \int_\Omega [A(\vec{u})\vec{v} - \vec{p}\vec{v}]\, d\Omega = 0 \tag{23}$$

so that (at least formally) :

$$J''(\vec{u})(\vec{v},\vec{w}) = \int_\Omega [A'(\vec{u})\vec{w}]\,\vec{v}\, d\Omega \,. \tag{24}$$

By comparison with (21), we can formulate an incremental step for the nonlinear thin shell problem as follows :

Given $\vec{u}(= \vec{u}^n)$, find $\vec{v}(= \vec{u}^{n+1} - \vec{u}^n)$ such that

$$J''(\vec{u})(\vec{v},\vec{w}) = \frac{1}{N}\int_\Omega \vec{p}\,\vec{w}\, d\Omega \,, \quad \forall\vec{w} \tag{25}$$

where, from (6), we have

$$\begin{aligned} J''(\vec{u})(\vec{v},\vec{w}) = \int_\Omega eE^{\alpha\beta\lambda\Gamma} \{&\gamma_{\alpha\beta}(\vec{u})(\gamma_{\lambda\Gamma})''(\vec{u})(\vec{v},\vec{w}) + [(\gamma_{\lambda\Gamma})'(\vec{u})\vec{w}][(\gamma_{\alpha\beta})'(\vec{u})\vec{v}] + \\ &+ \frac{e^2}{12}\{\rho_{\alpha\beta}(\vec{u})(\rho_{\lambda\Gamma})''(\vec{u})(\vec{v},\vec{w}) + [(\rho_{\lambda\Gamma})'(\vec{u})\vec{w}][(\rho_{\alpha\beta})'(\vec{u})\vec{v}]\}\}\, dS \,. \end{aligned} \tag{26}$$

Thus, it remains to compute the second Fréchet derivatives :

$[(\gamma_{\alpha\beta})''(\vec{u})](\vec{v},\vec{w})$ and $[(\rho_{\alpha\beta})''(\vec{u})](\vec{v},\vec{w})$.

For instance, in case of moderate rotations, we get

$$\begin{aligned} [(\gamma_{\alpha\beta})''(\vec{u})](\vec{v},\vec{w}) = {} & \frac{1}{2}\,\phi_\alpha(\vec{w})\,\phi_\beta(\vec{v}) + \frac{1}{2}\,\phi_\alpha(\vec{v})\,\phi_\beta(\vec{w}) + a_{\alpha\beta}\,\phi(\vec{w})\,\phi(\vec{v}) - \\ & - \frac{1}{2}\,[\theta^\lambda_\alpha(\vec{w})\,\omega_{\lambda\beta}(\vec{v}) + \theta^\lambda_\beta(\vec{w})\,\omega_{\lambda\alpha}(\vec{v}) + \theta^\lambda_\alpha(\vec{v})\,\omega_{\lambda\beta}(\vec{w}) + \theta^\lambda_\beta(\vec{v})\,\omega_{\lambda\alpha}(\vec{w})] \end{aligned}$$

and

$$[(\rho_{\alpha\beta})''(\vec{u})](\vec{v},\vec{w}) \equiv 0 \,.$$

Of course, convergence of such methods has to be analyzed. Some possibilities arise when $\Gamma_0 = \Gamma$, while great difficulties can be foresee when boundary conditions are of mixed type.

7. Some existence results for thin shell problems

The question of existence - and possibly uniqueness - of solutions for thin shell problems is an important one. Indeed, in order to construct appropriate linearization algorithm and to study suitable approximation methods, it is necessary to know the mathematical properties of the operators into consideration.

But, by contrast with the great number of linear or nonlinear shell models, there exist very few results concerning the mathematical analysis of theories governing the geometrically nonlinear behaviour of elastic shells. We mention, for example, NAUMANN [18], RUPPRECHT [22], VOROVICH and LEBEDEV [25] and WEINITSCHKE [26]. A review of some results can be found in DIKMEN [14]. In particular, the study of existence for models undergoing large or finite rotations seems to be completely open.

In the following, we just record three results concerning the cases of small rotations for general or shallow shells, and the case of moderate rotations for shallow shells ; these cases correspond to the equations (7), (7bis) and (12bis) in connection with variational formulation (5) and (6).

7.1. Existence and uniqueness for the small rotation case

Theorem 7.1.1. : The variational equation (5) (6) obtained in combination with (7) has one and only one solution in the admissible displacement space :

$$\vec{V} = \{\vec{v} \in (H^1(\Omega))^2 \times H^2(\Omega) \quad ; \quad \vec{v}|_{\Gamma_0} = \frac{\partial v_3}{\partial n}\Big|_{\Gamma_0} = 0\} .$$

Proof (BERNADOU and CIARLET [5])

The proof takes essentially three steps :

i) the square root of the strain energy

$$I(\vec{u}) = \frac{1}{2}\int_\Omega \{n^{\alpha\beta}(\vec{u})\ \gamma_{\alpha\beta}(\vec{u}) + m^{\alpha\beta}(\vec{u})\ \rho_{\alpha\beta}(\vec{u})\}\ \sqrt{a}\ d\xi^1\ d\xi^2$$

is a norm on the space $\vec{V}$. The main difficulty is to prove that $I(\vec{u}) = 0 \Rightarrow \vec{u} = \vec{0}$ in the Sobolev space $\vec{V}$. This is obtained by proving first that

$$\{I(\vec{u}) = 0 \ , \quad \vec{u} \in (H^1(\Omega))^2 \times H^2(\Omega)\}$$

implies $\vec{u}$ is an infinitesimal rigid body motion. Then, by adding boundary conditions, this rigid body motion is reduced to $\vec{0}$.

ii) the energy norm $\sqrt{I(\vec{u})}$ is equivalent to the usual norm on the product space $(H^1)^2 \times H^2$. This is completed by using arguments of lower weakly semi-continuity.

iii) then we apply the Lax-Milgram lemma.

□

7.2. Existence and uniqueness for shallow shells undergoing small rotations

Theorem 7.2.1. : The variational equation (5) (6) combined with (7bis) has one and only one solution in the admissible displacement $\vec{V}$ given by (27), as soon as the curvature of the middle surface of the shell is sufficiently small (i.e. "sufficiently shallow" shell).

Proof (BERNADOU and LALANNE [7]) :

The main ideas of the proof are to decompose the variational formulation into two parts : the first one concerns the integrals containing $\underline{v} = (v_1, v_2)$ and the second one concerns the integral in v_3, i.e.,

$$\left.\begin{aligned} &\mathcal{B}(\underline{u},\underline{v}) + b(u_3,\underline{v}) = \int_\Omega p^\alpha\, v_\alpha \sqrt{a}\, d\xi^1\, d\xi^2 \ , \quad \forall \underline{v} \in (V_1(\Omega))^2 \\ &\mathcal{A}(u_3,v_3) + c(\underline{u},v_3) = \int_\Omega p^3\, v_3 \sqrt{a}\, d\xi^1\, d\xi^2 \ , \quad \forall v_3 \in V_2(\Omega) \end{aligned}\right\} \qquad (27)$$

where the admissible displacement spaces are such that

$$V_1(\Omega) \subset H^1(\Omega) \quad \text{and} \quad V_2(\Omega) \subset H^2(\Omega) \ .$$

Then, we show that there exist constants c_1 and c_2 such that

$$\left.\begin{aligned} &\mathcal{B}(\underline{v},\underline{v}) \geq c_1 \|\underline{v}\|^2_{H^1} \ , \quad \forall \underline{v} \in (V_1(\Omega))^2 \\ &\mathcal{A}(v_3,v_3) \geq c_2 \|v_3\|^2_{H^2} \ , \quad \forall v \in V_2(\Omega) \end{aligned}\right\} \qquad (28)$$

In the same time, for sufficiently shallow shells, i.e., $|b^\beta_\alpha| \ , |b^\beta_{\alpha|\lambda}| \leq \varepsilon$, we have

$$\left.\begin{aligned} &|b(u_3,\underline{v})| \leq c_3 \varepsilon (\|u_3\|^2_{L^2} + \|\underline{v}\|^2_{H^1}) \ , \quad \forall v \in (V_1(\Omega))^2 \ , \quad \forall u_3 \in V_2(\Omega) \\ &|c(\underline{u},v_3)| \leq c_3 \varepsilon (\|\underline{u}\|^2_{H^1} + \|v_3\|^2_{L^2}) \ , \quad \forall \underline{u} \in (V_1(\Omega))^2 \ , \quad \forall v_3 \in V_2(\Omega) \end{aligned}\right\} \qquad (29)$$

Finally relations (28) (29) allow to show that the square root of the strain energy of the shallow shell is an equivalent norm to the product norm on $(H^1)^2 \times H^2$, i.e.,

$$a_S(\vec{v},\vec{v}) \geq (c_1-2c_3\varepsilon)\|\underline{v}\|^2_{H^1} + (c_2-2c_3\varepsilon)\|v_3\|^2_{H^2}$$

with

$$a_S(\vec{v},\vec{v}) = \mathcal{B}(\underline{v},\underline{v}) + b(v_3,\underline{v}) + \mathcal{A}(v_3 v_3) + c(\underline{v},v_3) .$$

It remains to apply Lax-Milgram's lemma to conclude.

□

7.3. Existence and uniqueness for shallow shells undergoing moderate rotations

Theorem 7.3.1. : For "sufficiently shallow" shells and for "sufficiently small" tangential loads, the problem associated to the variational formulation (5) (6) combined with (12bis) has at least one solution. There is uniqueness when the normal load p^3 is sufficiently small.

Proof (BERNADOU and ODEN [8]) :

For shallow shells the nonlinearity concerns only u_3. Hence

i) we can solve the problem with respect to u_1 and u_2 (Lax-Milgram lemma) ;

ii) we substitute u_1 and u_2 into the third equation and we obtain a nonlinear equation $\mathcal{A}u_3 = f$;

iii) then, we show that operator $\mathcal{A}$ is coercive and pseudomonotone so that (see LIONS [17]) $\mathcal{A}$ is surjective.

□

Unfortunately, we cannot expect to extend such a result to general shells undergoing moderate-and, of course, large or finite rotations. Indeed the previous proof is essentially based on the fact the equations are linear on u_1 and u_2 and nonlinear on u_3. In other words, shallow shell equations undergoing moderate rotations are very near from plate equations.

The situation concerning existence theorems for general shells undergoing moderate, large or finite rotations is quite open. Perhaps one should work first on regularity results for corresponding linear equations and next try to obtain existence results by using a process similar to that followed in the study of convergence of the total incremental method, like in BERNADOU, CIARLET and HU [6]. When looking for regularity results, one should find two difficulties :

- the first one to obtain regularity results for the linear equation under clamped boundary conditions. In this way, it seems that a direct application of results like those of AGMON, DOUGLIS and NIRENBERG [2] is difficult... ;

- the second one concerns the extension of possible such regularity results to the case of more general boundary conditions like, for instance, boundary conditions of mixed type.

References

1. ADAMS, R.A. [1975] : Sobolev spaces, Academic Press, New York.
2. AGMON, S. ; DOUGLIS, A. ; NIRENBERG, L. [1959] : Estimates near the boundary for solution of elliptic partial differential equations satisfying general boundary conditions (I). Commun. Pure Appl. Math. 12, pp. 623-727.
3. BERNADOU, M. [1980] : Convergence of conforming finite element methods for general shell problems. Int. J. Engng. Sci. 18, pp. 249-276.
4. BERNADOU, M. ; BOISSERIE, J.M. [1982] : The finite element method in thin shell theory : Application to arch dam simulations. Birkhaüser, Boston.
5. BERNADOU, M. ; CIARLET, P.G. [1976] : Sur l'ellipticité du modèle linéaire de coques de W.T. Koiter, in Computing Methods in Applied Sciences and Engineering (R. Glowinski and J.L. Lions, Ed.) pp.89-136, Lecture Notes in Economics and Mathematical Systems, Vol. 134, Springer-Verlag, Berlin.
6. BERNADOU, M. ; CIARLET, P.G. ; HU, J. [1984] : On the convergence of the semi-discrete incremental method in nonlinear, three-dimensional elasticity. J. Elasticity 14, pp. 425-440.
7. BERNADOU, M. ; LALANNE, B. [1985] : Sur l'approximation des coques minces par des méthodes "B-splines et éléments finis" ; Partie 1 : Formulation du problème et estimations d'erreur. Rapport de Recherche INRIA (to appear).
8. BERNADOU, M. ; ODEN, J.T. [1980] : An existence theorem for a class of nonlinear shallow shell problems. J. Math. Pures Appl. 60, (1981), pp. 285-308.
9. CIARLET, P.G. [1976] : Conforming finite element methods for the shell problems, in the Mathematics of Finite Elements and Applications II (J.R. Whiteman, Ed.), Academic Press, London, pp. 105-123.
10. CIARLET, P.G. [1978] : The Finite Element Method for Elliptic Problems. North-Holland, Amsterdam.
11. CIARLET, P.G. [1986] : Topics in Mathematical Elasticity. Vol. 1, North-Holland, Amsterdam.
12. DESTUYNDER, P. [1980] : Sur une justification mathématique des théories de plaques et de coques en élasticité linéaire. Thèse d'Etat, Université Pierre et Marie Curie, Paris.
13. DESTUYNDER, P. ; LUTOBORSKI, A. [1980] : A penalty method for the Budiansky-Sanders shell model. Rapport Interne 67, Centre de Mathématiques Appliquées de l'Ecole Polytechnique.
14. DIKMEN, M. [1982] : Theory of Thin Elastic Shells, Pitman, Boston.
15. KOITER, W.T. [1966] : On the nonlinear theory of thin elastic shells. Proc. Kon. Ned. Akad. Wetensch. B 69, pp. 1-54.

16. LIONS, J.L. ; MAGENES, E. [1968] : Problèmes aux limites non homogènes et applications. Vol.1, Dunod, Paris.

17. LIONS, J.L. [1969] : Quelques Méthodes de Résolution des Problèmes aux Limites Non Linéaires. Dunod et Gauthier-Villars, Paris.

18. NAUMANN, J. [1974] : On bifurcation buckling of thin elastic shells, J. Mécanique 13, n° 4, pp. 715-741.

19. PIETRASZKIEWICZ, W. [1980] : Finite rotations in the nonlinear theory of thin shells, in : Thin Shell Theory, New Trends and Applications, (Olszak, W. Ed.), CISM Courses and Lectures n° 240, Springer-Verlag, Wien-New York 1980, pp. 153-208.

20. PIETRASZKIEWICZ, W. ; SZWABOWICZ, M. [1981] : Entirely Lagrangian nonlinear theory of thin shells, Archives of Mechanics 33, n° 2, pp. 273-288.

21. PIETRASZKIEWICZ, W. ; SZWABOWICZ, M. [1982] : Hu-Washizu variational functional for the lagrangian geometrically nonlinear theory of thin elastic shells, ZAMM 62, T156-T158.

22. RUPPRECHT, G. [1981] : A singular perturbation approach to nonlinear shell theory, Rocky Mountain J. Math. 11, n° 1, pp. 75-98.

23. STEPHAN, E. ; WEISSGERBER, V. [1978] : Zur approximation von schalem mit hybriden elementen. Computing 20, n° 1, pp. 75-95.

24. VALID, R. [1981] : Mechanics of continuous media and analysis of structures. North-Holland, Amsterdam.

25. VOROVICH, I.I. ; LEBEDEV, L.P. [1972] : On the existence of solutions of the nonlinear theory of shallow shells. J. Appl. Math. and Mech. 36, n° 4, pp. 691.704.

26. WEINITSCHKE, H.J. [1976] : Some mathematical problems in the nonlinear theory of elastic membranes, plates and shells. Trends in Applications of Pure Mathematics to Mechanics, G. Fichera Ed., Pitman Publishing, London, pp. 409-424.

POSTCRITICAL DEFORMATIONS OF MERIDIONAL CROSS-SECTION OF ELASTIC TOROIDAL SHELLS SUBJECT TO EXTERNAL PRESSURE *

JAN BIELSKI
Institute of Mechanics and Machine Design
Technical University of Cracow
31-155 Kraków, ul. Warszawska 24, Poland

Abstract

A thin-walled, rotationally symmetric, elastic toroidal shell subject to external pressure and in-plane bending is considered. Geometrically nonlinear theory of finite displacements and rotations is applied. Both, symmetric as well as nonsymmetric forms of equilibrium of meridional cross-section are taken into account. The method of detecting of critical loadings is proposed and the analysis of post critical forms is performed. The influence of geometrical imperfections (jump-like thickness variation) on the equilibrium forms is discussed.

1. Introduction

Toroidal shells are a subdivided class of rotationally symmetric ones due to a specific geometry: one of principal curvatures changes its sign. Additionally, the negative influence of external pressure on the stability of the wall of a torus demands particular treatment.

The stability analysis, under assumption of a membrane precritical state of the shell has been performed by SOBEL and FLÜGGE [15]. On the basis of similar assumption, that is not confirmed by experiments, FEDOSOV has investigated the stability problem, too [6]. Using the semi-mem-

---

* The paper is a part of the lecture presented at the 197 Euromech Colloqium by J.BIELSKI and J.SKRZYPEK under the title 'Optimal design of a geometrically nonlinear, elastic toroidal shell subject to external pressure and bending.' The second part dealing with optimal design will be published separately.

brane shell theory AXELRAD [1,2] has analized various instability modes of a torus under external pressure. Forms, both, symmetric as well as asymmetric with respect to the symmetry plane of initial torus have been allowed for. JORDAN [10] has applied an asymptotic analysis of stability of a torus. Good accordance of both, theoretical and experimental results has been obtained. Nonlinear theory for precritical state of a toroidal shell under external pressure action has been used in stability analysis by BULYGIN [4,5]. Also nonlinear theory of large displacements but small rotations has been used by GAYDAYCHUK, GOTSULAK and GULAYEV [7,8] as well as by GULAYEV, BAZHENOV and GOTSULAK [9]. Symmetric as well as asymmetric forms of deformation have been considered there. Using nonlinear theory of finite displacements and finite rotations BIELSKI and SKRZYPEK [3] have discussed the instability forms of a toroidal shell subject to external pressure and in-plane bending. However, only the forms symmetric with respect to the symmetry plane of initial torus have been allowed for. Nonlinear theory combined with energetical approach to the stability analysis has been used by KRÄTZIG [11] and ZINTIKIS [16].

The influence of initial imperfections of shape of the shell on the equilibrium form and values of critical pressure is a separate problem. The influence of the thickness distribution and thickness imperfection has been discussed by GAYDAYCHUK, GOTSULAK and GULAYEV [7,9] as well as by JORDAN [10], KRÄTZIG [11] and POLISHCHUK [12]. On the other hand the influence of initial ovalization of the meridional section of the torus on the critical state has been considered by GAYDAYCHUK, GOTSULAK GULAYEV and BAZHENOV in previously quoted papers [7,8,9].

The purpose of the present paper is the analysis of both, symmetric and asymmetric, postcritical deformations of meridional section of a torus under external pressure, using geometrically nonlinear approach.

2. Assumptions

The geometrical relations as well as equilibrium equations are based on the papers [13],[14]. The main assumptions made there have been retained but, additionally, nonsymmetric deformations of meridional section have been allowed for.

A thin sandwich elastic toroidal shell with initially circular meridional section is analised. It is assumed that deformations obey the classical Love-Kirchhoff hypothesis: segments normal to the middle surface of the shell remain normal during the process. The distance between sandwich sheets, prescribed by a given function of meridional coordinate Φ , remains constant during the process. Also the Bernoulli hypothesis is applied: meridional cross sections of a tube remain plane and normal to the deformed circular axis of the torus. However, only rotationally symmetric deformations of a shell are allowed for, and hence the meridional and the circumferential directions are materially fixed principal directions. Geometrically nonlinear theory of finite displacements and finite rotations is applied.

Normal, uniform external pressure is carried by working sheets of the wall. Its intesity changes quasistatically. However, the unit toroidal angle α remains constant, and thus, the overall bending moment and the overall circumferential normal force are regarded as reactions.

Holding up the symmetry of initial shape and loadings with respect to the symmetry plane of initial torus both, symmetric as well as asymmetric forms of deformation are allowed for.

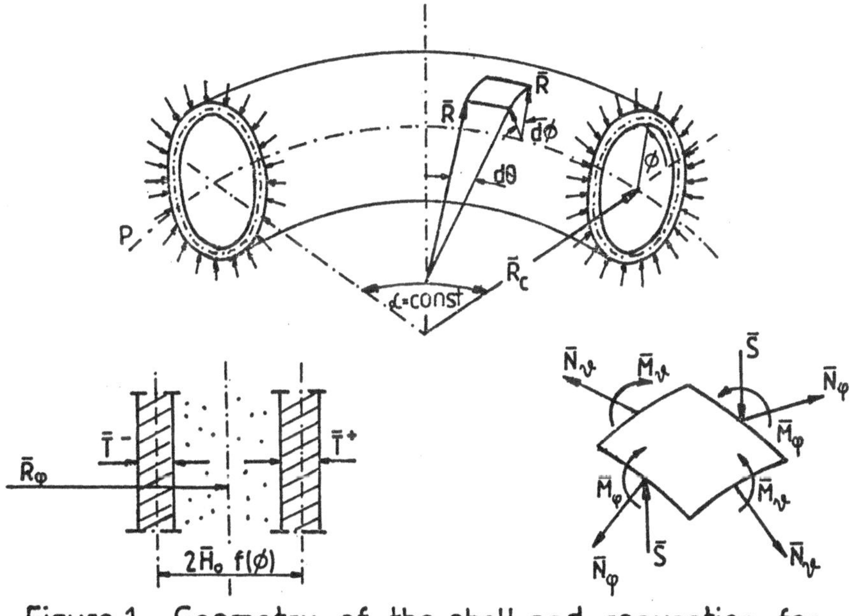

Figure 1 – Geometry of the shell and convention for generalized stresses

3. Notations and definitions

Capital letters denote quantities related to initial configuration whereas lower case letters - to current one. Dash over a symbol refers to dimensional quantity. Superscripts '+' or '-' refer to the outer or inner sandwich layers, respectively. Primes denote differentiation with respect to the material coordinate $\bar{\Phi}$. Subscript 'φ' refers to the radial direction whereas 'ϑ' - to the circumferential one.

The following notation is applied :

$\bar{\Phi}, \varphi$ - initial and current slope of normal with respect to the symmetry plane of initial torus,

$\bar{R}_\varphi, \bar{R}_\vartheta$ - meridional and circumferential principal curvature radii,

$\bar{R}=\bar{R}_\vartheta \cos\bar{\Phi}$ - distance of a middle surface point from the axis of revolution,

$\bar{U}_r, \bar{U}_z$ - radial and axial middle surface displacements, expressed in Cartesian spatial coordinate system,

$\bar{H}=f\bar{H}_o$ - half of the distance between working sheets,

$f(\bar{\Phi})$ - dimensionless distance function,

$\bar{T}^+=\bar{T}^-=\bar{T}$ - thicknesses of working sheets,

$\varkappa = \frac{d\vartheta}{d\Theta} - 1$ - change of the unit toroidal angle; in the present paper it is assumed $\varkappa = 0$,

$\varepsilon^{\pm}_\varphi, \varepsilon^{\pm}_\vartheta$ - principal strains in the working sheets

$$\varepsilon^{\pm}_{\varphi,\vartheta} = \frac{\ni_{\varphi,\vartheta} \pm \bar{H}\,\bar{\varkappa}_{\varphi,\vartheta}}{1 \pm \bar{H}/\bar{R}_{\varphi,\vartheta}} \quad , \qquad (3.1)$$

$\ni_\varphi, \ni_\vartheta$ - principal elongations of the middle surface (see [13])

$$\ni_\varphi = \frac{\sin\bar{\Phi} - \bar{U}_r'/\bar{R}_\varphi}{\sin\varphi} - 1 \quad , \qquad (3.2)$$

$$\ni_\vartheta = \frac{\bar{U}_r}{\bar{R}}(1+\varkappa) + \varkappa \quad , \qquad (3.3)$$

$\bar{\varkappa}_\varphi, \bar{\varkappa}_\vartheta$ - increments of middle surface curvatures

$$\bar{\varkappa}_\varphi = \frac{\varphi' - 1}{\bar{R}_\varphi} \quad , \qquad (3.4)$$

$$\bar{\varkappa}_\vartheta = \frac{\cos\varphi(1+\varkappa) - \cos\bar{\Phi}}{\bar{R}} \quad , \qquad (3.5)$$

$\bar{\sigma}^{\pm}_\varphi, \bar{\sigma}^{\pm}_\vartheta$ - principal stresses in working sandwich layers,

$\bar{N}_\varphi, \bar{M}_\varphi, \bar{N}_\vartheta, \bar{M}_\vartheta$ - generalized stresses

normal forces

$$\bar{N}_\varphi = [\,\bar{\sigma}^+_\varphi(1+\bar{H}/\bar{R}_\vartheta) + \bar{\sigma}^-_\varphi(1-\bar{H}/\bar{R}_\vartheta)\,]\bar{T} \quad , \qquad (3.6)$$

$$\bar{N}_\vartheta = [\,\bar{\sigma}^+_\vartheta(1+\bar{H}/\bar{R}_\varphi) + \bar{\sigma}^-_\vartheta(1-\bar{H}/\bar{R}_\varphi)\,]\bar{T} \quad , \qquad (3.7)$$

bending moments

$$\bar{M}_\varphi = [-\bar{\sigma}^+_\varphi(1+\bar{H}/\bar{R}_\vartheta) + \bar{\sigma}^-_\varphi(1-\bar{H}/\bar{R}_\vartheta)]\bar{T}\bar{H} , \quad (3.8)$$

$$\bar{M}_\vartheta = [-\bar{\sigma}^+_\vartheta(1+\bar{H}/\bar{R}_\varphi) + \bar{\sigma}^-_\vartheta(1-\bar{H}/\bar{R}_\varphi)]\bar{T}\bar{H} , \quad (3.9)$$

$\bar{S}$ - shearing force (reaction),

$\bar{p}$ - external pressure,

$\bar{E}, \nu$ - Young's modulus and Poisson's ratio (ν =0.3 in numerical calculations).

Dimensionless quantities are applied

$$\begin{aligned}
&R_j = \bar{R}_j/\bar{H}_o , \quad U_j = \bar{U}_j/\bar{H}_o ,\\
&\varkappa_{\varphi,\vartheta} = \bar{\varkappa}_{\varphi,\vartheta}\bar{H}_o ,\\
&N_{\varphi,\vartheta} = \bar{N}_{\varphi,\vartheta}/2\bar{T}\,\bar{\sigma}_o ,\\
&S = \bar{S}/2\bar{T}\,\bar{\sigma}_o ,\\
&M_{\varphi,\vartheta} = \bar{M}_{\varphi,\vartheta}/2\bar{T}\bar{H}_o\,\bar{\sigma}_o ,\\
&p = \bar{p}\bar{H}_o/2\bar{T}\,\bar{\sigma}_o ,\\
&\sigma_{\varphi,\vartheta} = \bar{\sigma}_{\varphi,\vartheta}/\bar{\sigma}_o ,\\
&E = \bar{E}/\bar{\sigma}_o ,
\end{aligned} \quad (3.10)$$

where

$2\bar{H}_o$ denotes a characteristical, chosen distance between working sheets,

$\bar{\sigma}_o$ is an arbitrary stress dimension quantity; in the paper it has been chosen $\bar{\sigma}_o = 0.001\bar{E}$

4. System of governing equations

Physical relations are assumed in the form of Hooke's law

$$\varepsilon^{\pm}_{\varphi,\vartheta} = \frac{1}{E}\,(\sigma^{\pm}_{\varphi,\vartheta} - \nu\,\sigma^{\pm}_{\vartheta,\varphi}) . \quad (4.1)$$

On the basis of presented definitions and after elimination of stresses as well as solving with respect to first order derivatives one can obtain the following system of differential equations in the dimensionless form :

$$\varphi' = 1+R_\varphi \varkappa_\varphi , \quad (4.2.1)$$

$$U_z' = R_\varphi[-\cos\Phi + (1+\ni_\varphi)\cos\varphi] , \quad (4.2.2)$$

$$U_r' = R_\varphi[\sin\Phi - (1+\ni_\varphi)\sin\varphi] . \quad (4.2.3)$$

Three equations of equlibrium must be added. They were derived in [13] on the basis of the principle of virtual work:

$$N'_\varphi = [-R'N_\varphi - R_\varphi \sin\varphi(1+\varkappa)N_\vartheta - \varphi' RS\,]/R \quad , \tag{4.2.4}$$

$$S' = [-R'S + R_\varphi \cos\varphi(1+\varkappa)N_\vartheta + \varphi' RN_\varphi - RR_\varphi p\,]/R \quad , \tag{4.2.5}$$

$$M'_\varphi = [-R'M_\varphi - R_\varphi \sin\varphi M_\vartheta - RR_\varphi S\,]/R \quad . \tag{4.2.6}$$

For the shorteness the following functions have been introduced:

$$\varepsilon_\vartheta = U_r(1+\varkappa)/R + \varkappa \quad , \tag{4.3.1}$$

$$\varkappa_\vartheta = [\cos\varphi(1+\varkappa) - \cos\Phi\,]/R \quad , \tag{4.3.2}$$

$$\varepsilon_\varphi = \frac{1-\nu^2}{E}[N_\varphi + M_\varphi(\frac{\cos\Phi}{R} - \frac{1}{R_\varphi})] - \nu\varepsilon_\vartheta + \nu f^2 \varkappa_\vartheta(\frac{\cos\Phi}{R} - \frac{1}{R_\varphi}) \quad , \tag{4.3.3}$$

$$\varkappa_\varphi = \nu\varepsilon_\vartheta(\frac{\cos\Phi}{R} - \frac{1}{R_\varphi}) - \nu\varkappa_\vartheta + \frac{1-\nu^2}{E}[N_\varphi(\frac{1}{R_\varphi} - \frac{\cos\Phi}{R}) - M_\varphi/f] \quad , \tag{4.3.4}$$

$$N_\vartheta = \frac{E}{1-\nu^2}[\varepsilon_\vartheta + f^2\varkappa_\vartheta(\frac{1}{R_\varphi} - \frac{\cos\Phi}{R}) + \nu\varepsilon_\varphi] \quad , \tag{4.3.5}$$

$$M_\vartheta = \frac{Ef^2}{1-\nu^2}[\varepsilon_\vartheta(\frac{\cos\Phi}{R} - \frac{1}{R_\varphi}) - \varkappa_\vartheta - \nu\varkappa_\varphi] \quad . \tag{4.3.6}$$

Equations (4.2) - (4.3) represent governing system of six quasilinear first order ordinary differential equations that describes the state of the shell. Suitable boundary conditions must be added so as to complete the problem.

5. Boundary conditions

According to the considered class of deformation of the cross-section of a torus one of two various types of boudary conditions has to be employed.

(1) In the case if the deformed shape of meridional section is symmetric with respect to the symmetry plane of initial torus the boundary conditions are equivalent to the necessary conditions of symmetry and may be written

$$\underline{F}_s(\underline{X}_s) = \underline{0} \quad , \tag{5.1}$$

where

$$\underline{X}_s = [M_\varphi(0), N_\varphi(0), U_r(0)] \quad , \tag{5.2}$$

$$\underline{F}_s = [\varphi(\pi) - \pi, S(\pi), U_z(\pi)] \quad . \tag{5.3}$$

The vector function $\underline{F}_s$ depends on the components of the vector $\underline{X}_s$, provided symmetry conditions are fulfilled at the point $\Phi=0$ i.e.

$$\varphi(0) = S(0) = U_z(0) = 0 \quad . \tag{5.4}$$

(2) In the more general case (not necessarily symmetric) the boundary conditions coincide with the necessary conditions of periodicity of the solution. They may be written in the form

$$\underline{F}_a(\underline{X}_a) = \underline{0} \ , \tag{5.5}$$

where

$$\underline{X}_a = [\varphi(0), S(0), M_\varphi(0), N_\varphi(0), U_r(0)] \ , \tag{5.6}$$

$$\underline{F}_a = [\varphi(0)-\varphi(2\pi), S(0)-S(2\pi), M_\varphi(0)-M_\varphi(2\pi), N_\varphi(0)-N_\varphi(2\pi), U_r(0)-U_r(2\pi)] \ . \tag{5.7}$$

The vector of unknowns $\underline{X}_a$ is an argument of the vector function $\underline{F}_a$. A rigid motion of the shell as a whole is eliminated by putting $U_z(0)=0$.

In both discussed cases vectors $\underline{X}_s$ or $\underline{X}_a$, that describe values of suitable functions for $\Phi=0$, are to be chosen so as to fulfill equations (5.1) or (5.5), respectively. So, we deal with two-point boundary value problem. It should be emphasized that problem of determination of the shell state is reduced to solving a set of algebraic equations (5.1) or (5.5).

The Newton-Raphson algorithm has been employed to solve the equations (5.1) or (5.5) , whereas the 4-th order Runge-Kutta method has been used for integration the system (4.2) so as to calculate values of the functions $\underline{F}_s$ or $\underline{F}_a$ for given vectors $\underline{X}_s$ or $\underline{X}_a$.

6. Detection of critical pressures

Values of vector functions $\underline{F}_s$ or $\underline{F}_a$ depend not only on boundary conditions at the point $\Phi=0$ but, simultaneously, on loading parameters: pressure p and change of the unit toroidal angle $\varkappa$. In the case under consideration the parameter $\varkappa$ is assumed to be identically equal to zero, so it is only the pressure to cause deformation of the shell. For particular values of pressure it happens to be impossible to solve systems (5.1) or (5.5) using Newton's procedure because of singularity of matrices of derivatives $[\partial\underline{F}_s/\partial\underline{X}_s]$ or $[\partial\underline{F}_a/\partial\underline{X}_a]$, respectively. It may be proved that such values of p correspond to critical pressures in the sense of stability of the shell.

Let the vectors $\underline{X}_{ao}$ and $\underline{X}_{so}$ satysfaying the equations (5.1) and (5.5) belong to the path of symmetric equilibrium of the shell. Then we may write

$$\underline{X}_{ao} = [0, 0, \underline{X}_{so}] \ , \tag{6.1}$$

because a symmetric form has to satisfy the continuity conditions, too.

The vector functions $\underline{F}_s$ and $\underline{F}_a$ are expanded into generelized Taylor series in the neighbourhood of points $\underline{X}_{so}$ and $\underline{X}_{ao}$. Restricting the expansions to the first terms we obtain

$$\underline{F}_s(\underline{X}_{so}+\Delta\underline{X}_s) \cong \underline{F}_s(\underline{X}_{so}) + [\partial\underline{F}_s/\partial\underline{X}_s]\Big|_{\underline{X}_{so}} \Delta\underline{X}_s \tag{6.2}$$

and

$$\underline{F}_a(\underline{X}_{ao}+\Delta\underline{X}_a) \cong \underline{F}_a(\underline{X}_{ao}) + [\partial\underline{F}_a/\partial\underline{X}_a]\Big|_{\underline{X}_{ao}} \Delta\underline{X}_a \ . \tag{6.3}$$

In both cases the center of expansion is the solution of (5.1) or (5.5), respectively. It means that necessary condition of existing adjacent solutions is singularity of the corresponding Jacobi matrix (obviously excluding the trivial cases $\Delta\underline{X}_s=\underline{0}$, $\Delta\underline{X}_a=\underline{0}$) .

If for a specific value of p it takes place

$$\det[\partial\underline{F}_a/\partial\underline{X}_a]\Big|_{\underline{X}_{ao}} = 0 \qquad \text{and} \qquad \det[\partial\underline{F}_s/\partial\underline{X}_s]\Big|_{\underline{X}_{so}} \neq 0 \ , \tag{6.4}$$

than we deal with such a critical state in the neighbourhood of which nonsymmetric postcritical form may exist. So it is a point of bifurcation in classical meaning and the value of parameter $p=p_{bif}$ is the bifurcation pressure.

If, on the other hand, it happens simultaneously

$$\det[\partial\underline{F}_s/\partial\underline{X}_s]\Big|_{\underline{X}_{so}} = 0 \qquad \text{and} \qquad \det[\partial\underline{F}_a/\partial\underline{X}_a]\Big|_{\underline{X}_{ao}} = 0 \ , \tag{6.5}$$

it means that the critical state results in symmetric postcritical equilibrium form (fulfills eq. (5.1)) . It is the point of the maximal carrying capacity, $p=p_{max}$.

As an example the relationship of critical pressures and geometry of shell is presented in Fig. 2 . The radius of meridional section as well as wall thickness are fixed and constant. One can see that difference between bifurcation and maximal pressures decreases with an increase of principle curvature radius and vanishes for the straight tube (R_c tends to infinity) .

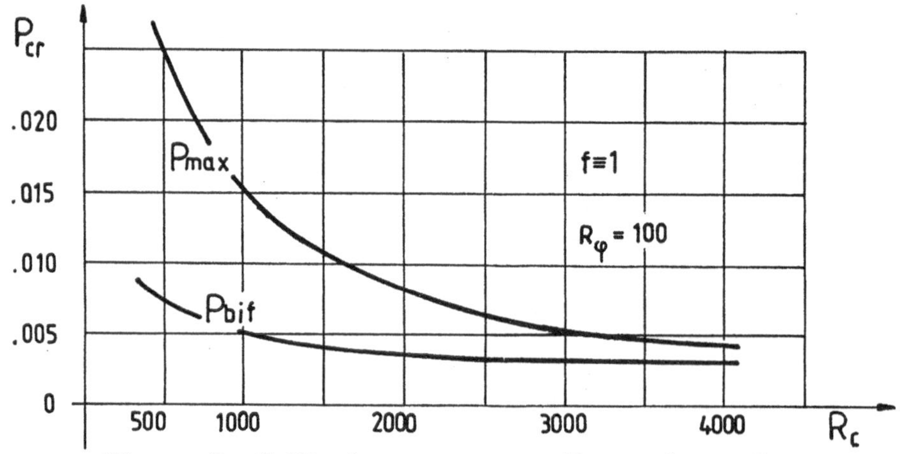

Figure 2 - Critical pressures v. the main radius

7. Analysis of postcritical forms

The path of symmetric equilibrium of the shell has been obtained by solving equation (5.1) for changing step by step the value of pressure p. Using the described above method both critical points, the maximal carrying capacity M and the bifurcation loss of stability B, have been detected. Equilibrium paths for both symmetric and nonsymmetric meri-

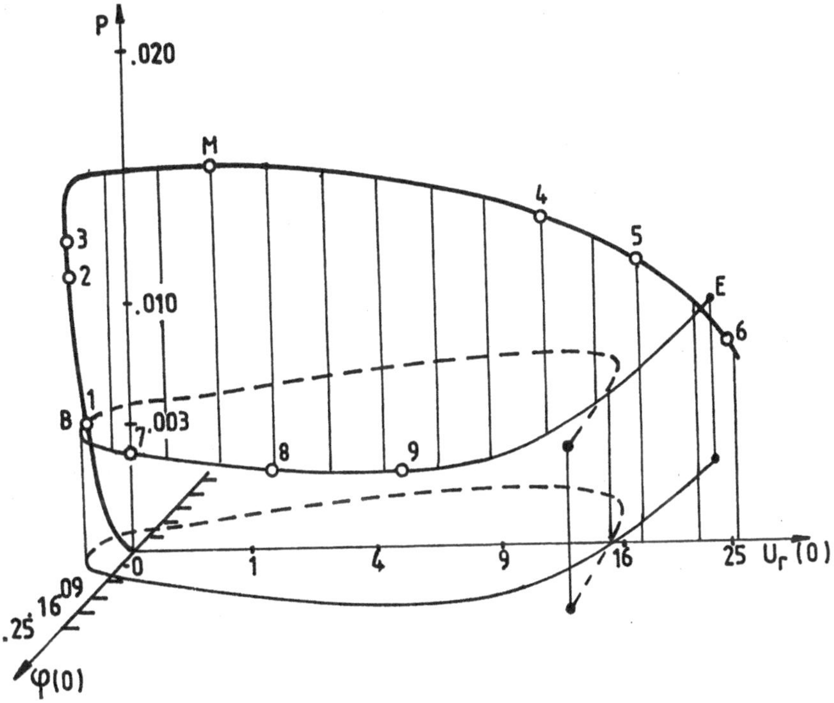

Figure 3 – Fundamental and postcritical paths.

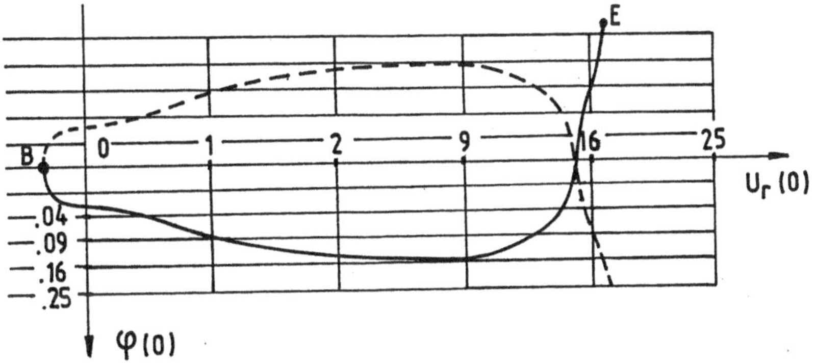

Figure 3a – Top view of the curve BE.

dional deformations are plotted in Fig.3. The axes of coordinates are: 1° pressure (p), 2° radial displacement of the point of meridional coordinate $\Phi = 0$ ($U_r(0)$) and 3° slope of normal at that point ($\varphi(0)$). Nonlinear, quadratic scales for both displacement and slope axes are applied because of wide range of variation.

At the point B the path of equilibrium bifurcates and the new, postbifurcation path emanates from the in-plane, fundamental curve OBM. Along the postbifurcation path the pressure value slightly increases, whereas the absolute value of the slope $\varphi(0)$ initialy increases, then decreases. Moreover, the path is symmetric with respect to the $\langle p - U_r(0) \rangle$ plane in the space under consideration. The analysis of the process of nonsymmetric deformation is terminated when the walls of the shell touch each other.

The evolution of shape of meridional section as well as the corresponding redistribution of stresses, both along the postbifurcation path BE are ilustrated in Fig.4. The angle of rotation of normal is marked for a chosen point of the middle surface. One may notice a strong flattening of the shape, the stronger the more asymmetric form is developed (Fig.4a). Meridional stress distribution in both working sheets of the sandwich wall (Fig.4b) indicates strongly bent regions that are visible in the diagram of shape. The last part, Fig.4c, shows a distribution of the circumferential normal force.

The evolution of a symmetric form near the maximal carrying capacity point M is presented in a similar way (Fig.5). Fig.5a shows the evolution of the shape before maximal pressure is reached. The scale of displacement is 100:1 so as to enable a quantitative illustration of maximal capacity effect: no higher pressure can be carried by the shell because of concavities that occur on its surface. It is possible to follow symmetric equilibrium path after the point M is reached, provided the pressure is no longer a control parameter. Strong increase of deformations is observed on the postmaximal, unstable part of the equilibrium path.

Development of postmaximal shape (using 1:1 scale for displacements) as well as distribution of meridional stresses and circumferential normal force is shown in Figs.5b, 5c and 5d, respectively.

All numerical results presented in diagrams 3,4,5 are computed for the shell of dimensionless main curvature radius $R_c = 1000$ and meridional section radius $R_\varphi = 100$.

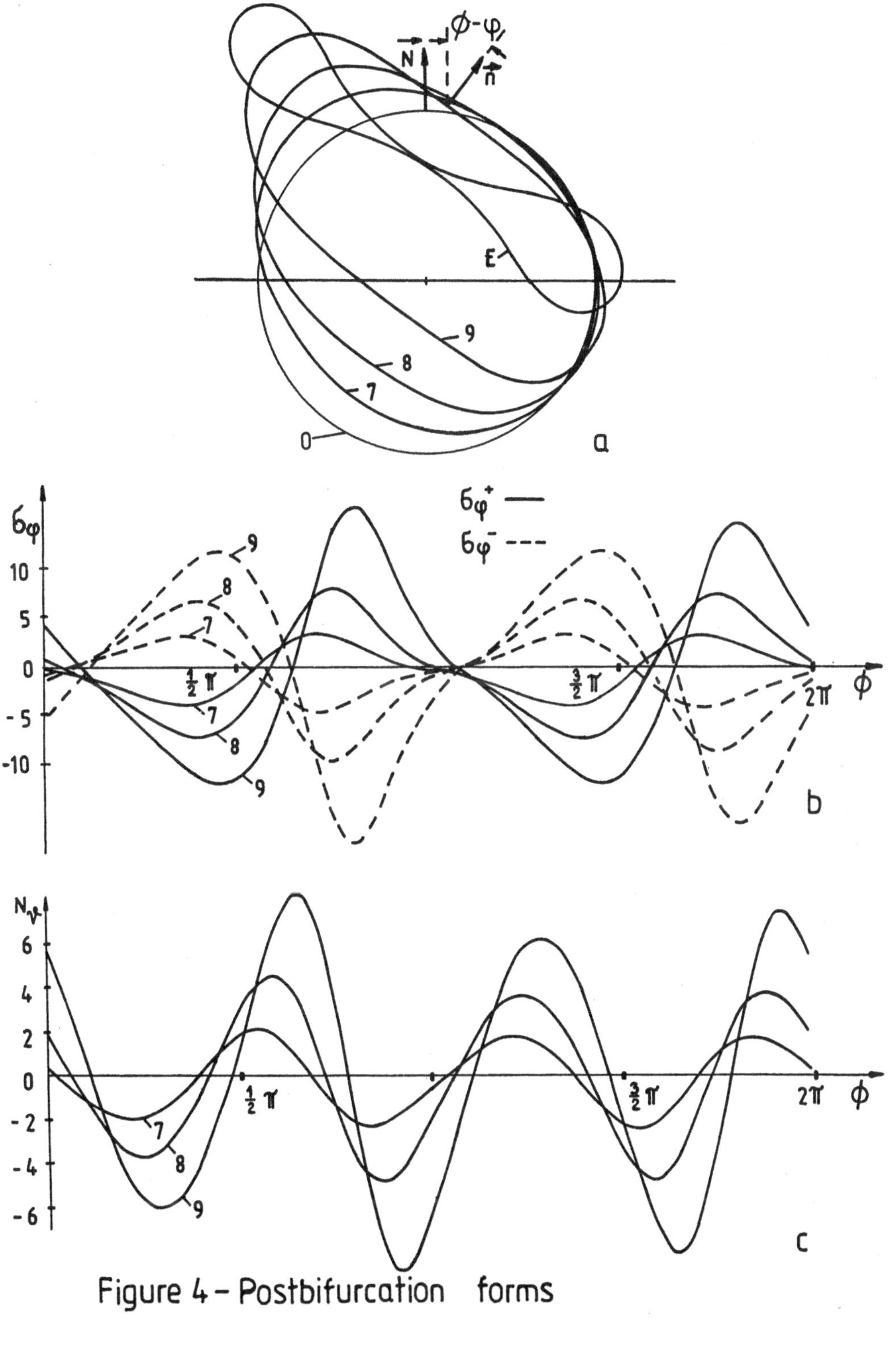

Figure 4 – Postbifurcation forms

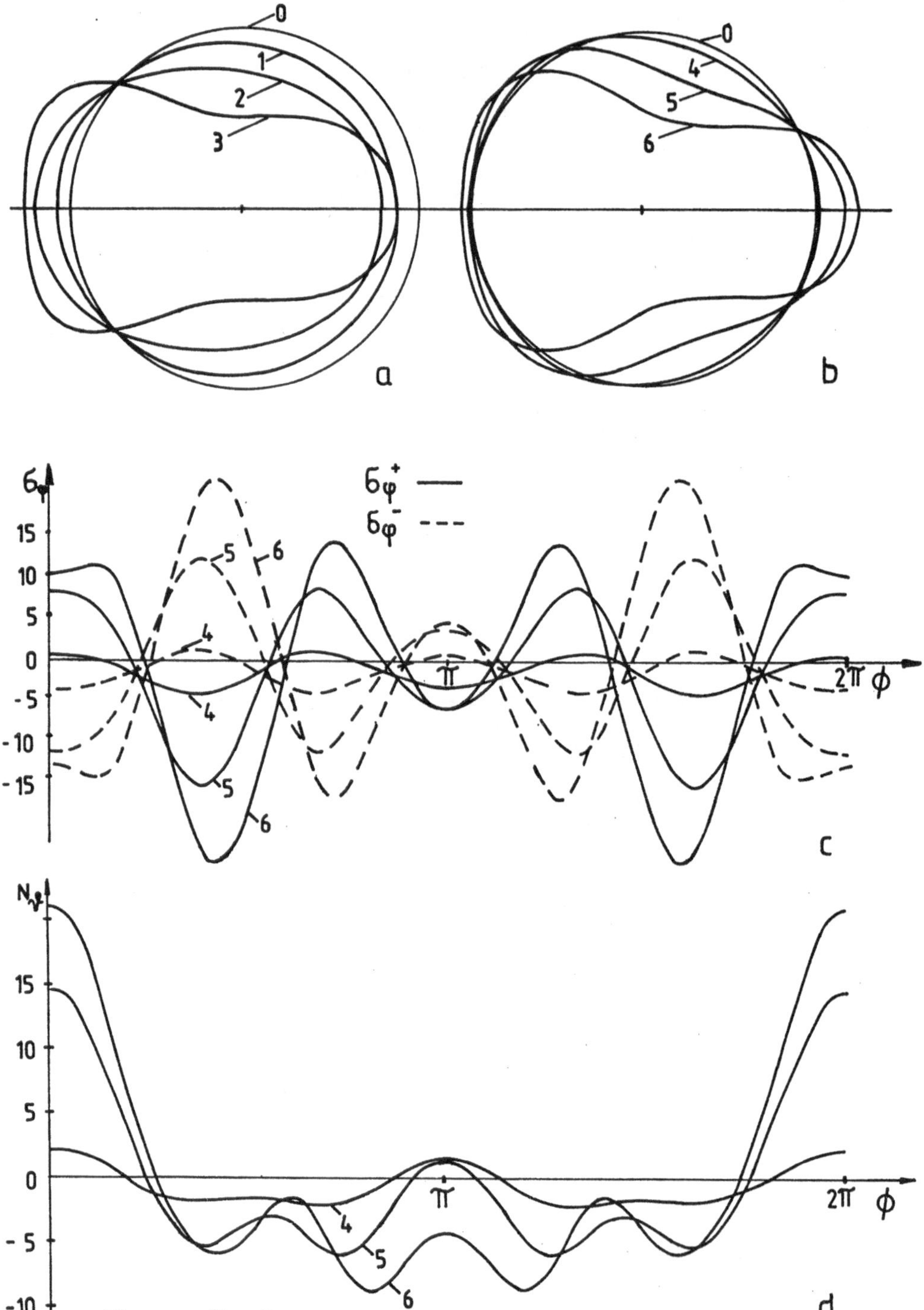

Figure 5 - Symmetric forms

8. Influence of shape imperfections

Symmetric equilibrium forms of meridional section of the shell may exist if the initial shape of the middle surface, the wall thickness and the pressure distribution are symmetric with respect to the symmetry plane of initial torus. Any deviation of symmetry results in non-symmetric form of deformation from the very begining of the loading process. For small imperfections, if a pressure is lower than bifurcation one (in the ideal shell sense) the difference between ideal and real forms of deformation is small, too. It increases, however, if the loading parameter comes closer to the critical value p_{bif}.

The influence of a jump-like variation of the distance between working sandwich sheets is presented. Dimensionless function f has been assumed to be discoutinous and chosen so as to assure nonsymmetric deformation (Fig. 6). The pressure versus the slope of the normal at the point $\Phi = 0$ is plotted in Fig. 6. Each curve corresponds to another value of imperfection parameter δ . The diagrams are symmetric with respect to the axis of pressure p for the reverse sign of δ. One can see the curves inscribe in a bifurcation one for a perfect shell. Thus, critical pressure as well as postbifurcation equilibrium path may be obtained using limit transition when the imperfection parameter δ tends to zero.

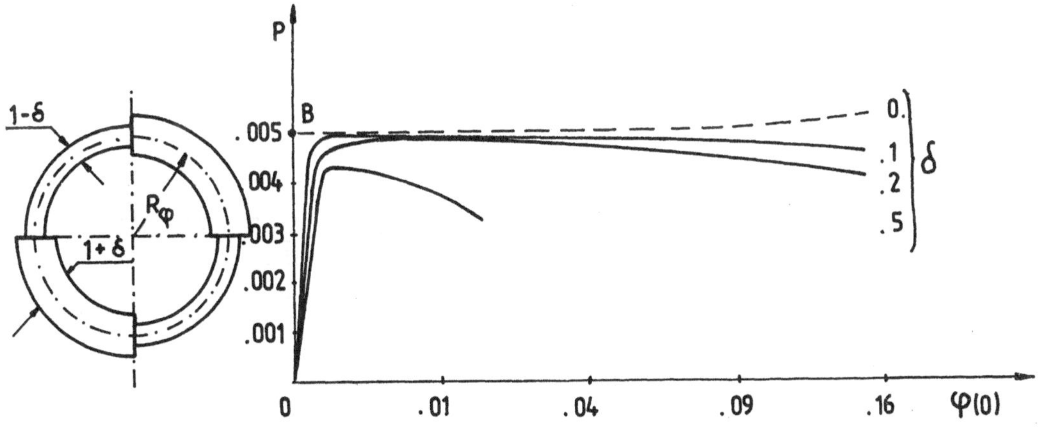

Figure 6 - Influence of thickness imperfections

9. Final remarks

Only rotationally symmetric bifurcation forms are cesidered. As it was shown in [7],[10],[15] they are most frequent and typical. In general, however, there may occur instability forms with some circumferential waves.

The diagrams of stresses suggest that elastis-plastic analysis should be performed. All stresses are related to one thousand fraction of E modulus and that quantity may approximate a yield point for many structural materials. On the other hand the resultant stresses in presented cases are greater than unity.

One can expect that there is possibility of optimal design of the function f so as to increase critical pressure value.

References

1. AXELRAD E.L., On stability of a curved pipe with circular cross section under external pressure (in Russian), Inzh. Zhurn. Mekh. Tverd. Tela, 2(1967), 117-120.
2. AXELRAD E.L., Flexible shell-theory and buckling of toroidal shells and tubes, Ing. Arch. 2(1979), 47, 95-104.
3. BIELSKI J., SKRZYPEK J., Forms of instability of an elastic toroidal shell subject to external pressure and in-plane bending (in Polish), Zesh. Nauk. WSI w Opolu, Bud., 18(1982), 31-36.
4. BULYGIN A.V., Stability of toroidal shell under external pressure action (in Russian), Trudy Kazan. Aviats. Inst. 160(1973), 16-23.
5. BULYGIN A.V., Stability and axisymmetric deformations of toroidal shell under external pressure (in Russian), Teoria Obolochek i Plast. 9th, Leningrad, 1973, Trudy Vses. Konf., (1975), 114-116.
6. FEDOSOV J., Stability of toroidal shell under external pressure (in Russian), Izvestya Vys. Uch. Zav., Aviats. Techn., 3(1971), 108-112.
7. GAYDAYCHUK W.W., GOTSULAK E.A. and GULAYEV W.I., Bifurcation of solutions of nonlinear equations of toroidal shells under external pressure (in Russian), Prikl. Mekh., 9(1978), 14, 38-45.
8. GAYDAYCHUK W.W., GOTSULAK E.A. and GULAYEV W.I., Nonlinear stability of toroidal shells with variable thickness under external pressure (in Russian), Mekh. Tver. Tela, Izv. Akad. Nauk SSSR, 3(1978),107-113
9. GULAYEV W.I., BAZHENOV W.A. and GOTSULAK E.A., Stability of nonlinear mechanical systems (in Russian), Lvov: Vishtsha Shkola, Izd. pri Lvov. Univ., 1982.
10. JORDAN P., Buckling of toroidal shells under hydrostatic pressure, AIAA Journal, 11(1973), 10, 1439-1441.
11. KRÄTZIG W.B., WITTEK U. and BASAR Y., Buckling of general shells--theory and numerical analysis, Collapse: Buckl. Struct. Theory and Pract. Symp., London, 31.08-3.09, 1982, Cambridge e.a. Univ. Press, 1983, 377-394.

12. POLISHCHUK T.I., Deformation of curved tubes of variable thickness, Sopr. Mat. i Teor. Soeruzh., Kiev, 42(1983), 124-127.

13. SKRZYPEK J., Kinematics and statics of symmetric finite deformations of an incomplete toroidal shell, Bull. Acad. Pol. Sci., Ser. Sci. Techn., 5-6(1980), 28, 113-120.

14. SKRZYPEK J. and ŻYCZKOWSKI M., Termination of processes of finite plastic deformations of incomplete toroidal shells, SM Archives, 8(1983), 39-98.

15. SOBEL L. and FLÜGGE W., Stability of toroidal shells under uniform external pressure, AIAA Journal 3(1967), 5, 425-431.

16. ZINTIKIS G.M., Pressure buckling of end supported shells of revolution, Croll. J.G.A. Eng. Struct., 4(1982), 4, 222-232.

The author would like to express his gratitude to prof. Jacek Skrzypek and prof. Michał Życzkowski for help and many invaluable remarks in preparing the paper.

The paper is prepared with a support of the Grant 05.12.

THE COMPLEMENTARY ENERGY PRINCIPLE IN FINITE ELASTOSTATICS AS A DUAL PROBLEM

W.R. BIELSKI[(1)] and J.J. TELEGA[(2)]
Polish Academy of Sciences
(1) Institute of Geophysics, ul. Pasteura 3, 02-093 Warsaw, Poland
(2) Institute of Fundamental Technological Research ul. Świętokrzyska 21, 00-049 Warsaw, Poland

1. Introduction

A revival of interest in a formulation of the principle of complementary energy for finitely deformed hyperelastic bodies should be attributed to LEVINSON [44]. Concerning earlier investigations the reader may refer to Refs [26,49,58,61,64]. A critical review of existing formulations is out of scope of the present paper. Here we shall only comment on various approaches to the problem.

Levinson's principle is based on the first (unsymmetric) Piola-Kirchhoff stress tensor and an inversion of the stress-deformation gradient relation is required. Such an approach has been revised and elucidated in many papers [3,13,15,25,31,34-36,49-51,64-68]. Invertibility is also used a priori by GRIOLI [29,30] without referring to existing contributions. In order to avoid the difficulty of inverting the nonlinear constitutive relation LEE and SHIELD [41,42] have proposed the variational principle of the complementary energy type and not the complementary energy itself. In this approach trial functions for the actual deformation gradient are used under the nonlinear constraints imposed by equilibrium equations and traction boundary conditions.

FRAEIJS DE VEUBEKE [25] has formulated the complementary energy principle which involves both the first Piola-Kirchhoff stress and a rigid rotation as independent variables, see also Refs [3,13,15,23,31,50,51,63,66].

In order to formulate the principles of the potential and complementary energies as dual extremum principles STUMPF [58] uses appropriate concepts of convex analysis. This approach is a straightforward extension of the method employed earlier by MOREAU [47] for small displacement theories, see also [48]. The basic hypothesis is very stringent since the convexity of a strain energy density is assumed, see also [38,39]. The complementary energy principle for incompressible materials is considered in [22] . The exterior penalty method is also applied.

A first attempt to formulate the complementary energy principle in the form of the maximum principle, provided that the stored energy function is not necessarily convex, is due to BERDIČEVSKII and MISIOURA [8]. Unsymmetric description is used. These authors were not able to prove that for nonconvex problems we may have, see the subsequent sections

$$\inf P = \sup P^{*} \qquad (1.1)$$

In [8] the hyperelastic body is subjected to a dead loading. Under similar assumption the relation (1.1) has been proved in [32] using Rockafellar's theory of duality. An extension to more general types of potential loadings is given in the next section.

The reader interested in variational formulations for nonlinear elastic plates and shells may consult Refs [3,9-11,40,56,57,59,60,63, 64] and the references cited therein. Non-potential elasticity has been dealt with in the papers [27,28] .

In the contributions [8,32] mentioned above only the unsymmetric approach, involving the first Piola-Kirchhoff stress tensor and the deformation gradient, has been studied. Then the choice of a linear and continuous operator Λ , playing an important role in Rockafellar's theory of duality, is straightforward, since $\Lambda = \nabla$ = grad [32]. In the case of the symmetric description, involving the second Piola-Kirchhoff stress tensor and its conjugate strain measure, the choice of Λ is more complicated due to the nonlinearity of strain--displacement relation (3.1) below. The same Λ as previously could be used, but then the complementary energy principle would be expressed in terms of the first Piola-Kirchhoff stress tensor solely. New formulations of the complementary energy principle involving a symmetric stress tensor are available provided that Λ is appropriately chosen. Two different formulations of this principle are derived in Section 4. The complementary energy principle for the two-dimensional, small strain and moderate rotation theory will be studied in a forthcoming paper.

2. Elements of convex analysis

In this section indispensable concepts of convex analysis are adduced. For details the reader may refer to [24,47,48,53,54] . However Th.2.1., relating the primal nonconvex problem P with the convex dual problem P^*, seems to be new. Particular case has been examined in [32] .

Let V be a real reflexive Banach space and V^* its topological dual. Let $\langle\cdot,\cdot\rangle : V^* \times V \to R$ be the duality pairing and $f : V \to \bar{R} = R \cup \{-\infty\} \cup \{+\infty\}$ a functional, not necessarily convex. The Fenchel transformation

$$f^*(u^*) = \sup\left\{\langle u^*,u\rangle - f(u) \,\middle|\, u \in V\right\}, \quad u^* \in V^* \tag{2.1}$$

defines the polar (conjugate) functional f^* of f. The polar functional f^* is convex and lower semicontinuous . The formula (2.1) implies

$$f^*(u^*) + f(u) \geqslant \langle u^*,u\rangle , \quad \forall u \in V, \quad \forall u^* \in V^* \tag{2.2}$$

An element $u^* \in V^*$ such that

$$f(v) \geqslant f(u) + \langle u^*,v-u\rangle , \quad \forall v \in V \tag{2.3}$$

is called a subgradient of f at u. The sets of all subgradients u^* satisfying (2.3) is denoted by $\partial f(u)$, and called the subdifferential. We write $u^* \in \partial f(u)$. Particularly we have $\partial f(u) = \emptyset$, if $f(u) = \infty$; here $\emptyset$ stands for the empty set. The following property is important

$$f^*(u^*)+f(u) = \langle u^*,u\rangle \Longleftrightarrow u^* \in \partial f(u), \text{ or } u \in \partial f^*(u^*) \tag{2.4}$$

Applying the Fenchel transformation to f^* we obtain the bipolar functional f^{**} of f, that is

$$f^{**} = (f^*)^* \tag{2.5}$$

The polar f^{***} of f^{**} is

$$f^{***} = (f^{**})^* \tag{2.6}$$

The functional f^* maps V^* into $\overline{R}$, and due to the reflexivity of V we have $f^{**} : V \rightarrow \overline{R}$. The bipolar functional f^{**} is the lower semi-continuous and convex envelope of f, that is $f^{**}(u) \leqslant f(u)$, $\forall u \in V$. Moreover we have

$$f^{***} = f^* \tag{2.7}$$

Now suppose that $\partial f(u) \neq \emptyset$. Then we obtain

$$f(u) = f^{**}(u) \tag{2.8}$$

The following minimization problem, which means evaluating

$$(P) \qquad \inf\{f(u) \mid u \in V\} \tag{2.9}$$

is called the primal problem. The dual problem of P, or P^*-problem, can be formulated using the approach proposed primarily by Rockafellar. Now it will be briefly presented.

Let $\Phi(u,q)$ be a so called perturbed functional defined on $V \times Y$ and such that $\Phi(u,0) = f(u)$. Hence we have $\inf f(u) = \inf \Phi(u,0)$. Here Y is a Hausdorff topological space, for instance a space L^p, $p \geqslant 1$. Then the dual problem P^* is formulated as follows

$$(P^*) \qquad \sup\{-\Phi^*(0,q^*) \mid q^* \in Y^*\} \tag{2.10}$$

In the calculus of variations the following functional occurs

$$f(u) = J(u, \Lambda u) , \tag{2.11}$$

where Λ is a continuous linear operator, $\Lambda : V \rightarrow Y$.
Then the problem P^* means evaluating

$$\sup\{-J^*(-\Lambda^* q^*, q^*) \mid q^* \in Y^*\} \tag{2.12}$$

In applications important is the specific case of (2.11)

$$J(u, \Lambda u) = G(\Lambda u) + L(u) , \tag{2.13}$$

Here G and L are given functionals. Then the following perturbed functional may be considered

$$\Phi(u,q) = G(\Lambda u+q) + L(u) \tag{2.14}$$

The dual problem P^* takes on the form

$$\sup\left\{- G^*(q^*) - L^*(- \Lambda^* q^*) \mid q^* \in Y^*\right\} \tag{2.15}$$

where $\Lambda^* : Y^* \longrightarrow V^*$ is the dual or conjugate operator of Λ.

In virtue of (2.2) we can write

$$\Phi(u,0)+ \Phi^*(0,q^*) \geqslant \langle (0,q^*),(u,0)\rangle =0, \quad \forall u \in V, \quad \forall q^* \in Y^* \tag{2.16}$$

The last inequality implies

$$\inf \ \Phi(u,0) \geqslant \sup\left\{- \Phi^*(0,q^*)\right\} \quad , \tag{2.17}$$

or in the usually assumed symbolic notation

$$\inf P \geqslant \sup P^* \tag{2.18}$$

Suppose that in (2.9) the infimum is attained at a point $\tilde{u} \in V$, that is

$$f(\tilde{u}) = \inf\left\{ f(u) \mid u \in V\right\} \tag{2.19}$$

The inequality $f^{**} \leqslant f$ implies

$$\inf f^{**}(u) \leqslant \inf f(u) = f(\tilde{u}) \ .$$

We observe that the constant function $l(u) = f(\tilde{u})$, $u \in V$, is a continuous affine minorant of f and then $f^{**}(u) \geqslant f(\tilde{u})$, $u \in V$. Thus we arrive at

$$f^{**}(\tilde{u})=\inf\{f^{**}(u) \mid u \in V\} =f(\tilde{u})=\inf\left\{ f(u) \mid u \in V\right\} \tag{2.20}$$

provided that in (2.9) a finite infimum is attained at $\tilde{u} \in V$. Obviously the same holds for specific cases (2.11) and (2.13).

The problems P^{**} and P^{***} can be treated as pair of dual convex problems. In virtue of (2.7) we can write $P^{***} = P^*$. The following theorem is an immediate consequence of (2.20) and Th.4.1, Chapter III, of [24] .

Theorem 2.1. Assume that the functional f has the form (2.11). Let $\inf P$ be attained at $\tilde{u} \in V$. Suppose that $u_0 \in V$ exists such that $J^{**}(u_0, \Lambda u_0) < \infty$ and the mapping $q \longrightarrow J^{**}(u_0,q)$ is continuous at Λu_0 . Then we have

$$\inf P = \inf P^{***} = \sup P^{****} = \sup P^{*} \tag{2.21}$$

and the problem P^{***}, and thus also P^{*}, has at least one solution $\tilde{q}^{*} \in Y^{*}$.

■

It is worth noting here that in [32] solely the specific case (2.13) is studied, provided that the functional L is linear.

As we know the functional J^{**} is convex and lower semicontinuous on $V \times Y$. Using Lemma 2.1 and Corollary 2.5 of [24, Chapter I] we deduce that the function $q \to J^{**}(u_0, q)$ is continuous at Λu_0 if the mapping $q \to J(u_0, q)$ is finite in a neighbourhood of Λu_0.

Further interconnexion between the primal and dual problems reveals

Theorem 2.2. Let the functional f be given by (2.11). Then the following conditions are equivalent

(i) $\tilde{u}$ is a solution of the primal problem (2.9), $\tilde{q}^{*}$ is a solution of the dual problem (2.12) and $\inf P = \sup P^{*}$.

(ii) $\tilde{u} \in V$ and $\tilde{q}^{*} \in Y^{*}$ satisfy the extremality condition

$$J(\tilde{u}, \Lambda\tilde{u}) + J^{*}(-\Lambda^{*}\tilde{q}^{*}, \tilde{q}^{*}) = 0, \tag{2.22}$$

which is equivalent to

$$(-\Lambda^{*}\tilde{q}^{*}, \tilde{q}^{*}) \in \partial J(\tilde{u}, \Lambda\tilde{u}) \tag{2.23}$$

■

If the functional J has the form (2.13) then (2.22) results in

$$G(\Lambda\tilde{u}) + G^{*}(\tilde{q}^{*}) = \langle \tilde{q}^{*}, \Lambda\tilde{u} \rangle, \tag{2.24}$$

$$L(\tilde{u}) + L^{*}(-\Lambda^{*}\tilde{q}^{*}) = \langle -\Lambda^{*}\tilde{q}^{*}, \tilde{u} \rangle \tag{2.25}$$

The extremality conditions (2.24) and (2.25) are equivalent to

$$\tilde{q}^{*} \in \partial G(\Lambda\tilde{u}), \tag{2.26}$$

$$-\Lambda^{*}\tilde{q}^{*} \in \partial L(\tilde{u}), \tag{2.27}$$

respectively.

In applications it is very important to find an explicit form of the polar functional of an integral functional, say M, defined on $L^{p}(\Omega, R^{m})$, $1 \leq p < \infty$. An accessible presentation of L^{p}-spaces and Sobolev spaces used throughout this paper may be found in [37].

Suppose that Ω is a bounded domain of R^n while $g: \Omega \times R^m \to \bar{R}$, $g \geqslant 0$, is a normal integrand, particularly a Carathéodory function. Let us consider the integral functional

$$M(u) = \int_{\Omega} g(x,u(x))dx , \quad u \in L^p(\Omega ,R^m) \tag{2.28}$$

The following theorem has been proved in [24, Chapter IX] , see also [54] .

Theorem 2.3. Let the functional M be given by (2.28) and satisfies the above assumptions. Assume that $u_o \in L^{\infty}(\Omega,R^m)$ exists such that $M(u_o) < \infty$. Then for each $u^* \in L^{p'}(\Omega,R^m)$ we have

$$M^*(u^*) = \int_{\Omega} g^*(x,u^*(x))dx \tag{2.29}$$

where $1/p + 1/p' = 1$.

The condition of nonnegativeness of the integrand g can be replaced by

$$g(x,y) \geqslant a(x) - b \sum_{i=1}^{m} |y_i|^p \tag{2.30}$$

where $a \in L^1(\Omega)$ and $b \geqslant 0$.

3. The primal problem for hyperelastic solids

Extremum principles which will be studied in the sequel correspond to a mixed boundary value problem for a finitely deformed hyperelastic bodies. In this section we shall discuss the minimum principle of the total potential energy for such a body.

Let Ω be a bounded and sufficiently regular domain of three--dimensional Euclidean space R^3 . Its closure $\bar{\Omega}$ is identified with an underformed configuration of the hyperelastic body B. Let S be the boundary of Ω . We assume that $S = \bar{S}_0 \cup \bar{S}_1$, $S_0 \cap S_1 = \emptyset$. Orthogonal Cartesian coordinates are used throughout. By $x = (x_A)$, $A = 1,2,3$, we denote a point of Ω . Next we set $u = (u_A)$, $E = (E_{AB})$, $T = (T_{AB})$ for the displacement vector, the Green strain tensor and the second (symmetric) Piola-Kirchhoff stress tensor, respectively, see [18,46] . The strain-displacement relation is given by

$$E_{AB}(u) = \frac{1}{2}(u_{A,B} + u_{B,A} + u_{C,A}u_{C,B}) \tag{3.1}$$

where $u_{A,B} = \partial u_A / \partial x_B$. It is worth noting here that the relation (3.1) is nonlinear. As we shall see just this nonlinearity strongly influences the manner of the formulation of the complementary energy principle.
By ϱ_0 and $n = (n_A)$ we denote the mass density in the reference configuration and the unit outward normal to S, respectively.

Let $W(E)$ be the stored energy function, not necessarily convex. For nonhomogeneous materials W depends explicity on x. We assume that on the part S_0 of the boundary the body is clamped while the surface traction $t = (t_A)$ is prescribed on S_1 . Then the space of kinematically admissible displacement fields u is defined by

$$V = \left\{ u=(u_A) \mid u_A \in W^{1,p}(\Omega), \quad u=0 \quad \text{on} \quad S_0 \right\} \tag{3.2}$$

where $W^{1,p}(\Omega)$ is the usual Sobolev space [37] . We observe that u restricted to S_0 or S_1 is understood in the sense of trace. Further we assume that t_A belongs to the dual space of $W^{1-(1/p),p}(S_0)$. Obviously, we have $u_{A,B} \in L^p(\Omega)$. Then Th.1 of [16] yields $u_{C,A}u_{C,B} \in L^{p/2}(\Omega)$. Therefore we set $p \geqslant 2$.

We assume that the virtual work of the potential external loading is given by the functional

$$L(u) = \int_{\Omega} g(x,u)dx - \int_{S_1} t_A u_A dx , \qquad u \in V , \tag{3.3}$$

where the integrand g obeys the conditions of Th.2.3 formulated in the preceding section. Then the total potential energy corresponding to a displacement field $u \in V$ is expressed by

$$J(u) = \int_{\Omega} W(E(u))dx + L(u) \tag{3.4}$$

The minimum principle of the total potential energy means evaluating

$$(P) \qquad \inf \left\{ J(u) \mid u \in V \right\} \tag{3.5}$$

Now we shall concisely discuss the existence of a minimizer solving the primal problem (3.5). Let $K: \bar{\Omega} \longrightarrow R^3$ be a sufficiently regular deformation of the body B, that is

$$\det F > 0 \quad , \tag{3.6}$$

where

$$K = Id + u \,, \qquad Id = \text{identity mapping} \,, \tag{3.7}$$

and

$$F = \nabla K = I + \nabla u \,, \tag{3.8}$$

is the deformation gradient. Here I stands for the identity matrix. We have [18,46]

$$E(F) = \tfrac{1}{2}(F^T F - I) = E(u) \tag{3.9}$$

The displacement boundary condition $u=0$ on S_0 is equivalent to $K(x) = x$, $x \in S_0$. The functional (3.4) yields

$$J_1(K) = \int_\Omega W(E(\nabla K))dx + \int_\Omega g(x,K(x)-x)dx - \int_{S_1} t_A\left[K_A(x)-x\right] dS \tag{3.10}$$

A minimizer $\tilde{K}$ of the functional J_1 gives the minimizer $\tilde{u}=\tilde{K}-Id$ of the problem (3.5). For details concerning the study of functionals of the type (3.10) the reader should refer to the fundamental papers by BALL [4–6] , see also [1,7,17,18,20,21,43,46,52] . This author investigates the problem of existence of minimizers of the total potential energy for various boundary conditions provided that the stored energy function

$$W_1(F) = W(E(F)) \tag{3.11}$$

is <u>polyconvex</u>, otherwise not necessarily convex.
The condition (3.6) is then incorporated into the definition of the space of kinematically admissible deformations K. As far as we know Ball's results are actually valid for isotropic materials.

When studying duality we may assume that the function (3.11) is such that

$$W_1(F) = +\infty \qquad \text{if} \qquad \det F \leqslant 0 \tag{3.12}$$

Polyconvexity is a specific case of the so called <u>quasi-convexity</u> [4,5,7,20,21,46] . An existence theorem concerning a minimizer of an integral functional has been given by MARCELLINI and SBORDONE [45] . These authors assume that the integrand is quasi-convex; moreover some additional conditions must be satisfied. Only the Dirichlet boundary problem has been dealt with. It seems that results obtained in the paper [45] are applicable to the corresponding boundary value problem of not necessarily isotropic finite elasticity. However, it is not clear whether the condition (3.6) will be satisfied.

Concerning questions of uniqueness in finite elasticity the reader should refer to the papers [2,33,52,62] and the references cited therein.

4. Dual problems or the complementary energy principle

We proceed to the formulation of two convex dual problems for the non-convex primal problem (3.5). A difference in those dual problems issues from the choice of Λ . Still another dual problem has been derived in our paper [12]. The latter dual problem depends explicity on the second Piola-Kirchhoff stress tensor whereas a kinematical field enters parametrically.

4.1. Natural choice of Λ

Section 2 suggests that in order to formulate the P^*-problem we must in the first place choose an appropriate operator Λ entering the functional (2.13). This choice is immediate in the case of the unsymmetric description since F depends linearly on K, see the formula (3.8). In the case of the symmetric description used in this paper the allowable form of Λ must be deduced from the strain-displacement relation (3.1).

In this subsection we take the following linear operator

$$\Lambda u = (Au, Bu) , \qquad \Lambda : V \to Y = Y_1 \times Y_2 \tag{4.1}$$

where

$$Au = e(u) = ((u_{A,B} + u_{B,A})/2) , \quad Bu = \nabla u \tag{4.2}$$

and

$$A : V \to Y_1 \subset L^p(\Omega, R^6), \; A^* : Y_1^* \to V^* , \; Y_1^* \subset L^{p'}(\Omega, R^6), \tag{4.3}$$

$$B : V \to Y_2 = L^p(\Omega, R^9), \quad B^* : Y_2^* = L^{p'}(\Omega, R^9) \to V^* \tag{4.4}$$

Here $1/p + 1/p' = 1$, Y_1 is the space of symmetric matrices $s = (s_{AB})$ such that $s_{AB} \in L^p(\Omega)$. The continuity of Λ is straightforward, since for instance

$$\|Au\|_{Y_1} = \|e(u)\|_{Y_1} \leqslant c\,\|v\|_V , \qquad c > 0 .$$

The strain tensor (3.1) can be written as follows

$$E(u) = Au + \tfrac{1}{2}(Bu)^T Bu \tag{4.5}$$

We now pass to the derivation of explicit forms of Λ^*, A^* and B^* . We have

$$\langle (s^*,q^*), \Lambda u\rangle = \langle (s^*,q^*),(Au,Bu)\rangle = \langle A^*s^*+B^*q^*,u\rangle =$$

$$= \langle \Lambda^*(s^*,q^*),u\rangle , \quad u \in V, \quad (s^*,q^*) \in Y^*,$$

or

$$\Lambda^*(s^*,q^*) = A^*s^*+B^*q^* , \quad (s^*,q^*) \in Y^* = Y_1^* \times Y_2^* \tag{4.6}$$

Next we obtain

$$\langle s^*, Au\rangle = \int_\Omega s^*_{AB} e_{AB}(u)dx = -\int_\Omega s^*_{AB,B} u_B dx + \int_{S_1} s^*_{AB} n_B u_A dS =$$
$$= \langle A^* p^*, v\rangle , \quad v \in V , \quad s^* \in Y_1^* . \tag{4.7}$$

Hence we have

$$A^*s^* = \begin{cases} -\operatorname{div} s^* , & \text{in } \Omega \\ s^*n , & \text{on } S_1 \end{cases} \tag{4.8}$$

where $(\operatorname{div} s^*)_A = s^*_{AB,B}$ and $(s^*n)_A = s^*_{AB} n_B$.

Similarly it can readily be proved that

$$B^*q^* = \begin{cases} -\operatorname{div} q^* , & \text{in } \Omega \\ q^*n , & \text{on } S_1 \end{cases} \tag{4.9}$$

The form (3.4) of the functional of the total potential energy yields

$$G(\Lambda u) = G(Au,Bu) = \int_{\Omega} W(E(u))dx \ , \quad u \in V \tag{4.10}$$

where (4.5) has to be taken into account.
The functional G^*, or the polar of (4.10), is given by

$$G^*(s^*,q^*) = \sup\{\langle s^*,s\rangle + \langle q^*,q\rangle - G(s,q) \mid (s,q) \in Y\} \tag{4.11}$$

where $(s^*,q^*) \in Y^*$.
Using the formulas (2.1), (4.6), (4.7) and (4.9) we write

$$L^*(-\Lambda^*(s^*,q^*)) = \sup\{\langle -A^*s^*-B^*q^*,u\rangle - L(u) \mid u \in V\} =$$
$$= \sup\Big\{\int_{\Omega}\big[\mathrm{div}(s^*+q^*)\big]\cdot u dx - \int_{\Omega} g(x,u)dx + \int_{S_1}\big[t - (s^*+q^*)n\big]\cdot u dS \mid u \in V\Big\} \ ,$$

where $u \cdot v = u_A v_A$.
Taking account of Th.2.3 the last relation readily furnishes

$$L^*(-\Lambda^*(s^*,q^*)) = \begin{cases} \int_{\Omega} g^*(x,\mathrm{div}(s^*+q^*))dx, & \text{if } (s^*+q^*)n=t, \text{ on } S_1 \\ +\infty & , \text{ otherwise} \end{cases} \tag{4.12}$$

since in virtue of our assumptions (2.29) holds.
The problem P^* reads

find

$$\sup\Big\{-G^*(s^*,q^*) - \int_{\Omega} g^*(x,\mathrm{div}(s^*+q^*))dx \mid (s^*,q^*) \in Y^*\Big\} \tag{4.13}$$

subject to

$$(s^*_{AB} + q^*_{AB})n_B = t_A \ , \quad \text{on } S_1 \tag{4.14}$$

The extremality condition (2.26) furnishes the global constitutive relations

$$(\tilde{s}^*,\tilde{q}^*) \in \partial G(\Lambda\tilde{u}) \Longleftrightarrow \tilde{s}^* \in \partial_1 G(\Lambda\tilde{u}) \text{ and } \tilde{q}^* \in \partial_2 G(\Lambda\tilde{u}) \tag{4.15}$$

provided that $\tilde{u}$ is a minimizer of the primal problem (3.5) whereas $(\tilde{s}^*,\tilde{q}^*)$ solves the P^*-problem (4.13) and (4.14). Here $\partial_1 G(\Lambda\tilde{u})$, etc., denotes the subdifferential of G with respect to Au at $\Lambda\tilde{u}$. Theorem 3 of [55] readily yields

$$\partial_1 G(\Lambda\tilde{u}) = \partial G(\tilde{E}) , \quad \partial_2 G(\Lambda\tilde{u}) = (\nabla\tilde{u})[\partial G(\tilde{E})] \tag{4.16}$$

where $\tilde{E} = E(\tilde{u})$. The subdifferential relations (4.15) are global. The corresponding local form is given by

$$\tilde{s}^*(x) \in \partial W(\tilde{E}(x)) , \quad \tilde{q}^*(x) \in [\nabla\tilde{u}(x)] [\partial W(\tilde{E}(x)] \tag{4.17}$$

and it holds for almost every $x \in \Omega$. To derive (4.17) the physically plausible assumptions of Th.1 due to CLARKE [19] are tacitly presupposed.

The second extremality relation, or (2.27), yields the global form of the equilibrium equations and the traction boundary condition on S_1. The local form is given by

$$\operatorname{div}[\tilde{s}^*(x) + \tilde{q}^*(x)] \in \partial_2 g(x,\tilde{u}(x)) , \quad \text{a.e. } x \in \Omega \tag{4.18}$$

$$[\tilde{s}^*(x) + \tilde{q}^*(x)]\, n(x) = t(x) , \quad \text{a.e. } x \in S_1 \tag{4.19}$$

where $\partial_2 g(x,\tilde{u}(x))$ is the subdifferential of the function $g(x,\cdot)$ at $\tilde{u}(x)$.
From (4.17) - (4.19) we conclude that s^* may be identified with the second Piola-Kirchhoff stress tensor T. From $(4.17)_2$ we infer that $Q \overset{df}{=} q^* = (\nabla u)T$. For further discussion the reader may refer to our papers [9-12] .

The P^*-problem (4.13) and (4.14) represents <u>the complementary energy principle as the really dual problem</u>. The supremum in (4.13) is taken over T <u>and</u> Q treated as independent static fields.

If the stored energy function W is differentiable then (4.17) takes on more familiar form

$$T = \frac{\partial W}{\partial e} = \frac{\partial W}{\partial E} , \qquad Q = \frac{\partial W}{\partial (Bu)} = (\nabla u)T \tag{4.20}$$

where for the sake of simplicity the argument x is dropped.

Let us assume that the body force potential g is differentiable with respect to the second argument. Then <u>the multivalued equation</u> (4.18) becomes

$$\operatorname{div}(T+Q) = \frac{\partial g}{\partial u} , \quad \text{almost everywhere in } \Omega. \tag{4.21}$$

Remark 1

BALL [6] discusses the Euler-Lagrange equations resulting from the study of the potential energy principle in the unsymmetric case provided that W_1 is differentiable. He shows that minimizers do not necessarily satisfy these equations. On the other hand, the Euler--Lagrange equations yield equilibrium equations and traction boundary conditions on S_1, provided that $S_1 \neq \emptyset$. Thus the Euler-Lagrange equations are equations of statics. As we have seen the equations of statics are satisfied via the duality. Though such an answer does not solve Ball's problem, yet we think that it is reasonable at least from the point of view of mechanics.

4.2. Choice of Λ useful in the study of finite and moderate rotations

CASEY and NAGHDI [14] have proposed various approximate theories of elasticity. For the purpose of performing approximations the displacement gradient $H = H(u) = \nabla u$ is decomposed into its symmetric and skew-symmetric parts

$$H = e+w \ , \quad e = (H+H^T)/2 \ , \quad w = (H-H^T)/2 = -w^T \tag{4.22}$$

where H^T is the transpose of H. The strain tensor E defined by (3.1) may be rewritten as follows

$$E = e + \tfrac{1}{2}H^TH = e + \tfrac{1}{2}(e^2 + ew - we - w^2) \tag{4.23}$$

The strain tensor E is thus the function of e and w. Consequently we take

$$\Lambda : V \longrightarrow X, \quad \Lambda u = (Au, Cu), \quad Cu = w(u) \tag{4.24}$$

where $X = Y_1 \times Y_3$. Y_3 is the space of skew-symmetric matrices $r = (r_{AB})$ such that $r_{AB} \in L^p(\Omega)$.

Let the functional J of the total potential energy be given by (2.13), provided that Λ is specified by (4.24) and

$$G(\Lambda u) = G(Au, Cu) = \int_{\Omega} W(E(u))dx \ . \tag{4.25}$$

Here the strain tensor E is given by (4.23).

The external potential loading is defined by

$$L(u) = \int_{\Omega} g(x,u)dx + \int_{S_1} h(x,u)dS \tag{4.26}$$

In the present subsection no restrictions are imposed on e and w. Small strain, moderate rotation theory will be dealt with in a separate paper.

The conjugate operator Λ^* of Λ can be derived similarly as in the preceding subsection. We now have

$$\langle (s^*,r^*), \Lambda u\rangle = \langle (s^*,r^*),(Au,Cu)\rangle =$$

$$= \langle A^*s^*+C^*r^*,u\rangle, \; u \in V, \; (s^*,r^*) \in X^* = Y_1^* \times Y_3^* \; . \tag{4.27}$$

Hence

$$\Lambda^*(s^*,r^*) = A^*s^* + C^*r^* \; . \tag{4.28}$$

It can be easily shown that

$$\Lambda^*(s^*,r^*) = \begin{cases} -\operatorname{div}(s^*+r^*) \; , & \text{in } \Omega \; , \quad (4.29) \\ (s^*+r^*)n \; , & \text{on } S_1 \quad (4.30) \end{cases}$$

The dual problem, or the complementary energy principle, reads

$$\left|\begin{array}{l} \text{find} \\ \qquad \sup\{-G^*(s^*,r^*) - L^*(-A^*s^*-C^*r^*) \mid (s^*,r^*) \in X^*\} \end{array}\right. \tag{4.31}$$

where

$$G(s^*,r^*) = \sup\{\langle s^*,s\rangle + \langle r^*,r\rangle - G_1(s,r) \mid (s,r) \in X\} \tag{4.32}$$

Let us consider the specific case when the body is subjected to a dead loading, that is

$$L_1(u) = -\int_{\Omega} \varrho_0 t_A u_A dx - \int_{S_1} t_A u_A dx \; , \quad u \in V \tag{4.33}$$

In virtue of (2.1) and (4.28) we obtain

$$L^*(-\Lambda^*(s^*,r^*)) = \sup\{\langle -A^*s^*-C^*r^*,u\rangle - L_1(u) \mid u \in V\} =$$

$$= \begin{cases} 0, & \text{if } \operatorname{div}(s^*+r^*)+\varrho_0 b=0 \text{ in } \Omega, \text{ and } (s^*+r^*)n=0, \text{ on } S_1 \\ +\infty \; , & \text{otherwise} \end{cases} \tag{4.34}$$

Taking account of (4.34) the complementary energy principle means evaluating

$$\sup\{-G(s^{*},r^{*}) \mid (s^{*},r^{*}) \in X^{*}\} \tag{4.35}$$

subject to

$$\operatorname{div}(s^{*}+r^{*}) + \rho_0 b = 0 , \quad \text{in } \Omega, \tag{4.36}$$

and

$$(s^{*}+r^{*})n = t , \quad \text{on } S_1 \tag{4.37}$$

We now proceed to a mechanical interpretation of the dual variables s^{*} and r^{*}. For the purpose we assume that $\tilde{u} \in V$ solves the primal problem whereas $(\tilde{s}^{*},\tilde{r}^{*})$ is a solution of the dual problem (4.35) - (4.37). If the stored energy function W is differentiable then the extremality relation (2.26) furnishes

$$\tilde{s}^{*}_{AB} = \frac{\partial W(E(\tilde{u}))}{\partial (Au)_{AB}} = \frac{\partial W(E(u))}{\partial E_{CD}} \frac{\partial E_{CD}}{\partial e_{AB}} =$$

$$= \tilde{T}_{AB} + \frac{1}{2}(\tilde{T}_{AC}\tilde{e}_{CB} + \tilde{e}_{AC}\tilde{T}_{CB}) + \frac{1}{2}(\tilde{w}_{AC}\tilde{T}_{CB} - \tilde{T}_{AC}\tilde{w}_{CB}) \tag{4.38}$$

where

$$\tilde{T} = \frac{\partial W(E(\tilde{u}))}{\partial E} , \quad \tilde{e} = e(\tilde{u}) , \quad \tilde{w} = w(\tilde{u}) .$$

Hence dropping tilde we may write

$$s^{*} = T + \frac{1}{2}\left[(e+w)T + T(e-w)\right] = T + (HT+TH^{T})/2 \tag{4.39}$$

In a similar manner using the second constitutive equation

$$\tilde{r}^{*} = \frac{\partial W(E(\tilde{u}))}{\partial (Cu)} \tag{4.40}$$

we deduce that

$$r^{*} = \frac{1}{2}(eT-Te) + \frac{1}{2}(wT+Tw) = \frac{1}{2}(HT-TH^{T}). \tag{4.41}$$

Thus the dual variable s^{*} is the symmetric part of the first Piola--Kirchhoff stress tensor and r^{*} is its skew-symmetric part. We observe that this interpretation is also suggested by (4.36) and (4.37).

References

1. ANTMAN S.S., The influence of elasticity on analysis: modern developments, Bull. Amer. Math. Soc., 9(1983), 3, 267-291.

2. ARON M., On the physical meaning of an uniqueness condition in finite elasticity, J. Elasticity, 8(1978), 111-115.

3. ATLURI S.N., Alternate stress and conjugate strain measures and mixed variational formulations involving rigid rotations, for computational analyses of finitely deformed solids, with application to plates and shells - I., Comp. Struct., 18(1983), 1, 93-116.

4. BALL J.M., Convexity conditions and existence theorems in nonlinear elasticity, Arch. Rat. Mech. Anal., 63(1977), 337-403.

5. BALL J.M., Constitutive inequalities and existence theorems in nonlinear elastostatics, in: Nonlinear Analysis and Mechanics, vol.1, Ed. by R.J. Knops, 187-241, Pitman Adv. Publ. Program, London 1977.

6. BALL J.M., Minimizers and the Euler-Lagrange equations, in: Trends and Applications of Pure Mathematics to Mechanics, Ed. by P.G. Ciarlet and M. Roseau, 1-4, Springer-Verlag, Berlin 1984.

7. BALL J.M., MURAT F., $W^{1,p}$-quasiconvexity and variational problems for multiple integrals, J. Funct. Analysis, 58(1984), 225-253.

8. BERDIČEVSKII V.L., MISIOURA V.A., On the dual variational principle in geometrically nonlinear theory of elasticity, Prikl. Mat. Mekh., 43(1979), 321-329.

9. BIELSKI W.R., TELEGA J.J., A note on duality for von Kármán plates in the case of the obstacle problem, Arch. Mech., 37(1985), 1, 135-141.

10. BIELSKI W.R., TELEGA J.J., A contribution to contact problems for a class of solids and structures, Arch. Mech., 37(1985), 4-5.

11. BIELSKI W.R., TELEGA J.J., On the obstacle problem for linear and nonlinear elastic plates, in: Variational Methods in Engineering, Ed. by C.A. Brebbia, (3-55) - (3-64), Springer--Verlag, Berlin 1985.

12. BIELSKI W.R., TELEGA J.J., On the complementary energy principle in finite elasticity, in: Proc. of the Int. Conf. on Nonlinear Mechanics, Shanghai, China, October 28-31, 1985.

13. BUFLER H., On the work theorems for finite and incremental elastic deformations with discontinuous fields: a unified treatment of different versions, Comp. Meth. Appl. Mech. Eng., 36 (1983), 95-124.

14. CASEY J., NAGHDI P.M., Physically nonlinear and related approximate theories of elasticity, and their invariance properties, Arch. Rat. Mech. Anal., 88(1985), 1, 59-82.

15. CHRISTOFFERSEN J., On Zubov's principle of stationary complementary energy and a related principle, DCAMM, Report No 44, April 1973, The Technical University of Denmark.

16. CHUNG-LIE WANG, Variants of Hölder inequality and its inverses, Canad. Math. Bull., 20(1977), 377-384.

17. CIARLET P.G., Quelques remarques sur les problèmes d'existence en élasticité non linéaire, in: Computing Methods in Applied Sciences and Engineering, Ed. by R. Glowinski and J.L. Lions, 235-254, North-Holland, Amsterdam 1982.

18. CIARLET P.G., Lectures on Three-Dimensional Elasticity, Springer -Verlag, Berlin 1983.

19. CLARKE F.H., Generalized gradients of Lipschitz functionals, Adv. in Math., 40(1981), 1, 52-67.

20. DACOROGNA B., Weak Continuity and Weak Lower Semicontinuity of Non-Linear Functionals, Springer-Verlag, Berlin 1982.

21. DACOROGNA B., Remarques sur les notions de polyconvexité, quasi convexité et convexité de rang 1, J. Math. Pures Appl., in print.

22. DE CAMPOS L.T., ODEN J.T., On the principle of stationary complementary energy in finite elastostatics, Int. J. Eng. Sci., 23 (1985), 1, 57-63.

23. DILL E.H., The complementary energy principle in nonlinear elasticity, Letters in Appl. Eng. Sci., 5(1977), 2, 95-106.

24. EKELAND I., TEMAM R., Convex Analysis and Variational Problems, North-Holland, Amsterdam 1976.

25. FRAEIJS DE VEUBEKE B., A new variational principle for finite elastic displacements, Int. J. Eng. Sci., 10(1972), 745-763.

26. FUNG Y.C., Foundations of Solid Mechanics, Prentice-Hall, Englewood Cliffs, New Jersey 1965.

27. GAŁKA A., TELEGA J.J., A variational method for finite elasticity in the case of non-potential loadings - I, First Piola-Kirchhoff stress tensor, Bull. Acad. Pol. Sci., Sér. Sci. Techn., 30(1982), 9-10, 471-478; II. Symmetric stress tensor and some comments, ibid., 479-485.

28. GAŁKA A., TELEGA J.J., On variational principles and conservation laws in finite non-potential elasticity, in: Variational Methods in Engineering, Ed. by C.A. Brebbia, (3-13) - (3-22), Springer -Verlag, Berlin 1985.

29. GRIOLI G., Aspetti variazionali nella meccanica dei continui con deformazioni finite, in: Trends in Applications of Pure Mathematics to Mechanics, Ed. by G. Fichera, 145-155, Pitman 1976.

30. GRIOLI G., A variational approach to finite elasticity, in: Finite Elasticity, Ed. by D.E. Carlson and R.T. Shield, 179-189, Martinus Nijhoff Publishers, The Hague 1982.

31. GUO ZHONG-HENG, The unified theory of variational principles in nonlinear elasticity, Arch. Mech., 32(1980), 4, 577-596.

32. HANYGA A., SEREDYŃSKA M., The complementary energy principle of nonlinear elasticity, Fisica Mat., Suppl. Boll. Un. Mat. Ital., 2(1983), 153-172.

33. KNOPS R.J., STUART C.A., Quasiconvexity and uniqueness of equilibrium solutions in nonlinear elasticity, Arch. Rat. Mech. Anal., 86(1984), 3, 233-249.

34. KOITER W.T., On the principle of stationary complementary energy in the nonlinear theory of elasticity, SIAM J. Appl. Math., 25(1973), 424-434.

35. KOITER W.T., On the complementary energy theorem in non-linear elasticity, in: Trends in Applications of Pure Mathematics to Mechanics, Ed. by G. Fichera, 207-232, Pitman, London 1976.

36. KOITER W.T., Complementary energy, neutral equilibrium and buckling, Meccanica, 19(1984), 52-56.

37. KUFNER A., JOHN O., FUČIK S., Function Spaces, Noordhoff Int. Publ., Leyden 1977.

38. LABISCH F.K., Error bounds in nonlinear elastostatics, Int. J. Eng. Sci., 18(1980), 389-407.

39. LABISCH F.K., On the dual formulation of boundary value problems in non-linear elastostatics, Int. J. Eng. Sci., 20(1982), 413-431.

40. LABISCH F.K., Approximation of non-unique solutions of a homogeneous elastic plate model, Zeitschr. Ang. Math. Mech., 63 (1983), 607-619.

41. LEE S.J., SHIELD R.T., Variational principles in finite elastostatics, Zeitschr. Ang. Math. Physik, 31(1980), 4, 437-453.

42. LEE S.J., SHIELD R.T., Applications of variational principles in finite elasticity, Zeitschr. Ang. Math. Physik, 31(1980), 4, 454-472.

43. LE TALLEC P., Existence and approximation results for nonlinear problems: application to incompressible finite elasticity, Numer. Math., 38(1982), 365-382.

44. LEVINSON M., The complementary energy theorem in finite elasticity, J. Appl. Mech., Trans. ASME, E 32 (1965), 4, 826-828.

45. MARCELLINI P., SBORDONE C., On the existence of minimum of multiple integrals of the calculus of variations, J. Math. Pures Appl., 62(1983), 1-9.

46. MARSDEN J.E., HUGHES T.J.R., Mathematical Foundations of Elasticity, Prentice Hall, Englewood Cliffs, New Jersey 1983.

47. MOREAU J.J., On unilateral constraints, friction and plasticity, in: New Variational Techniques in Mathematical Physics, Ed. by G. Capriz and G. Stampacchia, 173-322, Edizioni Cremonese, Roma 1974.

48. NAYROLES B., Elements of convex analysis in mechanics of solids (in Polish), in: Metody Analizy Funkcjonalnej w Plastyczności, Ed. by J.J. Telega, 127-235, Ossolineum, Wrocław 1981.

49. NEMAT-NASSER S., General variational principles in nonlinear and linear elasticity with applications, in: Mechanics Today, vol.1, Ed. by S. Nemat-Nasser, 214-261, Pergamon Press, New York 1972.

50. OGDEN R.W., A note on variational theorems in non-linear elastostatics, Math. Proc. Camb. Phil. Soc., 77(1975), 609-615.

51. OGDEN R.W., Extremum principles in non-linear elasticity and their application to composites - I, Int. J. Solids Structures, 14(1978), 265-282.

52. OSMOLOVSKII V.G., On the local solvability of a problem of the non-linear theory of elasticity (in Russian), in: Kraevye Zadači Matematičeskoy Fiziki i Smeznye Voprosy Teorii Funkcij. 13, 163-173, Zapiski Naučnykh Seminarov LOMI, t. 110, Nauka, Leningrad 1981.

53. ROCKAFELLAR R.T., Convex Analysis, Princeton University Press, Princeton 1970.

54. ROCKAFELLAR R.T., Integral functionals, normal integrands and measurable selections, in: Nonlinear Operators and the Calculus of Variations, Ed. by J.P. Gossez, E.J. Lami Dozo J. Mawhin and L. Waelbroeck, 157-207, Springer-Verlag, Berlin 1976.

55. ROCKAFELLAR R.T., Directionally Lipschitzian functions and subdifferential calculus, Proc. London Math. Soc., 39(1979), 331-355.

56. SCHMIDT R., PIETRASZKIEWICZ W., Variational principles in the geometrically non-linear theory of shells undergoing moderate rotations, Ing.-Archiv, 50(1981), 187-201.

57. STUMPF H., The principle of complementary energy in nonlinear plate theory, J. Elasticity, 6(1976), 1, 95-104.

58. STUMPF H., Dual extremum principles and error bounds in non-linear elasticity, J. Elasticity, 8(1978), 425-438.

59. STUMPF H., The derivation of dual extremum and complementary stationary principles in geometrical non-linear shell theory, Ing.-Archiv, 48(1979), 221-237.

60. STUMPF H., Stationary and extremal variational principles in nonlinear moderate rotation shell theory, in: Mechanics of Inelastic Media and Structures, Ed. by O. Mahrenholtz and A. Sawczuk, 267-281, PWN - Polish Scientific Publishers, Warszawa - Poznań 1982.

61. TRUESDELL C., NOLL W., The Non-Linear Field Theories of Mechanics, Encyclopedia of Physics, vol. III/3, Springer-Verlag, Berlin 1965 (pp. 324-328).

62. VALENT T., Sul problema di posto in elastostatica non lineare. Teoremi di esistenza e di unicità, Fisica Matematica, Suppl. Boll. Un. Mat. Ital., IV-5 (1985), 1, 281-295.

63. VALID R., Le principe de l'énergie complémentaire en théorie non linéaire des coques, La Recherche Aérospatiale, 1/1981, 43-53.

64. WASHIZU K., Complementary variational principles in elasticity and plasticity, in: Duality and Complementarity in Mechanics of Solids, Ed. by A. Borkowski, 7-93, Ossolineum, Wrocław 1979.

65. WASHIZU K., A note on the principle of stationary complementary energy in nonlinear elasticity, in: Mechanics Today, vol.5, Ed. by S. Nemat-Nasser, 509-522, Pergamon Press, New York 1980.

66. WEMPNER G., Complementary theorems of solid mechanics, in: Variational Methods in the Mechanics of Solids, Ed. by S. Nemat-Nasser, 127-135, Pergamon Press, Oxford 1980.

67. ZUBOV L.M., The stationary principle of complementary work in nonlinear theory of elasticity (in Russian), Prikl. Mat. Mekh., 34(1970), 241-245.

68. ZUBOV L.M., On the representation of the displacement gradient of an isotropic elastic body by the Piola stress tensor (in Russian), Prikl. Mat. Mekh., 40(1976), 1070-1077.

FINITE ROTATIONS AND COMPLEMENTARY EXTREMUM PRINCIPLES

H. Bufler
Institute of Mechanics (Civil Engineering) of the University
Pfaffenwaldring 7, D - 7000 Stuttgart 80, FRG

1. Introduction

The equations of nonlinear elasticity can be written in terms of the following conjugate quantities (BUFLER [1]): (a) Deformation (or displacement) gradients and 1.Piola-Kirchhoff stresses, (b) Green's deformations (or strains) and 2.Piola-Kirchhoff stresses, (c) stretches (or extensions) and Biot stresses. In connection with the complementary energy theorem in nonlinear elasticity frequently the first group of variables is used, see f.i. KOITER [2]. In this case (a), however, the "stress and strain quantities" (deformation gradients) are not objective and the inversion of the necessarily nonlinear constitutive equations generally yields severe difficulties (OGDEN [3]) which do not appear if restriction is made to moderate rotations (STUMPF [4]). In the formulations (b) and (c), on the other hand, objective quantities are involved and a linear material behaviour (Kirchhoff and semilinear material respectively) can be used. Formulation (c) seems to be advantageous for two further reasons: Firstly the stretches (or extensions) and the conjugate stresses (called engineering stresses by F. DE VEUBEKE [5]) are to be considered as "natural quantities" [3] - indeed these ones are frequently used for large deformation problems of plates, membranes and shells, and the assumption of a convex strain energy density with respect to the extensions is physically meaningful - and secondly the rotations are taken into account explicitely. Furthermore the associated most general variational principle does allow independent variations of the displacements, stretches, stresses, reactive forces and *rotations* the corresponding Euler equations representing the force equilibrium (including the statical boundary conditions and the reactive forces), the constitutive equations, the kinematical field equations, the kinematical boundary conditions and the *moment equilibrium*, see REISSNER [6] and BUFLER [7] [8]. On these grounds formulation (c) is used in this paper and no restriction concerning the magnitude of the rotation is made. In comparison to [7] and [8] the incremental formulations and the conditions for the existence of complementary *extremum* principles - minimum of the potential energy under the kinematical subsidiary conditions, and maximum of the complementary potential under the statical subsidiary conditions - are given additionally. When these ones are fulfilled, global (energetic) bounds of the solution can be constructed. The corresponding discrete problems have been investigated by the author in [9] and [10].

Explicit use of finite rotations is, among others, also made by F. DE VEUBEKE [5] WEMPNER [11], WUNDERLICH and OBRECHT [12] and ATLURI [13]. As shown for the first time in [5] the moment equilibrium equation comes out as Euler equation of a variational statement if the rotation is considered as an independent variable (as is done also in [12] and [13], for instance). For independent displacements and rotations, however, the stretches are no more symmetrical. Therefore their symmetry has to be postulated additionally (a point, which is usually overlooked): In [6] and [7] it is shown that the Euler equation corresponding to the variation of the antisymmetrical part of the Biot stresses yields this condition.

2. Kinematical and statical basic equations

We use Lagrangian's description with the (Cartesian) material coordinates x_A $(A = I,II,III)$ and the (Cartesian) spatial coordinates ξ_a $(a = 1,2,3)$. For clearness the finite rotation tensor α_{AB} which transforms the body attached vector $\overset{\circ}{\vec{v}} = \vec{e}_A \overset{\circ}{v}_A$ to the vector $\vec{v}$ owing to a finite rotation (axis $\vec{e} = \vec{e}_A c_A$, angle ϕ ; Fig.1), shall be derived shortly:

$$d\vec{v} = d\phi \vec{e}\times\vec{v} \quad \text{or} \quad dv_A = d\phi\, e_{ABC}c_B v_C = d\phi(-e_{ACB}c_B)v_C = d\phi c_{AC}v_C.$$

From $\frac{dv_A}{d\phi} = c_{AC}v_C$ or in matrix notation $\frac{d\underline{v}}{d\phi} = \underline{c}\,\underline{v}$ there follows by integration

$$\underline{v} = e^{\underline{c}\phi}\overset{\circ}{\underline{v}} = e^{\underline{\phi}}\overset{\circ}{\underline{v}} = \underline{\alpha}\overset{\circ}{\underline{v}} \qquad \text{or} \qquad v_A = \alpha_{AB}\overset{\circ}{v}_B \tag{2.1}$$

the eigenvalues of $\underline{\phi}$ being $0,\ i\phi,\ -i\phi$ $(i = \sqrt{-1})$. Via Cayley-Hamilton's theorem one obtains

$$\underline{\alpha} = e^{\underline{\phi}} = c_o\,\underline{1} + c_1\underline{\phi} + c_2\underline{\phi}^2 = \underline{1} + \frac{\sin\phi}{\phi}\,\underline{\phi} + \frac{2}{\phi^2}\sin^2(\tfrac{\phi}{2})\,\underline{\phi}^2\,, \quad \text{or}$$

$$\alpha_{AB} = \delta_{AB} - e_{ABC}\frac{\sin\phi}{\phi}\phi_C + \frac{2}{\phi^2}\sin^2(\tfrac{\phi}{2})(\phi_A\phi_B - \delta_{AB}\phi^2) = \delta_{AB} - e_{ABC}\phi_C + \tfrac{1}{2}(\phi_A\phi_B - \delta_{AB}\phi_C\phi_C) + \ldots \tag{2.2}$$

where $\phi_A = c_A\phi$ and $\phi = \sqrt{\phi_I^2 + \phi_{II}^2 + \phi_{III}^2}$. Further there is $\underline{\alpha}\underline{\alpha}^T = \underline{\alpha}^T\underline{\alpha} = \underline{1}$; $\underline{\alpha}^T = \underline{\alpha}^{-1}$; $\det\underline{\alpha} = 1$. For other (equivalent) representations of finite rotations see PIETRASZKIEWICZ and BADUR [14].

Next the polar decomposition of the deformation gradient is explained graphically in Fig.2. Here the following configurations are distinguished: Initial state A, final state D, intermediate states B and C. State C is produced from state D by a rigid translation (change of the basis) and state B from state C by a rigid rotation of the deformed element, *the rotation being assumed to be given for the present.* The pseudo stress vector $\vec{t}_I$ (force per unit of undeformed element acting on the deformed area element of D and C respectively whose orientation in the undeformed reference configuration A is $\vec{e}_I$) is also rigidly rotated with the element from state C to state B and is denoted by $\vec{r}_I$ in B. Then for the "generalized stretches" U_{BI} and the

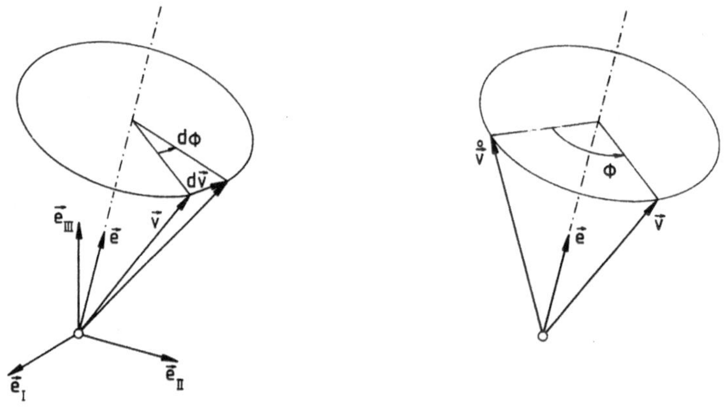

<u>Fig.1</u> Infinitesimal and finite rotation about a fixed axis

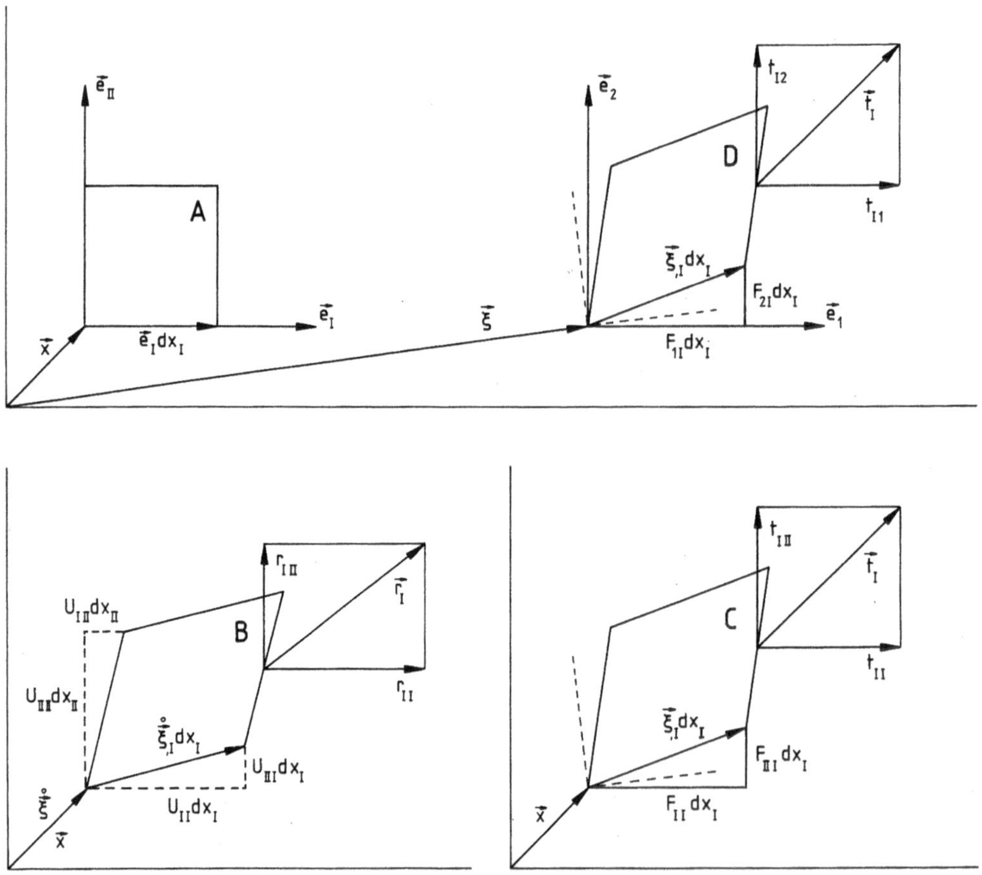

<u>Fig.2</u> "Generalized stretches" U_{AB} and "generalized referential Biot stresses" r_{AB}

deformation gradients F_{AI} and F_{aI} on the one side and the "generalized referential Biot stresses" r_{IB}, the Lagrange stresses t_{IA} and the Piola (1.Piola-Kirchhoff) stresses t_{Ia} on the other side the following relationships hold:

$$F_{AI} = \alpha_{AB}U_{BI} \qquad \Big| \qquad t_{IA} = \alpha_{AB}r_{IB}$$

$$F_{aI} = g_{aA}F_{AI} = \alpha_{aB}U_{BI} \qquad \Big| \qquad t_{Ia} = g_{aA}t_{IA} = \alpha_{aB}r_{IB}$$

where g_{aA} is a shifter and $\alpha_{aB} = g_{aA}\alpha_{AB}$. More generally one gets

$$F_{aB} = \alpha_{aC}U_{CB} \quad \text{or} \quad \underline{F} = \underline{\alpha}\underline{U} \qquad \Big| \qquad t_{Ba} = r_{BC}\alpha^T_{Ca} \quad \text{or} \quad \underline{t} = \underline{r}\underline{\alpha}^T \qquad (2.3)(2.4)$$

and

$$U_{AB} = \alpha_{cA}F_{cB} \quad \text{or} \quad \underline{U} = \underline{\alpha}^T\underline{F} \qquad \Big| \qquad r_{AB} = t_{Ac}\alpha_{cB} \quad \text{or} \quad \underline{r} = \underline{t}\underline{\alpha}. \qquad (2.5)(2.6)$$

As the rotation is so far considered to be independent of the deformation, $\underline{U}$ will be unsymmetric. Let us define the symmetrical and antisymmetrical part of it by

$$\underline{U}^s = \frac{1}{2}(\underline{\alpha}^T\underline{F} + \underline{F}^T\underline{\alpha}) \ ; \qquad \underline{U}^a = \frac{1}{2}(\underline{\alpha}^T\underline{F} - \underline{F}^T\underline{\alpha}). \qquad (2.7)(2.8)$$

In the case of symmetrical "generalized stretches", there holds $\underline{U}^a = \underline{0}$;

$$\underline{U} = \underline{U}^T = \underline{U}^* \ ; \ \underline{F}^T\underline{F} = \underline{U}^{*T}\underline{U}^* = \underline{U}^{*2} \ ; \ \underline{U}^* = \sqrt{\underline{F}^T\underline{F}} \ ; \quad \underline{\alpha}^* = \underline{F}\underline{U}^{*-1} \qquad (2.9)$$

showing that $\underline{\alpha}^*$ depends on the deformations and is not to be considered as an independent quantitiy. Further it turns out that $\underline{U}^*$ (contrarily to $\underline{U}$) is objective (frame indifferent) [1], i.e. the symmetry condition $\underline{U}^a = \underline{0}$ can be interpreted as condition of objectivity of the "generalized stretches". U^*_{AB} are the components of the right stretch tensor. In the special case $\underline{\alpha} = \underline{\alpha}^*$ the result (2.3) is the well known polar decomposition theorem $\underline{F} = \underline{\alpha}^*\underline{U}^*$ and (2.6) defines the referential Biot stresses $\underline{r}^* = \underline{t}\underline{\alpha}^*$. Contrarily to $\underline{r}$, $\underline{r}^*$ is objective. Further $\underline{\alpha}^*$ describes the rigid rotation of the principal axes of Green's strain tensor which coincide with those of $\underline{U}^*$. It should be noted that Reissner's approach [6] is equivalent to that based on the polar decomposition as has been shown in [7] [8].

The following symbols are introduced: Position vector $u = (\xi_a)$, load vector per unit volume and stress vector per unit surface of the stress free body $q = (q_a)$ in V and $g^1 = (g^1_a)$ on ∂V_1 as given quantities, and $g^2 = (g^2_a)$ on ∂V_2 as reactive force. We write

$$\underset{\sim}{u} = \begin{bmatrix} u \\ u \\ 0 \end{bmatrix} + \begin{bmatrix} 0 \\ 0 \\ u \end{bmatrix} \begin{matrix} \text{in } V \\ \text{on } \partial V_1 \\ \text{on } \partial V_2 \end{matrix} \ ; \qquad \underset{\sim}{q} = \begin{bmatrix} q \\ g^1 \\ 0 \end{bmatrix} + \begin{bmatrix} 0 \\ 0 \\ g^2 \end{bmatrix} \begin{matrix} \text{in } V \\ \text{on } \partial V_1 \\ \text{on } \partial V_2 \end{matrix}$$

$$\underset{\sim}{u} = \underset{\sim}{\bar{u}} + \underset{\sim}{\tilde{u}} \qquad\qquad \underset{\sim}{q} = \underset{\sim}{\bar{q}} + \underset{\sim}{\tilde{q}} \qquad (2.10)$$

and note that for finite element applications the ordered 3-tuples have to be enlarged by taking into account the interelement surfaces [1]. The *force equilibrium* equation $\vec{t}_{B,B} + \vec{q} = \vec{0}$ in V and the statical equation $\vec{t}_B\, n_B = \vec{g}$ on ∂V are with (2.4) in component form, $\vec{n} = \vec{e}_B n_B$ being the external normal of the undeformed surface,

$$- (\alpha_{aC,B} + \alpha_{aC}\partial_B) r_{BC} - q_a = 0 \quad \text{in } V \qquad n_B \alpha_{aC} r_{BC} - g^1_a = 0 \quad \text{on } \partial V_1 \qquad n_B \alpha_{aC} r_{BC} - g^2_a = 0 \quad \text{on } \partial V_2$$

or

$$\begin{bmatrix} \bar{A}(\phi)\sigma \\ n\bar{a}(\phi)\sigma \\ n\bar{a}(\phi)\sigma \end{bmatrix} - \begin{bmatrix} q \\ g^1 \\ 0 \end{bmatrix} - \begin{bmatrix} 0 \\ 0 \\ g^2 \end{bmatrix} = \begin{bmatrix} 0 \\ 0 \\ 0 \end{bmatrix}$$

$$\boxed{\underset{\sim}{\bar{A}}(\phi)\sigma - \underset{\sim}{\bar{q}} - \underset{\sim}{\tilde{q}} = \underset{\sim}{0}} \tag{2.11}$$

and accordingly the *moment equilibrium* equation $\vec{\xi}_{,A} \times \vec{t}_A = \vec{0}$ reads

$$e_{ead}\alpha_{eC}\xi_{a,B} r_{BC} = 0 \text{ in } V \quad \text{or} \quad \boxed{\bar{A}^m(u,\phi)\sigma = 0} \tag{2.12}$$

$\bar{A}(\phi)$ and $\bar{A}^m(u,\phi)$ being called *statical operators*. The *kinematical relations* together with the symmetry (or objectivity) condition are after (2.7) and (2.8)

$$U^s_{CB} = \frac{1}{2}(\alpha_{aC}\partial_B + \alpha_{aB}\partial_C)\xi_a \quad \text{in } V \quad \text{or} \quad \boxed{\varepsilon^s = B^s(u,\phi)} \tag{2.13}$$

$$U^a_{CB} = \frac{1}{2}(\alpha_{aC}\partial_B - \alpha_{aB}\partial_C)\xi_a = 0 \quad \text{in } V \quad \text{or} \quad \boxed{\varepsilon^a = B^a(u,\phi) = 0}\,. \tag{2.14}$$

By addition one gets $U_{CB} = \alpha_{aC}\partial_B\xi_a$ or $\varepsilon = B(u,\phi) = \bar{B}(\phi)\, u$ with $B(u,\phi)$ as the *kinematical operator*. The *kinematical boundary conditions* are

$$- \xi_a + \hat{\xi}_a = 0 \quad \text{on } \partial V_2 \quad \text{or} \quad \boxed{- \underset{\sim}{\tilde{u}} + \underset{\sim}{\hat{\tilde{u}}} = 0}\,. \tag{2.15}$$

Summarizing the statical and kinematical equations (2.11) - (2.15) they can be put into the following form

$$\begin{bmatrix} B^s(u,\phi) \\ B^a(u,\phi) \\ \underset{\sim}{\bar{A}}(\phi)\sigma \\ \bar{A}^m(u,\phi)\sigma \\ \underset{\sim}{0} \end{bmatrix} = \begin{bmatrix} \varepsilon^s \\ \varepsilon^a = 0 \\ \underset{\sim}{\bar{q}} + \underset{\sim}{\tilde{q}} \\ m = 0 \\ \underset{\sim}{\tilde{u}} - \underset{\sim}{\hat{\tilde{u}}} \end{bmatrix} \quad \text{or} \quad \boxed{\underset{\sim}{M}(\underset{\sim}{s}) = \underset{\sim}{e}} \quad \text{with} \quad \underset{\sim}{s} = \begin{bmatrix} \sigma^s \\ \sigma^a \\ u \\ \phi \\ \tilde{q} \end{bmatrix} \tag{2.16}$$

with the "supervectors" $\underset{\sim}{e}$ and $\underset{\sim}{s}$ and the "superoperator" $\underset{\sim}{M}(\underset{\sim}{s})$.

3. Mixed principle of virtual work and complementary variational principles

First it is necessary to introduce the following bilinear forms on the corresponding product spaces:

$$[\sigma,\varepsilon] = \int_V r_{AB}U_{BA}\,dV \;; \qquad \langle \underset{\sim}{q},\underset{\sim}{u}\rangle = \langle \underset{\sim}{\bar{q}},\underset{\sim}{\bar{u}}\rangle + \langle \underset{\sim}{\tilde{q}},\underset{\sim}{\tilde{u}}\rangle \;;$$

$$\langle \underset{\sim}{\bar{q}},\underset{\sim}{\bar{u}}\rangle = \int_V q_a\xi_a\,dV + \int_{\partial V_1} g^1_a\xi_a\,dS \;; \quad \langle \underset{\sim}{\tilde{q}},\underset{\sim}{\tilde{u}}\rangle = \int_{\partial V_2} g^2_a\xi_a\,dS \;; \quad \langle m,\phi\rangle_\phi = \int m_a\phi_a\,dV \;;$$

$$\{\underset{\sim}{s},\underset{\sim}{e}\} = [\sigma^s,\varepsilon^s] + [\sigma^a,\varepsilon^a] + \langle \underset{\sim}{\bar{q}},\underset{\sim}{\bar{u}}\rangle + \langle \underset{\sim}{\tilde{q}},\underset{\sim}{\tilde{u}}\rangle + \langle m,\phi\rangle_\phi + \langle \underset{\sim}{\tilde{u}}-\underset{\sim}{\hat{\tilde{u}}},\underset{\sim}{\tilde{q}}\rangle\,.$$

The Gâteaux differential (or variation) of the kinematical operator $B(u,\phi)$ at (u,ϕ) with increments (u',ϕ') is

$$\delta B(u,\phi;u',\phi') = \delta_u B(u,\phi;\, u',0) + \delta_\phi B(u,\phi;0,\phi') = B_u(u,\phi)u' + B_\phi(u,\phi)\phi'$$

where B_u and B_ϕ are called partial Gâteaux derivatives. Their adjoints are defined according to

$$[B_u(u,\phi)u',\sigma] = \langle \underset{\sim}{B}_u^*(u,\phi)\sigma,\underset{\sim}{u}' \rangle \tag{3.1}$$

$$\int_V \alpha_{aC}\xi'_{a,B}r_{BC}dV = -\int_V (\alpha_{aC}r_{BC})_{,B}\xi'_a dV + \int_{\partial V_1+\partial V_2} n_B\alpha_{aC}r_{BC}\xi'_a dS$$

and

$$[B_\phi(u,\phi)\phi',\sigma] = \langle B_\phi^*(u,\phi)\sigma,\phi' \rangle_\phi \tag{3.2}$$

$$\int_V \alpha'_{aC}\xi_{a,B}r_{BC}dV = \int_V e_{ead}\alpha_{eC}\xi_{a,B}r_{BC}\phi'_d dV$$

and the comparison with the statical operators shows that there holds

$$\boxed{\underset{\sim}{\bar{A}}(\phi) = \underset{\sim}{B}_u^*(u,\phi) = \underset{\sim}{B}_u^*(\phi)} \quad \boxed{\bar{A}^m(u,\phi) = B_\phi^*(u,\phi)} \quad \boxed{B_u(u,\phi) = \bar{B}(\phi)}\ . \tag{3.3}$$

The variation of the functional $P_M = [B(u,\phi),\sigma] = [B^s(u,\phi),\sigma^s] + [B^a(u,\phi),\sigma^a]$is $\delta P_M = [B^s(u,\phi),\delta\sigma^s] + [B^a(u,\phi),\delta\sigma^a] + [B_u(u,\phi)\delta u,\sigma] + [B_\phi(u,\phi)\delta\phi,\sigma]$ where, because of (3.1), (3.2) and (3.3), the third and forth term can be transformed to $[B_u(u,\phi)\delta u,\sigma] = \langle \underset{\sim}{\bar{A}}(\phi)\sigma,\delta\underset{\sim}{u}\rangle$ and $[B_\phi(u,\phi)\delta\phi,\sigma] = \langle \bar{A}^m(u,\phi)\sigma,\delta\phi \rangle_\phi$. As a result there follows with (2.16)

$$\delta P_M = \{ \frac{\delta P_M}{\delta \underset{\sim}{s}}, \delta\underset{\sim}{s} \} = \{ \underset{\sim}{M}(\underset{\sim}{s}),\delta\underset{\sim}{s} \}, \tag{3.4}$$

i.e.

$$\boxed{P_M = [B(u,\phi),\sigma]} \tag{3.5}$$

is the potential of the operator $\underset{\sim}{M}(\underset{\sim}{s})$ (and vice versa: $\underset{\sim}{M}(\underset{\sim}{s})$ is the gradient of P_M).

Putting (2.16) into its weak form $\{ \underset{\sim}{M}(\underset{\sim}{s}),\delta\underset{\sim}{s} \} - \{ \underset{\sim}{e},\delta\underset{\sim}{s} \} = 0$ one gets with (3.4) and (3.5) the *mixed principle of virtual work*

$$\boxed{-[\varepsilon^s,\delta\sigma^s] - \langle \underset{\sim}{\bar{q}},\delta\underset{\sim}{\bar{u}} \rangle - \delta\langle \underset{\sim}{\bar{u}} - \underset{\sim}{\hat{\bar{u}}},\underset{\sim}{\bar{q}} \rangle + \delta P_M = 0} \tag{3.6}$$

which does depend neither on the material nor on the load behaviour. It is, of course, equivalent to (2.16). From (3.6) further virtual work principles can be derived, for instance the principle of virtual displacements

$$[\sigma^s,\delta\varepsilon^s] - \langle \underset{\sim}{\bar{q}},\delta\underset{\sim}{\bar{u}} \rangle = 0 \tag{3.7}$$

with the kinematical conditions (2.13), (2.14) and (2.15) as constraints. Note that in (3.7) only the symmetrical part of the stress matrix is involved.

If the (configuration dependent) loading $\underset{\sim}{\bar{q}} = \underset{\sim}{C}(\underset{\sim}{\bar{u}})$ in V and on ∂V_1 respectively is *conservative*, then the gradient of the corresponding potential $P_C(\underset{\sim}{\bar{u}}) = -\int_V \pi_V^{(e)}dV - \int_{\partial V_1}\pi_S^{(e)}dS$ equals the *load operator* $\underset{\sim}{C}(\underset{\sim}{\bar{u}})$:

$$\delta P_C = \langle \frac{\delta P_C}{\delta\underset{\sim}{\bar{u}}},\delta\underset{\sim}{\bar{u}} \rangle = \langle \underset{\sim}{C}(\underset{\sim}{\bar{u}}),\delta\underset{\sim}{\bar{u}} \rangle = \langle \underset{\sim}{\bar{q}},\delta\underset{\sim}{\bar{u}} \rangle. \tag{3.8}$$

A *hyperelastic body* is characterized by the existence of a strain energy $P_D(\varepsilon) = \int_V \pi^{(i)}dV$ whose variation is the work of the applied forces (first law of thermodynamics for an adiabatic process) $\delta P_{D^-} = [\frac{\delta P_{D^-}}{\delta\varepsilon}, \delta\varepsilon] = \langle \tilde{q}, \delta\bar{u} \rangle$.

Together with (3.7) there follows $P_{D^-} = P_{D^-}(\varepsilon^s)$ and

$$\sigma^s = \frac{\delta P_{D^-}}{\delta\varepsilon^s} = D^{-1}(\varepsilon^s) \tag{3.9}$$

where $\varepsilon^a = 0$ guarantees the objectivity (frame indifference) of ε^s. If det $\frac{\partial\sigma^s}{\partial\varepsilon^s} \neq 0$, then (3.9) is locally invertible and the internal complementary potential $P_D(\sigma^s) = \int_V \tilde{\pi}^{(i)} dV$ can be calculated from

$$P_D(\sigma^s) + P_{D^-}(\varepsilon^s) = [\sigma^s, \varepsilon^s] \tag{3.10}$$

and

$$\varepsilon^s = \frac{\delta P_D}{\delta\sigma^s} = D(\sigma^s). \tag{3.11}$$

Further there holds

$$[\varepsilon^s, \delta\sigma^s] = [D(\sigma^s), \delta\sigma^s] = [\frac{\delta P_D}{\delta\sigma^s}, \delta\sigma^s] = \delta P_D \tag{3.12}$$

$D(\sigma^s)$ being called *material operator*. With the help of (3.8) and (3.12) the first three terms of (3.6) can be summarized:

$$-\delta P_N = -\{\frac{\delta P_N}{\delta \underset{\sim}{s}}, \delta \underset{\sim}{s}\} = -\{\underset{\sim}{N}(\underset{\sim}{s}), \delta \underset{\sim}{s}\} \tag{3.13}$$

where $\boxed{P_N = P_D(\sigma^s) + P_C(\bar{\underset{\sim}{u}}) + \langle \tilde{\underset{\sim}{u}} - \hat{\underset{\sim}{u}}, \tilde{\underset{\sim}{q}} \rangle}$. (3.14)

(3.6) and (3.13) yield the *mixed variational principle* of the type Hellinger-Reissner

$$\boxed{\bar{I}(\underset{\sim}{u}, \phi, \tilde{\underset{\sim}{q}}, \sigma^s, \sigma^a) = -P_N + P_M = -P_D(\sigma^s) - P_C(\bar{\underset{\sim}{u}}) - \langle \tilde{\underset{\sim}{u}} - \hat{\underset{\sim}{u}}, \tilde{\underset{\sim}{q}} \rangle + P_M(u, \phi, \sigma^s, \sigma^a) = \text{stat}} . \tag{3.15}$$

From its variation

$$\delta\bar{I} = [\frac{\delta\bar{I}}{\delta\sigma^s}, \delta\sigma^s] + [\frac{\delta\bar{I}}{\delta\sigma^a}, \delta\sigma^a] + \langle \frac{\delta\bar{I}}{\delta\tilde{\underset{\sim}{q}}}, \delta\tilde{\underset{\sim}{q}} \rangle + \langle \frac{\delta\bar{I}}{\delta\underset{\sim}{u}}, \delta\underset{\sim}{u} \rangle + \langle \frac{\delta\bar{I}}{\delta\phi}, \delta\phi \rangle_\phi = 0$$

one gets as Euler equations

$$\frac{\delta\bar{I}}{\delta\sigma^s} = -\frac{\delta P_D}{\delta\sigma^s} + B^s(u,\phi) = 0$$

$$\frac{\delta\bar{I}}{\delta\sigma^a} = B^a(u,\phi) = 0$$

$$\frac{\delta\bar{I}}{\delta\tilde{\underset{\sim}{q}}} = -\tilde{\underset{\sim}{u}} + \hat{\underset{\sim}{u}} = \underset{\sim}{0}$$

$$\frac{\delta\bar{I}}{\delta\underset{\sim}{u}} = \underset{\sim}{B}^*_u(u,\phi)\sigma - \frac{\delta P_C}{\delta\bar{\underset{\sim}{u}}} - \tilde{\underset{\sim}{q}} = \underset{\sim}{0}$$

$$\left.\begin{aligned}
&\frac{\partial\tilde{\pi}^{(i)}}{\partial r^s_{BC}} + \frac{1}{2}(\alpha_{aC}\partial_B + \alpha_{aB}\partial_C)\xi_a = 0 && \text{in } V\\
&\frac{1}{2}(\alpha_{aC}\partial_B - \alpha_{aB}\partial_C)\xi_a = 0 && \text{in } V\\
&-\xi_a + \hat{\xi}_a = 0 && \text{on } \partial V_2\\
&-(\alpha_{aC} r_{BC})_{,B} + \frac{\partial\pi_V^{(e)}}{\partial\xi_a} = 0 && \text{in } V\\
&n_B\alpha_{aC} r_{BC} + \frac{\partial\pi_S^{(e)}}{\partial\xi_a} = 0 && \text{on } \partial V_1\\
&n_B\alpha_{aC} r_{BC} - g_a^2 = 0 && \text{on } \partial V_2
\end{aligned}\right\} \tag{3.16}$$

$$\frac{\delta\bar{I}}{\delta\phi} = B^*_\phi(u,\phi)\sigma = 0 \qquad\qquad \Big| \quad e_{ead}\alpha_{eC}\xi_{a,B}r_{BC} = 0 \quad \text{in } V \quad \Big]$$

They are: Definition of the stretch plus material law, symmetry condition for the stretch, kinematical boundary condition, force-equilibrium in V and on ∂V_1 including the load behaviour, definition of the reactive stresses on ∂V_2 and moment equilibrium.

As shown by Sewell [15], each variational principle can be strengthened to *complementary extremum principles,* if the corresponding functional is a saddle one, i.e. convex with respect to the first and concave with respect to the second group of variables. This saddle condition reads for the functional (3.15)

$$\bar{I}|_2 - \bar{I}|_1 \underset{(<)}{(>)} \left\langle \frac{\delta\bar{I}}{\delta\underset{\sim}{u}}\Big|_{1\,(2)}, \underset{\sim}{u}_2 - \underset{\sim}{u}_1 \right\rangle + \left\langle \frac{\delta\bar{I}}{\delta\phi}\Big|_{1\,(2)}, \phi_2 - \phi_1 \right\rangle_\phi + \left\langle \frac{\delta\bar{I}}{\delta\underset{\sim}{\tilde{q}}}\Big|_{2\,(1)}, \underset{\sim}{\tilde{q}}_2 - \underset{\sim}{\tilde{q}}_1 \right\rangle + \left[\frac{\delta\bar{I}}{\delta\sigma^s}\Big|_{2\,(1)}, \sigma^s_2 - \sigma^s_1 \right] + \left[\frac{\delta\bar{I}}{\delta\sigma^a}\Big|_{2\,(1)}, \sigma^a_2 - \sigma^a_1 \right] \tag{3.17}$$

(convexity in $\underset{\sim}{u},\phi$; concavity in $\underset{\sim}{\tilde{q}},\sigma^s,\sigma^a$), where "1" and "2" stand for two different "points" $(\underset{\sim}{u},\phi,\underset{\sim}{\tilde{q}},\sigma^s,\sigma^a)$. Defining now the new functionals J_I and J_{II}

$$J_I = \bar{I} - \left[\frac{\delta\bar{I}}{\delta\sigma^s},\sigma^s\right] - \left[\frac{\delta\bar{I}}{\delta\sigma^a},\sigma^a\right] - \left\langle \frac{\delta\bar{I}}{\delta\underset{\sim}{\tilde{q}}},\underset{\sim}{\tilde{q}}\right\rangle = - P_D(\sigma^s) + \left[\frac{\delta P_D}{\delta\sigma^s},\sigma^s\right] - P_C(\underset{\sim}{\bar{u}}); \tag{3.18}$$

$$J_{II} = \bar{I} - \left\langle \frac{\delta\bar{I}}{\delta\underset{\sim}{u}},\underset{\sim}{u}\right\rangle - \left\langle \frac{\delta\bar{I}}{\delta\phi},\phi\right\rangle_\phi = - P_D(\sigma^s) + \langle \underset{\sim}{\bar{u}},\underset{\sim}{\tilde{q}}\rangle - P_C(\underset{\sim}{\bar{u}}) + \left\langle \frac{\delta P_C}{\delta\underset{\sim}{\bar{u}}},\underset{\sim}{\bar{u}}\right\rangle - \langle \bar{A}^m(u,\phi)\sigma,\phi\rangle_\phi \tag{3.19}$$

it can be seen that, if $\bar{I}$ has the property (3.17), the following inequalities hold:

$$J_I|_2 - J_I|_1 > \left\langle \frac{\delta\bar{I}}{\delta\underset{\sim}{u}}\Big|_1, \underset{\sim}{u}_2-\underset{\sim}{u}_1\right\rangle + \left\langle \frac{\delta\bar{I}}{\delta\phi}\Big|_1, \phi_2-\phi_1\right\rangle_\phi + \left[\frac{\delta\bar{I}}{\delta\sigma^s}\Big|_1 - \frac{\delta\bar{I}}{\delta\sigma^s}\Big|_2, \sigma^s_1\right] + \left[\frac{\delta\bar{I}}{\delta\sigma^a}\Big|_1 - \frac{\delta\bar{I}}{\delta\sigma^a}\Big|_2, \sigma^a_1\right] + \left\langle \frac{\delta\bar{I}}{\delta\underset{\sim}{\tilde{q}}}\Big|_1 - \frac{\delta\bar{I}}{\delta\underset{\sim}{\tilde{q}}}\Big|_2, \underset{\sim}{\tilde{q}}_1\right\rangle; \tag{3.20}$$

$$J_{II}|_2 - J_{II}|_1 < \left[\frac{\delta\bar{I}}{\delta\sigma^s}\Big|_1, \sigma^s_2-\sigma^s_1\right] + \left[\frac{\delta\bar{I}}{\delta\sigma^a}\Big|_1, \sigma^a_2-\sigma^a_1\right] + \left\langle \frac{\delta\bar{I}}{\delta\underset{\sim}{\tilde{q}}}\Big|_1, \underset{\sim}{\tilde{q}}_2-\underset{\sim}{\tilde{q}}_1\right\rangle + \left\langle \frac{\delta\bar{I}}{\delta\underset{\sim}{u}}\Big|_1 - \frac{\delta\bar{I}}{\delta\underset{\sim}{u}}\Big|_2, \underset{\sim}{u}_1\right\rangle + \left\langle \frac{\delta\bar{I}}{\delta\phi}\Big|_1 - \frac{\delta\bar{I}}{\delta\phi}\Big|_2, \phi_1\right\rangle_\phi . \tag{3.21}$$

If "1" denotes the solution point (which satisfies (3.16)), and "2" a "kinematically admissible point" (which satisfies (3.16/1), (3.16/2) and (3.16/3)) then we get from (3.20)

$$\boxed{J_I|_2 - J_I|_1 > 0} \tag{3.22}$$

under the kinematical side conditions. If, however, "2" denotes a "statically admissible point" (which satisfies (3.16/4) and (3.16/5)), then from (3.21) there follows

$$\boxed{J_{II}|_2 - J_{II}|_1 < 0} \tag{3.23}$$

under the statical side conditions. At the solution point there is

$$J_I = J_{II} = \bar{I} \, . \tag{3.24}$$

The inequalities (3.22) and (3.23) represent the *principle of (absolute) minimum potential energy* and the *principle of(absolute) maximum complementary energy* respectively. (In mechanics frequently $-J_{II}$ is considered as complementary energy rather than J_{II}). (3.22), (3.23) and (3.24) are called *complementary extremum principles*. It is clear, that because of (3.18) and (3.19) the inequalities (3.22) and (3.23) are equivalent to

$\bar{I}|_2 - \bar{I}|_1 > 0$ under the kinematical side conditions, and to

$\bar{I}|_2 - \bar{I}|_1 < 0$ under the statical side conditions.

Using a Taylor expansion with remainder at "point 1" in (3.17) one gets as a local saddle condition

$$\delta^2 P_D(\sigma^*;\delta\sigma) - \delta^2 P_C(\bar{\underset{\sim}{u}}^*;\delta\bar{\underset{\sim}{u}}) + [\delta^2 B(u^*,\phi^*;\delta u,\delta\phi),\sigma] > 0$$

or equivalently

$$[\delta D(\sigma^*;\delta\sigma),\delta\sigma] - \langle \delta \underset{\sim}{C}(\bar{\underset{\sim}{u}}^*;\delta\bar{\underset{\sim}{u}}),\delta\bar{\underset{\sim}{u}} \rangle + [\delta^2 B(u^*,\phi^*;\delta u,\delta\phi),\sigma] > 0 \tag{3.25}$$

where $\delta\sigma = \sigma_2 - \sigma_1, \quad \sigma^* = \sigma_1 + \gamma\delta\sigma, \qquad 0 < \gamma < 1 \ldots$

4. Incremental formulations

The fundamental state is produced by the load $\bar{q}$ and the prescribed displacements $\hat{\underset{\sim}{u}}$. The increments $\overset{\prime}{\bar{q}}$ and $\hat{\underset{\sim}{u}}'$ lead to the adjacent state $\underset{\sim}{u} + \underset{\sim}{u}'$, $\phi + \phi' \ldots$ The corresponding linearized and material independent basic equations are obtained from (2.16):

$$\boxed{\underset{\sim}{M}'(\underset{\sim}{s})\underset{\sim}{s}' = \underset{\sim}{e}' - (\underset{\sim}{M}(\underset{\sim}{s}) - \underset{\sim}{e})} \tag{4.1}$$

with $\underset{\sim}{M}'(\underset{\sim}{s}) = \underset{\sim}{M}_{\underset{\sim}{s}}(\underset{\sim}{s})$ as Gâteaux derivative. Since $\underset{\sim}{M}(\underset{\sim}{s})$ is a potential operator, the symmetry $\{\underset{\sim}{M}'(\underset{\sim}{s})\underset{\sim}{s}'_1,\underset{\sim}{s}'_2\} = \{\underset{\sim}{M}'(\underset{\sim}{s})\underset{\sim}{s}'_2,\underset{\sim}{s}'\}$ holds and the potential of the linear operator $\underset{\sim}{M}'\underset{\sim}{s}'$ (with respect to $\underset{\sim}{s}'$) is

$$\boxed{P'_M = \frac{1}{2}\{\underset{\sim}{M}'\underset{\sim}{s}',\underset{\sim}{s}'\} = [\delta B(u,\phi;u',\phi'),\sigma'] + \frac{1}{2}[\delta^2 B(u,\phi;u',\phi'),\sigma] = \frac{1}{2}\delta^2 P_M(\underset{\sim}{s};\underset{\sim}{s}')} \tag{4.2}$$

where $\delta B(u,\phi;u',\phi') = B_u u' + B_\phi \phi'$ and $\delta^2 B(u,\phi;u',\phi') = 2\,\bar{B}_{\phi u}\phi' u' + B_{\phi\phi}\phi'\phi'$. In (4.2) the following properties of the operators $\underset{\sim}{A}(\phi,\sigma) = \bar{\underset{\sim}{A}}(\phi)\sigma$, $A^m(u,\phi,\sigma) = \bar{A}^m(u,\phi)\sigma$, $B(u,\phi) = \bar{B}(\phi)u$ have been used:

$$\left.\begin{aligned} &[B_u u',\sigma'] = \langle \underset{\sim}{A}_\sigma \sigma',\underset{\sim}{u}'\rangle ; \ [B_{u\phi} u'\phi',\sigma] = [B_{\phi u}\phi' u',\sigma] = \langle \underset{\sim}{A}_\phi \phi',\underset{\sim}{u}'\rangle = \langle A^m_u u',\phi'\rangle_\phi ; \\ &[B_\phi \phi',\sigma'] = \langle A^m_\sigma \sigma',\phi'\rangle_\phi ; \ [B_{\phi\phi}\phi'_1\phi'_2,\sigma] = [B_{\phi\phi}\phi'_2\phi'_1,\sigma] = \langle A^m_\phi \phi'_2,\phi'_1\rangle_\phi . \end{aligned}\right\} \tag{4.3}$$

They are based on (3.3). Putting (4.1) into its weak form $\{\underset{\sim}{M}'\underset{\sim}{s}',\delta\underset{\sim}{s}'\} - \{\underset{\sim}{e}',\delta\underset{\sim}{s}'\} + \{\underset{\sim}{M} - \underset{\sim}{e},\delta\underset{\sim}{s}'\} = 0$ one obtains with (4.2) the *mixed principle of incremental virtual work*

$$\boxed{-[\acute{\varepsilon}^s,\delta\acute{\sigma}^s] - \langle \acute{q},\delta\acute{u} \rangle - \delta\langle \acute{u} - \acute{\hat{u}},\acute{q} \rangle + \delta\acute{P}_M + \delta\{\underline{M - e},\acute{s}\} = 0}\ . \tag{4.4}$$

Describing the load behaviour by $\bar{q} = C(\bar{u}) = \mu\bar{C}(\bar{u})$ with μ as the load parameter for the load increment there follows

$$\acute{\bar{q}} = \acute{\mu}\bar{C}(\bar{u}) + \mu\acute{\bar{C}}(\bar{u})\acute{\bar{u}} \quad \text{where} \quad \acute{\bar{C}}(\bar{u}) = \bar{C}_u(\bar{u})$$

is the Gâteaux derivative of $\bar{C}(\bar{u})$. If $\bar{C}(\bar{u})$ *is potential*, there is

$$\langle \acute{\bar{q}},\delta\acute{\bar{u}} \rangle = \delta\acute{P}_C \tag{4.5}$$

with

$$\acute{P}_C = \acute{\mu}\langle \bar{C},\acute{\bar{u}} \rangle + \frac{1}{2}\mu\langle \acute{\bar{C}}\acute{\bar{u}},\acute{\bar{u}} \rangle = \frac{1}{2}\delta^2 P_C(\mu,\bar{u};\acute{\mu},\acute{\bar{u}}). \tag{4.6}$$

For a *hyperelastic solid* $D(\sigma^s)$ in (3.11) is a potential operator and as a consequence its Gâteaux derivative $D_\sigma(\sigma^s) = \acute{D}(\sigma^s)$ is symmetric, the corresponding potential being

$$\acute{P}_D = \frac{1}{2}[\acute{D}\acute{\sigma}^s,\acute{\sigma}^s] = \frac{1}{2}\delta^2 P_D(\sigma^s;\acute{\sigma}^s). \tag{4.7}$$

Therefore with $\acute{\varepsilon}^s = \acute{D}(\sigma^s)\acute{\sigma}^s$ one gets

$$[\acute{\varepsilon}^s,\delta\acute{\sigma}^s] = \delta\acute{P}_D\ . \tag{4.8}$$

Using (4.5) and (4.8) in (4.4) there results the *mixed variational principle for the incremental quantities:*

$$\boxed{\begin{aligned}\acute{I}(\acute{u},\acute{\phi},\acute{q},\acute{\sigma}^s,\acute{\sigma}^a) = &- \acute{P}_D(\acute{\sigma}^s) - \acute{P}_C(\acute{\bar{u}}) - \langle \acute{\bar{u}} - \acute{\hat{\bar{u}}},\acute{\bar{q}} \rangle + \acute{P}_M(\acute{u},\acute{\phi},\acute{\sigma}^s,\acute{\sigma}^a) \\ &+ \langle \underline{\bar{A}(\phi)\sigma-\mu\bar{C}(\bar{u})-\bar{q}},\acute{u} \rangle + \langle \underline{\bar{A}^m(u,\phi)\sigma},\acute{\phi} \rangle_\phi + [\underline{B^s(u,\phi)-D(\sigma^s)},\acute{\sigma}^s] \\ &+ [\underline{B^a(u,\phi)},\acute{\sigma}^a] - \langle \underline{\tilde{u} - \hat{\tilde{u}}},\acute{\tilde{q}} \rangle = \text{stat}\,.\end{aligned}} \tag{4.9}$$

The underlined terms do vanish only if at "point" $(u,\phi,\tilde{q},\sigma^s,\sigma^a)$ the basic equations are satisfied. The saddle condition for the functional $\bar{I}'$ (see (3.17)) leads to

$$\boxed{[\acute{D}\delta\acute{\sigma}^s,\delta\acute{\sigma}^s] - \mu\langle \acute{\bar{C}}\delta\acute{\bar{u}},\delta\acute{\bar{u}} \rangle + [\delta^2 B(u,\phi;\delta\acute{u},\delta\acute{\phi}),\sigma] > 0} \tag{4.10}$$

with $\delta\acute{\sigma}^s = \acute{\sigma}^s_2 - \acute{\sigma}^s_1$,... and shows that, if (3.25) is satisfied, (4.10) holds also. In this case the following *complementary extremum principles for the incremental problem* are valid ("1" denotes the "solution point")

$$\boxed{\acute{J}_I\big|_2 - \acute{J}_I\big|_1 > 0} \tag{4.11}$$

under the kinematical side conditions for "point 2"

$$\left.\begin{aligned}\frac{\delta\acute{I}}{\delta\acute{\sigma}^s} &= -\acute{D}\acute{\sigma}^s + B^s_u\acute{u} + B^s_\phi\acute{\phi} - \underline{D(\sigma^s) + B^s(u,\phi)} = 0 \\ \frac{\delta\bar{I}'}{\delta\acute{\sigma}^a} &= B^a_u\acute{u} + B^a_\phi\acute{\phi} + \underline{B^a(u,\phi)} = 0\end{aligned}\right\} \tag{4.12}$$

$$\frac{\delta \dot{\bar{I}}}{\delta \dot{\underset{\sim}{q}}} = - \dot{\underset{\sim}{\bar{u}}} + \hat{\underset{\sim}{\bar{u}}}' - \underline{\underset{\sim}{\bar{u}} + \hat{\underset{\sim}{\bar{u}}}} = \underset{\sim}{0}$$

and

$$\boxed{\dot{J}_{II}\big|_2 - \dot{J}_{II}\big|_1 < 0} \qquad (4.13)$$

under the statical side conditions for "point 2"

$$\left.\begin{aligned}
\frac{\delta \dot{\bar{I}}}{\delta \dot{\underset{\sim}{u}}} &= \bar{A}(\phi)\dot{\sigma} + \bar{A}_{\phi}(\phi;\dot{\phi})\sigma - \dot{\mu}\bar{C}(\bar{u}) - \mu\dot{\bar{C}}(\bar{u})\dot{\bar{u}} - \dot{\underset{\sim}{q}} + \underline{\bar{A}(\phi)\sigma - \mu\bar{C}(\bar{u}) - \underset{\sim}{\bar{q}}} = \underset{\sim}{0} \\
\frac{\delta \dot{\bar{I}}}{\delta \dot{\phi}} &= \bar{A}^m(u,\phi)\dot{\sigma} + (\bar{A}^m_u \dot{u} + \bar{A}^m_{\phi}\dot{\phi})\sigma + \underline{\bar{A}^m(u,\phi)\sigma} = 0.
\end{aligned}\right\} \qquad (4.14)$$

The functionals in (4.11) and (4.13) are (see (3.18) and (3.19)):

$$\dot{J}_I = \dot{P}_D(\dot{\sigma}^s) - \dot{P}_C(\dot{\underset{\sim}{\bar{u}}}) + \frac{1}{2}[\delta^2 B(u,\phi;\dot{u},\dot{\phi}),\sigma] + \langle \underline{\bar{A}(\phi)\sigma - \mu\bar{C}(\bar{u}) - \underset{\sim}{\bar{q}}},\dot{\underset{\sim}{u}}\rangle + \langle \underline{\bar{A}^m(u,\phi)\sigma},\dot{\phi}\rangle_{\phi} \qquad (4.15)$$

and

$$\begin{aligned}
\dot{J}_{II} = &- \dot{P}_D(\dot{\sigma}^s) + \langle \hat{\underset{\sim}{\bar{u}}}',\dot{\underset{\sim}{q}}\rangle - \dot{P}_C(\dot{\underset{\sim}{\bar{u}}}) + \langle \dot{\mu}\bar{C}(\bar{u}),\dot{\underset{\sim}{\bar{u}}}\rangle + \mu\langle \dot{\bar{C}}(\bar{u})\dot{\bar{u}},\dot{\bar{u}}\rangle - \frac{1}{2}[\delta^2 B(u,\phi;\dot{u},\dot{\phi}),\sigma] \\
&+ [\underline{B^s(u,\phi) - D(\sigma^s)},\dot{\sigma}^s] + [\underline{B^a(u,\phi)},\dot{\sigma}^a] - \langle \underline{\underset{\sim}{\bar{u}} - \hat{\underset{\sim}{\bar{u}}}},\dot{\underset{\sim}{q}}\rangle.
\end{aligned} \qquad (4.16)$$

The potential $\dot{P}_M$ in (4.2) is explicitely given by

$$\begin{aligned}
\dot{P}_M = \int_V [&(\alpha_{aC}\dot{\xi}_{a,B} + e_{dae}\alpha_{dC}\xi_{a,B}\dot{\phi}_e)\dot{r}_{BC} \\
&+ (e_{dae}\alpha_{dC}\dot{\xi}_{a,B}\dot{\phi}_e + \frac{1}{2}\alpha_{dC}\xi_{a,B}(\dot{\phi}_a\dot{\phi}_d - \delta_{ad}\dot{\phi}_e\dot{\phi}_e))r_{BC}]dV
\end{aligned} \qquad (4.17)$$

where the formulas

$$\delta\alpha_{aC}(\phi;\dot{\phi}) = e_{dae}\alpha_{dC}\dot{\phi}_e; \quad \delta^2\alpha_{aC}(\phi;\dot{\phi}) = \alpha_{dC}(\dot{\phi}_a\dot{\phi}_d - \delta_{ad}\dot{\phi}_e\dot{\phi}_e) \qquad (4.18)$$

have been used. They are based on (see (2.2))

$$\alpha_{AC}(\phi) + \delta\alpha_{AC}(\phi;\dot{\phi}) + \frac{1}{2}\delta^2\alpha_{AB}(\phi;\dot{\phi}) + \ldots = \alpha_{AB}(\dot{\phi})\alpha_{BC}(\phi) \quad \text{and} \quad \alpha_{aC} = g_{aA}\alpha_{AC}.$$

5. Examples

In order to discretize an originally continuous problem one has to introduce suitable Ritz functions into (3.15), (3.18) or (3.19) the variations leading to a system of nonlinear algebraic equations for the unknown node quantities which is solved usually by the Newton-Raphson method. Alternatively (but equivalently) one could start with the functionals (4.9), (4.15) or (4.16) whose variation directly produces the linear algebraic equations for the increments of the unknown node quantities. The underlined terms in these functionals do vanish only if at the respective starting point of the iteration process the equilibrium and/or kinematical equations are fulfilled. The following examples are worked out on the basis of the stationarity of the functionals J_I and J_{II}.

5.1. First Example: Circular membrane

We consider a nonprestressed circular membrane under axisymmetrical dead loading the radial coordinate being denoted by r (Fig.3). Then with

$$\underset{\sim}{u} = \underset{\sim}{\bar{u}} + \underset{\sim}{\tilde{u}} \equiv \begin{bmatrix} \xi_1 \\ \xi_3 \\ \hline 0 \\ 0 \end{bmatrix} + \begin{bmatrix} 0 \\ 0 \\ \hline \xi_1(a) \\ \xi_3(a) \end{bmatrix} ; \quad \underset{\sim}{q} = \underset{\sim}{\bar{q}} + \underset{\sim}{\tilde{q}} \equiv \begin{bmatrix} q_1 \\ q_3 \\ \hline 0 \\ 0 \end{bmatrix} + \begin{bmatrix} 0 \\ 0 \\ \hline \tilde{g}_1 \\ \tilde{g}_3 \end{bmatrix},$$

the force equilibrium equation (2.11) reads with respect to polar coordinates

$$\underset{\sim}{\bar{A}}(\phi)\sigma - \underset{\sim}{\bar{q}} - \underset{\sim}{\tilde{q}} \equiv \begin{bmatrix} -\frac{1}{r}(rN_{11}\cos\phi)_{,1} + \frac{1}{r}N_{22} \\ -\frac{1}{r}(rN_{11}\sin\phi)_{,1} \\ \hline N_{11}(a)\cos\phi(a) \\ N_{11}(a)\sin\phi(a) \end{bmatrix} - \begin{bmatrix} q_1 \\ q_3 \\ \hline 0 \\ 0 \end{bmatrix} - \begin{bmatrix} 0 \\ 0 \\ \hline \tilde{g}_1 \\ \tilde{g}_3 \end{bmatrix} = \underset{\sim}{0}; \quad \tan\phi = \frac{\xi_{3,1}}{\xi_{1,1}} \tag{5.1}$$

where N_{11} and N_{22} are membrane forces and the moment equilibrium conditions are satisfied with $N_{13} = 0$. (For simplicity we use Arabic subscripts instead of Roman ones). The kinematical equations are (definition of the stretches and kinematical boundary conditions) with $B(u,\phi) = \bar{B}(\phi)u$:

$$\bar{B}(\phi)u - \varepsilon \equiv \begin{bmatrix} \xi_{1,1}\cos\phi + \xi_{3,1}\sin\phi \\ \frac{1}{r}\xi_1 \end{bmatrix} - \begin{bmatrix} U_{11} \\ U_{22} \end{bmatrix} = 0; \quad \tan\phi = \frac{\xi_{3,1}}{\xi_{1,1}} \tag{5.2}$$

$$\underset{\sim}{\tilde{u}} - \underset{\sim}{\hat{u}} \equiv \begin{bmatrix} 0 \\ 0 \\ \hline \xi_1(a) \\ \xi_3(a) \end{bmatrix} - \begin{bmatrix} 0 \\ 0 \\ \hline a \\ 0 \end{bmatrix} = \underset{\sim}{0}. \tag{5.3}$$

There follows $U_{31} = -\xi_{1,1}\sin\phi + \xi_{3,1}\cos\phi = 0$ and with $\vec{\xi} = \vec{\xi}(r)$ likewise $U_{13} = 0$; hence the symmetry condition (2.14) is fulfilled.

Using the bilinear forms

$$[\sigma,\varepsilon] = 2\pi\int_0^a (N_{11}U_{11} + N_{22}U_{22})r\,dr; \quad \langle \underset{\sim}{q}, \underset{\sim}{u} \rangle = 2\pi\int_0^a (q_1\xi_1 + q_3\xi_3)r\,dr + 2\pi a\tilde{g}_1\xi_1(a) + 2\pi a\tilde{g}_3\xi_3(a)$$

it can be easily proved that

$$[B_u(u,\phi)\delta u,\sigma] = \langle \underset{\sim}{B}_u^*(u,\phi)\sigma, \delta\underset{\sim}{u} \rangle \quad \text{with} \quad \underset{\sim}{B}_u^*(u,\phi)\sigma := \underset{\sim}{B}_u^*(\phi)\sigma = \underset{\sim}{\bar{A}}(\phi)\sigma .$$

Therefore according to (3.5) and (5.2) there follows after the elimination of ϕ

$$P_M = [B(u,\phi),\sigma] = 2\pi\int_o^a (\sqrt{\xi_{1,1}^2 + \xi_{3,1}^2}\; N_{11} + \frac{1}{r}\xi_1 N_{22})rdr \qquad (5.4)$$

and

$$\langle \tilde{u} - \hat{u}, \tilde{q} \rangle = 2\pi a(\xi_1(a) - a)\tilde{g}_1 + 2\pi a\xi_3(a)\tilde{g}_3. \qquad (5.5)$$

A *semilinear material* is characterized by (E, ν elastic moduli, t membrane thickness)

$$D(\sigma) - \varepsilon \equiv \begin{bmatrix} 1 + \frac{1}{Et}(N_{11} - \nu N_{22}) \\ 1 + \frac{1}{Et}(N_{22} - \nu N_{11}) \end{bmatrix} - \begin{bmatrix} U_{11} \\ U_{22} \end{bmatrix} = 0$$

leading to the internal complementary potential

$$P_D = 2\pi\int_o^a \{N_{11} + N_{22} + \frac{1}{2Et}(N_{11}^2 + N_{22}^2 - 2\nu N_{11}N_{22})\}rdr. \qquad (5.6)$$

The (negative) external potential in the case of *dead vertical loading* $q_3 = \hat{q}$ is given by

$$P_C = 2\pi \int_o^a \hat{q}\, \xi_3\, rdr. \qquad (5.7)$$

With the help of (5.4) - (5.7) the functionals (3.15), (3.18) and (3.19) can be constructed and there results

$$\frac{1}{2\pi}\bar{I} = -\int_o^a \{N_{11} + N_{22} + \frac{1}{2Et}(N_{11}^2 + N_{22}^2 - 2\nu N_{11}N_{22}) + \hat{q}\xi_3 - \sqrt{\xi_{1,1}^2 + \xi_{3,1}^2}\; N_{11} - \frac{\xi_1}{r}N_{22}\}rdr$$
$$- a\{\xi_1(a) - a\}\tilde{g}_1 - a\xi_3(a)\tilde{g}_3 = \text{stat} \qquad (5.8)$$

without side conditions;

$$\frac{1}{2\pi}J_I = \frac{1}{2\pi}\{\bar{I} - [\frac{\delta\bar{I}}{\delta\sigma},\sigma] - \langle \frac{\delta\bar{I}}{\delta\tilde{q}}, \tilde{q} \rangle\} = \int_o^a \{\frac{1}{2Et}(N_{11}^2 + N_{22}^2 - 2\nu N_{11}N_{22}) - \hat{q}\xi_3\}rdr = \text{stat} \qquad (5.9)$$

with the kinematical side conditions

$$\frac{1}{2\pi}\frac{\delta\bar{I}}{\delta\sigma} \equiv \begin{bmatrix} -1 - \frac{1}{Et}(N_{11} - \nu N_{22}) + \sqrt{\xi_{1,1}^2 + \xi_{3,1}^2} \\ -1 - \frac{1}{Et}(N_{22} - \nu N_{11}) + \frac{\xi_1}{r} \end{bmatrix} = \begin{bmatrix} 0 \\ 0 \end{bmatrix} \qquad (5.10)$$

and

$$\frac{1}{2\pi}\frac{\delta\bar{I}}{\delta\tilde{q}} = 0\,, \text{ that is } \xi_1(a) = a \text{ and } \xi_3(a) = 0; \qquad (5.11)$$

$$\frac{1}{2\pi}J_{II} = \frac{1}{2\pi}\{\bar{I} - \langle \frac{\delta\bar{I}}{\delta u}, u \rangle\} = -\int_o^a \{N_{11} + N_{22} + \frac{1}{2Et}(N_{11}^2 + N_{22}^2 - 2\nu N_{11}N_{22})\}rdr + a^2\tilde{g}_1 = \text{stat} \qquad (5.12)$$

with the statical side conditions

$$\frac{1}{2\pi}\frac{\delta\bar{I}}{\delta u} \equiv \begin{bmatrix} -\frac{1}{r}(rN_{11}\cos\phi)_{,1} + \frac{1}{r}N_{22} \\ -\frac{1}{r}(rN_{11}\sin\phi)_{,1} - \hat{q} \\ \hline N_{11}(a)\cos\phi(a) - \tilde{g}_1 \\ N_{11}(a)\sin\phi(a) - \tilde{g}_3 \end{bmatrix} = 0; \quad \tan\phi = \frac{\xi_{3,1}}{\xi_{1,1}}. \qquad (5.13)$$

Elimination of N_{11} and N_{22} in (5.9) by means of (5.10) yields the *principle of*

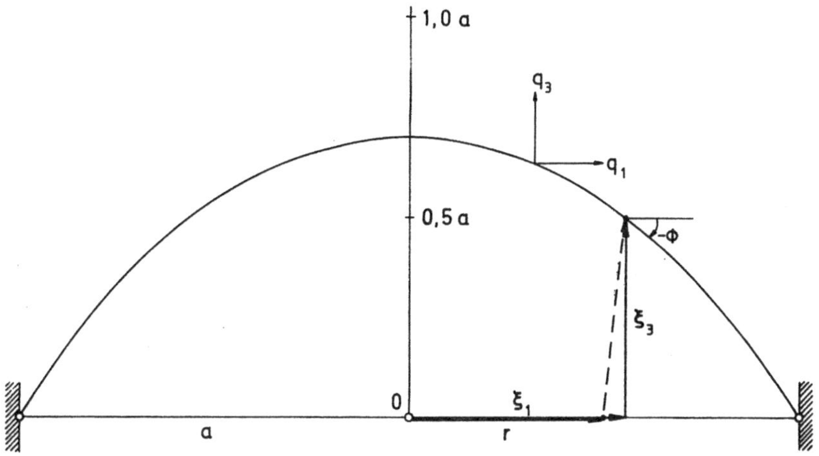

Fig.3 Circular membrane under constant loading $q_3 = \hat{q}$

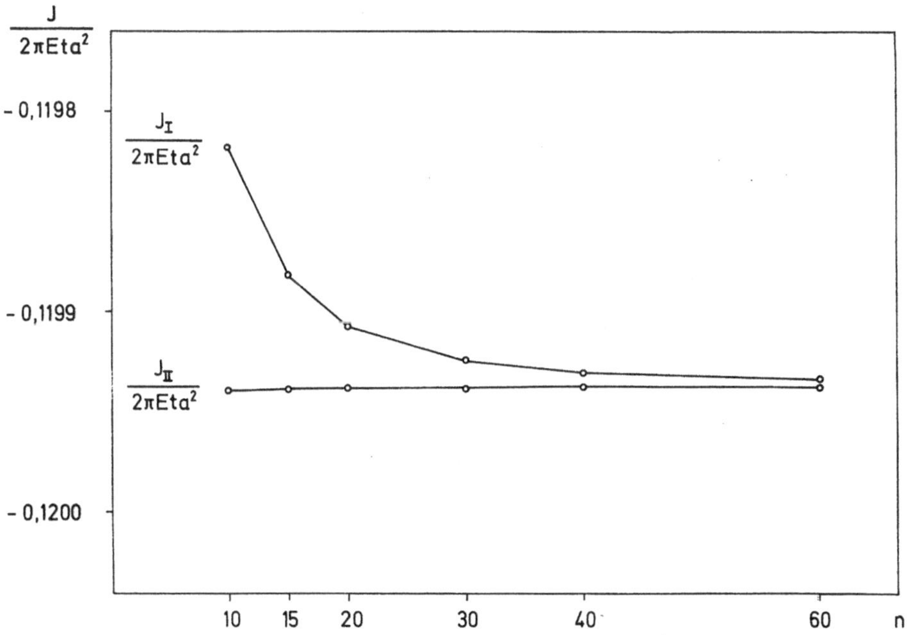

Fig.4 Potential energy and complementary potential for the circular membrane

stationary potential energy in the form

$$\frac{1}{2\pi}J_I = \int_o^a \{\frac{Et}{2(1-\nu^2)}[(\sqrt{\xi_{1,1}^2+\xi_{3,1}^2}-1)^2+(\frac{\xi_1}{r}-1)^2+2\nu(\sqrt{\xi_{1,1}^2+\xi_{3,1}^2}-1)(\frac{\xi_1}{r}-1)]-\hat{q}\xi_3\}rdr = \text{stat} \quad (5.14)$$

with the kinematical boundary conditions (5.11). In a similar manner N_{22} can be eliminated in (5.12) with the help of (5.13/1/2):

$$N_{11}\sin\phi = -\frac{1}{r}\int_o^r r\hat{q}dr \equiv -Q(r); \quad N_{22} = (r\sqrt{N_{11}^2 - Q^2})_{,1}.$$

Then one gets the *principle of stationary complementary potential* in the form

$$\frac{1}{2\pi}J_{II} = -\int\{\frac{1}{2Et}[N_{11}^2+(r\sqrt{N_{11}^2-Q^2})_{,1}^2 -2\nu N_{11}(r\sqrt{N_{11}^2-Q^2})_{,1}]+N_{11}-\sqrt{N_{11}^2-Q^2}\}rdr = \text{stat} \quad (5.15)$$

without any side conditions.

The local saddle condition is after (3.25)

$$\int_o^a \frac{1}{Et}\{\nu(\delta N_{11}-\delta N_{22})^2+(1-\nu)(\delta N_{11}^2+\delta N_{22}^2)\}rdr + \int_o^a \frac{(-\xi_{3,1}^*\delta\xi_{1,1}+\xi_{1,1}^*\delta\xi_{3,1})^2}{(\xi_{1,1}^{*2}+\xi_{3,1}^{*2})^{3/2}} N_{11}rdr > 0. \quad (5.16)$$

It is fulfilled, if the membrane stress is a tension stress: $N_{11} > 0$. This is always the case for the considered membrane. Numerical results are shown in Fig.3 and 4 for $\hat{q} = \frac{Et}{a}$ and $\nu = 0{,}5$. They are based on (5.14) and (5.15) respectively where piecewise linear functions have been used for ξ_1, ξ_3 and $T \equiv \sqrt{N_{11}^2 - Q^2}$ the number of intervals being n. The saddle property is confirmed by the numerical results in Fig.4.

5.2. Second Example: Thin beam under dead weight

We assume the Euler-Bernoulli theory for a thin beam with an inextensional axis and a linear moment-curvature relation, for simplicity. It turns out to be suitable to use the rotation ϕ instead of the deformation (ξ_1,ξ_3) and the shear force Q instead of the loading (q_1,q_3) as variables (Fig.5). For brevity we summarize the individual steps:

Definition of $\underset{\sim}{u}$ and $\underset{\sim}{q}$:

$$\underset{\sim}{u} = \underset{\sim}{\bar{u}} + \underset{\sim}{\tilde{u}} \equiv \begin{bmatrix} \phi(x) \\ \hline \phi(\ell) \\ 0 \end{bmatrix} + \begin{bmatrix} 0 \\ \hline 0 \\ \phi(0) \end{bmatrix}; \quad \underset{\sim}{q} = \underset{\sim}{\bar{q}} + \underset{\sim}{\tilde{q}} \equiv \begin{bmatrix} Q(x) \\ \hline M_\ell \\ 0 \end{bmatrix} + \begin{bmatrix} 0 \\ \hline 0 \\ -\tilde{M}_o \end{bmatrix}.$$

Moment equilibrium, statical boundary condition and definition of the reactive moment at x = 0:

$$\bar{A}\underset{\sim}{\sigma} - \underset{\sim}{\bar{q}} - \underset{\sim}{\tilde{q}} \equiv \begin{bmatrix} -\frac{dM}{dx} \\ \hline M(\ell) \\ -M(0) \end{bmatrix} - \begin{bmatrix} Q(x) \\ \hline M_\ell \\ 0 \end{bmatrix} - \begin{bmatrix} 0 \\ \hline 0 \\ -\tilde{M}_o \end{bmatrix} = \underset{\sim}{0}. \quad (5.17)$$

Definition of curvature and kinematical boundary condition at x = 0:

$$\bar{B}u - \epsilon \equiv \frac{d\phi}{dx} - \kappa = 0 \quad (5.18)$$

$$\tilde{\underset{\sim}{u}} - \hat{\underset{\sim}{u}} \equiv \begin{bmatrix} 0 \\ \hline 0 \\ \phi(0) \end{bmatrix} - \begin{bmatrix} 0 \\ \hline 0 \\ \hat{\phi}_o = 0 \end{bmatrix} = \underset{\sim}{0}. \tag{5.19}$$

Bilinear forms:

$$[\sigma,\varepsilon] = \int_0^\ell M\kappa dx; \quad \langle \underset{\sim}{q},\underset{\sim}{u}\rangle = \int_0^\ell Q\phi dx + M_\ell \phi(\ell) - \tilde{M}_o \phi(0).$$

Definition of the adjoint operator $\underset{\sim}{B}^*\sigma$:

$$[\bar{B}\delta u,\sigma] - \langle \bar{\underset{\sim}{B}}^*\sigma, \delta\underset{\sim}{u}\rangle \equiv \int_0^\ell \frac{d\delta\phi}{dx} M dx + \int_0^\ell \frac{dM}{dx}\delta\phi dx - M(\ell)\delta\phi(\ell) + M(0)\delta\phi(0) = 0.$$

Note that the statical and kinematical operators are linear ones and that $\bar{\underset{\sim}{A}}\sigma = \bar{\underset{\sim}{B}}^*\sigma$ holds. Functionals:

$$P_M = [\bar{B}u,\sigma] = \int_0^\ell \frac{d\phi}{dx} M dx; \quad \langle \tilde{\underset{\sim}{u}} - \hat{\underset{\sim}{u}}, \tilde{\underset{\sim}{q}}\rangle = -\phi(0)\tilde{M}_o. \tag{5.20}$$

Material behaviour and internal complementary potential (EJ bending stiffness):

$$D\sigma - \varepsilon \equiv \frac{1}{EJ} M - \kappa = 0; \quad P_D = \int_0^\ell \frac{M^2}{2EJ} dx. \tag{5.21}$$

Relationship between $\bar{\underset{\sim}{q}}$ and $\bar{\underset{\sim}{u}}$:

$$\bar{\underset{\sim}{q}} - \underset{\sim}{C}(\bar{\underset{\sim}{u}}) \equiv \begin{bmatrix} Q \\ \hline M_\ell \\ 0 \end{bmatrix} - \begin{bmatrix} \hat{q}(\ell-x)\cos\phi \\ \hline 0 \\ 0 \end{bmatrix} \tag{5.22}$$

(where $q_3 = \hat{q} = \text{const}$ and $q_1 = 0$ have been taken).

Potential of the operator $\underset{\sim}{C}(\bar{\underset{\sim}{u}})$:

$$P_C = \hat{q} \int_0^\ell (\ell - x)\sin\phi dx. \tag{5.23}$$

The first line of (5.22) is based on the force equilibrium equation and the second one specifies the given boundary moment whereas in (3.8) $\bar{\underset{\sim}{q}} = \underset{\sim}{C}(\bar{\underset{\sim}{u}})$ describes the load behaviour. Nevertheless P_C in (5.23) represents the negative external potential exactly as in (3.8). Indeed the external potential energy is

$$-\Pi^{(e)} = \hat{q}\int_0^\ell u_3 dx = \hat{q}\{u_3(\ell)\ell - \int_0^\ell x\frac{du_3}{dx}dx\} = \hat{q}\int_0^\ell (\ell-x)\sin\phi dx = P_C$$

where $u_3(x) = \int_0^x \sin\phi dx$.

The variational principles based on the functionals (3.15), (3.18) and (3.19) are, using (5.20), (5.21) and (5.23):

$$\bar{I} = \int_0^\ell \{-\frac{M^2}{2EJ} - \hat{q}(\ell-x)\sin\phi + \frac{d\phi}{dx}M\}dx + \phi(0)\tilde{M}_o = \text{stat} \tag{5.24}$$

without side conditions;

$$J_I = \bar{I} - \left[\frac{\delta\bar{I}}{\delta\sigma},\sigma\right] - \left\langle\frac{\delta\bar{I}}{\delta\bar{\underset{\sim}{q}}},\tilde{\underset{\sim}{q}}\right\rangle = \int_0^\ell\{\frac{M^2}{2EJ} - \hat{q}(\ell-x)\sin\phi\}dx = \text{stat} \tag{5.25}$$

with the kinematical side conditions

$$\frac{\delta \bar{I}}{\delta \sigma} \equiv - \frac{M}{EJ} + \frac{d\phi}{dx} = 0 ; \quad \frac{\delta \bar{I}}{\delta \tilde{q}} \equiv \begin{bmatrix} 0 \\ \hline 0 \\ -\phi(0) \end{bmatrix} = \begin{bmatrix} 0 \\ \hline 0 \\ 0 \end{bmatrix} ; \tag{5.26}$$

$$J_{II} = \bar{I} - \left\langle \frac{\delta \bar{I}}{\delta \underset{\sim}{u}}, \underset{\sim}{u} \right\rangle = \int_0^{\ell} \{ - \frac{M^2}{2EJ} - \hat{q}(\ell - x)(\sin\phi - \phi\cos\phi) \} dx = \text{stat} \tag{5.27}$$

with the statical side conditions

$$\frac{\delta \bar{I}}{\delta \underset{\sim}{u}} \equiv \begin{bmatrix} - \frac{dM}{dx} - \hat{q}(\ell - x)\cos\phi \\ \hline M(\ell) \\ -M(0) + \tilde{M}_o \end{bmatrix} = \underset{\sim}{0} . \tag{5.28}$$

Elimination of M in (5.25) by means of (5.26/1) leads to the *principle of stationary potential energy* in the form

$$J_I = \int_0^{\ell} \{ \frac{EJ}{2} (\frac{d\phi}{dx})^2 - \hat{q}(\ell - x)\sin\phi \} dx = \text{stat}; \quad \phi(0) = 0. \tag{5.29}$$

In the principle of *stationary complementary potential* (5.27), however, the elimination of ϕ by means of (5.28) is of less practical use.

The local saddle condition (3.25) reads

$$\int_0^{\ell} \frac{1}{EJ} (\delta M)^2 dx + \int_0^{\ell} \hat{q}(\ell - x)\sin\phi^{*} (\delta\phi)^2 dx > 0. \tag{5.30}$$

It is satisfied for positive normal forces $N^{*} = \hat{q}(\ell - x)\sin\phi^{*} > 0$.

In Fig.6 the numerical results are shown for a clamped beam under dead weight and $\hat{q} = - 50 \frac{EJ}{\ell^3}$, n being the number of elements. We used piecewise linear trial functions for ϕ in (5.29). In (5.27) ϕ has been taken piecewise constant and M piecewise quadratic in such a manner that the constraint (5.28) is satisfied identically. There exist three equilibrium configurations. The first one (A) is characterized by J_I = min and J_{II} = max ((5.30) is satisfied). In the second one (B) and the third one (C) (5.30) is not satisfied. An a posteriori calculation of the second variation of J_I and J_{II} respectively shows, that in case B there is J_I = min (stable equilibrium) and J_{II} = stat, and in case C J_I = stat (non-stable equilibrium) and J_{II} = stat. As a result there follows that only in case A global bounds can be constructed as is confirmed by Fig.6. The problem of a clamped beam under single end force has been discussed in a similar manner by Lautenbach [16].

Acknowledgement. Author thanks his coworker R. Lautenbach for his contributions to this paper and the Deutsche Forschungsgemeinschaft for its support.

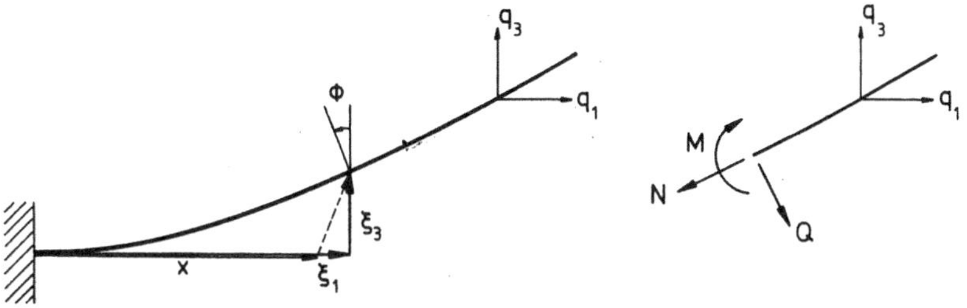

Fig.5 Clamped beam

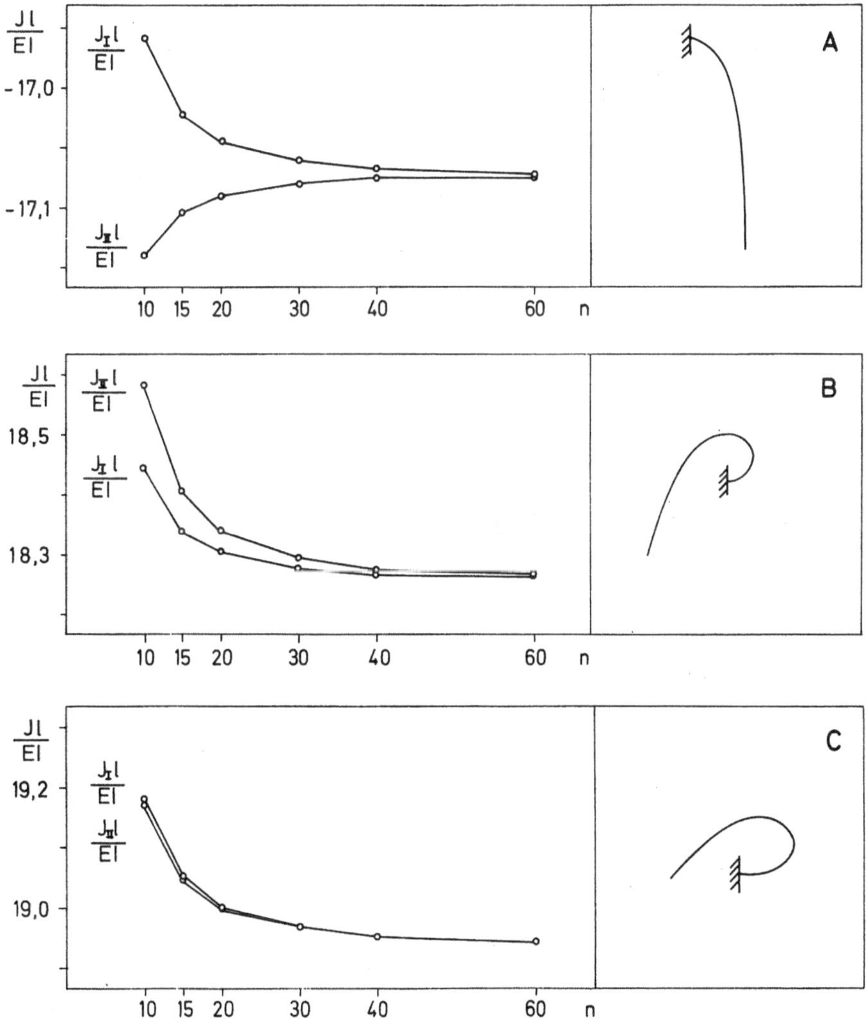

Fig.6 Potential energy and complementary potential for the clamped beam under dead weight. Three equilibrium configurations

References

1. BUFLER H., On the work theorems for finite and incremental elastic deformations with discontinuous fields: A unified treatment of different versions, Comp. Meth. Appl. Mech. Eng. 36 (1983), 95-124.

2. KOITER W.T., On the complementary energy theorem in nonlinear elasticity, in: Trends in applications of pure mathematics to mechanics, ed. by G. Fichera, Pitman, London 1976.

3. OGDEN R.W., Inequalities associated with the inversion of elastic stress-deformation relations and their implications, Math. Proc. Cambr. Phil. Soc. 81 (1977), 313-324.

4. STUMPF H., The derivation of dual extremum and complementary stationary principles in geometrical non-linear shell theory, Ing.-Arch. 48 (1979), 221-237.

5. FRAEIJS DE VEUBEKE B., A new variational principle for finite elastic displacements, Intern. J. Engng. Sc. 10 (1972), 745-763.

6. REISSNER E., Formulation of variational theorems in geometrically nonlinear elasticity, J. of Engng. Mech. 110 (1984), 1377-1390.

7. BUFLER H., The Biot stresses in nonlinear elasticity and the associated generalized variational principles, Ing.-Arch. 55 (1985), 450-462.

8. BUFLER H., Generalized variational principles in nonlinear elasticity using the Biot stresses and stretches as conjugate variables, in: Proc. of the intern.conf. on nonlinear mechanics Shanghai, ed. Chien Wei-zang, 219-224, Science press, Beijing 1985.

9. BUFLER H., NGUYEN-TUONG B., On the work theorems in nonlinear network theory, Ing.-Arch. 49 (1980), 275-286.

10. BUFLER H., Beitrag zu den komplementären Extremumsprinzipien in der nichtlinearen Elastostatik, Acta Mechanica 47 (1983), 153-159.

11. WEMPNER G., Complementary theorems of solid mechanics, in: Variational methods in the mechanics of solids, ed. by S. Nemat-Nasser, 127-135, Pergamon press, Oxford-New York 1980.

12. WUNDERLICH W., OBRECHT H., Large spatial deformations of rods using generalized variational principles, in: Nonlinear finite element analysis in structural mechanics, ed. by W. Wunderlich, E. Stein and K.J. Bathe, 185-216, Springer-Verlag Berlin 1981.

13. ATLURI S.N., Alternate stress and conjugate strain measures, and mixed variational formulations involving rigid rotations, for computational analysis of finitely deformed solids, with application to plates and shells - I theory, Computers and Structures 18 (1984), 93-116.

14. PIETRASZKIEWICZ W., BADUR J., Finite rotations in the description of continuum deformation, Int. J. Engng. Sc. 21 (1983) 1097-1115.

15. SEWELL M.J., On governing equations and extremum principles of elasticity and plasticity generated from a single functional, J. Struct. Mech. 2 (1973), 1-32 and 135-158.

16. LAUTENBACH R., Energetische Einschrankungen beim Balken mit großen elastischen Verformungen, ZAMM 65 (1985), T 275-T 276.

DEFORMATION OF THE SHELL BOUNDARY

K.F. CHERNYKH
Leningrad State University
Department of Applied Mathematics and Control Processes
Bibliotechnaya pl.,2, Leningrad 198904, USSR

Within the linear theory of shells, several problems associated with an infinitesimal deformation of the shell boundary were discussed by the author in [1,2]. In this paper the results are extended into the general case of large strains and rotations in thin shells.

1. Let us recall some basic relations which will be used in the following parts of the paper. The position vector $\boldsymbol{r}$ of the surface point is referred to the curvilinear coordinates α^1, α^2. The following relations hold [1]

$$
\begin{aligned}
&\boldsymbol{r}_i = \frac{\partial \boldsymbol{r}}{\partial \alpha^i}, \quad \boldsymbol{n} = (\boldsymbol{r}_1 \times \boldsymbol{r}_2)/|\boldsymbol{r}_1 \times \boldsymbol{r}_2|, \quad \boldsymbol{n}_i = \frac{\partial \boldsymbol{n}}{\partial \alpha^i}, \\
&a_{ij} = \boldsymbol{r}_i \cdot \boldsymbol{r}_j = a_{ji}, \quad b_{ij} = -\boldsymbol{r}_i \cdot \boldsymbol{n}_j = -\boldsymbol{r}_j \cdot \boldsymbol{n}_i = b_{ji}, \\
&a = a_{11}a_{22} - a_{12}^2, \quad a^{11} = a_{22}/a, \quad a^{12} = -a_{12}/a, \quad a^{22} = a_{11}/a, \\
&\boldsymbol{r}^i = a^{i\alpha}\boldsymbol{r}_\alpha, \quad \boldsymbol{r}_i = a_{i\alpha}\boldsymbol{r}^\alpha; \quad u_i = a_{i\alpha}u^\alpha, \quad u^j = a^{j\beta}u_\beta, \\
&t^{ij} = a^{i\alpha}t_\alpha{}^j = a^{j\beta}t^i{}_\beta = a^{i\alpha}a^{j\beta}t_{\alpha\beta}, \ \ldots; \\
&c_{12} = -c_{21} = \sqrt{a}, \quad c^{12} = -c^{21} = 1/\sqrt{a}, \quad c_{11} = c_{22} = c^{11} = c^{22} = 0; \\
&\boldsymbol{r}_i \times \boldsymbol{r}_j = c_{ij}\boldsymbol{n}, \quad \boldsymbol{n} \times \boldsymbol{r}_j = c_{i\alpha}\boldsymbol{r}^\alpha, \quad \boldsymbol{r}^i \times \boldsymbol{r}^j = c^{ij}\boldsymbol{n}, \quad \boldsymbol{n} \times \boldsymbol{r}^j = c^{j\alpha}\boldsymbol{r}_\alpha; \\
&\frac{\partial \boldsymbol{r}_i}{\partial \alpha^j} = \Gamma_{ij}^\alpha \boldsymbol{r}_\alpha + b_{ij}\boldsymbol{n}, \quad \frac{\partial \boldsymbol{r}^i}{\partial \alpha^j} = -\Gamma_{j\alpha}^i \boldsymbol{r}^\alpha + b_j^i \boldsymbol{n}, \quad \frac{\partial \boldsymbol{n}}{\partial \alpha^j} = -b_{j\alpha}\boldsymbol{r}^\alpha = -b_j^\alpha \boldsymbol{r}_\alpha; \\
&\Gamma_{ij}^h = \frac{1}{2}\left(\frac{\partial a_{j\alpha}}{\partial \alpha^i} + \frac{\partial a_{i\alpha}}{\partial \alpha^j} - \frac{\partial a_{ij}}{\partial \alpha^\alpha}\right) a^{\alpha h}, \\
&\nabla_j u^i = \frac{\partial u^i}{\partial \alpha^j} + \Gamma_{j\alpha}^i u^\alpha, \quad \nabla_k t^{ij} = \frac{\partial t^{ij}}{\partial \alpha^k} + \Gamma_{k\alpha}^i t^{\alpha j} + \Gamma_{k\alpha}^j t^{i\alpha}.
\end{aligned}
\tag{1.1}
$$

With the surface curve Γ (see Fig. 1) we associate vectors : $\boldsymbol{t}$ - the unit tangent, $\boldsymbol{n}$ - the unit normal and $\boldsymbol{\nu}$ - the outward unit normal. The following relations hold [1]

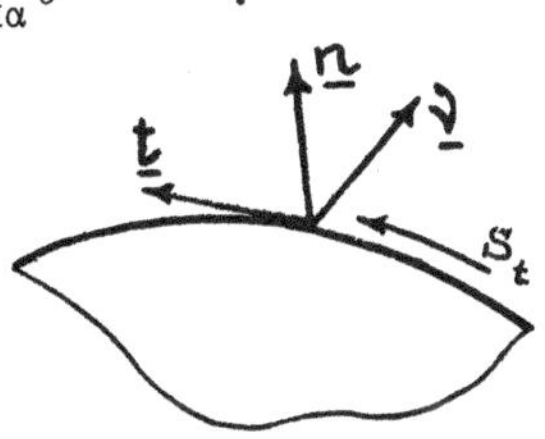

Fig. 1

$$\underset{\sim}{t}\times\underset{\sim}{n}=\underset{\sim}{\nu}\ ,\quad \underset{\sim}{n}\times\underset{\sim}{\nu}=\underset{\sim}{t}\ ,\quad \underset{\sim}{\nu}\times\underset{\sim}{t}=\underset{\sim}{n}\ ,$$
$$\underset{\sim}{t}\cdot\underset{\sim}{t}=\underset{\sim}{n}\cdot\underset{\sim}{n}=\underset{\sim}{\nu}\cdot\underset{\sim}{\nu}=1\ ,\quad \underset{\sim}{t}\cdot\underset{\sim}{n}=\underset{\sim}{n}\cdot\underset{\sim}{\nu}=\underset{\sim}{\nu}\cdot\underset{\sim}{t}=0\ ;$$
$$t^i=\frac{d\alpha^i}{ds_t}\ ,\quad \nu_i=c_{i\gamma}t^\gamma\ ,\quad \nu^j=a^{j\alpha}\nu_\alpha\ ,\quad d/ds_t=t^\alpha\nabla_\alpha\ ; \tag{1.2}$$
$$\frac{d\underset{\sim}{t}}{ds_t}=\sigma_t\underset{\sim}{n}-\rho_t\underset{\sim}{\nu}\ ,\quad \frac{d\underset{\sim}{n}}{ds_t}=\tau_t\underset{\sim}{\nu}-\sigma_t\underset{\sim}{t}\ ,\quad \frac{d\underset{\sim}{\nu}}{ds_t}=\rho_t\underset{\sim}{t}-\tau_t\underset{\sim}{n}\ ;$$
$$\sigma_t=b_{\alpha\beta}t^\alpha t^\beta\ ,\quad \tau_t=-b_{\alpha\beta}t^\alpha\nu^\beta\ ,\quad \rho_t=-\nu_\gamma t^\beta\nabla_\beta t^\gamma\ .$$

The parameters in the last line of (1.2) are called the normal curvature, the geodesic torsion and the geodesic curvature, respectively.

Let us associate the relations (1.1) and (1.2) with the middle surface of deformed shell. Analogous guantities associated with the undeformed shell middle surface will be marked with the nought ^{o}.

Position vectors of the material points of the shell before and after deformation can be given in the form [3-5](Fig. 2)

$$\overset{o}{\underset{\sim}{R}}(\alpha^1,\alpha^2;\zeta)=\overset{o}{\underset{\sim}{r}}(\alpha^1,\alpha^2)+\zeta\overset{o}{\underset{\sim}{n}}(\alpha^1,\alpha^2)\ , \tag{1.3}$$
$$\underset{\sim}{R}(\alpha^1,\alpha^2;\zeta)=\underset{\sim}{r}(\alpha^1,\alpha^2)+\lambda_\zeta(\alpha^1,\alpha^2)[\zeta+\tfrac{1}{2}\zeta^2\kappa_\zeta(\alpha^1,\alpha^2)]\underset{\sim}{n}(\alpha^1,\alpha^2)\ .$$

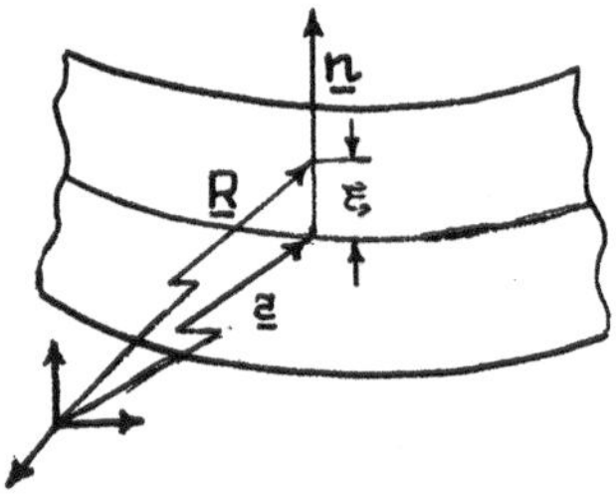

Fig. 2

The functions λ_ζ and κ_ζ describe an extension of the material fibre which is normal to the shell middle surface. When $\lambda_\zeta=1$ and $\kappa_\zeta=0$ the relations (1.3) reduce to the geometric Kirchhoff hypothesis. In case of incompressible material

$$\lambda_\zeta=(a/\overset{o}{a})^{-1/2}\ ,\quad \kappa_\zeta=-a^{\alpha\beta}\kappa_{\alpha\beta} \tag{1.4}$$

where

$$\kappa_{ij}=-\lambda_\zeta b_{ij}+\frac{1}{2}(a_{j\nu}\overset{o}{b}{}^{\nu}_{i}+a_{i\nu}\overset{o}{b}{}^{\nu}_{j})=\kappa_{ji} \tag{1.5}$$

are components of bending deformation.

2. Let us introduce the two-point tensor :

$$\underset{\sim}{Q}_t = \underset{\sim\sim}{n\overset{o}{n}} + \underset{\sim\sim}{\nu\overset{o}{\nu}} + \underset{\sim\sim}{t\overset{o}{t}} \tag{2.1}$$

associated with the material curve of the shell middle surface. According to (1.2), it has the following properties

$$\begin{aligned} &\underset{\sim}{Q}_t \cdot \overset{o}{\underset{\sim}{n}} = \underset{\sim}{n} \quad , \quad \underset{\sim}{Q}_t \cdot \overset{o}{\underset{\sim}{\nu}} = \underset{\sim}{\nu} \, , \quad \underset{\sim}{Q}_t \cdot \overset{o}{\underset{\sim}{t}} = \underset{\sim}{t} \quad , \\ &\underset{\sim}{n} \cdot \underset{\sim}{Q}_t = \overset{o}{\underset{\sim}{n}} \quad , \quad \underset{\sim}{\nu} \cdot \underset{\sim}{Q}_t = \overset{o}{\underset{\sim}{\nu}} \, , \quad \underset{\sim}{t} \cdot \underset{\sim}{Q}_t = \overset{o}{\underset{\sim}{t}} \quad . \end{aligned} \tag{2.2}$$

The tensor $\underset{\sim}{Q}_t$ rotates the triad of unit vectors $\overset{o}{\underset{\sim}{n}}$, $\overset{o}{\underset{\sim}{\nu}}$, $\overset{o}{\underset{\sim}{t}}$, which describe the normal cross-section of the undeformed shell, into the triad of unit vectors $\underset{\sim}{n}$, $\underset{\sim}{\nu}$, $\underset{\sim}{t}$ (Fig. 1), associated with the same material cross-section but in the deformed shell. Because of that the tensor $\underset{\sim}{Q}_t$ is called the rotation tensor of the normal element.

Let us introduce the tensor of change of curvature of the normal element

$$\underset{\sim}{K}_t = d\underset{\sim}{Q}_t \,/\, ds_t \tag{2.3}$$

If the extension of the middle-surface curve of the normal element is described by

$$\lambda_t = ds_t / d\overset{o}{s}_t \tag{2.4}$$

then

$$\frac{d}{ds_t} = \lambda_t^{-1} \frac{d}{d\overset{o}{s}_t} \quad . \tag{2.5}$$

Now from the relations (2.1)-(2.3) and (1.2) we obtain

$$\begin{aligned} \underset{\sim}{K}_t &= -\kappa_{t\nu}(\underset{\sim\sim}{n\overset{o}{\nu}} - \underset{\sim\sim}{\nu\overset{o}{n}}) + \kappa_{tn}(\underset{\sim\sim}{\nu\overset{o}{t}} - \underset{\sim\sim}{t\overset{o}{\nu}}) + \kappa_{tt}(\underset{\sim\sim}{t\overset{o}{n}} - \underset{\sim\sim}{n\overset{o}{t}}) = \\ &= [-\kappa_{t\nu}(\underset{\sim\sim}{n\nu} - \underset{\sim\sim}{\nu n}) + \kappa_{tn}(\underset{\sim\sim}{\nu t} - \underset{\sim\sim}{t\nu}) + \kappa_{tt}(\underset{\sim\sim}{tn} - \underset{\sim\sim}{nt})]\cdot \underset{\sim}{Q}_t = \\ &= \underset{\sim}{Q}_t \cdot [-\kappa_{t\nu}(\underset{\sim\sim}{\overset{o}{n}\overset{o}{\nu}} - \underset{\sim\sim}{\overset{o}{\nu}\overset{o}{n}}) + \kappa_{tn}(\underset{\sim\sim}{\overset{o}{\nu}\overset{o}{t}} - \underset{\sim\sim}{\overset{o}{t}\overset{o}{\nu}}) + \kappa_{tt}(\underset{\sim\sim}{\overset{o}{t}\overset{o}{n}} - \underset{\sim\sim}{\overset{o}{n}\overset{o}{t}})] \quad . \end{aligned} \tag{2.6}$$

Here the functions

$$-\kappa_{tt} = \sigma_t - \lambda_t^{-1}\overset{o}{\sigma}_t \quad , \quad \kappa_{t\nu} = \tau_t - \lambda_t^{-1}\overset{o}{\tau}_t \quad , -\kappa_{tn} = \rho_t - \lambda_t^{-1}\overset{o}{\rho}_t \tag{2.7}$$

may be regarded as components of the vectors of change of curvature of the normal element

$$\begin{aligned} \overset{o}{\underset{\sim}{\kappa}}_t &= -\kappa_{tn}\overset{o}{\underset{\sim}{n}} - \kappa_{tt}\overset{o}{\underset{\sim}{\nu}} + \kappa_{t\nu}\overset{o}{\underset{\sim}{t}} \quad , \\ \underset{\sim}{\kappa}_t &= -\kappa_{tn}\underset{\sim}{n} - \kappa_{tt}\underset{\sim}{\nu} + \kappa_{t\nu}\underset{\sim}{t} \quad , \end{aligned} \tag{2.8}$$

which are related, according to (2.2), by

$$\underset{\sim}{\kappa}_t = \underset{\sim}{Q}_t \cdot \overset{o}{\underset{\sim}{\kappa}}_t \quad , \quad \overset{o}{\underset{\sim}{\kappa}}_t = \underset{\sim}{\kappa}_t \cdot \underset{\sim}{Q}_t \tag{2.9}$$

Let us introduce the unit vectors of the principal normal $\overset{o}{\underset{\sim}{m}}$, $\underset{\sim}{m}$ and of binormal $\overset{o}{\underset{\sim}{b}}$, $\underset{\sim}{b}$, the spatial curvatures $\overset{o}{\sigma}\hat{}$ and $\sigma\hat{}$ as well as the spatial torsions $\overset{o}{\tau}\hat{}$ and $\tau\hat{}$ of the middle-surface curve of the normal element. Then, in analogy to transformations given above,

$$\begin{aligned} \underset{\sim}{K}_t &= \kappa_{tb}(\underset{\sim}{m}\overset{o}{\underset{\sim}{t}} - \underset{\sim}{t}\overset{o}{\underset{\sim}{m}}) + \kappa_{tm}(\underset{\sim}{b}\overset{o}{\underset{\sim}{m}} - \underset{\sim}{m}\overset{o}{\underset{\sim}{b}}) = \\ &= \underset{\sim}{Q}_t \cdot [\kappa_{tb}(\overset{o}{\underset{\sim}{m}}\overset{o}{\underset{\sim}{t}} - \overset{o}{\underset{\sim}{t}}\overset{o}{\underset{\sim}{m}}) + \kappa_{tm}(\overset{o}{\underset{\sim}{b}}\overset{o}{\underset{\sim}{m}} - \overset{o}{\underset{\sim}{m}}\overset{o}{\underset{\sim}{b}})] = \\ &= [\kappa_{tb}(\underset{\sim}{m}\underset{\sim}{t} - \underset{\sim}{t}\underset{\sim}{m}) + \kappa_{tm}(\underset{\sim}{b}\underset{\sim}{m}\ \underset{\sim}{m}\underset{\sim}{b})] \cdot \underset{\sim}{Q}_t \quad ; \\ \kappa_{tm} &= \tau\hat{} - \lambda_t^{-1}\overset{o}{\tau}\hat{} \quad , \quad \kappa_{tb} = \sigma\hat{} - \lambda_t^{-1}\overset{o}{\sigma}\hat{} \quad ; \\ \underset{\sim}{\kappa}_t &= \kappa_{tm}\underset{\sim}{m} + \kappa_{tb}\underset{\sim}{b} \quad , \quad \overset{o}{\underset{\sim}{\kappa}}_t = \kappa_{tm}\overset{o}{\underset{\sim}{m}} + \kappa_{tb}\overset{o}{\underset{\sim}{b}} \ . \end{aligned} \tag{2.10}$$

Application of (1.1) and (1.2) to the undeformed and deformed shell middle surfaces leads, with the help of (2.5), 'to

$$t^i = \lambda_t^{-1}\overset{o}{t}{}^i \quad , \quad \nu_j = \lambda_t^{-1}\sqrt{a/\overset{o}{a}}\ \overset{o}{\nu}_j \ . \tag{2.11}$$

If we multiply the representation

$$\underset{\sim}{t} = t^\alpha \underset{\sim}{r}_\alpha = \lambda_t^{-1}\overset{o}{t}{}^\alpha \underset{\sim}{r}_\alpha \tag{2.12}$$

by itself and take into account that $\underset{\sim}{t}\cdot\underset{\sim}{t} = 1$ and $\underset{\sim}{r}_i \cdot \underset{\sim}{r}_j = a_{ij}$ then

$$\lambda_t = \frac{1}{2}(a_{\alpha\beta}\overset{o}{t}{}^\alpha \overset{o}{t}{}^\beta) \tag{2.13}$$

Now with the help of (2.11) and (1.1)-(1.5) we obtain

$$\begin{aligned} \kappa_{tt} &= \lambda_t^{-1}[\lambda_t^{-1}\lambda_\zeta^{-1}\kappa_{\alpha\beta} - (\lambda_t^{-1}\lambda_\zeta^{-1}a_{\alpha\nu} - \overset{o}{a}_{\alpha\nu})\overset{o}{b}{}^\nu_\beta]\overset{o}{t}{}^\alpha\overset{o}{t}{}^\beta \quad , \\ \kappa_{t\nu} &= \lambda_t^{-1}[\lambda_t^{-1}\lambda_\zeta^{-1}\sqrt{a/\overset{o}{a}}\,a^{\gamma\beta}\kappa_{\alpha\beta} - \frac{1}{2}(\lambda_t^{-1}\lambda_\zeta^{-1}\sqrt{a/\overset{o}{a}} - 1)\overset{o}{b}{}^\gamma_\alpha - \\ &\quad - \frac{1}{2}(\lambda_t^{-1}\lambda_\zeta^{-1}\sqrt{a/\overset{o}{a}}\,a_{\alpha\nu}a^{\beta\gamma} - \overset{o}{a}_{\alpha\nu}\overset{o}{a}{}^{\beta\gamma})\overset{o}{b}{}^\nu_\beta]\overset{o}{t}{}^\alpha\overset{o}{\nu}{}^\gamma \quad , \\ \kappa_{tn} &= \lambda_t^{-1}[\lambda_t^{-1}\sqrt{a/\overset{o}{a}}\,\nabla_\beta \overset{o}{t}{}^\gamma - \overset{o}{\nabla}_\beta \overset{o}{t}{}^\gamma]\overset{o}{t}{}^\beta\overset{o}{\nu}_\gamma \ . \end{aligned} \tag{2.14}$$

3. Let us discuss the problem how to obtain the deformed configuration of the shell middle surface (or, equivalently, its translation) from the known a_{ij} and κ_{ij} . Note that in such a case quantities (2.13) and (2.14) are also known.

First according to (2.3),(2.6) and (2.5), the determination of the tensor $\underset{\sim}{Q}_t$ from given $\underset{\sim}{K}_t$ can be reduced to the solution of the tensor differential equation

$$\frac{d\underset{\sim}{Q}_t}{d\overset{o}{s}_t} = \underset{\sim}{Q}_t \cdot \underset{\sim}{A}(\overset{o}{s}_t) \tag{3.1}$$

where

$$\underset{\sim}{A}(\overset{o}{s}_t) = \lambda_t[-\kappa_{t\nu}(\overset{o}{\underset{\sim}{n}}\overset{o}{\underset{\sim}{\nu}} - \overset{o}{\underset{\sim}{\nu}}\overset{o}{\underset{\sim}{n}}) + \kappa_{tn}(\overset{o}{\underset{\sim}{\nu}}\overset{o}{\underset{\sim}{t}} - \overset{o}{\underset{\sim}{t}}\overset{o}{\underset{\sim}{\nu}}) + \kappa_{tt}(\overset{o}{\underset{\sim}{t}}\overset{o}{\underset{\sim}{n}} - \overset{o}{\underset{\sim}{n}}\overset{o}{\underset{\sim}{t}})] \tag{3.2}$$

is the skew-symmetric tensor, which is a known function of $\overset{o}{s}_t$. When $\underset{\sim}{Q}_t(\overset{o}{s}_t)$ is calculated (this is a difficult mathematical problem !) the position vector of the deformed middle surface point is determined, according to (2.2),(2.5) and (1.2), from the equation

$$\frac{d\underset{\sim}{r}}{d\overset{o}{s}_t} = \lambda_t \underset{\sim}{Q}_t \cdot \overset{o}{\underset{\sim}{t}} \tag{3.3}$$

An alternative approach to the solution of this (complex, in general) problem was giwen by SHAMINA [6] and PIETRASZKIEWICZ [7] .

4. Assuming the normal element to coincide with the boundary element, let us discuss again the relations (1.2). When the functions $\lambda_\zeta(\alpha^1,\alpha^2)$ $\kappa_\zeta(\alpha^1,\alpha^2)$ are determined by deformation of the middle surface (using (1.4), for example) the geometric boundary conditions require the assumption of two vectors

$$\underset{\sim}{r}[\alpha^1(\overset{o}{s}_t),\alpha^2(\overset{o}{s}_t)] = \underset{\sim}{r}(\overset{o}{s}_t) \quad , \quad \underset{\sim}{n}[\alpha^1(\overset{o}{s}_t),\alpha^2(\overset{o}{s}_t)] = \underset{\sim}{n}(\overset{o}{s}_t)$$

But according to (2.2) and (3.3)

$$\frac{d\underset{\sim}{r}}{d\overset{o}{s}_t} = \lambda_t(\overset{o}{s}_t)\underset{\sim}{t}(\overset{o}{s}_t).$$

As a result, in the second version of the geometric boundary conditions the guantities $\underset{\sim}{t}(\overset{o}{s}_t)$, $\underset{\sim}{n}(\overset{o}{s}_t)$, $\lambda_t(\overset{o}{s}_t)$ may be assumed. Here the vector $\underset{\sim}{\nu}(\overset{o}{s}_t) = \underset{\sim}{t}(\overset{o}{s}_t) \times \underset{\sim}{n}(\overset{o}{s}_t)$ is known as well. According to (2.2) , these three unit vectors are known if the tensor $\underset{\sim}{Q}_t(\overset{o}{s}_t)$ is given. Finaly, the relations (2.3),(2.6) and (2.8) show that it is possible to assume $\lambda_t(\overset{o}{s}_t)$ and $\underset{\sim}{\kappa}_t(\overset{o}{s}_t)$. The latter vector can be exchanged for the tensor $\underset{\sim}{K}_t(\overset{o}{s}_t)$.

Therefore, three following versions of the geometric boundary conditions can be formulated :

$$\text{I:} \quad \underset{\sim}{r} = \underset{\sim}{r}(s_t) \quad , \quad \underset{\sim}{n} = \underset{\sim}{n}(s_t) \quad ; \tag{4.1}$$

$$\text{II:}\ \underbrace{\underset{\sim}{t} = \underset{\sim}{t}(\overset{o}{s}_t)\ ,\quad \underset{\sim}{n} = \underset{\sim}{n}(\overset{o}{s}_t)\quad (\underset{\sim}{\nu} = \underset{\sim}{\nu}(\overset{o}{s}_t))}_{\underset{\sim}{Q}_t = \underset{\sim}{Q}_t(\overset{o}{s}_t)}\ ,\quad \lambda_t = \lambda_t(\overset{o}{s}_t)\ ; \tag{4.2}$$

$$\text{III:}\ \underbrace{\underset{\sim}{\kappa}_t = \underset{\sim}{\kappa}_t(\overset{o}{s}_t)}_{\underset{\sim}{K}_t = \underset{\sim}{K}_t(\overset{o}{s}_t)}\ ,\quad \lambda_t = \lambda_t(\overset{o}{s}_t)\ . \tag{4.3}$$

The first version is purely geometric, since in it the position of the boundary element in the deformed configuration is assumed. The third version may be regarded as deformational, since all the quantities appearing in it can be determined according to (2.13),(2.14) in terms of characteristics of the deformed middle surface. The second version is an intermediate one; it is expressed in terms of deformational quantity λ_t and rotation angles included implicitly into $\underset{\sim}{Q}_t$.

Note that the most interesting case of deformational boundary conditions is the condition of the rigid boundary

$$\lambda_t = 1\ ,\ \kappa_{tt} = \kappa_{t\nu} = \kappa_{tn} = 0\quad , \tag{4.4}$$

which differs from the geometric boundary condition for the clamped edge

$$\underset{\sim}{r} = \overset{o}{\underset{\sim}{r}}\quad ,\quad \underset{\sim}{n} = \overset{o}{\underset{\sim}{n}} \tag{4.5}$$

only by a rigid body translation and rotation. When the shell consists of several boundary contours (for a multi-connected middle surface) then the conditions (4.4) and (4.5) become totally different. Let us remind that the important feature of deformational conditions for a rigid boundary is that it is formulated entirely in terms of deformational characteristics of the middle surface.

5. When strains are small (but angles of rotation are arbitrary) the relations given above can be simplified. The relations (2.13) and (2.14) reduce to

$$\begin{aligned} \kappa_{tt} &= \overset{o}{t}{}^{\alpha}\overset{o}{t}{}^{\beta}\kappa_{\alpha\beta}\ ,\quad \kappa_{t\nu} = \overset{o}{t}{}^{\alpha}\overset{o}{\nu}{}^{\beta}\kappa_{\alpha\beta}\ , \\ \kappa_{tn} &= \overset{o}{t}{}^{\beta}\overset{o}{\nabla}_{\mu}(\overset{o}{c}{}^{\gamma\mu}\overset{o}{\varepsilon}_{\beta\gamma}) + \frac{d\varepsilon_{t\nu}}{d\overset{o}{s}_t} + \rho_t(\varepsilon_{\nu\nu} - \varepsilon_{tt})\quad ; \end{aligned} \tag{5.1}$$

$$\varepsilon_{tt} = \overset{o}{t}{}^{\alpha}\overset{o}{t}{}^{\beta}\overset{o}{\varepsilon}_{\alpha\beta}\quad . \tag{5.2}$$

Here

$$\begin{aligned} \varepsilon_{t\nu} &= \overset{o}{t}{}^{\alpha}\overset{o}{\nu}{}^{\beta}\overset{o}{\varepsilon}_{\alpha\beta}\ ,\quad \varepsilon_{\nu\nu} = \overset{o}{\nu}{}^{\alpha}\overset{o}{\nu}{}^{\beta}\overset{o}{\varepsilon}_{\alpha\beta}\quad ; \\ \overset{o}{\varepsilon}_{ij} &= -\beta_{ij} + \overset{o}{b}{}^{\gamma}_{i}\overset{o}{\varepsilon}_{j\gamma} + \overset{o}{b}{}^{\gamma}_{j}\overset{o}{\varepsilon}_{i\gamma}\ , \\ \overset{o}{E}_n\beta_{ij} &= -(\overset{o}{E}_n - 1)\overset{o}{b}_{ij} - (\overset{o}{\vartheta}_i\overset{o}{\vartheta}_j - \overset{o}{b}{}^{\alpha}_{i}\overset{o}{\omega}_{j\alpha}) + \overset{o}{P}_{ij,\beta}\overset{o}{\theta}{}^{\beta}\quad ; \end{aligned}$$

$$\overset{\circ}{P}_{ij,k} = \overset{\circ}{\nabla}_i \overset{\circ}{\varepsilon}_{jk} + \overset{\circ}{\nabla}_j \overset{\circ}{\varepsilon}_{ik} - \overset{\circ}{\nabla}_k \overset{\circ}{\varepsilon}_{ij} \quad ,$$

$$\overset{\circ}{E}_n = (1 + \overset{\circ}{\omega}_{1\cdot}^{\cdot 1})(1 + \overset{\circ}{\omega}_{2\cdot}^{\cdot 2}) - \overset{\circ}{\omega}_{1\cdot}^{\cdot 2} \overset{\circ}{\omega}_{2\cdot}^{\cdot 1} \quad ;$$

$$\overset{\circ}{e}_{ij} = \tfrac{1}{2}(\overset{\circ}{\nabla}_i \overset{\circ}{u}_j + \overset{\circ}{\nabla}_j \overset{\circ}{u}_i) - \overset{\circ}{b}_{ij} \overset{\circ}{w} \quad , \quad \overset{\circ}{\theta}_i = -\left(\frac{\partial \overset{\circ}{w}}{\partial \alpha^i} + \overset{\circ}{b}{}_i^{\alpha} \overset{\circ}{u}_{\alpha} \right) \quad ,$$

$$\overset{\circ}{\omega}_{ij} = \overset{\circ}{\nabla}_i \overset{\circ}{u}_j - \overset{\circ}{b}_{ij} \overset{\circ}{w} \quad , \quad \overset{\circ}{\omega}_{i\cdot}^{\cdot j} = \overset{\circ}{\nabla}_i \overset{\circ}{u}{}^j - \overset{\circ}{b}{}_i^j \overset{\circ}{w} \quad ;$$

and $\overset{\circ}{u}_i$, $\overset{\circ}{u}_j$, $\overset{\circ}{w}$ are components of the middle surface displacement vector $\underset{\sim}{u} = \overset{\circ}{u}{}^{\alpha} \overset{\circ}{\underset{\sim}{r}}_{\alpha} + \overset{\circ}{w} \overset{\circ}{\underset{\sim}{n}} = \overset{\circ}{u}_{\alpha} \overset{\circ}{\underset{\sim}{r}}{}^{\alpha} + \overset{\circ}{w} \overset{\circ}{\underset{\sim}{n}}$.

6. In the linear theory of shells, when the angles of rotation are small of the same order as the components of strain, the relations given above are further simplified. First, differences between the undeformed and deformed configurations may be ignored and the mark $^{\circ}$ in all the relations may be omitted. As a result, the relations given above reduce to the following

$$\kappa_{tt} = t^{\alpha} t^{\beta} \kappa_{\alpha\beta} \quad , \quad \kappa_{t\nu} = t^{\alpha} \nu^{\beta} \kappa_{\alpha\beta} \quad ,$$

$$\kappa_{tn} = t^{\mu} \nabla_{\beta} (c^{\gamma\beta} e_{\mu\gamma}) + \frac{de_{t\nu}}{ds_t} + \rho_t (e_{\nu\nu} - e_{tt}) \quad ,$$

$$e_{tt} = t^{\alpha} t^{\beta} e_{\alpha\beta} \quad ;$$

where

$$e_{t\nu} = t^{\alpha} \nu^{\beta} e_{\alpha\beta} \quad , \quad e_{\nu\nu} = \nu^{\alpha} \nu^{\beta} e_{\alpha\beta} \quad ;$$

$$\kappa_{ij} = - \beta_{ij} + b_i^{\gamma} e_{j\gamma} + b_j^{\gamma} e_{i\gamma} = \kappa_{ji} \quad ,$$

$$\beta_{ij} = - \nabla_i \theta_j + b_i^{\alpha} \omega_{j\alpha} = - \nabla_j \theta_i + b_j^{\alpha} \omega_{i\alpha} = \beta_{ji} \quad ;$$

$$e_{ij} = \tfrac{1}{2}(\nabla_i u_j + \nabla_j u_i) - b_{ij} w \quad ,$$

$$\omega_{ij} = \nabla_i u_j - b_{ij} w \quad , \quad \theta_i = -\left(\frac{\partial w}{\partial \alpha^i} + b_i^{\alpha} u_{\alpha}\right) \quad .$$

Apart from the displacement vector of the middle surface (Fig.3)

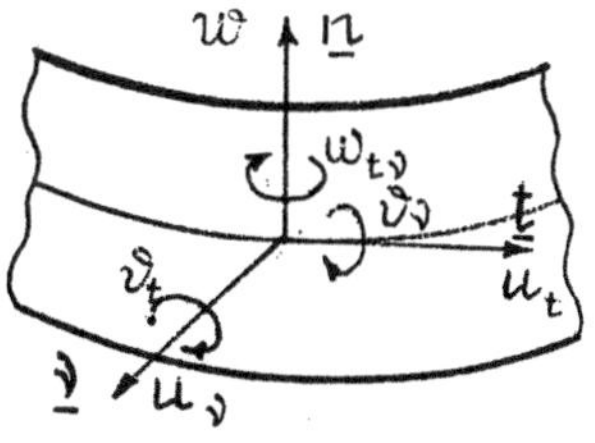

Fig. 3

$$\underset{\sim}{u} = u_\nu \underset{\sim}{\nu} + u_t \underset{\sim}{t} + w\underset{\sim}{n} \quad , \qquad (6.1)$$
$$u_\nu = \nu^\beta u_\beta \quad , \quad u_t = t^\beta u_\beta \quad ;$$

let us introduce the rotation vector of the normal element associated with the curve Γ (Fig.3)

$$\underset{\sim}{\Omega}_t = -\theta_t \underset{\sim}{\nu} + \theta_\nu \underset{\sim}{t} - \omega_{t\nu} \underset{\sim}{n} \qquad (6.2)$$
$$\theta_\nu = \nu^\beta \theta_\beta \quad , \quad \theta_t = t^\beta \theta_\beta \quad , \quad \omega_{t\nu} = t^\alpha \nu^\beta \omega_{\alpha\beta} \quad .$$

This vector should not be identified with the rotation vector of the neighbourhood of the middle surface points

$$\underset{\sim}{\Omega} = \mathbb{C}^{\alpha\beta} \theta_\alpha \underset{\sim}{r}_\beta + w\underset{\sim}{n}$$

which is connected to $\underset{\sim}{\Omega}_t$ by the relation

$$\underset{\sim}{\Omega} = \underset{\sim}{\Omega}_t + e_{t\nu} \underset{\sim}{n} \quad .$$

Using differential rules of the unit vectors (1.2) it is easy to obtain

$$\frac{d\underset{\sim}{u}}{ds_t} = e_{tt} \underset{\sim}{t} + \omega_{t\nu} \underset{\sim}{\nu} - \theta_t \underset{\sim}{n} = e_{tt} \underset{\sim}{t} + \underset{\sim}{\Omega}_t \times \underset{\sim}{t} \quad , \qquad (6.3)$$

$$\underset{\sim}{k}_t = \frac{d\underset{\sim}{\Omega}_t}{ds_t} = -k_{tt} \underset{\sim}{\nu} + k_{t\nu} \underset{\sim}{t} - k_{tn} \underset{\sim}{n} \quad ; \qquad (6.4)$$

where

$$k_{tt} = \frac{d\theta_t}{ds_t} + \rho_t \theta_\nu + \tau_t \omega_{t\nu} \quad ,$$

$$k_{t\nu} = \frac{d\theta_\nu}{ds_t} - \rho_t \theta_t + \sigma_t \omega_{t\nu} \quad ,$$

$$k_{tn} = \frac{d\omega_{t\nu}}{ds_t} - \tau_t \theta_t - \sigma_t \theta_\nu \quad .$$

It can be shown that the quantities given here are connected with the linearized quantities (2.14) by the relations

$$k_{tt} = \kappa_{tt} - \sigma_t e_{tt} \quad ,$$
$$k_{t\nu} = \kappa_{t\nu} + 2\sigma_t e_{\nu t} - \tau_t e_{\nu\nu} \quad ,$$
$$k_{tn} = \kappa_{tn} + \rho_t (e_{\nu\nu} - e_{tt}) \quad .$$

7. The formulae (6.1)-(6.4), given in earlier works of the author [1,2], allowed to express the displacement vector through $\underset{\sim}{k}_t$ and e_{tt}

in the form of the line integral

$$\underset{\sim}{u} = \underset{\sim}{U}^o + \underset{\sim}{\Omega}{}_t^o \times (\underset{\sim}{r} - \underset{\sim}{r}^o) + \left(\int_{s_o}^{s_t} \underset{\sim}{k}_t ds_t'\right) \times (\underset{\sim}{r} - \underset{\sim}{r}^o) +$$

$$+ \int_{s_o}^{s_t} [(\underset{\sim}{r}' - \underset{\sim}{r}^o) \times \underset{\sim}{k}_t + e_{tt}\underset{\sim}{t}'] \, ds_t' \quad . \qquad (7.1)$$

$$(\underset{\sim}{U}^o = \underset{\sim}{u}(s_o) \quad , \quad \underset{\sim}{\Omega}{}_t^o = \underset{\sim}{\Omega}{}_t(s_o) \quad , \quad \underset{\sim}{r}^o = \underset{\sim}{r}(s_o))$$

According to the static-geometric analogy, extended by the author to the boundary quantities, analogous to (7.1) formulae were obtained for the vector of stress function in terms of the principal force and moment resultant vectors of loads applied to the boundary contour.

Multivalued dislocational displacements were introduced into the shell theory. These quantities are connected with the single-valued components of the strain measures. The static-geometric analogy allowed to connect the non-uniqueness of the stress functions with the non-self-equilibrated boundary loads and to determine explicitly the non-self-equilibrated part of statically admissible stress functions.

For the shell with a non-asymptotic boundary contour the principal boundary-value problem (the membrane problem, in particular) was formulated together with associated boundary conditions. The relations for the simple edge effect were given in the explicit form. The conditions for connection of two shells as well as for shells with stiffners were also divided into the principal conditions and those associated with the simple edge effect.

The results listed above and some others are given in [1,2]. The extension of some results (apart from those valid only in the linear theory) to the large strains and large angles of rotation is given in [3--5].

References

1. CHERNYKH K.F., Linear theory of shells (in Russian), Leningrad Univ. Press, Leningrad 1964. English translation: Linear Theory of Shells, NASA - TT - F - 11 - 562, 1968.

2. CHERNYKH K.F., Some aspects of the linear theory of shells, in: Theory of Shells, 225-239, Ed. by W.T. Koiter and G.K. Mikhailov, North--Holland P. Co., Amsterdam 1980.

3. CHERNYKH K.F., Non-linear theory of thin isotropic elastic shells, Mechanics of Solids 1980, 2, 148-159.

4. CHERNYKH K.F., On the non-linear theory of thin elastic shells made of elastomers (in Russian), in: Deformation of Continuum and Regulation of Motion (in Russian), 3-25, Leningrad Univ. Press, Leningrad 1984.

5. CHERNYKH K.F., Theory of thin shells made of elastomers (rubber-like materials)(in Russian), Advances of Mechanics, 6 (1983), 1/2,111-147.

6. SHAMINA V.A., Some analogies in the problem of determination of the displacement vector from the deformation tensor, Mechanics of Solids, 1974, 3, 76-83.

7. PIETRASZKIEWICZ W., Introduction to the Non-Linear Theory of Shells, Ruhr-Universität, Mitt.Inst. für Mech. Nr 10, Bochum 1977.

COMPARISON OF NUMERICAL RESULTS FOR NONLINEAR FINITE ELEMENT ANALYSIS OF BEAMS AND SHELLS BASED ON 2-D ELASTICITY THEORY AND ON NOVEL FINITE ROTATION THEORIES FOR THIN STRUCTURES

J. Chróścielewski
Technical University of Gdańsk
Institute of Civil Engineering Mechanics
ul. Majakowskiego 11/12
PL-80-233 Gdańsk, Poland

R. Schmidt
University of Wuppertal
Institute of Civil Engineering Mechanics
Pauluskirchstrasse 7
D-5600 Wuppertal 2, Fed. Rep. of Germany

1. Introduction

In recent years a novel approach to the derivation of geometrically nonlinear Kirchhoff-Love type theories for thin elastic shells was initiated by PIETRASZKIEWICZ [10-13]. Based on the polar decomposition theorem he developed an exact theory of finite rotations in thin-walled structures. In [10-13] the rotational part of the shell deformation was described by a finite rotation vector, which in turn was expressed in terms of displacements of the shell middle surface and their gradients. This provides a sound basis for the derivation of constrained kinematic relations for thin elastic shells undergoing small strains (of $O(\eta)$, $\eta \ll 1$) accompanied by small ($O(\eta)$), moderate ($O(\eta^{1/2})$), large ($O(\eta^{1/4})$) or unrestricted rotations. Based on this approach a complete family of geometrically nonlinear Kirchhoff-Love type shell theories has been systematically derived in the works of PIETRASZKIEWICZ [11,13-18] and SCHMIDT [22-29]. In [22-29] special interest was focussed on the derivation of fully variationally consistent theories and associated energy principles. The aforementioned hierarchical set of shell equations consists of theories for unrestricted rotations, large rotations (with variants for large rotations of the normal only accompanied by moderate or small in-surface rotations) or moderate rotations. A shell theory for unrestricted rotations has been also developed by IURA and HIRASHIMA [4], while a variant for large rotations of the normal accompanied by small in-surface rotations has been also given by NOLTE and STUMPF [9]. STEIN [30] derived geometrically nonlinear theories for beams undergoing moderate or large rotations, respectively, while several other ones obtained by specializing the aforenamed shell theories for the one-dimensional case have been given recently by NOLTE [8]. For extensive references on the shell theories of the present approach we refer to the review of the state-of-the-art given in [29] , where also references to a large number of other geometrically nonlinear shell theories (e.g. to the important theories of REISSNER [20], LEONARD [6], SANDERS [21], KOITER [5] and many others) may be found.

First numerical results based on the novel shell theories mentioned above have been given by MAKOWSKI [7], and STEIN a.o. [31,32], who presented finite element results for moderate rotation shell problems. MAKOWSKI's work [7] has been extended into the range of large rotation beam and shell problems by NOLTE [8], who also carried out a numerical comparison of the present theories with a large number of other moderate and large rotation beam and shell theories available in literature. In some examples the results of MAKOWSKI [7] and NOLTE [8] confirm those obtained by other theories. For many problems, however, the present novel beam and shell theories predict considerable differences, e.g. in the load-deflection curves, especially in the post-buckling range of deformation where often large rotations occur. Sometimes not only quantitative differences are obtained, but even a qualitatively different post-buckling behaviour is predicted.

Therefore it was our interest to develop a method which is able to provide sound reference solutions for the problems under consideration without using explicitly beam or shell theory as such. Our starting-point was a variant of the NONSAP-code (see BATHE, RAMM, and WILSON [2]) which we have modified and extended essentially in order to make it applicable to problems of our specific interest. In particular the program has been made capable of calculating stability problems by introducing a displacement control method. Automatic change from load control to displacement control method and vice versa allows limit points to be passed and makes it possible to trace highly nonlinear load-deflection paths even in the deep post-buckling region. This control method is realized in the framework of standard or modified Newton-Raphson method. By means of this program we shall present 2-D elasticity theory solutions obtained by reduced integration technique for several typical highly nonlinear beam and shell problems. We shall compare our solutions with those obtained in [8] by means of the novel shell and beam theories described above. The result is a contribution to the numerical justification of these theories. We shall show that they yield accurate results in the range of moderate and large rotations, respectively, for which they have been developed.

2. Geometrically Nonlinear Shell Theories

In order to distinguish the various novel geometrically nonlinear shell theories under consideration we record here merely the respective strain-displacement relations and refer as to other details to the original literature mentioned in chapter 1.

(1) Unrestricted Rotation Shell Theory

$$\gamma_{\alpha\beta} = \frac{1}{2} \left(l^{\lambda}_{.\alpha} l_{\lambda\beta} + \varphi_\alpha \varphi_\beta - a_{\alpha\beta} \right) ,$$

$$\kappa_{\alpha\beta} = \left(m^{\lambda}|_{\beta} - b^{\lambda}_{\beta} m \right) l_{\lambda\alpha} + \left(b_{\lambda\beta} m^{\lambda} + m_{,\beta} \right) \varphi_\alpha + b_{\alpha\beta} \left(1 + \gamma^{\lambda}_{\lambda} \right) \tag{2.1}$$

Here, u_α and w are the tangential and normal components, respectively, of the displacement vector referred to the base vectors of the undeformed configuration and we have used the abbreviations $l_{\alpha\beta} = a_{\alpha\beta} + \varphi_{\alpha\beta}$, $\varphi_{\alpha\beta} = u_{\alpha|\beta} - b_{\alpha\beta}w$, $\omega_{\alpha\beta} = (1/2)(u_{\beta|\alpha} - u_{\alpha|\beta})$, $\varphi = (1/2)\,\varepsilon^{\alpha\beta}u_{\beta|\alpha}$, $\theta_{\alpha\beta} = (1/2)(u_{\alpha|\beta} + u_{\beta|\alpha}) - b_{\alpha\beta}w$, $\varphi_\alpha = w,_\alpha + b^\lambda_\alpha u_\lambda$, $m_\lambda = -(1+\theta^\mu_\mu)\varphi_\lambda + \varphi^\mu\varphi_{\mu\lambda}$, and $m = 1 + \theta^\lambda_\lambda - (1/2)(\theta^{\lambda\mu}\theta_{\lambda\mu} - \theta^\lambda_\lambda\theta^\mu_\mu) + \varphi^2$.

(2) Large Rotation Shell Theory

$$\gamma_{\alpha\beta} = \frac{1}{2}\left(l^{\lambda}_{.\alpha} l_{\lambda\beta} + \varphi_\alpha\varphi_\beta - a_{\alpha\beta} \right), \qquad \kappa_{\alpha\beta} = \frac{1}{2}\{ m^\lambda|_{(\alpha} l_{\lambda\beta)} + m,_{(\alpha}\varphi_{\beta)} - b^\lambda_{(\alpha}\varphi_{\lambda\beta)} +$$

$$+ b_{\alpha\beta}\varphi^\lambda\varphi_\lambda - b^\lambda_{(\alpha}\varphi_{\beta)}(\varphi_\lambda + \underline{\varphi^\kappa\omega_{\kappa\lambda}}) + \underline{b^\lambda_{(\alpha}\omega_{\lambda\beta)}(\theta^\kappa_\kappa + \varphi^2)} \} . \qquad (2.2)$$

Here, we have used the notation $A_{.(\alpha..}B_{.\beta)} = A_{.\alpha..}B_{.\beta} + A_{.\beta..}B_{.\alpha}$. The terms marked in (2.2) by a solid line may be dropped additionally, if the rotations about the normal remain moderate (of $O(\eta^{1/2})$), while the rotations of the normal are large.

(3) Large Rotations of the Normal Accompanied by Small In-Surface Rotations

For this case several variants have been proposed in recent literature which use the same expressions for the membrane strain tensor, but different bending strain-displacement relations (NOLTE and STUMPF [9] : (2.3), SCHMIDT [25,27] : (2.4) , PIETRASZKIEWICZ [17] : (2.5)). For a comparison of these variants we refer to [29] .

$$\gamma_{\alpha\beta} = \theta_{\alpha\beta} + \frac{1}{2}\varphi_\alpha\varphi_\beta + \frac{1}{2}\theta^\lambda_\alpha\theta_{\lambda\beta} - \frac{1}{2}\theta^\lambda_{(\alpha}\omega_{\lambda\beta)} ,$$

$$\kappa_{\alpha\beta} = -\frac{1}{2}\{ \varphi_{(\alpha|\beta)} + b^\lambda_{(\alpha}\theta_{\lambda\beta)} - b_{\alpha\beta}\varphi^\lambda\varphi_\lambda + b^\lambda_{(\alpha}\varphi_{\beta)}\varphi_\lambda + \varphi_{\lambda|(\alpha}\theta^\lambda_{\beta)} - \varphi_{(\alpha}\theta^\lambda_{\lambda|\beta)} \} , \qquad (2.3)$$

$$\kappa_{\alpha\beta} = -\frac{1}{2}\{ \varphi_{(\alpha|\beta)} + b^\lambda_{(\alpha}(\theta_{\lambda\beta)} - \omega_{\lambda\beta)}) - b_{\alpha\beta}\varphi^\lambda\varphi_\lambda + b^\lambda_{(\alpha}\varphi_{\beta)}\varphi_\lambda + \frac{1}{2}\varphi_{(\alpha}\varphi^\lambda\varphi_{\lambda|\beta)} \}, \qquad (2.4)$$

$$\kappa_{\alpha\beta} = -\frac{1}{2}\{ \varphi_{(\alpha|\beta)} + b^\lambda_{(\alpha}\theta_{\lambda\beta)} - b_{\alpha\beta}\varphi^\lambda\varphi_\lambda + b^\lambda_{(\alpha}\varphi_{\beta)}\varphi_\lambda + \varphi_{\lambda|(\alpha}\theta^\lambda_{\beta)} + \varphi_{(\alpha}\varphi^\lambda\varphi_{\lambda|\beta)} \}. \qquad (2.5)$$

(4) Moderate Rotation Shell Theory

$$\gamma_{\alpha\beta} = \theta_{\alpha\beta} + \frac{1}{2}\varphi_\alpha\varphi_\beta + \underline{\frac{1}{2}a_{\alpha\beta}\varphi^2 - \frac{1}{2}\theta^\lambda_{(\alpha}\omega_{\lambda\beta)}} , \qquad \kappa_{\alpha\beta} = -\frac{1}{2}\{ \varphi_{(\alpha|\beta)} + b^\lambda_{(\alpha}(\theta_{\lambda\beta)} - \omega_{\lambda\beta)}) \}. \qquad (2.6)$$

If only the rotations of the normal are moderate (of $O(\eta^{1/2})$) while the in-surface rotations remain small the terms underlined in (2.6) may be dropped as well.

3. Numerical Examples

The finite element formulation of 2-D elasticity theory has been described in detail by BATHE, RAMM and WILSON [2]. The following solutions for curved beam problems have been obtained with the 8-node plane stress element, while for the shell problems the 8-node axisymmetric element has been used. Reduced integration is realized by taking two integration points with stabilisation in direction of the midplane (see also [1]). Foundations of the displacement control method are described e.g. by RAMM [19] and WASZCZYSZYN [33], where also extensive references may be found. Our own displacement control method is described in [3]. With its help and, additionally, with automatic change to load control method and vice versa, the following highly nonlinear load-deflection curves have been traced. Our 2-D elasticity theory solutions will be compared with solutions obtained in [8] on the basis of the novel moderate and large rotation shell (and beam) theories , respectively, described in chapter 2. For a comparison of the latter results with additional ones obtained by a large number of other nonlinear shell theories proposed in literature we refer to [8]. In order to avoid misinterpretations we indicate here clearly for every point of the load-deflection curves the order of magnitude of the rotations occuring in the structure. The limits of applicability of the small ($O(\eta)$, $\eta \ll 1$), moderate ($O(\eta^{1/2})$), and large ($O(\eta^{1/4})$) rotation theories follow immediately from this clear classification. It is important to note that , speaking in other terms, they are at the most 2^0, 10^0, and 25^0, respectively.

As first example we have calculated the snap-through buckling of a moderately shallow arch under a point load given in Figure 1. The arch was idealized using twenty elements. In Figure 2 our load-deflection curve is compared with the result obtained in [8] using the large and unrestricted rotation theories (2.3) and (2.1), whose one-dimensional forms (for plane curved beams and cylindrical deformations of shells) can be easily shown to be identical. It follows from Figure 2 that the novel large (and unrestricted) rotation theories under consideration are indeed capable of predicting the highly nonlinear response of the arch in very good agreement with our 2-D elasticity theory reference solution.

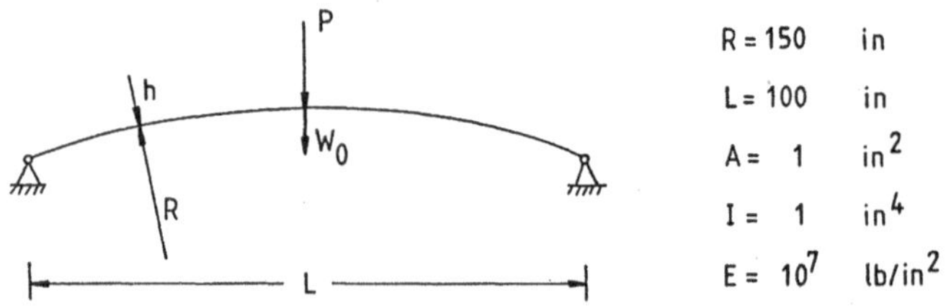

Fig. 1 : Moderately shallow arch under a point load

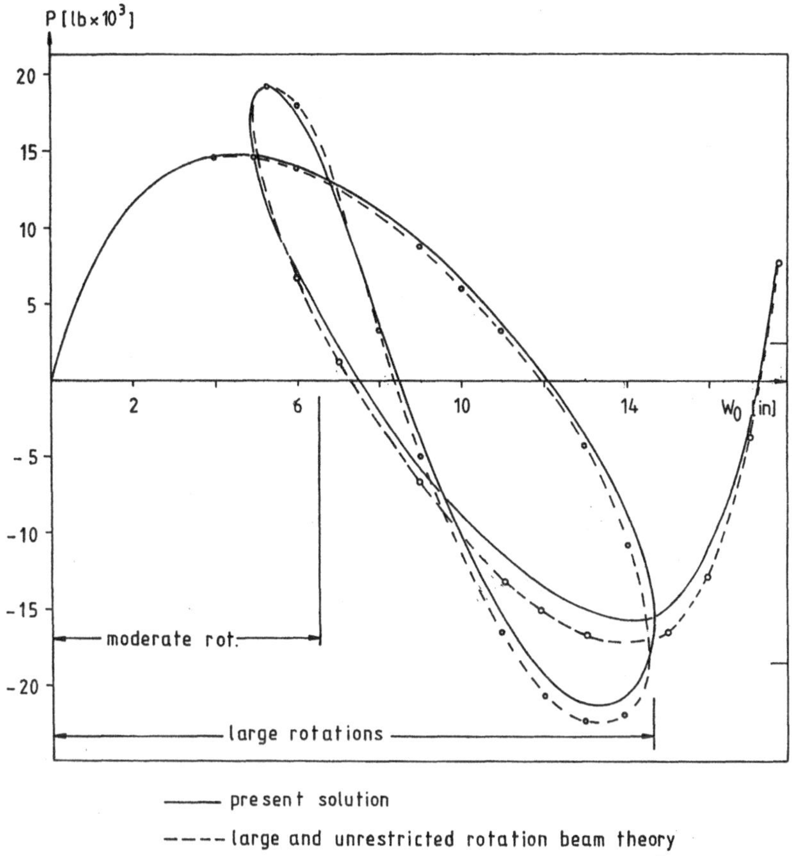

Fig. 2 : Load-deflection curves for a moderately shallow arch under a point load

Next we have studied the nonlinear response of a steep arch unsymmetrically loaded by a point load (Figure 3). The beam has been discretized using fourty elements.

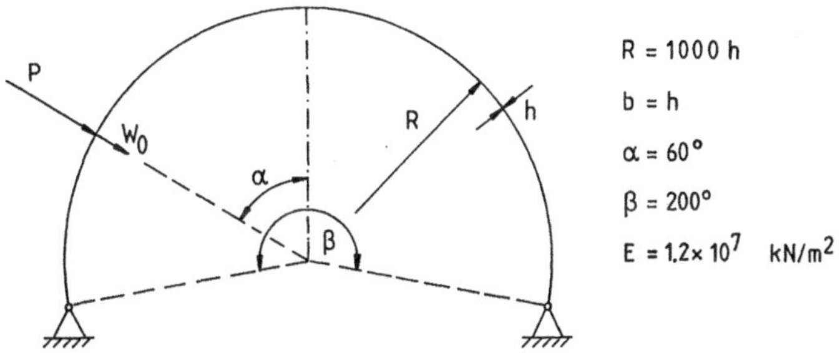

Fig. 3 : Curved beam subjected to an unsymmetric point load

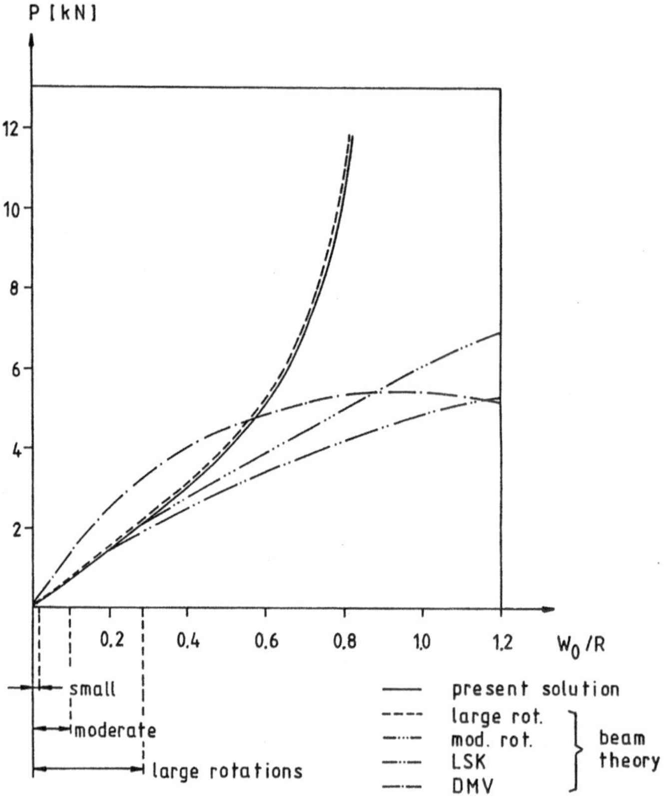

Fig. 4 : Load-deflection curves for an unsymmetrically loaded curved beam

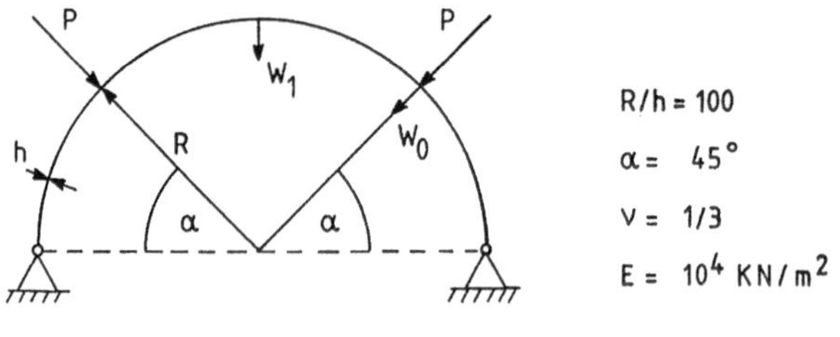

Fig. 5 : Spherical shell subjected to an axisymmetric line load

In Figure 4 our 2-D elasticity theory solution is compared with the results obtained in [8] using the one-dimensional versions of the large and unrestricted rotation theories (2.3) and (2.1) (which are identical in this case), of the moderate rotation theory (2.6), of the moderate rotation theory given by LEONARD [6], SANDERS [21], and KOITER [5] (LSK) and of the Donnell-Mushtari-Vlasov theory (DMV). It is clear that the latter theory cannot be applied here , because tangential displacements occur of the same order of magnitude as normal displacements. Both moderate rotation theories yield excellent results in the range of moderate rotations and depart from our reference solution in the large rotation range, what is not at all surprising. The large rotation theory matches our 2-D elasticity theory solution nearly exactly in the large rotation range. Even beyond this range the large rotation theory solution follows the reference solution nearly exactly. It should be mentioned, however, that this can only be expected for the variant (2.3) (not for the variants (2.4) and (2.5)) and only for the case of plane deformations of curved beams (and cylindrical deformations of shells), because in this case the one-dimensional version of (2.3) is identical to the one-dimensional version of the general theory (2.1). The variants (2.4) and (2.5) have been derived only for the range of large rotations and have in this range certain distinct advantages over the variant (2.3) (see e.g. [17], [18], [25 - 27], [29]), especially with respect to the consistency of variational principles

Next, we consider the stability problem of a spherical shell under axisymmetric line load given in Figure 5. The shell was discretized by 18 elements. This example has been calculated in [8] using the moderate rotation shell theory (2.6), the full large rotation theory (2.2), and its variant (2.3) which assumes small rotations about the normal. The results of the two latter theories are of course nearly identical in this example and are given in Figure 6 (and 7) by the broken line. A comparison with our 2-D elasticity theory solution shows that the novel moderate and large rotation shell theories under consideration yield very good results in the respective range of applicability. Disagreements between our reference solution and the large rotation shell theory solution can be found in the deep post-buckling region beyond the range of applicability of any large rotation shell theory. In particular we point out the dramatic nonlinear behaviour of the deflection w_1 , which we have observed beyond $w_1 = 12h$ and which we have drawn in Figure 7 in an enlarged scale. By very careful analysis choosing extremely small load and displacement increments, respectively, we found that the loop of the load-deflection curve involves two smaller loops with a local maximum located between them. Three corresponding local minima and maxima, respectively, can be also observed in the behaviour of the deflection w_0 at the same load values (near P = 9.17 , 11.54 and 12.72 kN/m) , so that in fact a local maximum exists where [8] predicts a local minimum. We assume that these details of the highly nonlinear structural response, which are pertinent to an extremely involved snap-through and snap-back behaviour, should be detected also, if the unrestricted rotation shell theory (2.1) is used.

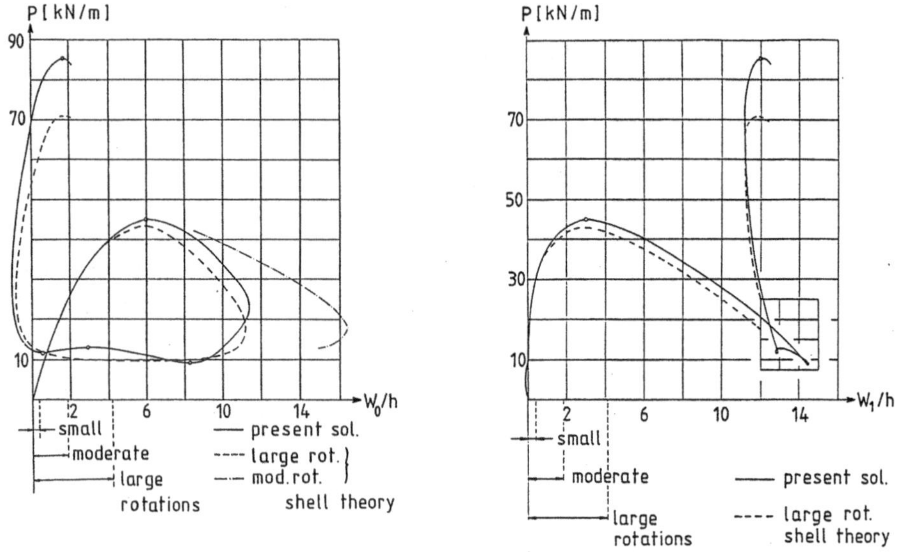

Fig. 6 : Load-deflection curves for a spherical shell under line load

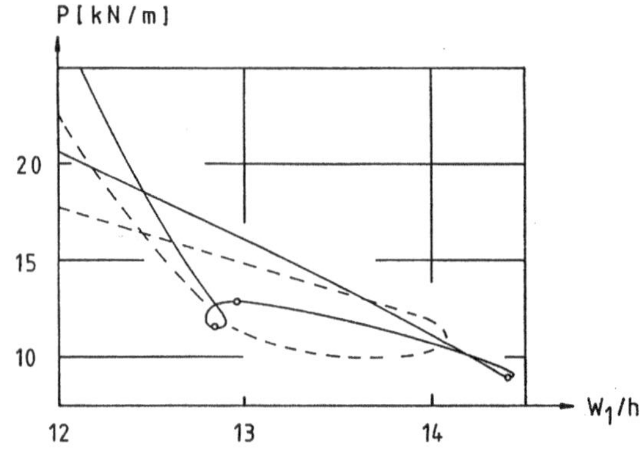

Fig. 7 : Detail of the load-deflection curve

Finally, we have analysed the behaviour of a torus subjected to an axisymmetric line load (see Figure 8). We have used 80 axisymmetric elements. In Figure 9 our 2-D elasticity theory solution is compared with the results obtained in [8] by the moderate rotation shell theory (2.6), the full large rotation shell theory (2.2), and its variant (2.3) for small rotations about the normal. Again the moderate rotation theory matches our reference solution exactly in the range of moderate rotations. Both large rotation theories (which yield identical results in this case) are in very good agreement with our 2-D elasticity theory solution in the large rotation range. Beyond that differences can be observed, e.g. near the second maximum of the load-deflection curve.

But this lies already far outside the range of applicability of large rotation shell theory and could be analysed correctly only by means of the general theory (2.1).

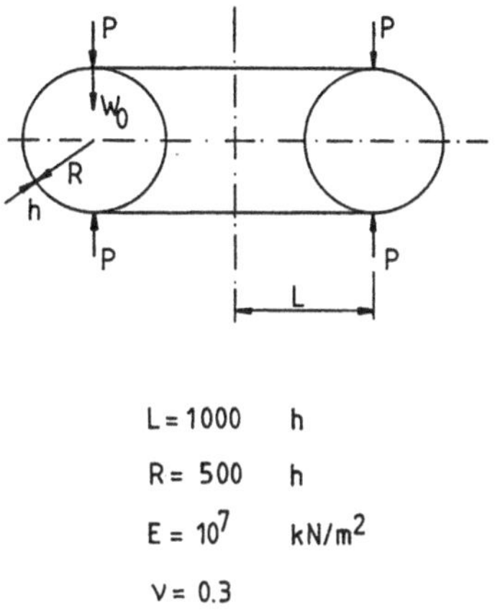

Fig. 8 : Torus under line load

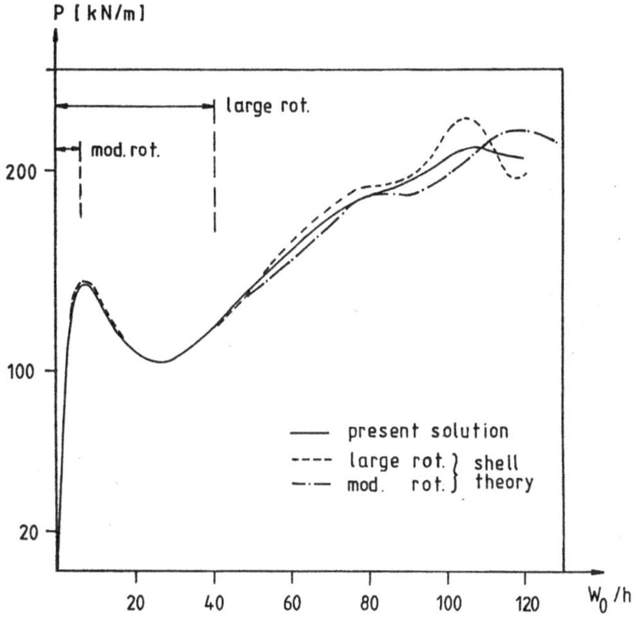

Fig. 9 : Load-deflection curves for Fig. 8

Acknowledgements

The second author gratefully acknowledges repeated support from the Deutsche Forschungsgemeinschaft (Grants 436 POL 111/2/84 and 111/8/85) and from the Polish Academy of Sciences for research visits to Gdańsk. This joint research was started, when both authors were staff members of the Institute of Mechanics of the Ruhr-University Bochum.

References

1. BADUR J. , CHRÓŚCIELEWSKI J. : Reduced Integration Technique in Nonlinear Shell Theory, Proc. 7th Conf. on Computer Meth. in Str. Mech., Gdynia 1985.
2. BATHE K.-J. , RAMM E. ; WILSON, E.L. : Int. J. for Numerical Methods in Engineering, 9 (1975), 353-386.
3. CHRÓŚCIELEWSKI J. , SCHMIDT R. : In: Proc. EUROMECH-Colloquium 200 "Post-Buckling of Elastic Structures", Mátrafüred (Hungary), 5.-7.10.85, to be published 1986.
4. IURA M. , HIRASHIMA M. : Proc. Jap. Soc. Civ. Eng., No. 344/I-1, 1984, 115-125.
5. KOITER W.T. : Proc. Kon. Ned. Ak. Wet. Ser. B 69 (1966), 1-54.
6. LEONARD R.W. : Nonlinear First Approximation Thin Shell and Membrane Theory, Ph.D. Thesis, Virginia Polytechnic Institute 1961.

7. MAKOWSKI J. : Linear and Nonlinear Analysis of Elastic Stability of Thin Shells (in Polish), Dissertation, Technical University of Gdańsk, 1981.

8. NOLTE L.-P. : Mitteilungen aus dem Institut für Mechanik, Nr. 39 (in German), Ruhr-Universität Bochum 1983.

9. NOLTE L.-P. , STUMPF, H. : Mech. Res. Comm. 10 (1983), 213-221.

10. PIETRASZKIEWICZ W. : Finite Rotations and Lagrangean Description in the Non-Linear Theory of Shells, Polish Scientific Publishers, Warszawa-Poznań 1979.

11. PIETRASZKIEWICZ W. : Mitteilungen aus dem Institut für Mechanik, Nr.10, Ruhr-Universität Bochum 1977.

12. PIETRASZKIEWICZ W. : In: Theory of Shells, eds. W.T. Koiter ; G.K. Mikhailov, 445-471, North-Holland Publ. Co., Amsterdam-New York-Oxford 1980.

13. PIETRASZKIEWICZ W. : In: Thin Shell Theory, New Trends and Applications, ed. W. Olszak, 153-208, Springer-Verlag, Wien-New York 1980.

14. PIETRASZKIEWICZ W. , SZWABOWICZ, M.L. : Archive of Mechanics 33 (1981), 273-288.

15. PIETRASZKIEWICZ W. : Mitteilungen aus dem Institut für Mechanik, Nr. 26, Ruhr-Universität Bochum 1981.

16. PIETRASZKIEWICZ W. : ZAMM 63 (1983), T200 - T202.

17. PIETRASZKIEWICZ W. : In: Flexible Shells, Theory and Applications, eds. E.L. Axelrad ; F.A. Emmerling, 106-123, Springer-Verlag, Berlin 1984.

18. PIETRASZKIEWICZ W. : Int. J. Non-Linear Mechanics 19 (1984), 115-140.

19. RAMM E. : In: Non-Linear Finite Element Analysis in Structural Mechanics, Proc. Europe-U.S. Workshop Bochum 1980, 63-89, Springer-Verlag, Berlin 1981.

20. REISSNER E. : Proc. 3rd U.S. Nat. Congr. Appl. Mech. 1958, 51-69.

21. SANDERS J.L. : Q. Appl. Math. 21 (1963), 21-36.

22. SCHMIDT R. : Variational Principles for Geometrically Nonlinear Theories of Shells Undergoing Moderate Rotations (in German), Dissertation, Bochum 1981.

23. SCHMIDT R. , PIETRASZKIEWICZ, W. : Ingenieur-Archiv 50 (1981), 187-201.

24. SCHMIDT R. : Proc. Int. Conf. on Finite Element Methods, Shanghai (China) 1982, 621-626, Gordon and Breach, New York 1982.

25. SCHMIDT R. : In: Flexible Shells, Theory and Applications, eds. E.L. Axelrad ; F.A. Emmerling, 76-90, Springer-Verlag, Berlin-Heidelberg-New York-Tokyo 1984.

26. SCHMIDT R. : Proc. 12th Southeastern Conf. on Theoretical and Applied Mechanics, Callaway Gardens, Georgia (USA), 1984, vol.1, 564-569, Auburn University Engineering Extension Service, Auburn, Alabama (USA), 1984.

27. SCHMIDT R. : Proc. 2nd Int. Conf. on Numerical Methods for Non-Linear Problems, Barcelona 1984, eds. C. Taylor, E. Hinton, D.R.J. Owen, vol. 2, 170-181, Swansea 1984.

28. SCHMIDT R. : ZAMM 85 (1985), 119-121.

29. SCHMIDT R. : Proc. of the Symp. on Advances and Trends in Structures and Dynamics, Washington 1984, eds. A.K. Noor ; R.J. Hayduk, 265-275, Pergamon Press, New York 1985, reprinted as Computers and Structures 20 (1985), 265-275.

30. STEIN E. : Proc. of the IUTAM Symposium on Finite Elasticity, Bethlehem (USA) 1980, eds. D.E. Carlson ; R.T. Shield, 379-400, Martinus Nijhoff Publ., The Hague 1982.

31. STEIN E., BERG A., WAGNER W. : In: Buckling of Shells, ed. E. Ramm, 91-136, Springer-Verlag, Berlin-Heidelberg-New York 1982.

32. STEIN E., WAGNER W., LAMBERTZ K.-H. : In: Flexible Shells, Theory and Applications, eds. E.L. Axelrad ; F.A. Emmerling, 22-43, Springer-Verlag, Berlin 1984.

33. WASZCZYSZYN Z. : In: Modern Methods for Stability Analysis of Structures, 341-380, Ossolineum, Wrocław 1981.

FUNDAMENTAL EQUATIONS AND EXTREMUM PRINCIPLES IN THE THEORY OF THIN SHELLS

R. DE BOER and W. WALTHER
Universität Essen, Institut für Mechanik
Universitätsstr. 15, 4300 Essen 1, FRG

1. Introduction

The present paper contributes to the non-linear theory of thin elastic shells in the frame of small strains but finite displacements and rotations (see [2],[3]). In describing the kinematic of deformations as well as in deducing the equilibrium conditions from the conservation laws of continuum mechanics the approximation is consequently applied that strains are small in every material point of the shell. The classification of occuring rotations as e.g. in [4] or restrictions in magnitude are not necessary.

Since large rotations with small strains in an elastic range are only imaginable in thin structures but not in a really three-dimensional continuum the product of shell thickness and tensor of curvature - $\Theta^3 \underset{\sim}{B}$ and $\Theta^3 \underset{\sim}{\bar{B}}$ - is neglected against the identity tensor in deducing the kinematics of deformation.

Proceeding from the equilibrium condition and the constitutive relations of the membrane forces and internal stress couples a basic differential equation for thin elastic shells is developed from which transitions to well-known non-linear theories - see e.g. [5] - can be shown.

Finally a minimum principle for the displacement rates is given taking into account the condition of stability. Notations refer to those in [1].

2. Geometry and Kinematics

In the reference shell placement the position vector to a material particle Y in shell space can be given by

$$\underset{\sim}{Y}(\Theta^i) = \underset{\sim}{X}(\Theta^\alpha) + \Theta^3 \underset{\sim}{a}_3 \quad , \quad \alpha = 1,\, 2 \quad , \qquad (2,1)$$

where $\underset{\sim}{X}$ is the position vector to the reference shell middle surface A, $\underset{\sim}{a}_3$ is the unit normal vector to A and Θ^3 is the distance from A, $-h/2 \le \Theta^3 \le h/2$, where h is the thickness of the shell.

With the middle surface A we associate the tangential base vectors $\underset{\sim}{a}_\alpha$, the unit normal vector $\underset{\sim}{a}_3$ and the symmetric curvature tensor $\underset{\sim}{B}$, which is described by the gradient of the unit normal vector with respect to the reference position vector $\underset{\sim}{X}$:

$$\underset{\sim}{B} = \nabla \underset{\sim}{a}_3 = \frac{\partial \underset{\sim}{a}_3}{\partial \underset{\sim}{X}} = b_{\alpha .}{}^{\beta} \underset{\sim}{a}^\alpha \otimes \underset{\sim}{a}_\beta \quad , \quad \underset{\sim}{B} = \underset{\sim}{B}^T \quad . \tag{2.2}$$

The transformation between the base systems of the middle surface $\underset{\sim}{a}_i$ and of the shell space $\underset{\sim}{h}_i$ is done by the shifter tensor $\underset{\sim}{Z}$:

$$\underset{\sim}{h}_i = \underset{\sim}{Z}\underset{\sim}{a}_i \quad , \quad \underset{\sim}{Z} = (\underset{\sim}{I} + \Theta^3 \underset{\sim}{B}) \quad , \quad \underset{\sim}{Z} = \underset{\sim}{Z}^T \quad , \quad \underset{\sim}{I} = \underset{\sim}{a}_i \otimes \underset{\sim}{a}^i \quad . \tag{2.3}$$

In some descriptions the inversion of the shifter tensor is needed. The inverse tensor and the determinant of the tensor contain the mean curvature H and the *Gaussian* curvature K:

$$\underset{\sim}{Z}^{-1} = \frac{1+2\Theta^3 H}{\det \underset{\sim}{Z}} (\underset{\sim}{I} - \Theta^3 \underset{\sim}{B}) + \frac{K(\Theta^3)^2}{\det \underset{\sim}{Z}} (\underset{\sim}{a}_3 \otimes \underset{\sim}{a}^3) \quad . \tag{2.4}$$

In the actual placement the base vectors $\bar{\underset{\sim}{a}}_i$ in a material point X of the deformed middle surface $\bar{A}$ can be determined by a linear mapping of the base vectors $\underset{\sim}{a}_i$ with the surface deformation gradient tensor $\bar{\underset{\sim}{F}}$, which can be splitted additively into two parts:

$$\bar{\underset{\sim}{a}}_i = \bar{\underset{\sim}{F}}\underset{\sim}{a}_i \quad , \quad \bar{\underset{\sim}{F}} = \underset{\sim}{F}' + \underset{\sim}{F}'' \tag{2.5}$$

with

$$\underset{\sim}{F}' = (\underset{\sim}{I}' + \nabla \underset{\sim}{u}) = \bar{\underset{\sim}{a}}_\alpha \otimes \underset{\sim}{a}^\alpha \quad , \quad \underset{\sim}{F}'' = \sqrt{\frac{a}{\bar{a}}}\, \overset{+}{\underset{\sim}{F}}{}' = \bar{\underset{\sim}{a}}_3 \otimes \underset{\sim}{a}^3 \quad . \tag{2.6}$$

The first part $\underset{\sim}{F}'$ including the reduced identity tensor $\underset{\sim}{I}' = \underset{\sim}{a}_\alpha \otimes \underset{\sim}{a}^\alpha$ and the gradient of the displacement field furnishes the linear mapping of the tangential vectors, whereas the second term transforms the unit normal vector from the reference to the actual placement. $\overset{+}{\underset{\sim}{F}}{}'$ is the adjoint tensor to $\underset{\sim}{F}'$.

We introduce the surface deformation tensor $\underset{\sim}{C}'$ and the surface strain tensor $\underset{\sim}{E}'$:

$$\underset{\sim}{C}' = \underset{\sim}{F}'^T \underset{\sim}{F}' \quad , \quad \underset{\sim}{E}' = \frac{1}{2}(\underset{\sim}{C}' - \underset{\sim}{I}') = \frac{1}{2}(\underset{\sim}{F}'^T \underset{\sim}{F}' - \underset{\sim}{I}') \quad . \tag{2.7}$$

The symmetric curvature tensor $\bar{\underset{\sim}{B}}$ in the actual placement is the gradient of the unit normal vector with respect to the actual position vector $\underset{\sim}{x}$:

$$\bar{\underset{\sim}{B}} = \text{grad } \bar{\underset{\sim}{a}}_3 = \frac{\partial \bar{\underset{\sim}{a}}_3}{\partial \underset{\sim}{x}} = \bar{b}_{\alpha .}{}^{\beta} \bar{\underset{\sim}{a}}^{\alpha} \otimes \bar{\underset{\sim}{a}}_{\beta} \quad , \quad \bar{\underset{\sim}{B}} = \bar{\underset{\sim}{B}}^T \quad . \tag{2.8}$$

Finally the gradient of the velocity field $\underset{\sim}{L}'$ is given by

$$\underset{\sim}{L}' = \text{grad } \dot{\underset{\sim}{x}} = \frac{\partial \dot{\underset{\sim}{x}}}{\partial \underset{\sim}{x}} = \dot{\bar{\underset{\sim}{a}}}_{\alpha} \otimes \bar{\underset{\sim}{a}}^{\alpha} \quad . \tag{2.9}$$

With the help of the *Love-Kirchhoff* hypothesis the position vector to the material point Y in the actual placement can be expressed by quantities of the middle surface

$$\underset{\sim}{y}(\Theta^i) = \underset{\sim}{Y}(\Theta^i) + \underset{\sim}{u}(\Theta^{\alpha}) + \Theta^3(\bar{\underset{\sim}{a}}_3 - \underset{\sim}{a}_3) \quad , \tag{2.10}$$

where $\underset{\sim}{u}(\Theta^{\alpha})$ is the displacement field of the shell middle surface and $\bar{\underset{\sim}{a}}_3$ the unit normal vector to the deformed surface $\bar{A}$.

The natural base vectors in shell space result from the partial derivation of the actual position vector $\underset{\sim}{y}$ and they again can be transformed from the base vectors of the middle surface by a linear mapping with the shifter tensor:

$$\bar{\underset{\sim}{h}}_i = \underset{\sim}{y}_{,i} \quad , \quad \bar{\underset{\sim}{h}}_i = \bar{\underset{\sim}{Z}} \bar{\underset{\sim}{a}}_i \quad , \quad \bar{\underset{\sim}{Z}} = \underset{\sim}{I} + \Theta^3 \bar{\underset{\sim}{B}} \quad , \quad \bar{\underset{\sim}{Z}} = \bar{\underset{\sim}{Z}}^T \quad . \tag{2.11}$$

The local deformation in a material point Y in shell space is given by the spatial deformation gradient tensor

$$\underset{\sim}{F} = \frac{\partial \underset{\sim}{y}}{\partial \underset{\sim}{Y}} = \bar{\underset{\sim}{h}}_i \otimes \underset{\sim}{h}^i \quad , \tag{2.12}$$

and with (2.3) and (2.11) it can be represented in the form

$$\underset{\sim}{F} = (\underset{\sim}{I} + \Theta^3 \bar{\underset{\sim}{B}})(\underset{\sim}{F}' + \underset{\sim}{F}'')(\underset{\sim}{I} + \Theta^3 \underset{\sim}{B})^{-1} \quad . \tag{2.13}$$

In order to make the description of the shell deformation significantly easier let us now state assumptions about the magnitude of some geometric and kinematic quantities.

On one hand we restrict ourselves to shells where the product of thickness and curvature - $\Theta^3 \underset{\sim}{B}$ and $\Theta^3 \bar{\underset{\sim}{B}}$, respectively - is very small. This postulation is for sure less restrictive than that of thin shells with small curvature which is often applied in the theory of shells.

On the other hand we assume the strains of the shell middle surface as well as those of the shell space to be very small and thus negligible against the identity tensor.

$$\underset{\sim}{E}' << \underset{\sim}{I}' \quad , \quad \underset{\sim}{E} << \underset{\sim}{I} \quad , \quad \underset{\sim}{E} = \frac{1}{2}(\underset{\sim}{F}^T \underset{\sim}{F} - \underset{\sim}{I}) \quad . \tag{2.14}$$

Both assumptions are consequently considered in the derivation of the kinematic relations. Finally we obtain the following equation for the *Cauchy-Green* strain tensor:

$$\underset{\sim}{E} = \underset{\sim}{E}' + \Theta^3 \Delta \underset{\sim}{B} \quad . \tag{2.15}$$

Herein, $\Delta \underset{\sim}{B}$ represents the symmetric tensor of change of curvature

$$\Delta \underset{\sim}{B} = \underset{\sim}{F}'^T \bar{\underset{\sim}{B}} \, \underset{\sim}{F}' - \underset{\sim}{B} = (\bar{b}_{\alpha.}^{\ \beta} - b_{\alpha.}^{\ \beta}) \underset{\sim}{a}^\alpha \otimes \underset{\sim}{a}_\beta \quad , \quad \Delta \underset{\sim}{B} = \Delta \underset{\sim}{B}^T \quad . \tag{2.16}$$

3. Equations of Equilibrium

The equations of equilibrium are derived from the conservation laws of continuum mechanics. With regard to the equilibrium of forces we start with the first *Cauchy* law, represented by the symmetric *Piola-Kirchhoff* stress tensor $\underset{\sim}{S}$ and other quantities of the reference placement, neglecting inertia forces and body forces

$$\int_P \mathrm{Div}(\underset{\sim}{F}\underset{\sim}{S})\, dV = \underset{\sim}{0} \quad , \tag{3.1}$$

where P denotes the body with the volume element dV in the reference placement. Splitting the volume integral into a double integral on the middle surface A and the thickness of the shell h

$$\int_A \int_h \mathrm{Div}(\underset{\sim}{F}\underset{\sim}{S})\, d\Theta^3 \, |d\underset{\sim}{A}| = \underset{\sim}{0} \tag{3.2}$$

leads to this local equilibrium equation

$$\int_h \mathrm{Div}(\underset{\sim}{F}\underset{\sim}{S})\, d\Theta^3 = \underset{\sim}{0} \quad . \tag{3.3}$$

For convenience we constitute the internal stress resultant tensor $\underset{\sim}{N}$ by integrating the stress tensor $\underset{\sim}{S}$ over the thickness of the shell

$$\underset{\sim}{N} = \int_h \underset{\sim}{S}\, d\Theta^3 \quad . \tag{3.4}$$

This tensor can be decomposed additively in the membrane forces $\underset{\sim}{N}'$, the shear forces $\underset{\sim}{Q}$ and its transposed $\underset{\sim}{Q}^T$:

$$\underset{\sim}{N} = \underset{\sim}{N}' + \underset{\sim}{Q} + \underset{\sim}{Q}^T \quad ,$$

$$\underset{\sim}{N}' = n^{\alpha\beta} \underset{\sim}{a}_\alpha \otimes \underset{\sim}{a}_\beta \quad , \qquad \underset{\sim}{Q} = n^{3\beta} \underset{\sim}{a}_3 \otimes \underset{\sim}{a}_\beta \quad . \tag{3.5}$$

The further evaluation of condition (3.3) finally leads to the equation of equilibrium of forces:

$$\mathrm{Div}(\underset{\sim}{F}'\underset{\sim}{N}') + \mathrm{Div}(\underset{\sim}{F}''\underset{\sim}{Q}) + \underset{\sim}{p} = \underset{\sim}{O} \tag{3.6}$$

with the loading vector $\underset{\sim}{p}$ and the definition

$$\mathrm{Div}(\ldots) = (\ldots)_{,\alpha} \underset{\sim}{a}^\alpha \quad . \tag{3.7}$$

In deriving the equilibrium of moments one proceeds from the conservation law of the moment of momentum:

$$\int_P \underset{\sim}{y} \times \mathrm{Div}(\underset{\sim}{F}\underset{\sim}{S})\,dV + \int_P \underset{\sim}{F} \times (\underset{\sim}{F}\underset{\sim}{S})\,dV = \underset{\sim}{O} \quad . \tag{3.8}$$

Again the volume integrals are splitted into double integrals and after introducing the internal stress couple tensor $\underset{\sim}{M}$

$$\underset{\sim}{M} = \int_h \underset{\sim}{S}\Theta^3 d\Theta^3 = \underset{\sim}{M}' + \underset{\sim}{M}'' + \underset{\sim}{M}''^T \quad ,$$

$$\underset{\sim}{M}' = m^{\alpha\beta} \underset{\sim}{a}_\alpha \otimes \underset{\sim}{a}_\beta, \qquad \underset{\sim}{M}'' = m^{3\beta} \underset{\sim}{a}_3 \otimes \underset{\sim}{a}_\beta \tag{3.9}$$

the local equation of the equilibrium of moments takes the form

$$\mathrm{Div}(\bar{\underset{\sim}{a}}_3 \times \underset{\sim}{F}'\underset{\sim}{M}') + \underset{\sim}{F}' \times (\underset{\sim}{F}'\underset{\sim}{N}') + \underset{\sim}{F}' \times (\underset{\sim}{F}''\underset{\sim}{Q}) = \underset{\sim}{O} \quad . \tag{3.10}$$

It is convenient to regard the components in the direction of the unit normal vector

$$(\bar{\underset{\sim}{B}}\underset{\sim}{F}') \times (\underset{\sim}{F}'\underset{\sim}{M}') + \underset{\sim}{F}' \times (\underset{\sim}{F}'\underset{\sim}{N}') = \underset{\sim}{O} \quad , \tag{3.11}$$

which is an identity - due to the symmetry of the stress tensor $\underset{\sim}{S}$ -, and in the direction of the tangential vectors:

$$\bar{\underset{\sim}{a}}_3 \times \mathrm{Div}(\underset{\sim}{F}'\underset{\sim}{M}') + \underset{\sim}{F}' \times (\underset{\sim}{F}''\underset{\sim}{Q}) = \underset{\sim}{O} \quad . \tag{3.12}$$

The equation (3.12) enables us to eliminate the shear forces in the equation of equilibrium of forces by the differential relation:

$$\underset{\sim}{F}''\underset{\sim}{Q} = \bar{\underset{\sim}{a}}_3 \otimes \mathrm{Div}\,\underset{\sim}{M}' \quad . \tag{3.13}$$

Finally, in the finite theory of thin shells under consideration of the specified assumptions (2.14) we can give the following equilibrium equation:

$$\mathrm{Div}(\underset{\sim}{F}'\underset{\sim}{N}' + \bar{\underset{\sim}{a}}_3 \otimes \mathrm{Div}\,\underset{\sim}{M}') + \underset{\sim}{p} = \underset{\sim}{O} \quad , \tag{3.14}$$

and the partial derivative with respect to the time yields the incremental formulation of this equilibrium condition:

$$\mathrm{Div}(\underset{\sim}{F}'\underset{\sim}{N}' + \bar{\underset{\sim}{a}}_3 \otimes \mathrm{Div}\,\underset{\sim}{M}')^{\bullet} + \dot{\underset{\sim}{p}} = \underset{\sim}{O} \quad . \tag{3.15}$$

4. Constitutive Equations and Fundamental Shell Equations

The elimination of the shear forces with relation (3.13) is necessary, because there is no material law available for these forces, as the following derivation will prove. In the derivation of constitutive equations between internal stress couple resultants and strains we start with the well-known linear mapping:

$$\underset{\sim}{S} = \overset{4}{\underset{\sim}{K}}\underset{\sim}{E} \quad , \quad \overset{4}{\underset{\sim}{K}} = \frac{E}{1+\nu}\,(\overset{4}{\underset{\sim}{I}} + \frac{\nu}{1-2\nu}\,\overset{4}{\underset{\sim}{\bar{I}}}) \tag{4.1}$$

which connects the symmetric *Piola-Kirchhoff* stress tensor with the symmetric *Cauchy-Green* strain tensor. The linear mapping of the strain tensor is done by the fourth-order elasticity tensor $\overset{4}{\underset{\sim}{K}}$, $\overset{4}{\underset{\sim}{I}}$ and $\overset{4}{\underset{\sim}{\bar{I}}}$ are fourth-order fundamental tensors and E and ν are *Young's* modulus and *Poisson's* ratio.

The constitutive equation (4.1) describes linear elastic material behaviour and is also valid in a finite theory under the assumption of small strains. In the theory of shells the usual assumptions of the vanishing stress components $s^{33} = 0$ leads to a material law in this form:

$$\underset{\sim}{S}' = \frac{E}{1-\nu^2}\,\overset{4}{\underset{\sim}{H}}{}'\underset{\sim}{E} \quad , \qquad \underset{\sim}{S}' = s^{\alpha\beta}\underset{\sim}{a}_\alpha \otimes \underset{\sim}{a}_\beta \quad . \tag{4.2}$$

Herein $\overset{4}{\underset{\sim}{H}}$ represents the fourth-order elasticity tensor

$$\overset{4}{\underset{\sim}{H}}{}' = (1-\nu)\overset{4}{\underset{\sim}{I}}{}' + \nu\overset{4}{\underset{\sim}{\bar{I}}}{}' \tag{4.3}$$

with the fourth-order fundamental tensor $\overset{4}{\underset{\sim}{I}}{}'$ and $\overset{4}{\underset{\sim}{\bar{I}}}{}'$, explained in [1].

After integrating on the thickness of the shell - e.g. (3.4) - one obtains this constitutive equation for the membrane forces

$$\underset{\sim}{N}' = D\,\overset{4}{\underset{\sim}{H}}{}'\underset{\sim}{E}' \quad , \qquad D = \frac{Eh}{1-\nu^2} \quad . \tag{4.4}$$

On the other side eq. (3.9) with the constitutive condition (4.3) yields the following constitutive equation for the internal stress couple tensor $\underset{\sim}{M}'$:

$$\underset{\sim}{M}' = B\,\overset{4}{\underset{\sim}{H}}{}'\Delta\underset{\sim}{B} \quad , \qquad B = \frac{Eh^3}{12(1-\nu^2)} \quad . \tag{4.5}$$

If we insert these constitutive equations (4.4) and (4.5) for the internal stress resultants tensor and the internal stress couple resultants tensor into the equilibrium condition (3.14) we obtain the fundamental equation for thin elastic shells under the assumption of small strains everywhere

$$D\,\mathrm{Div}(\underset{\sim}{F}'\overset{4}{\underset{\sim}{H}}{}'\underset{\sim}{E}') + B\,\mathrm{Div}[\underset{\sim}{\bar{a}}_3 \otimes \mathrm{Div}(\overset{4}{\underset{\sim}{H}}{}'\Delta\underset{\sim}{B})] + \underset{\sim}{p} = \underset{\sim}{O} \quad . \tag{4.6}$$

This differential equation contains, apart from the loading vector and the material properties, only parts of the deformation state of the middle surface.

Starting from this fundamental equation transitions to other known equations of linear and non-linear shell theory can be indicated. For example, the vanishing of the reference curvature yields the basic equation for thin elastic plates and the well-known *von Kármán* equations can be easily derived.

5. Extremum Principles

The development of extremum principles is of high relevance in numerical calculations. As an example we will give the minimum principle for the displacement rates. First we derive the principle of virtual displacements in an incremental form. We assume that an incremental step $\dot{\underset{\sim}{y}}Dt$ transfers the material point Y of the shell space into a placement close to the actual placement. As well the increment $\dot{\underset{\sim}{y}}^*Dt$ leads to a neighbouring comparative placement. Both placements have to be geometrically admissible. Then the following difference velocity fields can be chosen as virtual velocity fields

$$\dot{\hat{\underset{\sim}{x}}} = \dot{\underset{\sim}{x}}^* - \dot{\underset{\sim}{x}} \quad , \quad \dot{\hat{\underset{\sim}{Y}}} = \dot{\underset{\sim}{y}}^* - \dot{\underset{\sim}{y}} \quad , \tag{5.1}$$

and the virtual velocity gradient can be defined:

$$\hat{\underset{\sim}{L}}' = \underset{\sim}{L}^{*\prime} - \underset{\sim}{L}' \quad . \tag{5.2}$$

After some transformations by using the integral theorems the principle of virtual displacements in this incremental formulation can be given:

$$\oint_C (\bar{\underset{\sim}{F}}\underset{\sim}{N})^{\cdot} (\underset{\sim}{I} \times \underset{\sim}{a}_3) d\underset{\sim}{X} \cdot \dot{\hat{\underset{\sim}{x}}} + \oint_C (\underset{\sim}{F}'\underset{\sim}{M}')^{\cdot} (\underset{\sim}{I} \times \underset{\sim}{a}_3) d\underset{\sim}{X} \cdot \dot{\hat{\underset{\sim}{a}}}_3 -$$

$$\int_A (\hat{\underset{\sim}{L}}'\underset{\sim}{F}') \cdot (\underset{\sim}{F}'\underset{\sim}{N}')^{\cdot} \, |d\underset{\sim}{A}| - \int_A \nabla \dot{\hat{\underset{\sim}{a}}}_3 \cdot (\underset{\sim}{F}'\underset{\sim}{M}')^{\cdot} \, |d\underset{\sim}{A}| + \int_A \dot{\underset{\sim}{p}} \cdot \dot{\hat{\underset{\sim}{x}}} \, |d\underset{\sim}{A}| = 0 \ . \tag{5.3}$$

It contains the virtual work done by the rates of the loading vector and the forces and couples acting along the boundary curve. Moreover, two integrals for the virtual internal work of membrane forces and stress couple resultants occur.

In the frame of a finite theory the proof of a unique solution for a boundary value problem can be furnished only in connection with the condition of stability. In deriving the uniqueness theorem we first proceed from the assumption that a neutral equilibrium exists and consequently any neighbouring placements are possible. With the help of the principle of virtual displacements one finally obtains this integral equation built by parts of the internal work of membrane forces and moments:

$$J = \int_A [(\overset{*}{\underset{\sim}{F}}{}'\underset{\sim}{N}')^{\cdot} - (\underset{\sim}{F}'\underset{\sim}{N}')^{\cdot}] \cdot (\underset{\sim}{L}^{*}{}'\underset{\sim}{F}' - \underset{\sim}{L}'\underset{\sim}{F}') \, |d\underset{\sim}{A}| \; +$$

$$+ \int_A [(\overset{*}{\underset{\sim}{F}}{}'\underset{\sim}{M}')^{\cdot} - (\underset{\sim}{F}'\underset{\sim}{M}')^{\cdot}] \cdot (\nabla \dot{\underset{\sim}{a}}{}_3^{*} - \nabla \dot{\bar{\underset{\sim}{a}}}_3) \, |d\underset{\sim}{A}| \quad , \tag{5.4}$$

$$(\overset{*}{\underset{\sim}{F}}{}'\underset{\sim}{N}')^{\cdot} \neq (\underset{\sim}{F}'\underset{\sim}{N}')^{\cdot} \quad , \quad (\overset{*}{\underset{\sim}{F}}{}'\underset{\sim}{M}')^{\cdot} \neq (\underset{\sim}{F}'\underset{\sim}{M}')^{\cdot}$$

$$\underset{\sim}{L}^{*}{}'\underset{\sim}{F}' \neq \underset{\sim}{L}'\underset{\sim}{F}' \quad , \quad \nabla \dot{\underset{\sim}{a}}{}_3^{*} \neq \nabla \dot{\bar{\underset{\sim}{a}}}_3 \quad . \tag{5.5}$$

This integral equation becomes zero, if there is a neutral equilibrium and thus a bifurcation. The equilibrium state for $J > 0$ is stable. The satisfaction of this postulation is a sufficient condition for a unique neighbouring state.

Finally the minimum principle of displacements can be deduced from the principle of virtual displacements under consideration of the condition of stability. It leads to this integral form, which always has to be positive:

$$G^* - G = \frac{1}{2} \int_A [(\overset{*}{\underset{\sim}{F}}{}'\underset{\sim}{N}')^{\cdot} \cdot (\underset{\sim}{L}^{*}{}'\underset{\sim}{F}') - (\underset{\sim}{F}'\underset{\sim}{N}')^{\cdot} \cdot (\underset{\sim}{L}'\underset{\sim}{F}')] \, |d\underset{\sim}{A}| \; +$$

$$+ \frac{1}{2} \int_A [(\overset{*}{\underset{\sim}{F}}{}'\underset{\sim}{M}')^{\cdot} \cdot \nabla \dot{\underset{\sim}{a}}{}_3^{*} - (\underset{\sim}{F}'\underset{\sim}{M}')^{\cdot} \cdot \nabla \dot{\bar{\underset{\sim}{a}}}_3] \, |d\underset{\sim}{A}| \; -$$

$$- \oint_C (\bar{\underset{\sim}{F}}\underset{\sim}{N})^{\cdot} (\underset{\sim}{I} \times \underset{\sim}{a}_3) d\underset{\sim}{X} \cdot (\dot{\underset{\sim}{x}}^{*} - \dot{\underset{\sim}{x}}) \; -$$

$$- \oint_C (\underset{\sim}{F}'\underset{\sim}{M}')^{\cdot} (\underset{\sim}{I} \times \underset{\sim}{a}_3) d\underset{\sim}{X} \cdot (\dot{\underset{\sim}{a}}{}_3^{*} - \dot{\underset{\sim}{a}}_3) \; -$$

$$+ \int_A \dot{\underset{\sim}{p}} \cdot (\dot{\underset{\sim}{x}}^{*} - \dot{\underset{\sim}{x}}) \, |d\underset{\sim}{A}| \tag{5.6}$$

$$G^* - G \geq 0 \quad . \tag{5.7}$$

The statical quantities $(\overset{*}{\underset{\sim}{F}}{}'\underset{\sim}{N}')^{\cdot}$ and $(\overset{*}{\underset{\sim}{F}}{}'\underset{\sim}{M}')^{\cdot}$ in general do not fulfill the statical boundary conditions and the equation of equilibrium (3.15). On the other hand, however, we have to postulate in using the minimal principle for numerical calculations that for each further step the previous state has to be an equilibrium state. For this reason the chosen displacement fields must be examined considering the question whether the equilibrium conditions and the statical boundary conditions are adequately fulfilled.

References

1. DE BOER, R.: Vektor- und Tensorrechnung für Ingenieure, Springer-Verlag, Berlin-Heidelberg-New York 1982.

2. NAGHDI, P.M.: The theory of shells and plates, in: Encyclopedia of Physics, Vol. VI a/2, Mechanics of Solids II, Ed. by S. Flügge, 425-640, Springer-Verlag, Berlin-Heidelberg-New York 1972.

3. STEIN, E., BERG, A., WAGNER, W.: Different levels of nonlinear shell theory in finite element stability analysis, in: Buckling of shells, Ed. by E. Ramm, Springer-Verlag, Berlin-Heidelberg-New York 1982.

4. PIETRASZKIEWICZ, W.: Lagrangian description and incremental formulation in the non-linear theory of thin shells, Int. J. Non-linear Mechanics 19, No.2, 115-140, 1984.

5. VON KÁRMÁN, Th.:Festigkeitsprobleme im Maschinenbau, in: Encyclopädie der Mathematischen Wissenschaften, Vol. 4/4, 311-385, 1910.

INHOMOGENEITY AND ROTATION

MARCELO EPSTEIN
Department of Mechanical Engineering
University of Calgary
Calgary, Alberta, Canada

An elastic body B is said to be materially uniform [1,2] if all its points are made of the same elastic material. Formally, there exists a smooth distribution of local configurations, i.e. a smooth distribution of linear maps of the form

$$P(X) : V \longrightarrow T_X B ,$$

such that, when referred to those local configurations, the elastic response in terms of the stress tensor

$$T = T(F,X) ,$$

is identical for all points X of the body B. In other words, the following condition holds for all pairs of points X,Y of B :

$$T(FP^{-1}(X),X) = T(FP^{-1}(Y),Y) .$$

In the above formulae V is the standard Euclidean translation space, $T_X B$ is the tangent space of the body manifold B at the point X, and F denotes an arbitrary deformation gradient. The composition

$$P_X(Y) = P(Y)\ P^{-1}(X) \quad : \quad T_X B \longrightarrow T_Y B ,$$

called a material isomorphism between the points X and Y, represents a deformation of a neighbourhood of X which will render it, as far as the mechanical response is concerned, indistinguishable from a neighbourhood of Y. Obviously,

$$T(F,Y) = T(FP_X(Y),X) .$$

A body is locally homogeneous if for every point , Y , there exists a global configuration such that, on some neighbourhood of Y, the material isomorphisms $P_X(Y)$ can be chosen as a constant (independently of X in the neighbourhood). If the neighbourhood with this property can be extended to the whole interior of B, the body is (globally) homogeneous. A uniform elastic body which is not homogeneous is called inhomogeneous.

An important particular case of inhomogeneity is the so-called contorted aeolotropy [1], for which a global configuration exists such that the material isomorphisms between all pairs of points can be chosen as pure rotations. In this paper we restrict our analysis to this kind of inhomogeneity for solid crystals in two dimensions and we explore the question of whether more than one state of contorted aeolotropy can be found in such a uniform body. The notation is akin to that employed in [3,4].

1. Some geometrical results

Let a frame field K in two-dimensional Euclidean space E_2 be given by assigning smoothly at each point of E_2 a right-handed orthonormal basis e_I (I = 1,2) . A tensor field on E_2 is said to be K-constant (or K-parallel) if its components relative to the bases e_I are constant. Given a second order symmetric K-constant tensor field on E_2, under which conditions can it have a vanishing divergence or a vanishing curvature? The answer to these questions is provided, without proof, in the following two propositions.

PROPOSITION 1 : except in the trivial case of an Euclidean frame field, a second order symmetric K-constant tensor field in E_2 has a vanishing divergence if and only if it is spherical.

PROPOSITION 2 : the Riemann-Christoffel (curvature) tensor derived from a second order symmetric positive definite K-constant tensor G in E_2 vanishes identically if and only if

i) G is spherical; or

ii) the frame of eigenvectors of G is determined by a rotation angle w, relative to a cartesian coordinate system x,y , satisfying the PDE :

$$(\cos 2w)_{,xx} - (\cos 2w)_{,yy} + 2(\sin 2w)_{,xy} = 0 \quad .$$

A non-trivial example of a frame of the type prescribed in Proposition 2 is given by

$$w = \tan^{-1}(y/x) + \pi/4 \quad ,$$

whose integral curves are logarithmic spirals.

2. Discussion of contorted aeolotropy

In the case of solid crystals it is convenient to regard the uniformity maps P(X) as maps from a standard configuration of a reference crystal to the tangent spaces $T_X B$. For any chosen reference basis in the reference crystal the maps P(X) determine a basis field in a neighbourhood of each point (sometimes called a crystallographic basis). Assume that a solid crystal body is in a state of contorted aeolotropy, i.e., the material isomorphisms between all pairs of points in the given configuration are pure rotations. The uniformity maps P(X), however, will be rotations only if the reference crystal is appropriately chosen (as, say, one of the points of the body). In general, the uniformity maps will be given by

$$P(X) = R(X)\ H \ ,$$

where R(X) is a point dependent rotation tensor and H is a fixed tensor.

For a given configuration of a solid crystal body we define a Riemannian metric C by

$$G(X) = P^{T^{-1}}(X)\ P^{-1}(X) \ ,$$

where a T superscript denotes transposition. It follows from this definition that if a body is in a state of contorted aeolotropy its Riemannian metric G is of the form

$$G(X) = R(X)\ C\ R^T(X),$$

where C is the positive definite symmetric second order constant tensor

$$C = (\ H\ H^T\ )^{-1} \ .$$

This fact can be reinterpreted by stating that there exists a frame field

K such that the Riemannian metric G is K-constant and, therefore, the results of Propositions 1 and 2 can be used.

Consider, for instance, thè states of stress necessary to maintain a state of contorted aeolotropy of a uniform solid crystal body in two dimensions. In the absence of body forces the stress tensor T must satisfy the static equation

$$\operatorname{div} T = 0.$$

On the other hand, assuming an elastic constitutive law

$$T = T(F),$$

it follows that in a state of contorted aeolotropy the stress tensor is given by an expression of the form

$$T(X) = R(X)\, T_0\, R^T(X) \ ,$$

where T_0 is the stress at a fixed point (arbitrarily chosen). A straightforward application of Proposition 1 yields :

PROPOSITION 3 : except in the trivial case (homogeneity), the only states of stress compatible with a state of contorted aeolotropy in a two-dimensional solid crystal body in the absence of body forces are hydrostatic, i.e.

$$T = p\, I,$$

where p is a constant and I is the identity.

A deeper result can be obtained from Proposition 2 applied to the Riemannian metric defined above. This result is summarized in the following

PROPOSITION 4 : a uniform elastic body can be brought to a state of contorted aeolotropy if and only if the curvature of its Riemannian metric vanishes for some reference crystal (cf. NOLL [1]). For a solid crystal body in two dimensions there are just three possibilities :

i) there are no states of contorted aeolotropy (the curvature of the Riemannian metric cannot be made to vanish, no matter which reference crystal is used); or

ii) there exists essentially one state of contorted aeolotropy and its scalar multiples (for no orthonormal crystallographic basis the PDE of Proposition 2 is satisfied); or

iii) a whole class of states of contorted aeolotropy can be found. The strains of typical crystals in any two such states must differ by arbitrary stretches along two preferred orthogonal directions (which in each case satisfy the PDE of Proposition 2) .

The expression "scalar multiple" used above is short for states of strain that differ by a spherical dilatation. Thus ii) is a particular case of iii) when the two stretches are equal.

References

1. NOLL W., Materially uniform simple bodies with inhomogeneities, Arch. Rat. Mech. & Anal. (1967), 27, 1-32.
2. WANG C.-C., On the geometric structure of simple bodies, a mathematical foundation for the theory of continuous distributions of dislocations, Arch. Rat. Mech. & Anal. (1967), 27, 33-94.
3. COHEN H., EPSTEIN M., Remarks on uniformity in hyperelastic materials, Int. J. Solids & Structures (1984), 20, 233-243.
4. ELŻANOWSKI M., EPSTEIN M., Geometric characterization of hyperelastic uniformity, Arch. Rat. Mech. & Anal. (1985), 88, 347-357.

Acknowledgement : This work has been supported through a grant of the Natural Sciences and Engineering Research Council of Canada.

ON A GENERAL THEORY OF LARGE ROTATIONS AND SMALL STRAIN WITH APPLICATION TO THREE-DIMENSIONAL BEAM STRUCTURES

TH. HINKELMANN*, G. LUMPE** and H. ROTHERT*

* Institut für Statik
Universität Hannover
Callinstr. 32
D-3000 Hannover 1

** NOELL GmbH
Alfred-Nobel-Str. 20
D-8700 Würzburg 1

1. Introduction

For economical calculations of slender three-dimensional (3D) beam structures it becomes increasingly necessary to apply a more accurate method than the so-called 2nd order theory in structural engineering. To this kind of structures belong f.e. tower cranes, dredgers, masts and antennas. Because of the application of high-tensile steel (with yield stresses 800 - 1000 N/mm^2) very slender constructions may undergo large displacements and rotations without coming close to the yield limit at any point. Therefore the strain will also be limited by the order of 10^{-3}.

This paper wants to introduce a general theory of large rotations and small strain with application to complex three-dimensional beam structures which has been derived by the second-named author in [1]. Its efficiency will be proved by several examples.

2. Concept of the General Beam Theory

2.1 Preliminaries

For a scientifically founded beam theory a general derivation from continuum mechanics of elastic bodies is required. However, applying the complete 3D theory the effort of computation will be immense and become uneconomical. Therefore consistent approximations and error estimations are necessary for the simplification of the governing equations.

With reference to Fig. 1 the formulation of the elastic body motion is accomplished by LAGRANGEan coordinates. All functions are given in material coordinates orientated at the fixed cartesian base vector system $\overset{R}{\tilde{G}}_i$. The configuration of the elastic body is described by the convected coordinate system $\overset{o}{\tilde{G}}_i$ at the initial state and by $\tilde{g}_i$ at a current state.

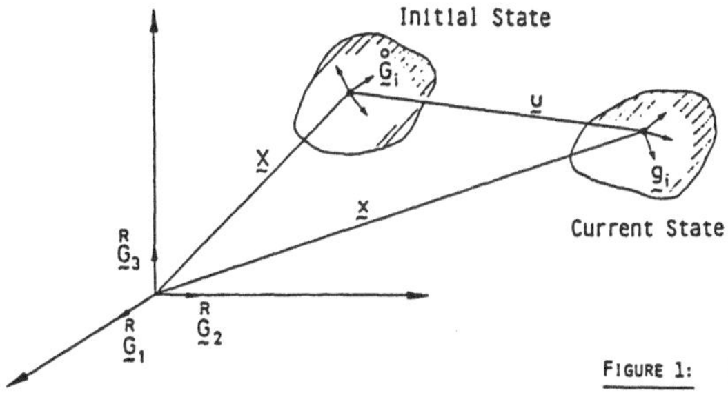

FIGURE 1: Reference Systems

2.2 Assumptions

The essential assumptions of this beam theory are:

- straight prismatic beam with constant cross section,
- cross section dimensions versus element length $D/L \lesssim 10^{-1}$,
- linear elastic, homogeneous and isotropic material, largest principal extension $\varepsilon \lesssim 10^{-3}$,
- cross section under deformation keeps its shape (exception: torsion),
- stress at any point of the cross-section is limited by the value of the yield stress.

The reduction of the general 3D continuum to the special one-dimensional continuum of a beam can be carried out within the following three steps:

2.3 Kinematic equation

The first step is to constitute the kinematic equation. Looking at Fig. 2 with $\tilde{\tilde{g}}_i$ as the strained base system of the current state in the center of gravity one finds the total displacement $\tilde{u}$ of any point P described by the convective coordinates θ^i consisting of the displacement

- of the centroidal axis $\tilde{\tilde{u}}(\theta^3)$,
- caused by the rotation of the reference cross section $\theta^\alpha(\tilde{\tilde{g}}_\alpha - \overset{o}{\tilde{\tilde{G}}}_\alpha)$,

- due to warping of the cross section $\underset{\sim}{u}_{\omega}$.

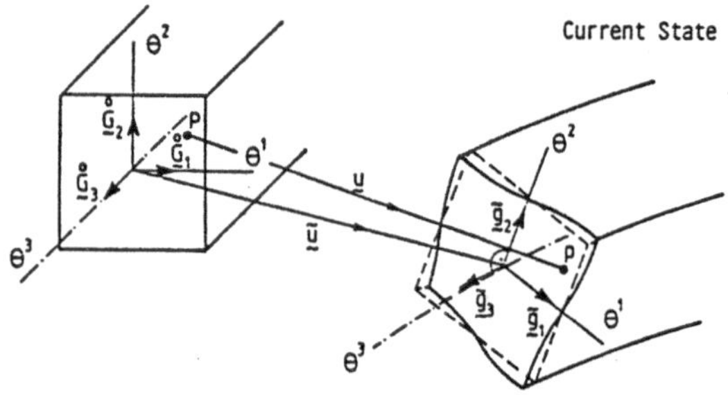

$$\underline{u}(\Theta^1,\Theta^2,\Theta^3) = \bar{\underline{u}}(\Theta^3) + \Theta^\alpha(\bar{\underline{g}}_\alpha - \mathring{\underline{G}}_\alpha) + \Omega(\Theta^1,\Theta^2)\,\underset{\sim}{\kappa}^3(\Theta^3)$$

FIGURE 2: KINEMATIC RELATION

Concerning the warping $\underset{\sim}{u}_{\omega} = \Omega(\theta^1,\theta^2)\ \underset{\sim}{K}^3\ (\theta^3)$ it can be specified that
- $\underset{\sim}{K}^3$ is the DARBOUX rotation vector which specifies the relative rotation respectively the twist of two neighboring cross sections and that
- $\Omega(\theta^1,\theta^2)$ is the unity warping referred to $K^3 = 1$.

2.4 Approximations of stress and strain measures

In order to deal with approximations and error estimations it is important to know the order of magnitude of the rotation of the cross section. Regarding Fig. 3 the rotation of the cross section caused by a bending moment can be easily determined. If the stress of the edge fiber has the value of the yield stress σ_F which correlates with the normal strain ε_F, one gets for an increment dx

$$d\varphi = \frac{2\varepsilon_F}{D}\,dx$$

and for a beam element with the length L

$$\varphi max = 2\varepsilon_F\ L/D.$$

Applying LANDAU's symbol 0 the order of magnitude of the rotation is

$O(\varepsilon_F L/D)$.

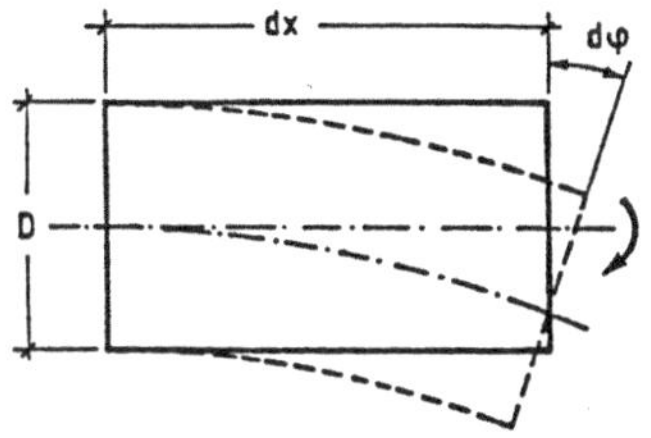

FIGURE 3:

Cross Section Rotation Caused by a Bending Moment

In the same way the rotations caused by torsion and shear can be computated. Since the bending moment and shear are generally coupled this rotation angle consists of two parts. It has to be pointed out that the shear deformation depending on support, element length and element structure cannot generally be neglected.

2.5 Separation of rigid body motion and actual deformation

It may be quite understandable that the strain energy does not depend on the rigid body modes (Fig. 4, angle Ψ) but only on the relative modes (Fig. 4, $\Delta\varphi$). This means that both the beam elements shown in Fig. 4 possess exactly the same strain energy.

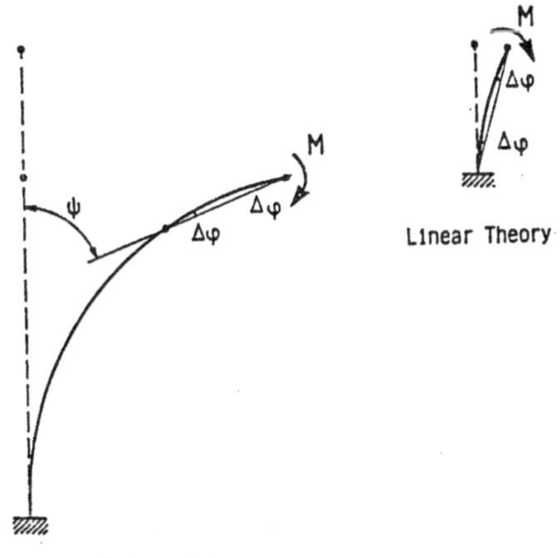

FIGURE 4: Rigid Body- and Relative Rotations

By choosing a properly limited element length the relative mode $\Delta\varphi$ (Fig. 5) will be reduced in such a way that the strain energy is a quadratic function of the relative modes. After having specified a set of

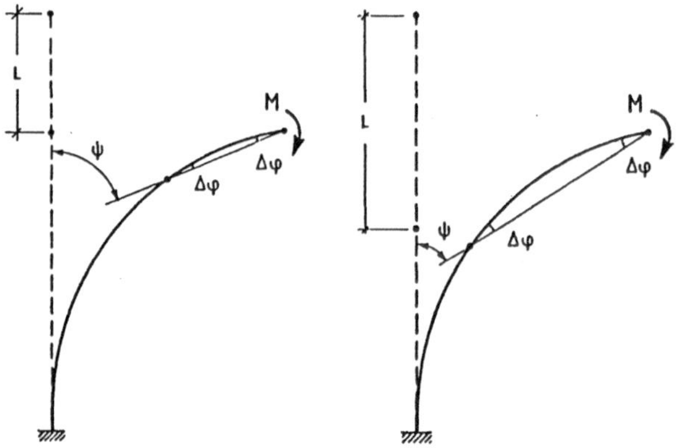

FIGURE 5: Dependence Between Relative Rotation $\Delta\varphi$ and Element Length L

functions for the displacement and rotation modes, on the one hand side an exact integration is possible and on the other hand the relations between element forces and relative modes are of a very simple character. Therefore any hinge connection can be taken into account.

The twelve kinematic element modes (Fig. 6) can be subdivided into six rigid body modes (here: $\underset{\sim}{\underline{v}}_s = \{\tilde{u}_a^1, \tilde{u}_a^2, \tilde{u}_a^3, \Psi^1, \Psi^2, \Psi^3\}$ which are the spatial displacement $\underset{\sim}{u}_a$ and the spatial angle of rotation $\underset{\sim}{\Psi}$) and into six so-called relative modes (here: $\Delta\underline{v} = \{\Delta\varphi_a^1, \Delta\varphi_a^2, \Delta\varphi_a^3, \Delta\varphi_b^1, \Delta\varphi_b^2, \Delta\varphi_b^3\}$). This form of separation has been well proved by comparative computations. The consistency between rigid body and relative modes has to be exactly determined and described by tensor notation.

One has to observe that the components of the rigid body rotation can be described by vectors but that they do not conform to the rule of commutativity. By the way, a tensorial formulation of the spatial rotation is absolutely necessary. LUMPE [1] expresses the large spatial angle of rotation $\underset{\sim}{\Psi}$ by EULERian angles and the spatial rotation of the cross section by an orthogonal tensor as a rotation about one skew axis in space.

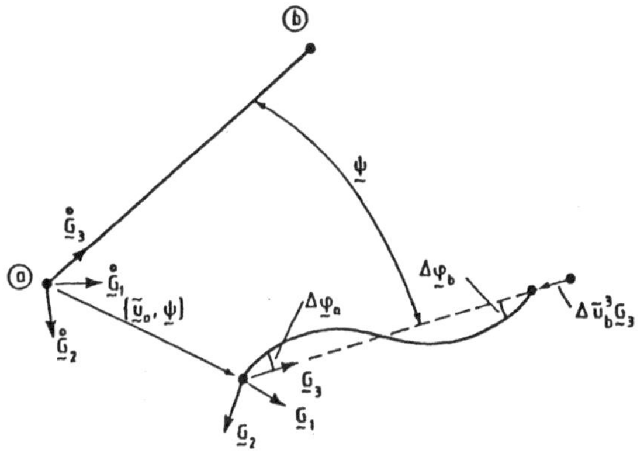

FIGURE 6: Definition of Rigid Body- and Relative Modes

3. Examples

The efficiency of the beam theory shall now be demonstrated by several examples. The following load-displacement diagrams belong to structures with moderate or large displacements and rotations.

3.1 COX-Problem (Fig. 7, 8)

The COX-beam model has the same buckling behaviour as an axially compressed cylindrical shell. The model consists of a vertical beam with a geometric imperfection e. This beam is supported in its center by two inclined bars in a plane which is normal to the plane of beam imperfection. The height of the slope is d, the load F is applied at point ⓐ Fig. 7 shows the load F versus the displacement w of point ⓑ in y-direction and Fig. 8 gives the load F versus the displacement u of point ⓐ in x-direction. In both cases the critical load F_{Kri} is the buckling load. And the solution of each loading state is found after three to four iterations.

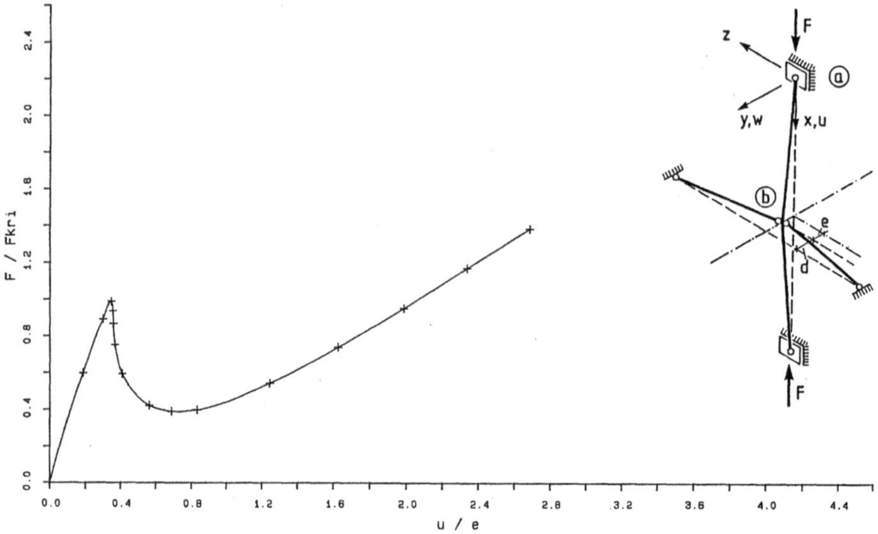

UNIVERSITAET HANNOVER INSTITUT FUER STATIK PROF. DR.-ING. H. ROTHERT	LOAD-DISPLACEMENT DIAGRAM OF C O X - PROBLEM e/d = 2/20	FIGURE 7

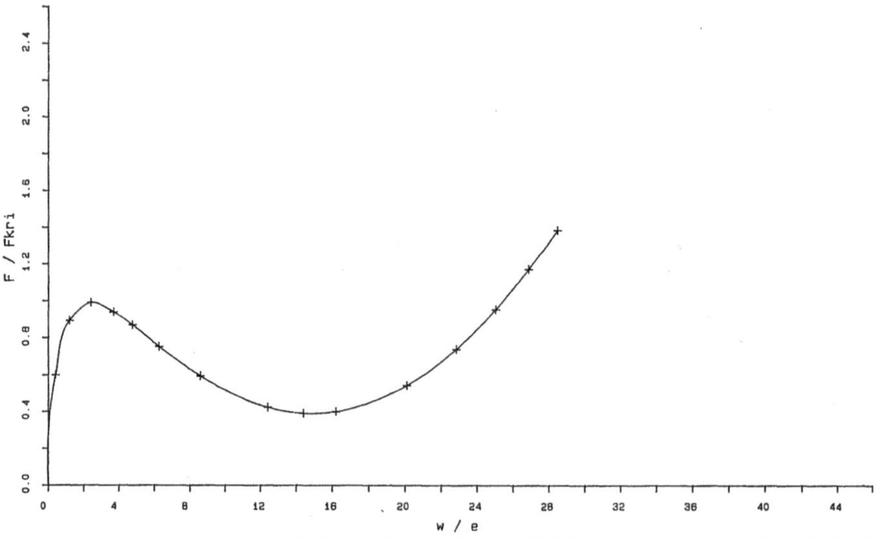

UNIVERSITAET HANNOVER INSTITUT FUER STATIK PROF. DR.-ING. H. ROTHERT	LOAD-DISPLACEMENT DIAGRAM OF C O X - PROBLEM e/d = 2/20	FIGURE 8

3.2 v.MISES-Truss (Fig. 9)

The next example is the well-known snap-through problem of a v.MISES-truss with an angle of three degrees subjected to a vertical load. On the ordinate one finds the ratio of actual load to the critical snap-through load, on the abscissa one has the displacement w of point ⓐ with respect to the heigth h of the structure. With three to four equilibration iterations the solution of each loading state can be found very quickly.

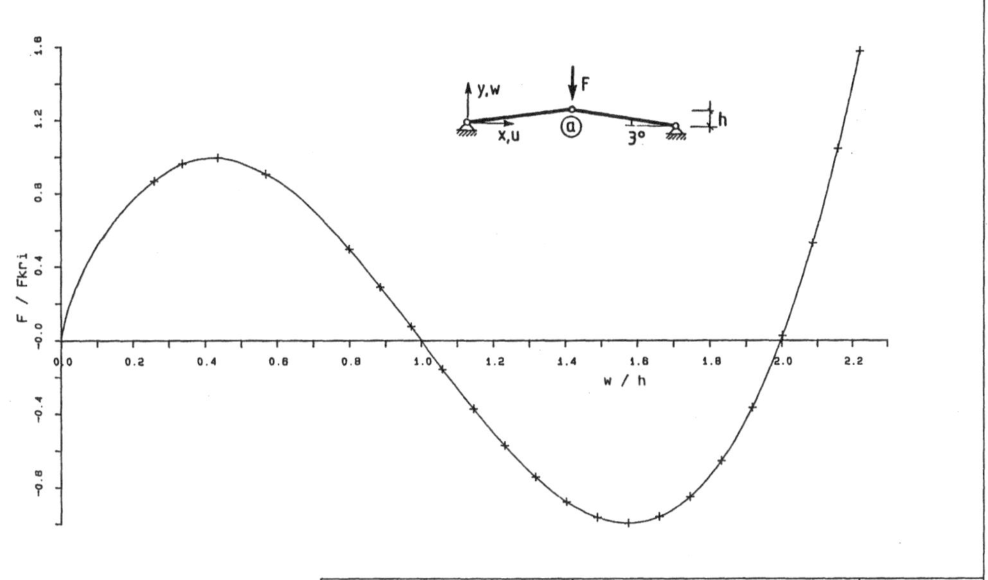

UNIVERSITAET HANNOVER INSTITUT FUER STATIK PROF. DR.-ING. H. ROTHERT	LOAD-DISPLACEMENT DIAGRAM OF A M I S E S - TRUSS BY VERTICAL LOAD	FIGURE 9

3.3 Lateral buckling of an angle girder (Fig. 10)

The angle girder shown in Fig. 10 with a width b = 30 mm and the thickness t = 0,6 mm is loaded by forces F_x and F_z. The calculation has been carried out with 10 elements and a relation of $F_z/F_x = 10^{-3}$. In this diagram one finds the displacements in z-direction of points ⓐ and ⓑ. The displacement w(b) given in the right lower curve describes a well-known lateral buckling behaviour. The upper curve has been calculated by ARGYRIS [2] applying 20 elements. The displacement w(a) has a quite interesting pattern. First one notes an also positive deflection caused by F_z; then the displacement w(a) decreases and reaches a negativ value caused by the twist of the horizontal angle-side. Finally caused by the lateral buckling of the angle-tip also the corner gets large displacements.

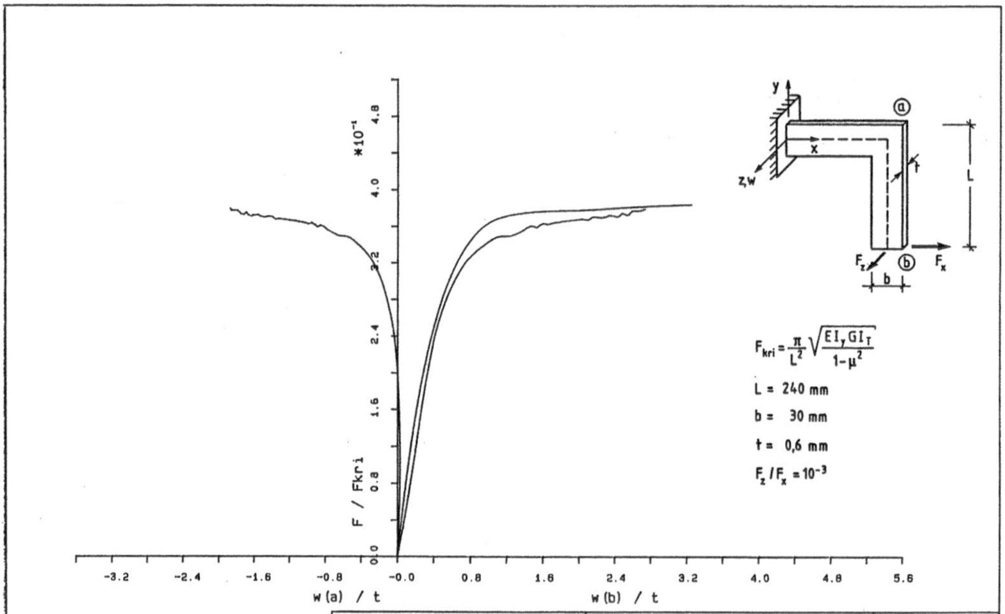

UNIVERSITAET HANNOVER INSTITUT FUER STATIK PROF. DR.-ING. H. ROTHERT	LOAD-DISPLACEMENT DIAGRAM OF A LATERAL BUCKLING ANGLE GIRDER	FIGURE 10

3.4 Mobil Rotary Tower Crane (Fig. 11)

Fig. 11 gives the side view of a mobil rotary tower crane with a maximum height of 115 m. At first the characteristics of its bearing behaviour should be explained. By vertical forces as dead- and weight-lifting-load a bending moment is caused which correlates with very great normal forces in the tower shaft and the cables. At the same time the outrigger can be loaded by side forces due to wind and rotational acceleration which cause a strong twist of the tower shaft. Therefore in a realistic computation the coupling of normal forces, bending and torsion in the tower shaft has to be taken into account.

An efficient and economical calculation of such a complex space structure (here: about 1500 bars and 700 nodes with 6 DOF) requires an idealisation of the outrigger and the shaft by makro-elements. As these elements have truss character the influence of shear deformation must be considered. The discretisation of the crane by an FE-model consisting 23 makro-elements is given in Fig. 12. The three displacement components of the outrigger tip (a) as a function of the load factor K are mapped in Fig.13.

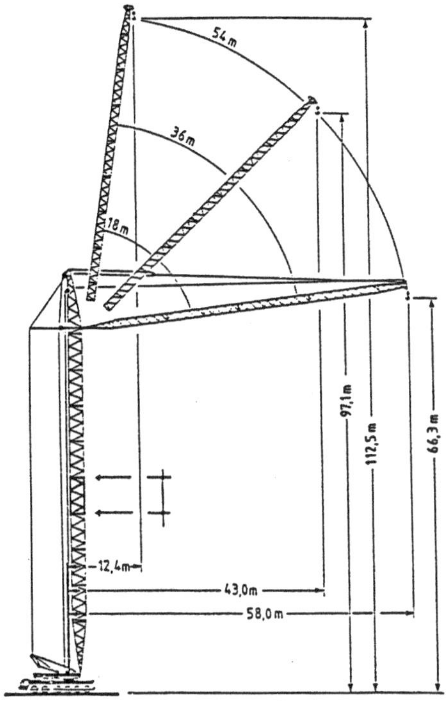

FIGURE 11:
Rotary Tower Crane

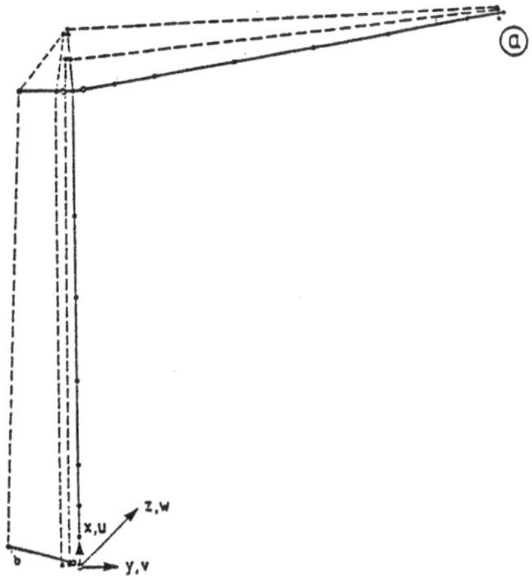

FIGURE 12:
FE - Model of a Rotary Tower Crane with Makro - Elements

One has to consider that the scale of the lateral displacement w_a is ten times larger than that of u_a and v_a. The load case weight-lifting load F_{xa} = -80 kN and side load F_{za} = 2,7 kN demonstrates quite well the influence of torsion of the tower shaft. The lateral displacement w_a amounts to about 0,50 m at the load state K = 1,6 and is more than five times the vertical displacement u_a and nearly thirty times the horizontal displacement v_a. The reverse behaviour of v_a is caused by the large rotation about the tower axis.

In addition it has to be mentioned that by calculating this crane according to the German standard DIN 15018 the lateral displacement w_a amounts to 15 m. This means that 70 to 80 % of this value result from the twist of the tower shaft.

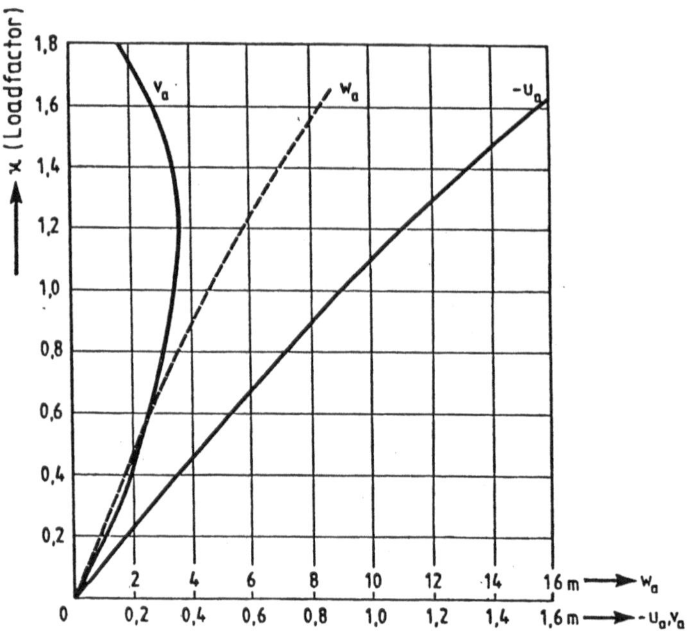

FIGURE 13: Weight - Lifting Load F_{xa} = -80,0 kN; Side Load F_{za} = 2,7 kN; by $\varkappa = 1{,}0$

4. Conclusions

It has been tried to give a general view on a 3D beam theory which has been successfully modelled by finite elements including large displacements and rotations; however, small strain. The computer program, which has been developed allows economical calculations even of complex and very large structures. By the separation of the rigid body motion and by an accurate formulation of the large spatial rotation combined with consistent error estimations all geometrically nonlinear effects can be taken into account to any desired accuracy. The magnitude of error is known as a function of the element length ($\Delta = O(L/D)$ with $\varepsilon = 10^{-3}$ and $D/L = 10^{-1} \rightarrow \Delta = O(10^{-2})$).

Additionally it may be mentioned that the FE program enables to consider

- any hinge fitting combinations,
- shear deformation,
- hindered torsion,
- cable-roll-connections,
- elastic supports,
- displacements of supports,
- support in any direction,

and to determine load-displacement diagrams up to buckling and post-buckling loads.

5. Literature

[1] LUMPE, G.: Geometrisch nichtlineare Berechnung von räumlichen Stabwerken. Mitt.d.Inst.f.Statik der Universität Hannover, Mitt.Nr. 28, 1982.

[2] ARGYRIS, J.H.; DUNNE, P.C.; MALEJANAKIS, G.A.; SCHARPF, D.W.: On Large Displacement - Small Strain Analysis of Structures with Rotational Degrees of Freedom. Comp.Meth.Appl.Mech.Eng. 14 (1978) 401; 15 (1978) 99.

FINITE DISPLACEMENT THEORY OF NATURALLY CURVED AND TWISTED BEAMS WITH FINITE ROTATIONS

M. IURA
Tokyo Denki University
Hatoyama, Hiki-gun
Saitama 350-03, Japan

1. Introduction

A finite rotation is not a quantity in a vector space. Various approaches, therefore, exist for evaluating the finite rotation. Euler angles [1-4], finite rotation tensors [5] and finite rotation vectors [6-8] have been employed as a measure for finite rotations. The exact finite displacement functions of beams have been derived in terms of the three translation and three rotation parameters. It is widely accepted, however, that four parameters are necessary and sufficient for formulating a beam theory under the Bernoulli-Euler hypotheses. When the number of parameters decreases from six to four, some approximations are introduced in the most of existing literature. Consequently, the most of existing equations are available only for the analyis of geometrically nonlinear behavior of beams with moderate rotations.

In general, there is a difference between the two angles which the deformed principal axes make with the undeformed ones. If the displacements of beams are assumed to be small, this difference can be neglected. In case of a small displacement theory, therefore, the fourth parameter is defined as the angle of rotation of cross sections. When the displacements of beams become finite, this difference can not be neglected. As a result, in case of a finite displacement theory, the fourth parameter associated with the rotation of cross sections has been defined in a various form.

In this paper, a new approach is introduced to investigate the coupling of finite rotations of beams under the Bernoulli-Euler hypotheses. As a measure for finite rotations, we introduce the finite rotation vector $\underset{\sim}{\theta}$. When we employ the Euler angles or finite rotation vector $\underset{\sim}{\omega}$ to derive the finite displacement field, we must pay attention to the order of rotations. However, in this paper, the exact finite displacement field is obtained without taking into account the order of rotations. With the use of a new parameter introduced herein, the displacement field and the twist and the curvatures after the deformation are derived without restricting the magnitude of strains, displacements or rotations. When we utilize the stress tensors, the rigorous treatment is available for deriving the equilibrium equations. It is found, however, that as far as the physical components of stress tensors are utilized, small-strain assumptions are necessary for deriving the equilibrium equations from the principle of virtual work. In the existing equilibrium equations [3-7,9-10], the third and higher order terms with respect to displacement components are neglected. While the present equilibrium equations and the associated boundary conditions are obtained without restricting the magnitude of displacements or rotation angles.

The fundamental hypotheses introduced here are itemized as follows:

(1) The cross section of beams is constant along the beam axis.

(2) The beam axis before the deformation is a smooth space curve.

(3) The transverse plane is assumed to remain plane and normal to the beam axis throughout the deformation; that is, the Bernoulli-Euler hypotheses hold and the warping of cross section is neglected.

Throughout this paper, the summation rule is adopted and Latain indices will have the range 1,2,3.

2. Preliminaries

Consider a naturally curved and twisted beam as shown in Fig.1. Two convected coordinate systems are introduced to describe the motion of the beam. One of these is the orthogonal curvilinear coordinate ξ^m with the base vectors $\underset{\sim}{e}_m$. The coordinate ξ^1 is taken along the beam axis. The coordinates ξ^2 and ξ^3 are taken along principal axes of a cross section. The other is the local Cartesian coordinates Z^m with the base vectors $\underset{\sim}{b}_m$. The coordinates Z^m are chosen so that the base vectors $\underset{\sim}{b}_m$ of Z^m coincide with the base vectors $\underset{\sim}{e}_m$ of ξ^m.

With the aid of the Frenet-Serret formulae, the following differential formula for the base vectors $\underset{\sim}{e}_m$ holds:

$$\underset{\sim}{e}_{m,1}=\underset{\sim}{\kappa}\times\underset{\sim}{e}_m,\qquad \underset{\sim}{\kappa}=\kappa_m\underset{\sim}{e}_m, \tag{1.a,b}$$

where κ_1 is the initial twist, and κ_2 and κ_3 are the components of the initial curvature. The base vectors at an arbitrary material point are obtained by differentiating the position vector, $\underset{\sim}{r}=\underset{\sim}{x}+\xi^2\underset{\sim}{e}_2+\xi^3\underset{\sim}{e}_3$, with respect to ξ^m,

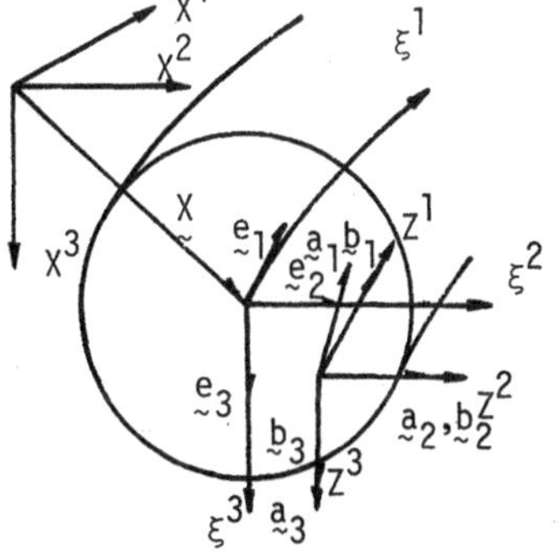

Fig.1 Coordinate Syttems

$$\underset{\sim}{a}_1=\sqrt{g}\,\underset{\sim}{e}_1-\kappa_1\xi^3\underset{\sim}{e}_2+\kappa_1\xi^2\underset{\sim}{e}_3,\quad \underset{\sim}{a}_2=\underset{\sim}{e}_2,\quad \underset{\sim}{a}_3=\underset{\sim}{e}_3, \tag{2.a-c}$$

where $\sqrt{g}=1+\kappa_2\xi^3-\kappa_3\xi^2$.

3. Finite Rotation Vector

According to the hypothesis (3), the orthonormal triad $\mathbf{e}_m$ is transformed to the orthonormal triad $\bar{\mathbf{e}}_m$ during the deformation. This transformation is, generally, accompanied with finite rotations. In this paper, the finite rotation vector $\theta=2\tan\frac{1}{2}\omega\,\phi$, where ϕ is the unit vector, the direction of which coincides with that of rotation axis, and ω the angle of rotation about its axis, is introduced as a measure for finite rotations. With the use of θ, we have the following relationship between $\mathbf{e}_m$ and $\bar{\mathbf{e}}_m$ [11]:

$$\bar{\mathbf{e}}_m=\mathbf{e}_m+\tfrac{1}{2}(1+\cos\omega)\theta\times\mathbf{e}_m+\tfrac{1}{4}(1+\cos\omega)\theta\times(\theta\times\mathbf{e}_m). \tag{3}$$

The components of the finite rotation vector θ, expressed by $\theta^m=\theta\cdot\mathbf{e}^m$, are not independent since the following relationship holds:

$$(\theta^1)^2+(\theta^2)^2+(\theta^3)^2=4(1-\cos\omega)/(1+\cos\omega). \tag{4}$$

Using the rotation parameters θ^i, we have

$$\bar{\mathbf{e}}_m=R_m{}^n\mathbf{e}_n, \tag{5}$$

where

$$\begin{aligned}
&R_1{}^1=1-\tfrac{1}{4}(1+\cos\omega)[(\theta^2)^2+(\theta^3)^2], && R_1{}^2=\tfrac{1}{2}(1+\cos\omega)(\theta^3+\tfrac{1}{2}\theta^1\theta^2),\\
&R_1{}^3=\tfrac{1}{2}(1+\cos\omega)(-\theta^2+\tfrac{1}{2}\theta^3\theta^1), && R_2{}^1=\tfrac{1}{2}(1+\cos\omega)(-\theta^3+\tfrac{1}{2}\theta^1\theta^2),\\
&R_2{}^2=1-\tfrac{1}{4}(1+\cos\omega)[(\theta^3)^2+(\theta^1)^2], && R_2{}^3=\tfrac{1}{2}(1+\cos\omega)(\theta^1+\tfrac{1}{2}\theta^2\theta^3),\\
&R_3{}^1=\tfrac{1}{2}(1+\cos\omega)(\theta^2+\tfrac{1}{2}\theta^3\theta^1), && R_3{}^2=\tfrac{1}{2}(1+\cos\omega)(-\theta^1+\tfrac{1}{2}\theta^2\theta^3),\\
&R_3{}^3=1-\tfrac{1}{4}(1+\cos\omega)[(\theta^1)^2+(\theta^2)^2].
\end{aligned} \tag{6.a-i}$$

To confirm the difference between the two angles which the deformed principal axes make with the undeformed ones, we take the dot product of $\mathbf{e}_2$ and $\mathbf{e}_3$ into both sides of $\bar{\mathbf{e}}_2$ and $\bar{\mathbf{e}}_3$ in Eqs.(5) respectively, and so obtain

$$\mathbf{e}_2\cdot\bar{\mathbf{e}}_2=1-\tfrac{1}{4}(1+\cos\omega)[(\theta^3)^2+(\theta^1)^2],\quad \mathbf{e}_3\cdot\bar{\mathbf{e}}_3=1-\tfrac{1}{4}(1+\cos\omega)[(\theta^1)^2+(\theta^2)^2]. \tag{7.a,b}$$

It concludes from Eqs.(7) that $\mathbf{e}_2\cdot\bar{\mathbf{e}}_2\neq\mathbf{e}_3\cdot\bar{\mathbf{e}}_3$ except for the following especial case: $\theta^2=\theta^3$.

4. Finite Displacement Field

According to the hypothesis (3), the displacement vector at an arbitrary material point, $\underset{\sim}{\mathbf{U}}=U^m \underset{\sim}{\mathbf{e}}_m^{\prime}$, is given by

$$\underset{\sim}{\mathbf{U}}=\underset{\sim}{\mathbf{u}}+\xi^2(\underset{\sim}{\bar{\mathbf{e}}}_2-\underset{\sim}{\mathbf{e}}_2)+\xi^3(\underset{\sim}{\bar{\mathbf{e}}}_3-\underset{\sim}{\mathbf{e}}_3), \tag{8}$$

where $\underset{\sim}{\mathbf{u}}=u^m \underset{\sim}{\mathbf{e}}_m$ is the displacement vector at the beam axis. Three translation parameters u^m and three rotation parameters θ^m are used to express the displacement vector $\underset{\sim}{\mathbf{U}}$ in Eq.(8). It is well known that four parameters are necessary and sufficient for describing the mechanical behavior of beams under the Bernoulli-Euler hypotheses. Therefore we must consider the way to decrease the number of parameters from six to four.

Since the tangent base vector to the beam axis after the deformation is obtained by differentiating the deformed position vector with respect to ξ^1, another expression for $\underset{\sim}{\bar{\mathbf{e}}}_1$ is given by

$$\underset{\sim}{\bar{\mathbf{e}}}_1=(\underset{\sim}{\mathbf{x}}+\underset{\sim}{\mathbf{u}})_{,1}/|(\underset{\sim}{\mathbf{x}}+\underset{\sim}{\mathbf{u}})_{,1}|. \tag{9}$$

By expressing the vector $\underset{\sim}{\bar{\mathbf{e}}}_1$ into the component form, we have

$$\underset{\sim}{\bar{\mathbf{e}}}_1=u^m\|_1 \underset{\sim}{\mathbf{e}}_m,\quad u^1\|_1=(1+u^1|_1)/\sqrt{G_0},\quad u^2\|_1=u^2|_1/\sqrt{G_0},\quad u^3\|_1=u^3|_1/\sqrt{G_0}, \tag{10.a-d}$$

where $(\ )|_1$ denotes the covariant differentiation with respect to the metric tensor $e_{ij}=\underset{\sim}{\mathbf{e}}_i\cdot\underset{\sim}{\mathbf{e}}_j$ and

$$G_0=(1+u^1|_1)^2+(u^2|_1)^2+(u^3|_1)^2. \tag{11}$$

Comparison of the components of the base vector $\underset{\sim}{\bar{\mathbf{e}}}_1$ between Eqs.(5) and (10.a) leads to

$$1-\tfrac{1}{4}(1+\cos\omega)[(\theta^2)^2+(\theta^3)^2]=u^1\|_1,\quad \tfrac{1}{2}(1+\cos\omega)(\theta^3+\tfrac{1}{2}\theta^1\theta^2)=u^2\|_1,$$
$$\tfrac{1}{2}(1+\cos\omega)(-\theta^2+\tfrac{1}{2}\theta^3\theta^1)=u^3\|_1. \tag{12.a-c}$$

If the translation parameters u^i are regarded as known variables, the expressions (12) become the equations for the unknowns θ^i. With the use of Eq.(4), it is easy to verify that there exist only two independent expressions in Eqs.(12). Therefore, because of redundancy of unknowns, it is impossible to express the rotation parameters θ^i by means of u^i. This fact indicates that four parameters are necessary and sufficient for describing the mechanical behavior of beams under the Bernoulli-

Euler hypotheses. It seems natural, herein, to introduce the new variable ψ, as the fourth parameter, defined as

$$\theta^1=2\tan\tfrac{1}{2}\psi. \tag{13}$$

The physical meaning of ψ will be discussed later.

From Eqs.(4), (12) and (13), we have

$$\theta^2=\lambda[u^2{}_{\|1}\sin\psi/(1+\cos\psi)-u^3{}_{\|1}],\quad \theta^3=\lambda[u^2{}_{\|1}+u^3{}_{\|1}\sin\psi/(1+\cos\psi)], \tag{14.a,b}$$

where $\lambda=2/(1+u^1{}_{\|1})$. By substituting Eqs.(14) into Eqs.(5) and (6), we have the following expressions for the base vectors $\bar{\mathbf{e}}_2$ and $\bar{\mathbf{e}}_3$ in terms of four parameters:

$$\begin{aligned}\bar{\mathbf{e}}_2=&-(u^2{}_{\|1}\cos\psi+u^3{}_{\|1}\sin\psi)\mathbf{e}_1+[\{1-\tfrac{1}{2}\lambda(u^2{}_{\|1})^2\}\cos\psi-\tfrac{1}{2}\lambda u^2{}_{\|1}u^3{}_{\|1}\sin\psi]\mathbf{e}_2\\ &+[\{1-\tfrac{1}{2}\lambda(u^3{}_{\|1})^2\}\sin\psi-\tfrac{1}{2}\lambda u^2{}_{\|1}u^3{}_{\|1}\cos\psi]\mathbf{e}_3,\\ \bar{\mathbf{e}}_3=&(u^2{}_{\|1}\sin\psi-u^3{}_{\|1}\cos\psi)\mathbf{e}_1+[\{-1+\tfrac{1}{2}\lambda(u^2{}_{\|1})^2\}\sin\psi-\tfrac{1}{2}\lambda u^2{}_{\|1}u^3{}_{\|1}\cos\psi]\mathbf{e}_2\\ &+[\{1-\tfrac{1}{2}\lambda(u^3{}_{\|1})^2\}\cos\psi+\tfrac{1}{2}\lambda u^2{}_{\|1}u^3{}_{\|1}\sin\psi]\mathbf{e}_3.\end{aligned} \tag{15.a,b}$$

When Eqs.(15) are introduced into Eq.(8), the finite displacement field is expressed in terms of the four parameters, u_i and ψ.

We take the dot products of $\mathbf{e}_2$ and $\mathbf{e}_3$ into both sides of Eqs.(15.a) and (15.b) respectively, and so obtain

$$\begin{aligned}\mathbf{e}_2\cdot\bar{\mathbf{e}}_2&=[1-\tfrac{1}{2}\lambda(u^2{}_{\|1})^2]\cos\psi-\tfrac{1}{2}\lambda u^2{}_{\|1}u^3{}_{\|1}\sin\psi,\\ \mathbf{e}_3\cdot\bar{\mathbf{e}}_3&=[1-\tfrac{1}{2}\lambda(u^3{}_{\|1})^2]\cos\psi+\tfrac{1}{2}\lambda u^2{}_{\|1}u^3{}_{\|1}\sin\psi.\end{aligned} \tag{16.a,b}$$

With the aid of Eqs.(10),(11) and (16), we obtain the following relationship:

$$\cos\psi=\tfrac{1}{2}\lambda(\mathbf{e}_2\cdot\bar{\mathbf{e}}_2+\mathbf{e}_3\cdot\bar{\mathbf{e}}_3). \tag{17}$$

The expression (17) indicates the physical meaning of the present fourth parameter ψ.

In accordance with Eqs.(1), we introduce the vector $\bar{\kappa}$ satisfying the following differential formula:

$$\bar{\mathbf{e}}_{m,1}=\bar{\kappa}\times\bar{\mathbf{e}}_m,\qquad \bar{\kappa}=\bar{\kappa}_m\mathbf{e}_m. \tag{18.a,b}$$

The notations $\bar{\kappa}_m$ are not exactly the twist and curvatures after the deformation since the deformed line is extended. Under the small strain assumptions, however, $\bar{\kappa}_1$ denotes the twist after the deformation, and

$\bar{\kappa}_2$ and $\bar{\kappa}_3$ denote the components of the curvature after the deformation. The explicit forms of $\bar{\kappa}_m$ are given in [10].

The base vectors at an arbitrary material point after the deformation are given by differentiating the position vector at an arbitrary material point with respect to ξ^m as

$$\bar{\mathbf{a}}_1=(\sqrt{G_0}-\bar{\kappa}_3\xi^2+\bar{\kappa}_2\xi^3)\bar{\mathbf{e}}_1-\bar{\kappa}_1\xi^3\bar{\mathbf{e}}_2+\bar{\kappa}_1\xi^2\bar{\mathbf{e}}_3,\quad \bar{\mathbf{a}}_2=\bar{\mathbf{e}}_2,\quad \bar{\mathbf{a}}_3=\bar{\mathbf{e}}_3. \tag{19}$$

5. Internal and External Virtual Works

The Green strain tensors associated with the directions of the ξ^i lines are given by

$$\gamma_{ij}=\tfrac{1}{2}(\bar{\mathbf{a}}_i\cdot\bar{\mathbf{a}}_j-\mathbf{a}_i\ \mathbf{a}_j). \tag{20}$$

With the use of the second Piola-Kirchhoff stress tensors, τ^{ij}, associated with the directions of the ξ^i lines, the internal virtual work is written as [12]

$$IVW=\int_V(\tau^{11}\delta\gamma_{11}+2\tau^{12}\delta\gamma_{12}+2\tau^{13}\delta\gamma_{13})dV, \tag{21}$$

where $dV=\sqrt{g}\ d\xi^1 d\xi^2 d\xi^3$.

Since the base vectors $\mathbf{a}_i$ are not orthogonal, the constitutive equations may become complicated. Therefore, for the latter convenience, we introduce the strain tensors ε_{ij} and the stress tensors σ^{ij} associated with the coordinates Z^i. The relationship between γ_{ij} and ε_{ij} is given by [12]

$$\varepsilon_{ij}=\partial\xi^m/\partial z^i\cdot\partial\xi^n/\partial z^j\cdot\gamma_{mn}. \tag{22}$$

Introducing Eqs.(19) and (20) into Eq.(22) leads to

$$\begin{aligned}&\varepsilon_{11}=(\varepsilon_0-\xi^2\kappa_3^*+\xi^3\kappa_2^*)[1/\sqrt{g}+(\varepsilon_0-\xi^2\kappa_3^*+\xi^3\kappa_2^*)/(2g)]+\rho^2(\kappa_1^*)^2/(2g),\\ &\varepsilon_{12}=\varepsilon_{21}=-\xi^3\kappa_1^*/(2\sqrt{g}),\qquad \varepsilon_{13}=\varepsilon_{31}=\xi^2\kappa_1^*/(2\sqrt{g}),\\ &\varepsilon_{22}=\varepsilon_{23}=\varepsilon_{32}=\varepsilon_{33}=0,\quad \varepsilon_0=\sqrt{G_0}-1,\quad \kappa_i^*=\bar{\kappa}_i-\kappa_i.\end{aligned} \tag{23,a-f}$$

where $\rho^2=(\xi^2)^2+(\xi^3)^2$. The relationship between τ^{ij} and σ^{ij} is given by [12]

$$\sigma^{ij}=\partial z^i/\partial\xi^m\cdot\partial z^j/\partial\xi^n\cdot\tau^{mn}. \tag{24}$$

From the practical point of view, it is necessary to distinguish the physical components from the tensor ones. When the stress resultants and moments are defined, the physical components of stress tensors are utilized in [9,10]. This is because we are not concerned with the tensors but with the physical components of tensors. The physical components of stress tensors σ^{ij} are given by [9]

$$\tilde{\sigma}^{ij}=\sigma^{ij}\sqrt{\bar{b}_{ii}\bar{b}_{jj}} \qquad (\text{ i,j not summed }), \tag{25}$$

where $b_{ij}=\mathbf{b}_i\cdot\mathbf{b}_j$ and $\bar{b}_{ij}=\bar{\mathbf{b}}_i\cdot\bar{\mathbf{b}}_j$ are metric tensors and

$$\bar{\mathbf{b}}_1=(\sqrt{g}+\varepsilon_0-\xi^2\kappa_3^*+\xi^3\kappa_2^*)\bar{\mathbf{e}}_1/\sqrt{g},\quad \bar{\mathbf{b}}_2=\bar{\mathbf{e}}_2,\quad \bar{\mathbf{b}}_3=\bar{\mathbf{e}}_3. \tag{26.a-c}$$

Substituting Eqs.(22),(23),(24) and (25) into Eq.(21) leads to

$$\begin{aligned}IVW=\int_V[\tilde{\sigma}^{11}/\sqrt{g}\ \{\delta\varepsilon_0-\xi^2\delta\kappa_3^*+\xi^3\delta\kappa_2^*+\underline{\rho^2\kappa_1^*\delta\kappa_1^*/(\sqrt{g}+\varepsilon_0-\xi^2\kappa_3^*+\xi^3\kappa_2^*)}\}\\ +(-\tilde{\sigma}^{12}\xi^3+\tilde{\sigma}^{13}\xi^2)\delta\kappa_1^*/\sqrt{g}\ ]dV.\end{aligned} \tag{27}$$

Let the beam be in equilibrium under the distributed force vector, $\mathbf{p}=p_i\mathbf{e}_i$, along the beam axis and the distributed surface force vector, $\bar{\mathbf{p}}=\bar{P}_i\mathbf{e}_i$ at the end-cross sections. The directions of force vectors are assumed to remain constant during the deformation. Then the external virtual work takes the form [12]

$$EVW=\int_V P_i\delta U^i dV + \Big[\int_A \bar{P}_i\delta U_i dA\ \Big]_{\xi^1=\xi^1_*}^{\xi^1=\xi^1_{**}}, \tag{28}$$

where ξ^1_* and ξ^1_{**} denote the coordinates of ξ^1 at both ends of the beam and $dA=d\xi^2 d\xi^3$.

6. Equilibrium Equations and Boundary Conditions

In the preceding section, the exact internal and external virtual works are obtained under the Bernoulli-Euler hypotheses. When Eq.(27) is used for evaluating the IVW, an important difficulty arises from the fact that the appropriate stress resultants and moments can not be defined. Because not only the coordinates but also the extension and the curvatures after the deformation are included in the denominator of the underlined term in Eq.(27). In view of this fact, we introduce the following small-strain assumptions:

$$\sqrt{g}\gg\varepsilon_0,\quad \xi^2\kappa_3^*,\quad \xi^3\kappa_2^*. \tag{29}$$

Although the small-strain assumptions are introduced here, it should be emphasized that the present theory places no restrictions on the magnitude of displacements or angles of rotations.

With the use of Eq.(29), the IVW takes the form

$$
\begin{aligned}
IVW=\int_V[&\tilde{\sigma}^{11}(\delta\varepsilon_0-\xi^2\delta\kappa_3^*+\xi^3\delta\kappa_2^*+\rho^2\kappa_1^*\delta\kappa_1^*/\sqrt{g})/\sqrt{g}\\
&+(-\tilde{\sigma}^{12}\xi^3+\tilde{\sigma}^{13}\xi^2)\delta\kappa_1^*/\sqrt{g}\ ]dV.
\end{aligned}
\tag{30}
$$

The equilibrium equations and the associated boundary conditions are obtained from the principle of virtual work, which asserts that IVW=EVW for any additional virtual displacement subject to geometric constrains. Using Eqs.(28) and (30) and integrating by parts, we obtain the following equilibrium equations:

$$
\begin{aligned}
&[N\bar{\underset{\sim}{e}}_1-(M_{3,1}+\bar{\kappa}_1M_2-\bar{\kappa}_2T)\bar{\underset{\sim}{e}}_2+(M_{2,1}-\bar{\kappa}_1M_3+\bar{\kappa}_3T)\bar{\underset{\sim}{e}}_3\\
&+\tfrac{1}{4}\lambda^2\{T_\eta^*+g_1(M_\eta^*+T^*+T_\xi^*)\}(\underset{\sim}{e}_1-u^1{}_{\|1}\bar{\underset{\sim}{e}}_1)-\{M_\psi^*-\tfrac{1}{2}\lambda g_2(M_\eta^*+T^*+T_\xi^*)\}(\underset{\sim}{e}_2-u^2{}_{\|1}\bar{\underset{\sim}{e}}_1)\\
&+\{M_\phi^*+\tfrac{1}{2}\lambda g_3(M_\eta^*+T^*+T_\xi^*)\}(\underset{\sim}{e}_3-u^3{}_{\|1}\bar{\underset{\sim}{e}}_1)]_{,1}+\underset{\sim}{Q}^*=0,\\
\\
&T_{,1}+\bar{\kappa}_2M_3-\bar{\kappa}_3M_2+M_\eta^*+T^*+T_\xi^*=0,
\end{aligned}
\tag{31.a,b}
$$

where stress resultants and moments are defined by

$$
\begin{aligned}
&N=\int\tilde{\sigma}^{11}dA,\quad M_2=\int\tilde{\sigma}^{11}\xi^3dA,\quad M_3=-\int\tilde{\sigma}^{11}\xi^2dA,\quad M_t=\int(-\tilde{\sigma}^{12}\xi^3+\tilde{\sigma}^{13}\xi^2)dA,\\
&M_\rho=\int\tilde{\sigma}^{11}\rho^2dA,\quad T=M_t+M_\rho\kappa_1^*,
\end{aligned}
\tag{32.a,b}
$$

and the external forces are defined by

$$
\begin{aligned}
&N^*=\int P_1\sqrt{g}dA,\quad M_2^*=\int P_1\xi^3\sqrt{g}dA,\quad M_3^*=-\int P_1\xi^2\sqrt{g}dA,\quad Q_2^*=\int P_2\sqrt{g}dA,\\
&Q_3^*=\int P_3\sqrt{g}dA,\quad T_{22}^*=-\int P_2\xi^2\sqrt{g}dA,\quad T_{23}^*=-\int P_2\xi^3\sqrt{g}dA,\quad T_{32}^*=\int P_3\xi^2\sqrt{g}dA,\\
&T_{33}^*=-\int P_3\xi^3\sqrt{g}dA,\qquad T_{xx}^*=T_{22}^*\sin\psi+T_{23}^*\cos\psi,\quad T_{xy}^*=T_{22}^*\cos\psi-T_{23}^*\sin\psi,\\
&T_{yx}^*=T_{33}^*\cos\psi-T_{32}^*\sin\psi,\quad T_{yy}^*=T_{32}^*\cos\psi+T_{33}^*\sin\psi,\quad M_x^*=M_2^*\cos\psi-M_3^*\sin\psi,\\
&T^*=T_{xx}^*[1-\tfrac{1}{2}\lambda(u^2{}_{\|1})^2]+T_{yy}^*[1-\tfrac{1}{2}\lambda(u^3{}_{\|1})^2],\quad M_y^*=M_2^*\sin\psi+M_3^*\cos\psi,\\
&M_\phi^*=M_x^*-\tfrac{1}{2}\lambda T_{xx}^*u^2{}_{\|1}-\lambda(u^3{}_{\|1}T_{yx}^*-\tfrac{1}{2}u^2{}_{\|1}T_{yy}^*),\quad M_\eta^*=M_x^*u^2{}_{\|1}+M_y^*u^3{}_{\|1},\\
&M_\psi^*=M_y^*-\tfrac{1}{2}\lambda T_{yy}^*u^3{}_{\|1}+\lambda(u^2{}_{\|1}T_{xy}^*+\tfrac{1}{2}u^3{}_{\|1}T_{xx}^*),\quad T_\xi^*=\tfrac{1}{2}\lambda u^2{}_{\|1}u^3{}_{\|1}(T_{xy}^*-T_{yx}^*),\\
&T_\eta^*=(u^2{}_{\|1})^2T_{xy}^*+(u^3{}_{\|1})^2T_{yx}^*+u^2{}_{\|1}u^3{}_{\|1}(T_{xx}^*-T_{yy}^*),\quad \underset{\sim}{Q}^*=N^*\underset{\sim}{e}_1+Q_2^*\underset{\sim}{e}_2+Q_3^*\underset{\sim}{e}_3,
\end{aligned}
\tag{33.a-v}
$$

and

$$
\begin{aligned}
&g_1=u^1{}_{\|1}(u^2{}_{\|1}\cos\psi+u^3{}_{\|1}\sin\psi)(u^3{}_{\|1}\cos\psi-u^2{}_{\|1}\sin\psi),\\
&g_2=u^3{}_{\|1}+\tfrac{1}{2}\lambda(u^2{}_{\|1})^2u^3{}_{\|1}\cos2\psi-\tfrac{1}{4}\lambda u^2{}_{\|1}[(u^2{}_{\|1})^2-(u^3{}_{\|1})^2]\sin2\psi,\\
&g_3=-u^2{}_{\|1}+\tfrac{1}{2}\lambda(u^3{}_{\|1})^2u^2{}_{\|1}\cos2\psi-\tfrac{1}{4}\lambda u^3{}_{\|1}[(u^2{}_{\|1})^2-(u^3{}_{\|1})^2]\sin2\psi.
\end{aligned}
\tag{34.a-c}
$$

The boundary conditions at both end-cross sections are given by

$$
\begin{aligned}
&N\bar{\mathbf{e}}_1-(M_{3,1}+\bar{\kappa}_1M_2-\bar{\kappa}_2T)\bar{\mathbf{e}}_2+(M_{2,1}-\bar{\kappa}_1M_3+\bar{\kappa}_3T)\bar{\mathbf{e}}_3\\
&=\bar{\mathbf{Q}}^*-\tfrac{1}{4}\lambda[T_\eta^*+g_1(M_\eta^*+T^*+T_\xi^*)](\mathbf{e}_1-u^1\|_1\bar{\mathbf{e}}_1)+[M_\psi^*-\tfrac{1}{2}g_2\lambda(M_\eta^*+T^*+T_\xi^*)](\mathbf{e}_2-u^2\|_1\bar{\mathbf{e}}_1)\\
&-[M_\phi^*+\tfrac{1}{2}g_3\lambda(M_\eta^*+T^*+T_\xi^*)](\mathbf{e}_3-u^3\|_1\bar{\mathbf{e}}_1) \qquad \text{or} \quad \mathbf{u}=\bar{\mathbf{u}},\\
&c_1M_3-c_2M_2+c_3T=\bar{M}_\psi^*+\lambda^2\bar{T}^*u^2\|_1/(4u^1\|_1) \qquad \text{or} \quad u^2\|_1=\bar{u}^2\|_1,\\
&d_1M_2-d_2M_3-d_3T=\bar{M}_\phi^*-\lambda^2\bar{T}^*u^3\|_1/(4u^1\|_1) \qquad \text{or} \quad u^3\|_1=\bar{u}^3\|_1,\\
&T=\bar{T}^*+\bar{T}_\xi^*+\bar{M}_\eta^* \qquad \text{or} \quad \psi=\bar{\psi}, \qquad (35.\text{a-d})
\end{aligned}
$$

where $\mathbf{u}$, ψ, $u^2\|_1$ and $u^3\|_1$ are prescribed values at both end-cross sections and the boundary forces, denoted by $\overline{(\ )}^*$, are given by introducing p_i into Eqs.(33) in place of p_i and putting g=1. The notations c_i and d_i are defined by

$$
\begin{aligned}
&c_1=[1-\tfrac{1}{2}\lambda(u^2\|_1)^2]\cos\psi-\tfrac{1}{2}\lambda u^2\|_1u^3\|_1\sin\psi+(u^2\|_1\cos\psi+u^3\|_1\sin\psi)u^2\|_1/u^1\|_1,\\
&c_2=[-1+\tfrac{1}{2}\lambda(u^2\|_1)^2]\sin\psi-\tfrac{1}{2}\lambda u^2\|_1u^3\|_1\cos\psi-(u^2\|_1\sin\psi-u^3\|_1\cos\psi)u^2\|_1/u^1\|_1,\\
&c_3=-\cos\psi(u^2\|_1\sin\psi-u^3\|_1\cos\psi)-[\lambda(u^2\|_1\cos\psi+\tfrac{1}{2}u^3\|_1\sin\psi)\\
&\quad+\tfrac{1}{4}\lambda^2(u^2\|_1\cos\psi+u^3\|_1\sin\psi)(u^2\|_1)^2/u^1\|_1]\times[\{-1+\tfrac{1}{2}\lambda(u^2\|_1)^2\}\sin\psi\\
&\quad-\tfrac{1}{2}\lambda u^2\|_1u^3\|_1\cos\psi]-[\tfrac{1}{2}\lambda u^3\|_1\cos\psi+\tfrac{1}{4}\lambda^2(u^2\|_1\cos\psi+u^3\|_1\sin\psi)u^2\|_1u^3\|_1/u^1\|_1]\\
&\quad\times[\{1-\tfrac{1}{2}\lambda(u^3\|_1)^2\}\cos\psi+\tfrac{1}{2}\lambda u^2\|_1u^3\|_1\sin\psi],\\
&d_1=[1-\tfrac{1}{2}\lambda(u^3\|_1)^2]\cos\psi+\tfrac{1}{2}\lambda u^2\|_1u^3\|_1\sin\psi-(u^2\|_1\sin\psi-u^3\|_1\cos\psi)u^3\|_1/u^1\|_1,\\
&d_2=[1-\tfrac{1}{2}\lambda(u^3\|_1)^2]\sin\psi-\tfrac{1}{2}\lambda u^2\|_1u^3\|_1\cos\psi+(u^2\|_1\cos\psi+u^3\|_1\sin\psi)u^3\|_1/u^1\|_1,\\
&d_3=-\sin\psi(u^2\|_1\sin\psi-u^3\|_1\cos\psi)-[\lambda(u^3\|_1\sin\psi+\tfrac{1}{2}u^2\|_1\cos\psi)\\
&\quad+\tfrac{1}{4}\lambda^2(u^2\|_1\cos\psi+u^3\|_1\sin\psi)(u^3\|_1)^2/u^1\|_1]\times[\{1-\tfrac{1}{2}\lambda(u^3\|_1)^2\}\cos\psi\\
&\quad+\tfrac{1}{2}\lambda u^2\|_1u^3\|_1\sin\psi]-[\tfrac{1}{2}\lambda u^2\|_1\sin\psi+\tfrac{1}{4}\lambda^2(u^2\|_1\cos\psi+u^3\|_1\sin\psi)u^2\|_1u^3\|_1/u^1\|_1]\\
&\quad\times[\{-1+\tfrac{1}{2}\lambda(u^2\|_1)^2\}\sin\psi-\tfrac{1}{2}\lambda u^2\|_1u^3\|_1\cos\psi]. \qquad (36.\text{a-f})
\end{aligned}
$$

To complete the beam theory, we seek the constitutive equations which relate the generalized stresses and strains. We postulate that the linear elastic relationships between strains and stresses hold. Then we have

$$
\tilde{\sigma}^{11}=E\tilde{\varepsilon}_{11}, \qquad \tilde{\sigma}^{12}=2G\tilde{\varepsilon}_{12}, \qquad \tilde{\sigma}^{13}=2G\tilde{\varepsilon}_{13} \qquad (37.\text{a-c})
$$

where E is the Young modulus, G the shear modulus and $\tilde{\varepsilon}_{ij}(=\varepsilon_{ij})$ are the physical components of strain tensors. Substituting Eqs.(23) and (37) into Eqs.(32) and integrating over cross sections leads to

$$
N=E[F_1\varepsilon_o-F_2\overset{*}{\kappa}_3+F_3\overset{*}{\kappa}_2+\tfrac{1}{2}F_\rho(\overset{*}{\kappa}_1)^2], \qquad M_2=E[F_3\varepsilon_o-I_{23}\overset{*}{\kappa}_3+I_{33}\overset{*}{\kappa}_2+\tfrac{1}{2}F_{\rho 3}(\overset{*}{\kappa}_1)^2],
$$

$$M_3=E[-F_2\varepsilon_o+I_{22}\kappa_3^*-I_{23}\kappa_2^*+\tfrac{1}{2}F_{\rho 2}(\kappa_1^*)^2],\quad M_t=GJ\kappa_1^*,$$
$$M_\rho=E[F_\rho\varepsilon_o-F_{\rho 2}\kappa_3^*+F_{\rho 3}\kappa_2^*+\tfrac{1}{2}F_{\rho\rho}(\kappa_1^*)^2], \qquad (38.a\text{-}e)$$

where

$$F_1=\int 1/\sqrt{g}\,dA,\quad F_2=\int \xi^2/\sqrt{g}\,dA,\quad F_3=\int \xi^3/\sqrt{g}\,dA,\quad F_\rho=\int \rho^2/g\,dA,$$
$$F_{\rho 2}=\int \rho^2\xi^2/g\,dA,\quad F_{\rho 3}=\int \rho^2\xi^3/g\,dA,\quad F_{\rho\rho}=\int \rho^4/(g\sqrt{g})\,dA,$$
$$I_{22}=\int (\xi^2)^2/\sqrt{g}\,dA,\quad I_{23}=\int \xi^2\xi^3/\sqrt{g}\,dA,\quad I_{33}=\int (\xi^3)^2/\sqrt{g}\,dA,$$
$$J=\int \rho^2/\sqrt{g}\,dA. \qquad (39.a\text{-}k)$$

REFERENCES

1. Love A.E.H., A Treatise on the Mathematical Theory of Elasticity, Dover Pub., New York (1944).
2. Novozhilov V.V., Foundation of the Nonlinear Theory of Elasticity, Graylock Press, New York (1953).
3. Rosen A., Friedmann P., The nonlinear behavior of elastic slender straight beams undergoing small strains and moderate rotations, J. Appl. Mech., Trans. ASME 46, 161-168 (1979).
4. Besseling J.F., Non-linear theory for elastic beams and rods and its finite element representation, Comput. Meths. Appl. Mech. Engrg. 31, 205-220 (1982).
5. Maeda Y., Hayashi M., Finite displacement analysis of space framed structures, Proc. JSCE No.253, 13-27 (1976).
6. Schroeder F.H., Allgemeine Stabtheorie des dunnwandigen raumlich vorgekrummten und vorgewundenen Tragers mit grossen Verformungen, Ing.-Arch. 39, 87-103 (1970).
7. Hirashima M., Iura M., Yoda T., Finite displacement theory of naturally curved and twisted thin-walled members, Proc. JSCE No.292, 13-27 (1979).
8. Stein E., Incremental methods in finite elasticity, especially for rods, Proc. of the IUTAM Symposium on Finite Elasticity, Bethlehem (USA) 1980, pp.379-400, Martinus Nijhoff Pub. (1982).
9. Kurakata Y., Nishino F., Finite displacement field and governing equations of solid curved beams, Proc. JSCE No.317, 15-30 (1982).
10. Iura M., Hirashima M., Geometrically nonlinear theory of naturally curved and twisted rods with finite rotations, Proc. of JSCE, Struct. Eng./Earthq. Eng., Vol.2, No.2, 107-117 (1985).
11. Pietraszkiewicz W., Finite Rotations and Lagrangian Description in The Non-linear Theory of Shells, Polish Scientific Publishers, Warszawa-Poznan (1979).
12. Washizu K., Variational Methods in Elasticity and Plasticity, 3rd. edition, Pergamon Press (1982).

HIGHER-ORDER MODERATE ROTATION THEORIES FOR ELASTIC ANISOTROPIC PLATES

Liviu Librescu
Tel Aviv University
Dept. of Solid Mechanics, Materials and Structures
Tel Aviv 69978, Israel

Rüdiger Schmidt
University of Wuppertal
Institute of Civil Engineering Mechanics
Pauluskirchstrasse 7
D-5600 Wuppertal 2, Fed. Rep. of Germany

1. Introduction

A great deal of interest in the substantiation of refined theories of elastic anisotropic plates and shells has been manifested in the specialized literature in the last two decades. This interest is largely due to the need for more adequate methods of analysis of structural elements exposed to severe and complex operational conditions in various branches of the advanced technology. In addition, the increased use of new exotic composite materials has provided a new impetus for such refined theories. As it was conclusively shown, the classical methods of analysis based on the Kirchhoff-Love assumptions are inadequate in many important instances. This is especially true whenever the material of the structure exhibits high degrees of anisotropy in its physical and mechanical properties. Such features are typical for the composite and refractory type materials used with increased frequency in the aerospace, naval, nuclear industries, etc. In such cases, refined models allowing a more adequate description of the structural response are needed. They should include transverse shear and transverse normal deformations and should account for the higher-order effects.

In this framework, the substantiation of the geometrically nonlinear theory of elastic anisotropic plates is of an evident theoretical and practical interest. Such a theory could provide inter alia the basic ingredients for the study of the associated stability problems (performed either in a linearized or nonlinear formulation). However, the geometrically nonlinear theory of plates (and shells) approached in its full generality by NAGHDI [1], LIBRESCU [2,3], and YOKOO and MATSUNAGA [4] or in the special case of first-order transverse shear approximation by HABIP [5,6], HABIP and EBCIOGLU [7], AINOLA [8], GALIMOV [9-10], PIETRASZKIEWICZ [11] and others, illustrates the high complexity of the problem. In order to obtain simpler field equations without impairing the accuracy of the results, suitable approximations are to be introduced. Such approximate theories of plates (and shells) may be obtained by using e.g. the concept of small strains accompanied by moderate (or large) rotations. This concept was largely employed with useful results in the framework of the classical theory of plates and shells (see e.g. SANDERS [12], KOITER [13], REISSNER [14], GALIMOV [9], PIETRASZKIEWICZ [11,15,16], SCHMIDT[17,18], and for a review of the state-of-the-art and extensive references see SCHMIDT [19]). For the refined theories of plates (and shells) including transverse normal deformation effects this concept was used only in a very few papers (see e.g. KAUL [2o], WEMPNER [21], GALIMOV [9,1o], NAGHDI and VONGSARNPIGOON [22], LIBRESCU [3]).

The present work is devoted to the substantiation of the higher-order theory of anisotropic plates accounting for small strains accompanied by moderate rotations. Appropriate estimations for the order of magnitude of linearized strain and rotation components are introduced. On this basis, two approximate variants of the strain-displacement relations are considered and their implications in the remaining field equations are analysed. By using the virtual work principle of 3-D elastokinetics and by postulating a higher-order representation of the displacement field across the wall thickness, the pertinent set of equations of motions and boundary conditions in a complete Lagrangian description is derived. In addition, by assuming that the material of the plate is (elastically and thermally) anisotropic, the appropriate constitutive equations are given. It is shown that the theory presented here includes as special cases a variety of nonlinear refined plate theories available in the literature. In the light of the present theory these approximate variants may be derived in a unified manner and related to each other on the basis of order-of-magnitude considerations, which accordingly makes it possible to clarify their range of validity. It is also shown that under Kirchhoff constraints the present theory reduces to the well known classical moderate rotation plate theory.

2. Preliminaries

Let σ denote the volume of the plate in the undeformed (reference) configuration. By S^+ and S^- we denote the upper and lower surfaces of σ, symmetrically located with respect to the midplane M of the undeformed plate. Let B denote the lateral boundary surface of σ generated by the normals to M along its boundary curve C (of arc length s). By B_f and B_v ($B = B_f \cup B_v$) we denote the parts of B, where stresses and displacements, respectively, are prescribed. Let us introduce a normal coordinate system x^i, $i = 1,2,3$, consisting of the set of curvilinear in-plane coordinates x^α, $\alpha = 1,2$, on M, and the coordinate x^3 normal to M. The constant plate thickness is denoted by h. The components of the metric tensor of the undeformed normal space are

$$g_{\alpha\beta} = \underset{\sim}{g}_\alpha \cdot \underset{\sim}{g}_\beta \ , \ g_{\alpha 3} = \underset{\sim}{g}_\alpha \cdot \underset{\sim}{g}_3 = 0 \ , \ g_{33} = \underset{\sim}{g}_3 \cdot \underset{\sim}{g}_3 = 1 \ ;$$
$$g^{\alpha\beta} = \underset{\sim}{g}^\alpha \cdot \underset{\sim}{g}^\beta \ , \ g^{\alpha 3} = \underset{\sim}{g}^\alpha \cdot \underset{\sim}{g}^3 = 0 \ , \ g^{33} = \underset{\sim}{g}^3 \cdot \underset{\sim}{g}^3 = 1 \qquad (2.1)$$

where $\underset{\sim}{g}_i$ and $\underset{\sim}{g}^i$, respectively, denote the covariant and contravariant base vectors of the plate continuum. For this specific case

$$\underset{\sim}{g}_\alpha = \underset{\sim}{a}_\alpha \ , \ \underset{\sim}{g}^\alpha = \underset{\sim}{a}^\alpha \ , \ \underset{\sim}{g}_3 = \underset{\sim}{g}^3 = \underset{\sim}{a}_3 = \underset{\sim}{a}^3 \ (\equiv \underset{\sim}{n}) \tag{2.2}$$

where $\underset{\sim}{a}_i$ and $\underset{\sim}{a}^i$ are the covariant and contravariant base vectors of M with surface metric tensors $a_{\alpha\beta}$ and $a^{\alpha\beta}$, respectively. Here also $g_{\alpha\beta} = a_{\alpha\beta}$, $g^{\alpha\beta} = a^{\alpha\beta}$, $g = a$, where $g \equiv \det(g_{ij})$ and $a \equiv \det(a_{\alpha\beta})$ are the determinants of the metric tensors.

At the boundary curve C of M we introduce unit tangent and outward normal vectors $\underset{\sim}{t}$ and $\underset{\sim}{\nu}$, respectively, by

$$\underset{\sim}{t} = t^\alpha \underset{\sim}{a}_\alpha \ , \qquad \underset{\sim}{\nu} = \nu^\alpha \underset{\sim}{a}_\alpha = \underset{\sim}{t} \times \underset{\sim}{n} \ . \tag{2.3}$$

Partial differentiation will be denoted by a comma, $(\ldots)_{,i} \equiv \partial(\ldots)/\partial x^i$, while the notation $(\ldots)|_\alpha$ stands for covariant differentiation with respect to the metric of M. It is mentioned, that for the case of flat plates, by virtue of (2.2), there is no distinction between space and surface differentiations in σ and on M, respectively. Throughout the paper the Einsteinian summation convention will be used, with Latin indices ranging from 1 to 3 and Greek indices ranging from 1 to 2.

3. Geometric Equations

3.1. Geometric Equations for the 3-D Plate Continuum

The displacement vector $\underset{\sim}{V}(x^\alpha, x^3, t)$ of any point in the 3-D plate space may be represented as

$$\underset{\sim}{V} = V^\alpha \underset{\sim}{g}_\alpha + V^3 \underset{\sim}{g}_3 = V^\alpha \underset{\sim}{a}_\alpha + V^3 \underset{\sim}{n} \tag{3.1}$$

while the covariant components of the associated Lagrangian strain tensor are given by

$$e_{ij} = \frac{1}{2} (V_{i|j} + V_{j|i} + V^k{}_{|i} V_{k|j}) \ . \tag{3.2}$$

For convenience, eq.(3.2) will be expressed exactly as

$$e_{ij} = \eta_{ij} + \frac{1}{2} \Omega_{ri} \Omega^r{}_{.j} + \frac{1}{2} (\eta_{rj} \Omega^r{}_{.i} + \eta_{ri} \Omega^r{}_{.j}) + \frac{1}{2} \eta_{rj} \eta^r{}_{.i} \tag{3.3}$$

where we have introduced the linearized strain and rotation tensors η_{ij} and Ω_{ij}, respectively, given by

$$\eta_{ij} = \frac{1}{2} (V_{i|j} + V_{j|i}), \tag{3.4}$$

$$\Omega_{ij} = \frac{1}{2} (V_{i|j} - V_{j|i}). \tag{3.5}$$

In the ensuing developments the following representation for the components of the displacement vector (3.1) across the plate thickness will be used

$$V_\alpha(x^\omega,x^3,t) = \sum_{r=o}^{R} (x^3)^r \overset{(r)}{V_\alpha} (x^\omega,t) , \tag{3.6}$$

$$V_3(x^\omega,x^3,t) = \sum_{s=o}^{S} (x^3)^s \overset{(s)}{V_3} (x^\omega,t) . \tag{3.7}$$

Here R and S stand for two natural numbers defining the level of truncation in the series expansions, while r and s are summation indices. For the sake of simplicity we shall assume, without any loss of generality, that $R=S\equiv N$. In cases for which $R\neq S$, i.e. when $R<N$ or $S<N$, the components $\overset{(r)}{V_\alpha}$, for $r>R$, or $\overset{(s)}{V_3}$, for $s>S$, respectively, are to be considered as zero quantities.

3.2. Geometric Equations for Moderate Rotation Plate Theory

As outlined in the introduction, the purpose of this paper is to derive a general higher-order theory of physically linear anisotropic, elastic plates. We shall assume that the rotations about the normal to the midplane are small and of the order of magnitude of the strains, while the rotations of the normal are moderate. These assumptions are consistent with the real features of these structures which exhibit a large in-plane rigidity and simultaneously a high transverse flexibility. In mathematical terms, the assumption of small strains means

$$e_{ij} = o(\theta^2)$$

where $\theta^2 \ll 1$. Furthermore the assumption that the rotations of the normal are moderate, while the in-surface rotations about the normal remain small, implies

$$\Omega_{\alpha 3} = o(\theta) \quad , \quad \Omega_{\alpha\beta} = o(\theta^2) . \tag{3.8}$$

Then, with the help of (3.3) we obtain the following order of magnitude for the linearized strains

$$\eta_{ij} = o(\theta^2) . \tag{3.9}$$

It is pointed out that the estimates (3.8) refer to the linearized rotations (3.5) as the principal part of their full nonlinear expressions. This point of view has been adopted in several works dealing with the classical theory of shells (see e.g. SANDERS [12], KOITER [13]),or with the first-order transverse shear approximation theory of plates and shells (see e.g. WEMPNER [21], GALIMOV [9,1o]), and will be used here, too. Under the estimates (3.8) - (3.9) the exact expression for the strain tensor (3.3) may be simplified to within a relative error $o(\theta^2)$ as

$$e_{\alpha\beta} = \eta_{\alpha\beta} + \frac{1}{2}\Omega_{3\beta}\Omega_{3\alpha} + \underline{\frac{1}{2}(\eta_{3\beta}\Omega_{3\alpha} + \eta_{3\alpha}\Omega_{3\beta})} , \tag{3.1o}$$

$$e_{\alpha 3} = \eta_{\alpha 3} + \underline{\frac{1}{2}\,\Omega_{\lambda 3}\Omega^{\lambda}_{.\alpha} + \frac{1}{2}\left(\eta_{\lambda\alpha}\Omega^{\lambda}_{.3} + \eta_{33}\Omega^{3}_{.\alpha}\right)} \,, \tag{3.11}$$

$$e_{33} = \eta_{33} + \frac{1}{2}\,\Omega_{\lambda 3}\Omega^{\lambda}_{.3} + \underline{\eta_{\lambda 3}\Omega^{\lambda}_{.3}} \,. \tag{3.12}$$

Here, the terms marked by solid lines are of the order θ^3. Their neglection would involve a relative error $o(\theta)$ in the strain-displacement relations. In the following, the two variants based on either retaining or discarding of the underlined terms will be considered. In the remaining field equations the terms generated by those underlined in (3.1o) - (3.12) will be also identified by a solid line. It may be observed that by discarding the underlined terms the transverse shearing strains remain expressed by linear strain measures only.

Here a remark concerning the strain-displacement relations used by YU [23] and KAUL [2o] is in order. In contrast to their approach, the present one allows to obtain appropriate strain-displacement relations on the basis of the postulated estimates. In this way it is possible to elaborate consistently approximate theories, which ensure that all principal terms are taken into account while the negligible small terms are disregarded.

Consideration of (3.4) - (3.7) in the strain-displacement relations (3.1o) - (3.12) yields the following representation of the strain components in the 3-D medium of the plate:

$$e_{\alpha\beta} = \sum_{n=o}^{2N} (x^3)^n \overset{(n)}{e}_{\alpha\beta}(x^1,x^2,t) \,,$$

$$e_{\alpha 3} = \sum_{n=o}^{2N} (x^3)^n \overset{(n)}{e}_{\alpha 3}(x^1,x^2,t) \,,$$

$$e_{33} = \sum_{n=o}^{2N} (x^3)^n \overset{(n)}{e}_{33}(x^1,x^2,t) \,, \tag{3.13}$$

where the two-dimensional n-th order strain measures are given by

$$\overset{(n)}{e}_{\alpha\beta} = \overset{(n)}{\eta}_{\alpha\beta} + \sum_{m=o}^{n} \left(\frac{1}{2}\,\overset{(m)}{\Omega}_{3\alpha}\overset{(n-m)}{\Omega}_{3\beta} + \underline{\frac{1}{2}\left(\overset{(m)}{\eta}_{3\alpha}\overset{(n-m)}{\Omega}_{3\beta} + \overset{(m)}{\eta}_{3\beta}\overset{(n-m)}{\Omega}_{3\alpha}\right)}\right) \,,$$

$$\overset{(n)}{e}_{\alpha 3} = \overset{(n)}{\eta}_{\alpha 3} + \sum_{m=o}^{n} \left(\underline{\frac{1}{2}\,\overset{(m)}{\Omega}_{\lambda 3}\overset{(n-m)}{\Omega}{}^{\lambda}_{.\alpha} + \frac{1}{2}\left(\overset{(m)}{\eta}_{\lambda\alpha}\overset{(n-m)}{\Omega}{}^{\lambda}_{.3} + \overset{(m)}{\eta}_{33}\overset{(n-m)}{\Omega}{}^{3}_{.\alpha}\right)}\right) \,,$$

$$\overset{(n)}{e}_{33} = \overset{(n)}{\eta}_{33} + \sum_{m=o}^{n} \left(\frac{1}{2}\,\overset{(m)}{\Omega}_{\lambda 3}\overset{(n-m)}{\Omega}{}^{\lambda}_{.3} + \underline{\overset{(m)}{\eta}_{\lambda 3}\overset{(n-m)}{\Omega}{}^{\lambda}_{.3}}\right) \,, \tag{3.14}$$

and where

$$\overset{(n)}{\eta}_{\alpha\beta} = \frac{1}{2}(\overset{(n)}{V}_{\alpha|\beta} + \overset{(n)}{V}_{\beta|\alpha}) \quad , \qquad \overset{(n)}{\Omega}_{\alpha\beta} = \frac{1}{2}(\overset{(n)}{V}_{\alpha|\beta} - \overset{(n)}{V}_{\beta|\alpha}) \quad ,$$

$$\overset{(n)}{\eta}_{\alpha 3} = \frac{1}{2}((n+1)\overset{(n+1)}{V}_{\alpha} + \overset{(n)}{V}_{3,\alpha}) \quad , \qquad \overset{(n)}{\Omega}_{\alpha 3} = \frac{1}{2}((n+1)\overset{(n+1)}{V}_{\alpha} - \overset{(n)}{V}_{3,\alpha}) \quad .$$

$$\overset{(n)}{\eta}_{33} = (n+1)\overset{(n+1)}{V}_{3} \quad , \tag{3.15}$$

In all the previous equations $\overset{(n)}{V}_{\alpha}$ and $\overset{(n)}{V}_{3}$ are to be considered zero whenever $n<o$ or $n>N$. Let us now consider a special case of the above relations which is referred to in the literature as the first-order transverse shear approximation theory. In this case a linear representation of the displacement field across the plate thickness is assumed

$$V_{\alpha} = \overset{(o)}{V}_{\alpha} + x^3\overset{(1)}{V}_{\alpha} \quad , \quad V_3 = \overset{(o)}{V}_{3} + x^3\overset{(1)}{V}_{3} \; . \tag{3.16}$$

As a result the strain components in the 3-D continuum of the plate take the form

$$e_{\alpha\beta} = \overset{(o)}{e}_{\alpha\beta} + x^3\overset{(1)}{e}_{\alpha\beta} + (x^3)^2\overset{(2)}{e}_{\alpha\beta} \quad ,$$

$$e_{\alpha 3} = \overset{(o)}{e}_{\alpha 3} + x^3\overset{(1)}{e}_{\alpha 3} + (x^3)^2\overset{(2)}{e}_{\alpha 3} \quad ,$$

$$e_{33} = \overset{(o)}{e}_{33} + x^3\overset{(1)}{e}_{33} + (x^3)^2\overset{(2)}{e}_{33} \quad , \tag{3.17}$$

where the two-dimensional n-th order strain measures are now given by

$$\overset{(o)}{e}_{\alpha\beta} = \overset{(o)}{\eta}_{\alpha\beta} + \frac{1}{2}\overset{(o)}{\Omega}_{3\alpha}\overset{(o)}{\Omega}_{3\beta} + \underline{\frac{1}{2}(\overset{(o)}{\eta}_{3\alpha}\overset{(o)}{\Omega}_{3\beta} + \overset{(o)}{\eta}_{3\beta}\overset{(o)}{\Omega}_{3\alpha})} \quad ,$$

$$\overset{(1)}{e}_{\alpha\beta} = \overset{(1)}{\eta}_{\alpha\beta} + \frac{1}{2}(\overset{(o)}{\Omega}_{3\alpha}\overset{(1)}{\Omega}_{3\beta} + \overset{(1)}{\Omega}_{3\alpha}\overset{(o)}{\Omega}_{3\beta}) + \underline{\frac{1}{2}(\overset{(o)}{\eta}_{3\alpha}\overset{(1)}{\Omega}_{3\beta} + \overset{(1)}{\eta}_{3\alpha}\overset{(o)}{\Omega}_{3\beta} + \overset{(o)}{\eta}_{3\beta}\overset{(1)}{\Omega}_{3\alpha} + \overset{(1)}{\eta}_{3\beta}\overset{(o)}{\Omega}_{3\alpha})} \quad ,$$

$$\overset{(2)}{e}_{\alpha\beta} = \frac{1}{2}\overset{(1)}{\Omega}_{3\alpha}\overset{(1)}{\Omega}_{3\beta} + \underline{\frac{1}{2}(\overset{(1)}{\eta}_{3\alpha}\overset{(1)}{\Omega}_{3\beta} + \overset{(1)}{\eta}_{3\beta}\overset{(1)}{\Omega}_{3\alpha})} \quad ; \tag{3.18}$$

$$\overset{(o)}{e}_{\alpha 3} = \overset{(o)}{\eta}_{\alpha 3} + \underline{\frac{1}{2}\overset{(o)}{\Omega}_{\lambda 3}\overset{(o)}{\Omega}{}^{\lambda}_{.\alpha}} + \underline{\frac{1}{2}(\overset{(o)}{\eta}_{\lambda\alpha}\overset{(o)}{\Omega}{}^{\lambda}_{.3} + \overset{(o)}{\eta}_{33}\overset{(o)}{\Omega}{}^{3}_{.\alpha})} \quad ,$$

$$\overset{(1)}{e}_{\alpha 3} = \overset{(1)}{\eta}_{\alpha 3} + \underline{\frac{1}{2}(\overset{(o)}{\Omega}_{\lambda 3}\overset{(1)}{\Omega}{}^{\lambda}_{.\alpha} + \overset{(1)}{\Omega}_{\lambda 3}\overset{(o)}{\Omega}{}^{\lambda}_{.\alpha})} + \underline{\frac{1}{2}(\overset{(o)}{\eta}_{\lambda\alpha}\overset{(1)}{\Omega}{}^{\lambda}_{.3} + \overset{(1)}{\eta}_{\lambda\alpha}\overset{(o)}{\Omega}{}^{\lambda}_{.3} + \overset{(o)}{\eta}_{33}\overset{(1)}{\Omega}{}^{3}_{.\alpha})} \quad ,$$

$$\overset{(2)}{e}_{\alpha 3} = \frac{1}{2}\overset{(1)}{\Omega}_{\lambda 3}\overset{(1)}{\Omega}{}^{\lambda}{}_{.\alpha} + \frac{1}{2}\overset{(1)}{\eta}_{\lambda\alpha}\overset{(1)}{\Omega}{}^{\lambda}{}_{.3} \quad ; \tag{3.19}$$

$$\overset{(o)}{e}_{33} = \overset{(o)}{\eta}_{33} + \frac{1}{2}\overset{(o)}{\Omega}_{\lambda 3}\overset{(o)}{\Omega}{}^{\lambda}{}_{.3} + \overset{(o)}{\eta}_{\lambda 3}\overset{(o)}{\Omega}{}^{\lambda}{}_{.3} \quad ,$$

$$\overset{(1)}{e}_{33} = \overset{(o)}{\Omega}_{\lambda 3}\overset{(1)}{\Omega}{}^{\lambda}{}_{.3} + \overset{(o)}{\eta}_{\lambda 3}\overset{(1)}{\Omega}{}^{\lambda}{}_{.3} + \overset{(1)}{\eta}_{\lambda 3}\overset{(o)}{\Omega}{}^{\lambda}{}_{.3} \quad ,$$

$$\overset{(2)}{e}_{33} = \frac{1}{2}\overset{(1)}{\Omega}_{\lambda 3}\overset{(1)}{\Omega}{}^{\lambda}{}_{.3} + \overset{(1)}{\eta}_{\lambda 3}\overset{(1)}{\Omega}{}^{\lambda}{}_{.3} \quad . \tag{3.20}$$

Let us compare the above results with their counterparts derived by WEMPNER (see chapters 7.14 and 7.16 of [21]). The results are obtainable from our equations (3.17) - (3.19) by retaining the strain measures $\overset{(o)}{e}_{\alpha\beta}$, $\overset{(1)}{e}_{\alpha\beta}$, $\overset{(o)}{e}_{\alpha 3}$ only, so yielding the equations

$$e_{\alpha\beta} = \overset{(o)}{e}_{\alpha\beta} + x^{3}\overset{(1)}{e}_{\alpha\beta} \; , \quad e_{\alpha 3} = \overset{(o)}{e}_{\alpha 3} \; , \quad e_{33} = 0. \tag{3.21}$$

It is argued in [21] that eqs.(3.21) are valid for thin plates. If additionally in the expressions for $\overset{(1)}{e}_{\alpha\beta}$ and $\overset{(o)}{e}_{\alpha 3}$ all nonlinear terms are disregarded and if the non-linear part of $\overset{(o)}{e}_{\alpha\beta}$ is simplified by omitting the strains $\overset{(o)}{\eta}_{\alpha 3}$ in the relation $\overset{(o)}{\Omega}_{\alpha 3} = -\overset{(o)}{V}_{3,\alpha} + \overset{(o)}{\eta}_{\alpha 3}$, so yielding $\overset{(o)}{\Omega}_{\alpha 3} = -\overset{(o)}{V}_{3,\alpha}$ the strain-displacement relations reduce to those obtained in [21]. They read:

$$\overset{(o)}{e}_{\alpha\beta} = \frac{1}{2}(\overset{(o)}{V}_{\alpha|\beta} + \overset{(o)}{V}_{\beta|\alpha}) + \frac{1}{2}\overset{(o)}{V}_{3,\alpha}\overset{(o)}{V}_{3,\beta} \quad ,$$

$$\overset{(1)}{e}_{\alpha\beta} = \frac{1}{2}(\overset{(1)}{V}_{\alpha|\beta} + \overset{(1)}{V}_{\beta|\alpha}) \; , \quad \overset{(o)}{e}_{\alpha 3} = \frac{1}{2}(\overset{(1)}{V}_{\alpha} + \overset{(o)}{V}_{3,\alpha}) \quad . \tag{3.22}$$

The form of $\overset{(o)}{e}_{\alpha\beta}$ reminds the one referred to as von Kārmān membrane strain tensor. It is worth remarking that equations (3.22) may be derived also in a more direct manner (see LIBRESCU [2], chapter IV), by starting with the representation of the displacement field in the form (3.16), specialized for $\overset{(1)}{V}_{3} \equiv 0$, and by considering a partially nonlinear variant

$$e_{ij} = \frac{1}{2}(V_{i|j} + V_{j|i} + V_{3|i}V_{3|j}) \tag{3.23}$$

instead of the full nonlinear strain-displacement relations (3.2). The equations in the form (3.22) have been used in the literature in treating various problems of plates with transverse shear deformations (see e.g. VINSON and CHOU [24], NOOR [25], REDDY [26], REDDY and CHAO [27]). It should also be mentioned that the linearized counterpart of eqs.(3.13) - (3.14) specialized for the case of flat plates leads to the relations given by BRULL and LIBRESCU [28].

3.3. Geometric Equations for Moderate Rotation Plate Theory Under Kirchhoff Contraints

In this instance, the pertinent strain-displacement relations will be obtained as a special case of eqs.(3.16) - (3.2o). We shall restrict ourselves to the case when in the strain-displacement relations the terms of $o(\theta^2)$ are retained only. Therefore the terms of $o(\theta^3)$ marked by a solid line will be dropped. Then, under the Kirchhoff constraints $e_{\alpha 3}\equiv o$ and $e_{33}\equiv o$, we get

$$\overset{(o)}{e}_{\alpha 3} + x^3 \overset{(1)}{e}_{\alpha 3} = o \quad , \tag{3.24}$$

$$\overset{(o)}{e}_{33} + x^3 \overset{(1)}{e}_{33} + (x^3)^2 \overset{(2)}{e}_{33} = o \quad . \tag{3.25}$$

The condition (3.24) in terms of displacements reads

$$\frac{1}{2} (\overset{(o)}{V}_{3,\alpha} + \overset{(1)}{V}_{\alpha}) + x^3 \frac{1}{2} \overset{(1)}{V}_{3,\alpha} = o \quad , \tag{3.26}$$

thus allowing to obtain:

$$\overset{(1)}{V}_{\alpha} = - \overset{(o)}{V}_{3,\alpha} \quad , \quad \overset{(1)}{V}_{3,\alpha} = o \quad . \tag{3.27}$$

A similar analysis of condition (3.25) yields

$$\overset{(1)}{V}_{3} = - \frac{1}{2} \overset{(o)}{V}_{3|\lambda} \overset{(o)}{V}_{3|}{}^{\lambda} \quad . \tag{3.28}$$

As a result of these conditions the non-vanishing two-dimensional strain measures become

$$\overset{(o)}{e}_{\alpha\beta} = \frac{1}{2} (\overset{(o)}{V}_{\alpha|\beta} + \overset{(o)}{V}_{\beta|\alpha} + \overset{(o)}{V}_{3,\alpha} \overset{(o)}{V}_{3,\beta}) \quad , \quad \overset{(1)}{e}_{\alpha\beta} = - \overset{(o)}{V}_{3|\alpha\beta} \quad . \tag{3.29}$$

These are nothing but the well known strain-displacement relations of classical plate theory expressed here in general coordinates. They allow to infer that the relations (3.17) - (3.2o) based on the representation (3.16) of the displacement field and derived as per the concept of small strains and moderate rotations, yield, under Kirchhoff constraints, the strain-displacement relations associated to the von Kármán theory of plates.

4. Equations of Motion and Boundary Conditions

4.1. Equations of Motion and Boundary Conditions for Moderate Rotation Plate Theory

The equations of motion for the higher-order moderate rotation plate theory will be derived with the help of the virtual work principle of 3-D elastokinetics

$$\int_{\sigma}(s^{ij}\delta e_{ij} - \rho(\underset{\sim}{H}-\underset{\sim}{h}) \cdot \delta \underset{\sim}{V})\, d\sigma - \int_{F} \overset{*}{\underset{\sim}{s}} \cdot \delta \underset{\sim}{V}\, dF = o \tag{4.1}$$

where the strain-displacement relations (3.13) - (3.15) and the geometric boundary conditions on B_v

$$V_\nu = V^\alpha \nu_\alpha = \overset{*}{V}_\nu \quad , \quad V_t = V^\alpha t_\alpha = \overset{*}{V}_t \quad , \quad V_3 = \overset{*}{V}_3 \tag{4.2}$$

have to be imposed as subsidiary conditions. Here F denotes those parts of the total undeformed boundary surface $S \cup B$, where the stress vector $\overset{*}{\underset{\sim}{s}}$ is prescribed; s^{ij} denotes the (symmetric) second Piola-Kirchhoff stress tensor; ρ stands for the mass density while $\underset{\sim}{H}$ and $\underset{\sim}{h}$ denote respectively, the body force and acceleration vectors, per unit volume of the undeformed body.

In order to derive the two-dimensional counterpart of the virtual work principle we shall make use in (4.1) of the representation of e_{ij} given by (3.13) - (3.15); of the expression for the volume element $d\sigma = dx^3 dM$; of the expression for the stress vector $\overset{*}{\underset{\sim}{s}} = {}^*t^i \underset{\sim}{g}_i$, ${}^*t^i = {}^*t^{ji} n_j$, where $t^{ij} = s^{ir}(\delta^j_r + V^j|_r)$ is the first Piola-Kirchhoff stress tensor and n_j are the components of the unit outward normal vector to F, as well as of the relations expressing the area elements of the bounding surfaces, i.e. $dS = dM$, $n_\alpha dB = \nu_\alpha dx^3 ds$.

Furthermore, following [2,3] we shall define:

a) the n-th order stress couples

$$\overset{(n)}{L}{}^{\alpha\beta} = \int_{-h/2}^{h/2} s^{\alpha\beta}\,(x^3)^n\, dx^3 \quad ,$$

$$\overset{(n)}{L}{}^{\alpha 3} = \int_{-h/2}^{h/2} s^{\alpha 3}\,(x^3)^n\, dx^3 \quad ,$$

$$\overset{(n)}{L}{}^{33} = \int_{-h/2}^{h/2} s^{33}\,(x^3)^n\, dx^3 \quad , \tag{4.3}$$

b) the n-th order body couples

$$\overset{(n)}{F}{}^{\alpha} = \int_{-h/2}^{h/2} \rho H^{\alpha}\,(x^3)^n\, dx^3 \quad ,$$

$$\overset{(n)}{F}{}^{3} = \int_{-h/2}^{h/2} \rho H^{3}\,(x^3)^n\, dx^3 \quad , \tag{4.4}$$

c) the n-th order inertia couples

$$\overset{(n)}{f}{}^{\alpha} = \sum_{q=o}^{N} \overset{(q+n)}{m} \overset{(q)}{\ddot{V}}{}^{\alpha} ,$$

$$\overset{(n)}{f}{}^{3} = \sum_{q=o}^{N} \overset{(q+n)}{m} \overset{(q)}{\ddot{V}}{}^{3} , \tag{4.5}$$

where the generalized mass is defined as

$$\overset{(r)}{m} = \rho\, \eta\,(r+1) \tag{4.6}$$

while $\eta(r)$ is given by [2,3]

$$\eta(r) = \frac{h^r}{r2^{r-1}} \qquad \text{if } r \text{ is odd,}$$

$$\eta(r) = o \qquad \text{if } r \text{ is even,} \tag{4.7}$$

d) the n-th order couples of the surface loads on $S^{\pm}$

$$\overset{(n)}{p}{}^{\alpha} = [\,{}^{*}t^{3\alpha}\,(x^3)^n]\Big|_{-h/2}^{h/2} ,$$

$$\overset{(n)}{p}{}^{3} = [\,{}^{*}t^{33}\,(x^3)^n]\Big|_{-h/2}^{h/2} , \tag{4.8}$$

e) the n-th order couples of the boundary loads prescribed on $\mathcal{B}$

$$\overset{(n)}{{}^{*}L}{}^{\alpha\beta} = \int_{-h/2}^{h/2} {}^{*}t^{\beta\alpha}\,(x^3)^n\, dx^3 ,$$

$$\overset{(n)}{{}^{*}L}{}^{3\beta} = \int_{-h/2}^{h/2} {}^{*}t^{\beta 3}\,(x^3)^n\, dx^3 . \tag{4.9}$$

Using the above definitions the 3-D virtual work principle (4.1) may be converted into its 2-D counterpart

$$\sum_{n=o}^{2R} \Big\{ \int_{M} [\, \overset{(n)}{L}{}^{\alpha\beta} \overset{(n)}{\delta e}_{\alpha\beta} + 2\overset{(n)}{L}{}^{\alpha 3} \overset{(n)}{\delta e}_{\alpha 3} + \overset{(n)}{L}{}^{33} \overset{(n)}{\delta e}_{33}$$

$$- (\overset{(n)}{F}{}^{\alpha} - \overset{(n)}{f}{}^{\alpha}) \overset{(n)}{\delta V}_{\alpha} - (\overset{(n)}{F}{}^{3} - \overset{(n)}{f}{}^{3}) \overset{(n)}{\delta V}_{3} - \overset{(n)}{p}{}^{\alpha} \overset{(n)}{\delta V}_{\alpha} - \overset{(n)}{p}{}^{3} \overset{(n)}{\delta V}_{3}]\, dM$$

$$- \int_{C_f} [\, \overset{(n)}{{}^{*}L}{}^{\alpha\beta} \nu_{\alpha}\nu_{\beta} \overset{(n)}{\delta V}_{\nu} + \overset{(n)}{{}^{*}L}{}^{\alpha\beta} t_{\alpha}\nu_{\beta} \overset{(n)}{\delta V}_{t} + \overset{(n)}{{}^{*}L}{}^{3\beta} \nu_{\beta} \overset{(n)}{\delta V}_{3}]\, ds \Big\} = o . \tag{4.1o}$$

The subsidiary conditions of (4.1o) are the 2-D strain-displacement relations as given by (3.14) - (3.15) and the geometric boundary conditions on C_v

$$\overset{(n)}{V}_\nu = \overset{*(n)}{V}_\nu \quad , \quad \overset{(n)}{V}_t = \overset{*(n)}{V}_t \quad , \quad \overset{(n)}{V}_3 = \overset{*(n)}{V}_3 \quad . \tag{4.11}$$

Here, we remind the convention that $\overset{(n)}{V}_\alpha$ and $\overset{(n)}{V}_3$ are zero for n>N. Introducing the above mentioned subsidiary conditions in (4.1o), transforming the result by means of partial integration and of the divergence theorem, and considering the variations of $\overset{(n)}{V}_\alpha$ and $\overset{(n)}{V}_3$ as arbitrary, we obtain the equations of motion for the theory of plates accounting for small strains and moderate rotations

$$\begin{aligned} \delta\overset{(n)}{V}_\alpha: &\quad \overset{(n)}{T}{}^{\alpha\beta}|_\beta + \overset{(n)}{Q}{}^{\alpha 3} + \overset{(n)}{F}{}^{\alpha} - \overset{(n)}{f}{}^{\alpha} + \overset{(n)}{p}{}^{\alpha} = o \quad , \\ \delta\overset{(n)}{V}_3: &\quad \overset{(n)}{T}{}^{3\beta}|_\beta + \overset{(n)}{Q}{}^{33} + \overset{(n)}{F}{}^{3} - \overset{(n)}{f}{}^{3} + \overset{(n)}{p}{}^{3} = o \quad , \end{aligned} \tag{4.12}$$

and the associated static boundary conditions on C_f

$$\begin{aligned} \delta\overset{(n)}{V}_\nu: &\quad \overset{(n)}{T}{}^{\alpha\beta}\nu_\alpha\nu_\beta = {}^{*}\overset{(n)}{L}{}^{\alpha\beta}\nu_\alpha\nu_\beta \quad , \\ \delta\overset{(n)}{V}_t: &\quad \overset{(n)}{T}{}^{\alpha\beta}t_\alpha\nu_\beta = {}^{*}\overset{(n)}{L}{}^{\alpha\beta}t_\alpha\nu_\beta \quad , \\ \delta\overset{(n)}{V}_3: &\quad \overset{(n)}{T}{}^{3\beta}\nu_\beta = {}^{*}\overset{(n)}{L}{}^{3\beta}\nu_\beta \quad , \end{aligned} \tag{4.13}$$

where

$$\overset{(n)}{T}{}^{\alpha\beta} = \overset{(n)}{L}{}^{\alpha\beta} + \sum_{p=o}^{N} \underline{\overset{(p)}{\Omega}{}^{\alpha}_{.3}\, \overset{(p+n)}{L}{}^{\beta 3}} \quad ,$$

$$\begin{aligned} \overset{(n)}{T}{}^{3\beta} = \overset{(n)}{L}{}^{3\beta} + \frac{1}{2}\sum_{p=o}^{N} \Big[& (\overset{(p)}{\Omega}_{3\lambda} + \underline{\overset{(p)}{\Omega}_{3\lambda} + \overset{(p)}{\eta}_{3\lambda}})\, \overset{(p+n)}{L}{}^{\beta\lambda} + \underline{\overset{(p)}{\eta}_{33}\, \overset{(p+n)}{L}{}^{\beta 3}} \\ & \underline{- (\overset{(p)}{\eta}{}^{\beta}_{.\lambda} + \overset{(p)}{\Omega}{}^{\beta}_{.\lambda})\, \overset{(p+n)}{L}{}^{3\lambda}} + (-\overset{(p)}{\Omega}{}^{\beta}_{.3} + \underline{\overset{(p)}{\Omega}{}^{\beta}_{.3} - \overset{(p)}{\eta}{}^{\beta}_{.3}})\, \overset{(p+n)}{L}{}^{33} \Big] \quad , \end{aligned}$$

$$\begin{aligned} \overset{(n)}{Q}{}^{\beta 3} = -n \Big\{ \overset{(n-1)}{L}{}^{\beta 3} + \frac{1}{2}\sum_{p=o}^{N} \Big[& (- \overset{(p)}{\Omega}_{3\lambda} + \underline{\overset{(p)}{\Omega}_{3\lambda} - \overset{(p)}{\eta}_{3\lambda}})\, \overset{(p+n-1)}{L}{}^{\beta\lambda} - \underline{\overset{(p)}{\eta}_{33}\, \overset{(p+n-1)}{L}{}^{\beta 3}} \\ & \underline{+ (\overset{(p)}{\eta}{}^{\beta}_{.\lambda} + \overset{(p)}{\Omega}{}^{\beta}_{.\lambda})\, \overset{(p+n-1)}{L}{}^{3\lambda}} + (\overset{(p)}{\Omega}{}^{\beta}_{.3} + \underline{\overset{(p)}{\Omega}{}^{\beta}_{.3} + \overset{(p)}{\eta}{}^{\beta}_{.3}})\, \overset{(p+n-1)}{L}{}^{33} \Big] \Big\} \quad , \end{aligned}$$

$$\overset{(n)}{Q}{}^{33} = -n \Big\{ \overset{(n-1)}{L}{}^{33} + \sum_{p=o}^{N} \underline{\overset{(p)}{\Omega}_{3\lambda}\, \overset{(p+n-1)}{L}{}^{\lambda 3}} \Big\} \quad . \tag{4.14}$$

In the above equations the underlined terms represent the induced counterpart of those terms identified in the strain-displacement relations by a solid line. Being of the order of magnitude θ^3, these latter terms are generally not negligible as compared to the remaining terms which are of the order of magnitude θ^2 (see (3.1o) - (3.12)). It may also be remarked that in the equations of motion some of those terms (i.e. products of stress couples with linearized rotations of the normal) generated by the $o(\theta^3)$-terms have the same order of magnitude as those induced by the $o(\theta^2)$-terms. This, in fact, constitutes another argument for not neglecting ab initio those $o(\theta^3)$-terms.

On the other hand it may be observed that among the underlined terms in the equations of motion there are several terms (as e.g. products of stress couples with linearized strains or in-surface rotations) which could be neglected consistently with the postulated relative error margin $o(\theta^2)$ of the theory considered here. Such a simplification would yield the equations of motion and the static boundary conditions in the same form as the one exhibited in eqs. (4.12) - (4.13), but now we have the following representations for $\overset{(n)}{T}{}^{3\beta}$ and $\overset{(n)}{Q}{}^{\beta 3}$:

$$\overset{(n)}{T}{}^{3\beta} = \overset{(n)}{L}{}^{3\beta} + \sum_{p=o}^{N} \overset{(p)}{\Omega}_{3\lambda} \overset{(p+n)}{L}{}^{\beta\lambda} \quad , \quad \overset{(n)}{Q}{}^{\beta 3} = -n\,\{ \overset{(n-1)}{L}{}^{\beta 3} + \sum_{p=o}^{N} \overset{(p)}{\Omega}{}^{\beta}{}_{.3} \overset{(p+n-1)}{L}{}^{33} \}. \quad (4.15)$$

This latter variant of the equations of motion (4.12), $(4.14)_{1,4}$, (4.15) and boundary conditions (4.13) agrees with that obtained by KAUL [2o], when in his equations the terms involving the in-surface rotations are dropped. It is to be noted that the present analysis was undertaken in the framework of the moderate rotation concept. Therefore, KAUL's [2o] assertion that his equations hold true for arbitrary unrestricted rotations is not valid.

Equations of motion and boundary conditions pertinent to the first-order transverse shear approximation theory (and consistent with the representation (3.16) of the displacement field) are obtainable through the appropriate specialization of eqs. (4.12) - (4.14). This may be done by taking in (4.12) - (4.14) the truncation index N as $N = 1$, the summation index n ranging from o to 1. In the same manner, by making use of the simplified expressions (4.15) instead of $(4.14)_{2,3}$, a simplified version of the equations of motion and associated boundary conditions for moderate rotation first-order transverse shear approximation plate theory may be obtained. In this context it may be noticed that consistent with the simplified kinematic relations (3.22), the equations of motion and static boundary conditions obtained by WEMPNER [21] may be viewed as special cases of the present equations (4.12) - (4.14). They may be obtained by taking $n = o,1$ in $(4.12)_1$ and (4.13), $n = o$ in $(4.12)_2$, and by retaining only one of the nonlinear terms, namely $-\overset{(o)}{V}_{3,\lambda}\overset{(o)}{L}{}^{\beta\lambda}$ in $(4.14)_2$. As it is the case with (3.22), also equations of motion and boundary conditions identical to those given by WEMPNER [21] have been obtained by LIBRESCU in chapter IV of [2], by using the partially nonlinear strain-displacement relations (3.23) together with the representation of the displacement field (3.16), specialized for $\overset{(1)}{V}_3 \equiv o$. Moreover, in [2], pp.484 - 488, the associated field equations have been reduced to a form yielding (a) the von Kármán plate equations in the case of infinite transverse shear rigidity, (b) upon linearization a refined small deflection plate theory, where the equations governing bending (Reissner's equations) and stretching appear in a decoupled form.

4.2. Equations of Motion and Boundary Conditions for Moderate Rotation Plate Theory Under Kirchhoff Constraints

Substituting (3.29) into (4.1o) and further paralleling the general procedure outlined in the preceding section, we obtain the equations of motion

$$\delta\overset{(o)}{V}_\alpha: \quad \overset{(o)}{L}{}^{\alpha\beta}|_\beta + \overset{(o)}{F}{}^\alpha - \overset{(o)}{f}{}^\alpha + \overset{(o)}{p}{}^\alpha = o \quad ,$$

$$\delta\overset{(o)}{V}_3: [\overset{(1)}{L}{}^{\alpha\beta}|_\alpha + \overset{(o)}{V}_{3,\alpha}\overset{(o)}{L}{}^{\alpha\beta} + \overset{(1)}{F}{}^\beta - \overset{(1)}{f}{}^\beta + \overset{(1)}{p}{}^\beta]|_\beta + \overset{(o)}{F}{}^3 - \overset{(o)}{f}{}^3 + \overset{(o)}{p}{}^3 = o \quad , \qquad (4.16)$$

the static boundary conditions on C_f

$$\delta\overset{(o)}{V}_\nu: \overset{(o)}{L}{}^{\alpha\beta}\nu_\alpha\nu_\beta = \overset{*(o)}{L}{}^{\alpha\beta}\nu_\alpha\nu_\beta \quad ,$$

$$\delta\overset{(o)}{V}_t: \overset{(o)}{L}{}^{\alpha\beta}t_\alpha\nu_\beta = *\overset{(o)}{L}{}^{\alpha\beta}t_\alpha\nu_\beta \quad ,$$

$$\delta\overset{(o)}{V}_3: (\overset{(1)}{L}{}^{\alpha\beta}|_\alpha + \overset{(o)}{V}_{3,\alpha}\overset{(o)}{L}{}^{\alpha\beta} + \overset{(1)}{F}{}^\beta - \overset{(1)}{f}{}^\beta + \overset{(1)}{p}{}^\beta)\nu_\beta + \frac{\partial}{\partial s}(\overset{(1)}{L}{}^{\alpha\beta}t_\alpha\nu_\beta)$$

$$= *\overset{(o)}{L}{}^{3\beta}\nu_\beta + \frac{\partial}{\partial s}(*\overset{(1)}{L}{}^{\alpha\beta}t_\alpha\nu_\beta) \quad ,$$

$$\delta\left(\frac{\partial \overset{(o)}{V}_3}{\partial s_\nu}\right): \overset{(1)}{L}{}^{\alpha\beta}\nu_\alpha\nu_\beta = *\overset{(1)}{L}{}^{\alpha\beta}\nu_\alpha\nu_\beta \quad , \qquad (4.17)$$

and the static corner conditions at each corner point P_j of C_f located at $s = s_j$, $j = 1,2,\ldots$,

$$\delta\overset{(o)}{V}_3: (\overset{(1)}{L}{}^{\alpha\beta}(s_j+o) - \overset{(1)}{L}{}^{\alpha\beta}(s_j-o))t_\alpha\nu_\beta = (*\overset{(1)}{L}{}^{\alpha\beta}(s_j+o) - *\overset{(1)}{L}{}^{\alpha\beta}(s_j-o))t_\alpha\nu_\beta \qquad (4.18)$$

In (4.17) $\partial/\partial s$ and $\partial/\partial s_\nu$ denote the derivatives along the tangent and normal to the boundary curve C, respectively. These equations constitute the well known geometrically nonlinear variant for the classical moderate rotation theory of thin plates.

5. Constitutive Equations

At this point it is necessary to supplement the preceding equations by constitutive relations. They will be deduced by assuming the plate material as being homogeneous and elastically and thermally anisotropic. We shall make use of the fact that for a physically linear but geometrically nonlinear theory a linear relation may be established between second Piola-Kirchhoff stress and Lagrange strain tensors. If in addition a nonuniform temperature field $T \equiv T\,(x^\alpha, x^3)$ is taken into consideration we may write

$$s^{ij} = E^{ijkl}\, e_{kl} + \lambda^{ij} T \tag{5.1}$$

where E^{ijkl} and λ^{ij} stand for the spatial tensors of elasticity moduli and thermal expansion coefficients, respectively, exhibiting the well-known symmetry properties. Restricting the anisotropy properties to the case of elastic and thermal symmetry with respect to the plane $x^3 = o$, taking the representation of the temperature field in the form

$$T(x^\alpha, x^3) = \sum_{n=o}^{N} (x^3)^n \overset{(n)}{T}(x^\alpha) \,,$$

inserting (5.1) into the expression for the stress couples (4.3) and having in view the representations (3.13) for the strain measures we obtain (see LIBRESCU [2])

$$\overset{(n)}{L}{}^{\alpha\beta} = \sum_{r=o}^{2N} \eta(r+n+1)\,\{E^{\alpha\beta\lambda\mu}\overset{(r)}{e}_{\lambda\mu} + E^{\alpha\beta 33}\overset{(r)}{e}_{33} + \lambda^{\alpha\beta}\overset{(r)}{T}\}\,,$$

$$\overset{(n)}{L}{}^{\alpha 3} = 2\sum_{r=o}^{2N} \eta(r+n+1)\, E^{\alpha 3\beta 3}\overset{(r)}{e}_{\beta 3}\,,$$

$$\overset{(n)}{L}{}^{33} = \sum_{r=o}^{2N} \eta(r+n+1)\,\{E^{33\alpha\beta}\overset{(r)}{e}_{\alpha\beta} + E^{3333}\overset{(r)}{e}_{33} + \lambda^{33}\overset{(r)}{T}\}\,. \tag{5.2}$$

These constitutive equations may be specialized for several types of elastic anisotropy (including e.g. orthotropy and transverse isotropy) and finally for isotropic materials. For each of these cases the pertinent expressions for the tensors of elastic moduli and thermal expansion coefficients may be found in [2]. In addition, (5.2) may provide by appropriate specialization the constitutive equations of the Kirchhoff theory of plates. For a straightforward deduction of these equations see e.g.[2].

6. Conclusions

On the basis of the full set of field equations, i.e. of the constitutive equations, the geometric equations, the equations of motion and the boundary conditions developed in the preceding sections, it is possible to derive the governing equations valid for the theory of anisotropic plates accounting for small strains and moderate rotations, which incorporates transverse shear and transverse normal deformation effects, higher-order effects as well as the kinetic and thermal ones. The generality of the present approach makes it possible to extract a variety of intermediate theories which would correspond to various special representations of the displacement field across the plate thickness. It was shown that the present theory includes also several variants of nonlinear higher-order and first-order transverse shear deformation theories obtained in literature in different ways. The theory derived here provides a common basis for all these approximate variants. Furthermore, the present approach, which is based on proper estimations of the order of magnitude of the quantities involved, makes it possible to classify these various variants and to clarify their limits of applicability. In addition, the linear higher-order theory of anisotropic plates [2] may also be obtained as a special case of the theory developed in the present paper. It is worth mentioning that the derived field equations may constitute an appropriate basis for the approach of the stability problem (either in linear or nonlinear formulation) as well as of the vibration problem of anisotropic plates. Some results allowing to infer the influence of the transverse shear deformation effect in these problems may be found e.g. in the works of LIBRESCU [2], VINSON and CHOU [24], NOOR [25], REDDY [26], REDDY and CHAO [27], AMBARTSUMIAN [29], BRUNELLE and ROBERTSON [3o]. It should also be mentioned that the 2-D variational formulation of the present theory as given in (4.1o) provides a basis for a numerical approach of the problems considered here e.g. by means of finite element or finite difference methods.

Acknowledgement

The present research was carried out when the second author was associated as a visiting research fellow with the Department of Solid Mechanics, Materials and Structures of the Tel Aviv University for the period of one year, during which he was supported by a MINERVA-Scholarship of the Max-Planck-Gesellschaft, Heidelberg. Both authors wish to express their sincerest thanks for this support, which made their cooperation possible.

References

1. NAGHDI P.M., The Theory of Shells and Plates, in "Handbuch der Physik", vol. VI a/2, ed. S. Flügge, 425-64o, Springer-Verlag, Berlin - Heidelberg - New York, 1972.

2. LIBRESCU L., Elastostatics and Kinetics of Anisotropic and Heterogeneous Shell-Type Structures, Noordhoff Int. Publ., Leyden, 1975.

3. LIBRESCU L., Refined Geometrically Non-Linear Theories of Anisotropic Laminated Shells, Quarterly of Applied Mathematics, (in print).

4. YOKOO Y., MATSUNAGA H., A General Theory of Elastic Shells, Int. J. Solids Structures 1o (1974), 261-274.

5. HABIP L.M., Theory of Elastic Shells in the Reference State, Ingenieur-Archiv 34 (1965), 228-237.

6. HABIP L.M., Theory of Elastic Plates in the Reference State, Int. J. Solids Structures 2 (1966), 157-166.

7. HABIP L.M., EBCIOGLU I.K., On the Equations of Motion of Shells in the Reference State, Ingenieur-Archiv 34 (1965), 28-32.

8. AINOLA L. IA., Nonlinear Timoshenko Type Theory of Elastic Shells (in Russian), Izv. Akad. Nauk. Eston. SSR 14 (1965), 337-344.

9. GALIMOV K.Z., Foundations of the Nonlinear Theory of Shells (in Russian), Kazan' University Press, Kazan' 1975.

1o. GALIMOV K.Z., Theory of Shells with Transverse Shear Deformation Effect (in Russian), Kazan' University Press, Kazan' 1977.

11. PIETRASZKIEWICZ W., Finite Rotations and Lagrangean Description in the Non-Linear Theory of Shells, Polish Scientific Publishers, Warszawa - Poznan, 1979.

12. SANDERS J.L., Nonlinear Theories for Thin Shells, Quart. Appl. Math. 21 (1963), 21-36.

13. KOITER W.T., On the Nonlinear Theory of Thin Elastic Shells, Proc. Kon. Ned. Ak. Wet., Series B, 69 (1966), 1-54.

14. REISSNER E., Rotationally Symmetric Problems in the Theory of Thin Elastic Shells, Proc. 3rd U.S. Nat. Congr. of Appl. Mech., 1958, 51-69.

15. PIETRASZKIEWICZ W., Finite Rotations in the Non-Linear Theory of Thin Shells, in "Thin Shell Theory, New Trends and Applications", ed. W. Olszak, 153-2o8, Springer-Verlag, Wien - New York, 198o.

16. PIETRASZKIEWICZ W., Lagrangian Description and Incremental Formulation in the Non-Linear Theory of Thin Shells, Int. J. Non-Linear Mechanics 19 (1984), 115-14o.

17. SCHMIDT R., Variational Principles for General and Restricted Kirchhoff-Love Type Shell Theories, Proc. Int. Conf. on Finite Element Methods, Shanghai (China) 1982, eds. He Guangqian, Y.K. Cheung, 621-626, Gordon and Breach, New York, 1982.

18. SCHMIDT R., On Geometrically Nonlinear Theories for Thin Elastic Shells, in "Flexible Shells, Theory and Applications", Proc. EUROMECH-Colloquium Nr. 165, Munich (Germany) 1983, eds. E.L. Axelrad, F.A. Emmerling, 76-9o, Springer-Verlag, Berlin - Heidelberg - New York - Tokyo, 1984.

19. SCHMIDT R., A Current Trend in Shell Theory: Constrained Geometrically Nonlinear Kirchhoff-Love Type Theories Based on Polar Decomposition of Strains and Rotations, Proc. Symp. on Advances and Trends in Structures and Dynamics, Arlington, Virginia (USA), 1984, eds. A.K. Noor, R.J. Hayduk, 265-275, Pergamon Press, New York, 1985, reprinted in Computers and Structures 2o (1985), 265-275.

2o. KAUL R.K., Finite Thermal Oscillations of Thin Plates, Int. J. Solids Structures 2 (1966), 337-35o.

21. WEMPNER G.A., Mechanics of Solids with Applications to Thin Bodies, McGraw-Hill, New York, 1973.

22. NAGHDI P.M., VONGSARNPIGOON L., A Theory of Shells with Small Strains Accompanied by Moderate Rotations, Arch. for Rational Mechanics and Analysis 83 (1983), 245-283.

23. YU Y.-Y., Generalized Hamilton's Principle and Variational Equation of Motion in Nonlinear Elasticity Theory, with Application to Plate Theory, J. of the Acoustical Society of America 36 (1964), 111-12o.

24. VINSON J.R., CHOU T.W., Composite Materials and Their Use in Structures, J. Wiley and Sons, New York - Toronto, 1974.

25. NOOR A.K., Stability of Multilayered Composite Plates, Fibre Science and Technology 8 (1975), 81-89.

26. REDDY J.N., Analysis of Layered Composite Plates Accounting for Large Deflections and Transverse Shear Strains, in "Recent Advances in Non-Linear Computational Mechanics", eds. E. Hinton, D.R.J. Owen and C. Taylor, Pineridge Press Ltd., Swansea, U.K., 1982.

27. REDDY J.N., CHAO W.C., Nonlinear Oscillations of Laminated Anisotropic Rectangular Plates, Journal of Applied Mechanics, Trans. ASME, 49 (1982), 396-4o2.

28. BRULL M.A., LIBRESCU L., Strain Measures and Compatibility Equations in the Linear High-Order Shell Theories, Quarterly of Appl. Mathematics, 4o (1982), 15-25.

29. AMBARTSUMIAN S.A., Theory of Anisotropic Plates (English Transl.), Technomic, Stamford, Conn., 197o.

3o. BRUNELLE E.J., ROBERTSON S.R., Initially Stressed Mindlin Plates, AIAA Journal 12 (1974), 1o36-1o45.

FINITE STRAINS AND ROTATIONS IN SHELLS

J. MAKOWSKI* and H. STUMPF
Ruhr - University, Universitätstr. 150 IA 3
4630 Bochum (FRG)

1. Introduction

In this article we present an exact and systematic derivation of a theory of finite strain deformation of shells from the principles of classical continuum mechanics. We assume that the three-dimensional deformation of the shell satisfies the following kinematical constraints; I) material fibres initially normal to the reference surface of the shell remain straight during the deformation, II) deformation is isochoric (volume preserving). These are the only assumptions made in our developments out which the first one is of simplifying nature while the other one reflects merely the real property of many materials. We show that the resulting theory is characterized by the following features; a) a rational incorporation of transverse shear deformation and exact incorporation of transverse normal deformation, b) the constitutive equations for the stress resultants and stress couple as nonlinear functions of appropriate strain measures and their surface derivatives, c) a sufficient geometric structure to account for a non-uniform change in the shell thickness at the boundary.

To handle this generality of the shell theory we find it convenient to formulate the basic equations without explicite reference to the displacement field. In our development the basic kinematical variables are the position vector of the reference surface and some proper orthogonal tensor $\underset{\sim}{Q}$. We show that this tensor may be decomposed into $\underset{\sim}{R}$ describing the local rotations of the reference surface and a rotation tensor $\underset{\sim}{\Gamma}$ entailing the transverse shear deformation. The corresponding strain measures and other relevant relations for a finitely deformed shell are next derived in terms of these alternate rotations. Finally, the conjugate stress measures are defined through the principle of virtual work.

---

*) Technical University, ul. ZSP 5, 45-223 Opole (Poland). Part of the paper has been written during the author's stay at Ruhr - University

2. Preliminaries

Consider a three-dimensional body in Euclidean space $\mathbb{E}^3$ and let Θ^i, $i=1,2,3$, be material (convected) coordinates assigned to each material point. Without loss in generality we may identify a body with the domain $\mathbb{B}$ in $\mathbb{E}^3$ that it occupies in the fixed initial configuration. Let $\underset{\sim}{X} = \underset{\sim}{X}(\Theta^i)$ be the position vector of any point $P \in \mathbb{B}$. Then the base vectors, the reciprocal base vectors and the components of the metric tensor at P are defined as usual

$$\underset{\sim}{g}_i = \underset{\sim}{X}_{,i} \ , \quad \underset{\sim}{g}^i \cdot \underset{\sim}{g}_j = \delta^i_j \ , \quad g_{ij} = \underset{\sim}{g}_i \cdot \underset{\sim}{g}_j \ , \quad g^{ik} g_{kj} = \delta^i_j \ , \qquad (2.1)$$

where $(\ )_{,i}$ indicate partial derivatives with respect to Θ^i. A deformation of $\mathbb{B}$ is a diffeomorphism $\chi\colon \mathbb{B} \to \mathbb{E}^3$ and we write $\underset{\sim}{x} = \underset{\sim}{\chi}(\underset{\sim}{X})$, wherein $\underset{\sim}{x} = \underset{\sim}{x}(\Theta^i)$ is the position vector of a place $\bar{P} \in \bar{\mathbb{B}} = \chi(\mathbb{B})$ in the deformed configuration of the body that is occupied by the particle P. Following the conventional notations we shall distinguish the geometric quantities defined at $\bar{P}$ by the bar over symbols, e.g. $\bar{\underset{\sim}{g}}_i$, $\bar{g}_{ij}$, etc.

A shell is a three-dimensional body $\mathbb{B}$ whose boundary surface $\partial\mathbb{B}$ has special features. To describe its geometry we take $(\Theta^i) = (\Theta^\alpha, \xi)$, $\alpha=1,2$, to be normal coordinates in $\mathbb{B}$. Then $\xi = 0$ defines a material surface Π in $\mathbb{B}$, called the undeformed reference surface, and the position vector of any point $P \in \mathbb{B}$ takes the form

$$\underset{\sim}{X}(\Theta^\alpha, \xi) = \underset{\sim}{r}(\Theta^\alpha) + \xi \underset{\sim}{n}(\Theta^\alpha) , \qquad \xi \in [-h_o^-, +h_o^+] . \qquad (2.2)$$

Here $\underset{\sim}{r}$ is the position vector of Π, $\underset{\sim}{n}$ denotes the unit normal vector to this surface and $h_o(\Theta^\alpha) = h_o^- + h_o^+$ is the initial (variable) shell thickness. We emphasize that the reference surface need not to be the middle surface in $\mathbb{B}$. Its geometry is determind by the formulas [1,2]

$$\begin{aligned} &\underset{\sim}{a}_\alpha = \underset{\sim}{r}_{,\alpha} \ , && \underset{\sim}{a}^\alpha \cdot \underset{\sim}{a}_\beta = \delta^\alpha_\beta \ , && \underset{\sim}{n} = \tfrac{1}{2}\varepsilon^{\alpha\beta} \underset{\sim}{a}_\alpha \times \underset{\sim}{a}_\beta \ , \\ &a_{\alpha\beta} = \underset{\sim}{a}_\alpha \cdot \underset{\sim}{a}_\beta \ , && a^{\alpha\beta} = \underset{\sim}{a}^\alpha \cdot \underset{\sim}{a}^\beta \ , && a = \det a_{\alpha\beta} > 0 \ , \qquad (2.3) \\ &b_{\alpha\beta} = \underset{\sim}{a}_{\alpha,\beta} \cdot \underset{\sim}{n} \ , && H = \tfrac{1}{2} b^\alpha_\beta \ , && K = \tfrac{1}{2}\varepsilon^{\alpha\lambda}\varepsilon^{\beta\varkappa} b_{\alpha\beta} b_{\lambda\varkappa} \ . \end{aligned}$$

Furthermore, owing to the particular choice of the material coordinates the base vectors at $P \in \mathbb{B}$ can be expressed in the form [1,2]

$$\begin{aligned} &\underset{\sim}{g}_\alpha = \mu^\beta_\alpha \underset{\sim}{a}_\beta \ , && \underset{\sim}{g}^\alpha = (\mu^{-1})^\alpha_\beta \underset{\sim}{a}^\beta \ , && \underset{\sim}{g}_3 = \underset{\sim}{g}^3 = \underset{\sim}{n} \ , \\ &\mu^\beta_\alpha = \delta^\beta_\alpha - \xi b^\beta_\alpha \ , && (\mu^{-1})^\alpha_\beta = \tfrac{1}{\mu}[\delta^\alpha_\beta - \xi(2H\delta^\alpha_\beta - b^\alpha_\beta)] \ , && \qquad (2.4) \end{aligned}$$

$$\mu = \det\mu_{\alpha}^{\beta} = \sqrt{\frac{g}{a}} = 1 - 2\xi H + \xi^2 K .$$

For a detailed explanation of notations used in (2.3) and (2.4) the reader is referred to [1,2]. Here we only add that throughout the paper Latin indices have the range 1, 2, 3 and Greek indices have the range 1, 2. These indices obey the summation convention.

A shell may be thin or thick. In general it is assumed that $(h_o/R)^p \ll 1$ for some positive rational number p, where R denotes some characteristic length of the undeformed reference surface (minimum radius of curvature, smallest dimension). The number p characterizes thinness of the shell.

The boundary $\partial\mathbb{B}$ of the shell consists of the bounding surfaces Π^+ and Π^-, their position vectors being $\underset{\sim}{X}^+ = \underset{\sim}{X}(\Theta^\alpha, h_o^+)$ and $\underset{\sim}{X}^- = \underset{\sim}{X}(\Theta^\alpha, -h_o^-)$, respectively and the lateral surface $\partial\mathbb{B}^*$. For simplicity we assume that $\partial\mathbb{B}^*$ is a ruled surface given by

$$\underset{\sim}{X}(s,\xi) = \underset{\sim}{r}(s) + \xi\underset{\sim}{n}(s) , \qquad \xi \in |-h_o^-, +h_o^+| , \tag{2.5}$$

where s denotes the arc length along piecewise smooth boundary curve $\partial\Pi$. At any regular point $M \in \partial\Pi$ the orthonormal triad $(\underset{\sim}{\nu}, \underset{\sim}{t}, \underset{\sim}{n})$ is defined by

$$\underset{\sim}{t}(s) = \frac{d}{ds}\underset{\sim}{r} = t^\alpha \underset{\sim}{a}_\alpha , \quad \underset{\sim}{\nu}(s) = \underset{\sim}{t}\times\underset{\sim}{n} = \nu^\alpha \underset{\sim}{a}_\alpha , \quad \underset{\sim}{n} = \underset{\sim}{\nu}\times\underset{\sim}{t} . \tag{2.6}$$

Its geometric meaning is clear from the definition (see [1]).

Under a smooth, but otherwise arbitrary deformation $\chi : \mathbb{B} \to \bar{\mathbb{B}}$ of the shell, the reference surface Π deforms into a surface $\bar{\Pi} = \chi(\Pi)$ in $\bar{\mathbb{B}}$ having position vector $\bar{\underset{\sim}{r}} = \bar{\underset{\sim}{r}}(\Theta^\alpha)$. We shall denote by $\bar{\underset{\sim}{a}}_\alpha$, $\bar{\underset{\sim}{a}}^\alpha$, $\bar{\underset{\sim}{n}}$, $\bar{a}_{\alpha\beta}$, $\bar{b}_{\alpha\beta}$, etc. the base vectors, their duals, the unit normal vector, the components of metric and curvature tensors and other geometric quantities that are defined on $\bar{\Pi}$ in the same way as their unbarred counterparts (2.3) are defined on Π. The position vector to any point $\bar{P} \in \bar{\mathbb{B}}$ can now be expressed in the form

$$\underset{\sim}{x}(\Theta^\alpha,\xi) = \bar{\underset{\sim}{r}}(\Theta^\alpha) + \underset{\sim}{\zeta}(\Theta^\alpha,\xi) . \tag{2.7}$$

For further reference we define the field of unit vectors $\underset{\sim}{N}$ over $\bar{\Pi}$ by

$$\underset{\sim}{N}(\Theta^\alpha) = \lambda_\xi^{-1} \underset{\sim}{\zeta}_{,\xi}|_{\xi=0} , \qquad \lambda_\xi(\Theta^\alpha) = |\underset{\sim}{\zeta}_{,\xi}|_{\xi=0} , \tag{2.8}$$

where $(\ )_{,\xi}$ stands for the derivative with respect to the normal coordinate ξ.

3. Geometry of deformation

Our aim in this chapter is to examine the structure of a three-dimensional deformation of the shell within the single

Assumption 1: Material fibres normal to the reference surface in the initial configuration remain straight during the shell deformation.

Accordingly, the position vector to any point $\bar{P} \in \bar{\mathbb{B}}$ must be of the form

$$\underset{\sim}{x}(\Theta^\alpha, \xi) = \underset{\sim}{\bar{r}}(\Theta^\alpha) + \zeta(\Theta^\alpha, \xi)\underset{\sim}{N}(\Theta^\alpha) , \tag{3.1}$$

where $\underset{\sim}{N}$ is a field of unit vectors over $\bar{\Pi}$ (its geometric meaning follows from (2.8)) and an unspecified, at this stage of considerations, function $\zeta = \zeta(\Theta^\alpha, \xi)$ incorporates an arbitrary transverse normal deformation of initially normal fibres. The material coordinates employed in the analysis imply that

$$\zeta(\Theta^\alpha, 0) = \zeta,_\beta(\Theta^\alpha, 0) = 0 . \tag{3.2}$$

Moreover, to ensure that the deformation be locally invertible and orientation-preserving the following conditions have to be satisfied

$$\underset{\sim}{\bar{n}}.\underset{\sim}{N} > 0 , \qquad \zeta,_\xi(\Theta^\alpha, \xi) > 0 \quad \text{for} \quad \xi \in [-h_\circ^-, +h_\circ^+] . \tag{3.3}$$

Our task now is to derive the expressions for the components of metric tensor at any point $\bar{P} \in \bar{\mathbb{B}}$ in an appropriate form for further analysis.

Differentiating (3.1) with respect to the material coordinates one gets

$$\underset{\sim}{\bar{g}}_\alpha = \underset{\sim}{\bar{a}}_\alpha + \zeta\underset{\sim}{N},_\alpha + \zeta,_\alpha\underset{\sim}{N} , \qquad \underset{\sim}{\bar{g}}_3 = \zeta,_\xi\underset{\sim}{N} . \tag{3.4}$$

To proceed further we introduce the following bases $\underset{\sim}{\bar{A}}_i(\Theta^\alpha) = \underset{\sim}{\bar{g}}_i(\Theta^\alpha, 0)$ and $\underset{\sim}{\bar{A}}^i(\Theta^\alpha) = \underset{\sim}{\bar{g}}^i(\Theta^\alpha, 0)$ that are defined on $\bar{\Pi}$. From (3.4) it follows that

$$\underset{\sim}{\bar{A}}_\alpha = \underset{\sim}{\bar{a}}_\alpha = \underset{\sim}{\bar{r}},_\alpha , \qquad \underset{\sim}{\bar{A}}_3 = \lambda_\xi\underset{\sim}{N} , \qquad \lambda_\xi(\Theta^\alpha) = \zeta,_\xi|_{\xi=0} . \tag{3.5}$$

Solving next the system of equations $\underset{\sim}{\bar{A}}^i.\underset{\sim}{\bar{A}}_j = \delta^i_j$ we obtain

$$\underset{\sim}{\bar{A}}^\alpha = \bar{\varphi}_n^{-1}\bar{\varepsilon}^{\alpha\beta}\underset{\sim}{\bar{a}}_\beta \times \underset{\sim}{N} = \underset{\sim}{\bar{a}}^\alpha - \bar{\varphi}_n^{-1}\bar{\varphi}^\alpha\underset{\sim}{\bar{n}} , \qquad \underset{\sim}{\bar{A}}^3 = \lambda_\xi^{-1}\bar{\varphi}_n^{-1}\underset{\sim}{\bar{n}} , \tag{3.6}$$

where

$$\underset{\sim}{N} = \bar{\varphi}^\alpha\underset{\sim}{\bar{a}}_\alpha + \bar{\varphi}_n\underset{\sim}{\bar{n}} = \bar{\varphi}_\alpha\underset{\sim}{\bar{a}}^\alpha + \bar{\varphi}_n\underset{\sim}{\bar{n}} , \qquad \bar{\varphi}_n = \sqrt{1 + \bar{\varphi}^\lambda\bar{\varphi}_\lambda} . \tag{3.7}$$

The base vectors (3.4) can now be expressed in the form

$$\bar{g}_\alpha = \bar{\mu}^\beta_\alpha \bar{a}_\beta + \bar{\mu}_\alpha N , \qquad \bar{g}_3 = \zeta,_\xi N , \tag{3.8}$$

$$\bar{\mu}^\beta_\alpha = \delta^\beta_\alpha - \zeta\bar{B}^\beta_\alpha , \qquad \bar{\mu}_\alpha = \zeta,_\alpha + \zeta\bar{B}^\beta_\alpha\varphi_\beta , \qquad \bar{B}^\beta_\alpha = -\bar{A}^\beta . N,_\alpha . \tag{3.9}$$

Here $\bar{B}^\beta_\alpha$ is defined in such a way that whenever $N = \bar{n}$, i.e. if the initially normal fibres remain normal to the deformed reference surface, we have $\bar{B}^\beta_\alpha = \bar{b}^\beta_\alpha$ and the relations (3.8) are reduced to a form obtained in [3] for this particular case. Moreover, they take the known form [1,2] if in addition we assume $\zeta = \xi$. Let us note that (3.3) implies that $\bar{\mu}^\beta_\alpha$ is non-singular for $\xi \in [-h_o^-, +h_o^+]$ and we can define its inverse by $(\bar{\mu}^{-1})^\alpha_\lambda \bar{\mu}^\lambda_\beta = \delta^\alpha_\beta$ The solution of this system of equations is (it should be kept in mind that in general $\bar{B}^1_2 \neq \bar{B}^2_1$)

$$(\bar{\mu}^{-1})^\alpha_\beta = \frac{1}{\bar{\mu}}\begin{bmatrix} 1 - \zeta\bar{B}^2_2 & \zeta\bar{B}^1_2 \\ \zeta\bar{B}^2_1 & 1 - \zeta\bar{B}^1_1 \end{bmatrix} , \tag{3.10}$$

$$\bar{\mu} = \det \bar{\mu}^\beta_\alpha = 1 - 2\zeta\tilde{H} + \zeta^2\tilde{K} , \tag{3.11}$$

$$\tilde{H} = \tfrac{1}{2}\bar{B}^\lambda_\lambda , \qquad \tilde{K} = \det \bar{B}^\beta_\alpha . \tag{3.12}$$

It is interesting to note that if $N = \bar{n}$ then $\tilde{H} = \bar{H}$ and $\tilde{K} = \bar{K}$, where $\bar{H}$, $\bar{K}$ are the mean and Gaussian curvatures of the deformed reference surface $\bar{\Pi}$. Finally, solving the system of equations $\bar{g}^i . \bar{g}_j = \delta^i_j$ we obtain

$$\bar{g}^\alpha = (\bar{\mu}^{-1})^\alpha_\beta \bar{A}^\beta , \qquad \bar{g}^3 = (\zeta,_\xi)^{-1}[-(\zeta,_\alpha + \bar{\varphi}_\alpha)(\bar{\mu}^{-1})^\alpha_\beta \bar{A}^\beta + N] , \tag{3.13}$$

where $\bar{A}^\beta$ are given by (3.5). The determination of the components of metric tensor in $\bar{\mathbb{B}}$ becomes now a simple matter of multiplications of the base vectors (3.8) and (3.13). The results are

$$\begin{aligned} \bar{g}_{\alpha\beta} &= \bar{\mu}^\lambda_\alpha \bar{\mu}^\varkappa_\beta (\bar{a}_{\lambda\varkappa} - \bar{\varphi}_\lambda \bar{\varphi}_\varkappa) + (\zeta,_\alpha + \bar{\varphi}_\alpha)(\zeta,_\beta + \bar{\varphi}_\beta) , \\ \bar{g}_{\alpha 3} &= \zeta,_\xi (\zeta,_\alpha + \bar{\varphi}_\alpha) , \\ \bar{g}_{33} &= (\zeta,_\xi)^2 , \end{aligned} \tag{3.14}$$

$$\begin{aligned} \bar{g}^{\alpha\beta} &= (\bar{\mu}^{-1})^\alpha_\lambda (\bar{\mu}^{-1})^\beta_\varkappa (\bar{a}^{\lambda\varkappa} - \bar{\varphi}_n^{-2} \bar{\varphi}^\alpha \bar{\varphi}^\beta) , \\ \bar{g}^{\alpha 3} &= -(\zeta,_\xi)^{-1}(\zeta,_\beta + \bar{\varphi}_\beta)\bar{g}^{\alpha\beta} , \\ \bar{g}^{33} &= (\zeta,_\xi)^{-2}[(\zeta,_\alpha + \bar{\varphi}_\alpha)(\zeta,_\beta + \bar{\varphi}_\beta)\bar{g}^{\alpha\beta} + 1] . \end{aligned} \tag{3.15}$$

In view of (3.11) and (2.4) the local change in volume of the shell is given by

$$J = \det \underset{\sim}{F} = \sqrt{\frac{\bar{g}}{g}} = j\bar{\varphi}_{n\mu}^{\bar{\mu}}\zeta_{,\xi}\ , \quad \bar{g} = \det\bar{g}_{ij}\ , \quad j = \sqrt{\frac{\bar{a}}{a}}\ , \tag{3.16}$$

where $\underset{\sim}{F} = \operatorname{grad}\chi = \bar{\underset{\sim}{g}}_i \otimes \underset{\sim}{g}^i$ is the deformation gradient. In a similar manner we can now obtain other relevant kinematical relations.

The main results of this chapter may be stated as

Remark 1: A three-dimensional deformation of the shell consistent with the kinematical assumption I is entirely specified by the position vector $\bar{\underset{\sim}{r}} = \bar{\underset{\sim}{r}}(\Theta^\alpha)$ describing the deformation of the reference surface, the unit vector $\underset{\sim}{N} = \underset{\sim}{N}(\Theta^\alpha)$ entailing the transverse shear deformation in the form of average (over shell thickness) rotations and the scalar function $\zeta = \zeta(\Theta^\alpha,\xi)$ reflecting an arbitrary transverse normal deformation of the initially normal fibres.

Various theories of shells can now be formulated in two basic ways. We may postulate a particular form of the function ζ, in general

$$\zeta(\Theta^\alpha,\xi) = \zeta(\xi;\beta_k(\Theta^\alpha))\ , \qquad k = 1,2,\ldots,K \tag{3.17}$$

where $\beta_k = \beta_k(\Theta^\alpha)$ are regarded as additional independent kinematical variables. Alternatively, we can impose additional constraints on the shell deformation or/and stresses in the shell that allow to determine $\beta_l = \beta_l(\Theta^\alpha)$, $l = L+1, L+2, \ldots, K$, in terms of $\bar{\underset{\sim}{r}}$, $\underset{\sim}{N}$ and β_k, $k = 1,2,\ldots, L < K$, and their derivatives. Examples of these approaches are listed in Tab. 1. In this paper we adopt the latter one.

Tab. 1

assumptions			theory (independent kinematical variables*), references
$\underset{\sim}{N}$	ζ	other	
$\underset{\sim}{N} = \bar{\underset{\sim}{n}}$	$\zeta = \xi$	none	classical Kirchhoff-Love's ($\bar{\underset{\sim}{r}}$)
none	$\zeta = \xi$	none	classical Reissner's ($\bar{\underset{\sim}{r}}$, $\underset{\sim}{N}$)
none	$\zeta = \beta_1\xi$	none	linear distribution of displacements across thickness ($\bar{\underset{\sim}{r}}$, $\underset{\sim}{N}$, β_1), see [1]
$\underset{\sim}{N} = \bar{\underset{\sim}{n}}$	$\zeta = \beta_1\xi$	$J = 1$	Biricikoglu, Kalnins ($\bar{\underset{\sim}{r}}$), [4]
$\underset{\sim}{N} = \bar{\underset{\sim}{n}}$	$\zeta = \beta_1\xi+\beta_2\xi^2$	$J = 1$	Chernykh ($\bar{\underset{\sim}{r}}$), [5,6]
$\underset{\sim}{N} = \bar{\underset{\sim}{n}}$	none	$J = 1$	Stumpf, Makowski ($\bar{\underset{\sim}{r}}$), [3]

* or any other equivalent kinematical variables

4. Isochoric deformation

The second assumption made in our considerations is

Assumption 2: The deformation of the shell is isochoric (volume preserving).

In view of (3.16) the assumption II leads to the following first order differential equation for the function ζ

$$(\tilde{K}\zeta^2 - 2\tilde{H}\zeta + 1)\zeta_{,\xi} = j^{-1}\bar{\varphi}_n^{-1}(K\xi^2 - 2H\xi + 1) \ . \tag{4.1}$$

It is seen now that $\lambda_\xi = \zeta_{,\xi|\xi=0} = j^{-1}\bar{\varphi}_n^{-1}$. Moreover, keeping in mind that $\zeta = 0$ for $\xi = 0$ the equation (4.1) can be integrated to give

$$(\tilde{K}\zeta^3 - 3\tilde{H}\zeta^2 + 3\zeta) = \lambda_\xi(K\xi^3 - 3H\xi^2 + 3\xi) \ . \tag{4.2}$$

It remains to solve for ζ the cubic algebraic equation. To this end three particular cases must be distinguished: 1) $\tilde{K} = \tilde{H} = 0$, 2) $\tilde{K} = 0$, $\tilde{H} \neq 0$, 3) $\tilde{K} \neq 0$. For the lack of space we shall not go into the details of this problem (compare [3]). For our purposes here it is enough to note that

Remark 2: Whenever the assumption II holds the function $\zeta = \zeta(\Theta^\alpha, \xi)$ is completely determind by five surface invariants; λ_ξ, $\tilde{H}$ and $\tilde{K}$ characterizing the shell deformation and H, K characterizing the undeformed reference surface.

The analogous results have been recently obtained in [3,7] for more restricted cases.

Obviously, instead of λ_ξ, $\tilde{H}$, $\tilde{K}$ we may choose another three mutually independent invariants. It may be convenient to choose the ones defined by

$$\lambda_\xi = \zeta_{,\xi|\xi=0}, \qquad \varkappa_\xi = 2(\lambda_\xi\tilde{H} - H) \ , \qquad \chi_\xi = \lambda_\xi^2\tilde{K} - K \ , \tag{4.3}$$

so that in general case we may write

$$\zeta(\Theta^\alpha, \xi) = \zeta(\xi; \lambda_\xi, \varkappa_\xi, \chi_\xi, H, K) \ . \tag{4.4}$$

This choice is dictated by the fact that λ_ξ, $\varkappa_\xi$ and χ_ξ appear in a natural way in the power expansion of the function ζ with respect to ξ. Indeed, from (4.2) we obtain

$$\zeta(\Theta^\alpha, \xi) = \lambda_\xi(\xi + \frac{1}{2}\varkappa_\xi\xi^2 + \sum_{k=1}^{\infty} \frac{1}{(k+2)!} \beta_k \xi^{(k+2)}) , \tag{4.5}$$

where

$$\begin{aligned} \beta_1 &= -2\chi_\xi + 3\varkappa_\xi(\varkappa_\xi + 2H) \ , \\ \beta_2 &= [-8\chi_\xi + 3\varkappa_\xi(5\varkappa_\xi + 8H)](\varkappa_\xi + 2H) - 12\varkappa_\xi(\chi_\xi + K) \ , \\ \beta_3 &= 5(\beta_2 + 2\varkappa_\xi\beta_1)(\varkappa_\xi + 2H) - 10(2\beta_1 + 3\varkappa_\xi^2)(\chi_\xi + K) \ , \ \dots \end{aligned} \tag{4.6}$$

We note that it is not known a priori how many terms in power expansion (4.5) should be retained to attain required accuracy of the resulting shell equations. Moreover, in some cases the power expansion need not be easier to handle than the exact solution of (4.2) itself (see [7]). For these reasons we rely in our considerations on (4.4) regarded as the exact solution of the equation (4.2).

There is another remarkable implication of the result obtained in this chapter. Namely, to ensure the existence of a unique solution, satisfying the conditions (3.2) and (3.3), of the equation (4.2) we must impose some restrictions on numerical values of λ_ξ, $\tilde{H}$, $\tilde{K}$ and $\frac{h_o}{R}$. In this way, we are able to establish the restrictions that must be imposed on the deformation and the initial thickness to ensure that the deformation of the shell consistent with the assumption I and II can take place at all. However, in general these restrictions have a quite complicated form and there is no need to present them here.

5. Strain measures

The results of the previous chapters show that the reference surface endowed with a structure formed by the field of unit vectors $\underset{\sim}{N} = \underset{\sim}{N}(\Theta^\alpha)$ creates an adequate kinematic model for the shell theory that is consistent with the assumption I and II. Our task now is to introduce suitable strain measures.

Following [8,9] we associate with each point $\bar{M} \in \bar{\Pi}$ a rigidly rotated (during shell deformation) triad $(\underset{\sim}{A}_i) = (\underset{\sim}{A}_\alpha, \underset{\sim}{N})$ that coincides with the base $(\underset{\sim}{a}_i) = (\underset{\sim}{a}_\alpha, \underset{\sim}{n})$ on the undeformed reference surface. Then there exists a rotation tensor $\underset{\sim}{Q} = \underset{\sim}{Q}(\Theta^\alpha)$ so that

$$\underset{\sim}{A}_i = \underset{\sim}{Q}\underset{\sim}{a}_i , \quad \underset{\sim}{A}^i = \underset{\sim}{Q}\underset{\sim}{a}^i, \quad \underset{\sim}{A}^i.\underset{\sim}{A}_j = \delta^i_j , \quad \underset{\sim}{Q}^T\underset{\sim}{Q} = \underset{\sim}{1} . \tag{5.1}$$

It should be noted that

$$\underset{\sim}{A}_\alpha.\underset{\sim}{A}_\beta = a_{\alpha\beta} , \quad \underset{\sim}{A}^\alpha.\underset{\sim}{A}^\beta = a^{\alpha\beta}, \quad \underset{\sim}{A}_\alpha.\underset{\sim}{N} = \underset{\sim}{A}^\alpha.\underset{\sim}{N} = 0 , \quad \underset{\sim}{A}^3 = \underset{\sim}{N} . \tag{5.2}$$

The strain measures introduced in [8,9] are

$$\underset{\sim}{E}_\alpha = \bar{\underset{\sim}{a}}_\alpha - \underset{\sim}{A}_\alpha = E_{\alpha\beta}\underset{\sim}{A}^\beta + E_\alpha\underset{\sim}{N} , \quad \underset{\sim}{K}_\alpha = \varepsilon^{\lambda\beta}K_{\alpha\beta}\underset{\sim}{A}_\lambda + K_\alpha\underset{\sim}{N} , \tag{5.3}$$

where the bending vector $\underset{\sim}{K}_\alpha$ is defined as the axial vector of the skew-symmetric tensor $\underset{\sim}{Q},_\alpha\underset{\sim}{Q}^T$ so that

$$\underset{\sim}{K}_\beta\times\underset{\sim}{A}_\alpha = \underset{\sim}{A}_{\alpha|\beta} - b_{\alpha\beta}\underset{\sim}{N} , \qquad \underset{\sim}{K}_\beta\times\underset{\sim}{N} = \underset{\sim}{N},_\beta + b^\alpha_\beta\underset{\sim}{A}_\alpha . \tag{5.4}$$

Here the vertical stroke indicates the covariant derivative in the metric of the undeformed reference surface Π. By virtue of (5.2) the indices at the components of strain measures (5.3) are raised or lowered in the metric of Π. We also note that in general $E_{\alpha\beta} \neq E_{\beta\alpha}$ and $K_{\alpha\beta} \neq K_{\beta\alpha}$.

We now proceed to derive the expressions for the deformation invariants λ_ξ, $\varkappa_\xi$ and χ_ξ in terms of the strain measures (5.3). Recalling that $\bar{\varepsilon}^{\alpha\beta} = \overset{\xi}{j}{}^{-1}\varepsilon^{\alpha\beta}$ and $\lambda_\xi = \overset{\xi}{j}{}^{-1}\bar{\varphi}_n^{-1}$ from (3.6) and $(5.3)_1$ we obtain

$$\bar{\underset{\sim}{A}}{}^{\alpha} = \lambda_\xi(\underset{\sim}{A}^{\alpha} + \varepsilon^{\alpha\beta}\underset{\sim}{E}_\beta \times \underset{\sim}{N}) = \lambda_\xi(a^{\alpha\beta} + \bar{\bar{E}}{}^{\alpha\beta})\underset{\sim}{A}_\beta \;, \qquad \bar{\underset{\sim}{A}}{}^{3} = \lambda_\xi \bar{\underset{\sim}{n}} \;. \tag{5.5}$$

Here and henceforth, the "$(\bar{\bar{\ }})$" operation on any tensor $T^{\alpha\beta}$ is defined by $\bar{\bar{T}}{}^{\alpha\beta} = a^{\alpha\beta}T^{\lambda}_{\lambda} - T^{\beta\alpha}$ [11]. Substituting now (5.5) and $(5.4)_2$ into the definition $(3.9)_3$ we find that

$$\bar{B}^{\alpha}_{\beta} = \lambda_\xi(a^{\alpha\lambda} + \bar{\bar{E}}{}^{\alpha\lambda})(b_{\beta\lambda} + K_{\beta\lambda}) \,. \tag{5.6}$$

As the consequences of (5.5), (5.6) and $(5.3)_1$ we have the following identities

$$\begin{aligned} &\lambda_\xi(a^{\alpha\lambda} + \bar{\bar{E}}{}^{\alpha\lambda})(a_{\beta\lambda} + \bar{\bar{E}}_{\beta\lambda}) = \delta^{\alpha}_{\beta} \;, \\ &\bar{B}^{\lambda}_{\alpha}(a_{\lambda\beta} + E_{\lambda\beta}) = b_{\alpha\beta} + K_{\alpha\beta} \;. \end{aligned} \tag{5.7}$$

Now it is easy to show that λ_ξ and $\tilde{H}$, $\tilde{K}$ defined by (3.12) can be expressed in the form

$$\begin{aligned} &\lambda_\xi^{-1} = 1 + E^{\lambda}_{\lambda} + \frac{1}{2}\bar{\bar{E}}{}^{\alpha\beta}E_{\alpha\beta} \;, \\ &\tilde{H} = \frac{1}{2}\lambda_\xi[2H + K^{\lambda}_{\lambda} + \bar{\bar{E}}{}^{\alpha\beta}(b_{\alpha\beta} + K_{\alpha\beta})] \;, \\ &\tilde{K} = \lambda_\xi[K + \bar{\bar{K}}{}^{\alpha\beta}(b_{\alpha\beta} + \frac{1}{2}K_{\alpha\beta})] \;, \end{aligned} \tag{5.8}$$

with the aid of which the invariants (4.3) can be expressed in terms of the strain measures $\underset{\sim}{E}_\alpha$ and $\underset{\sim}{K}_\alpha$. It is interesting to observe that

<u>Remark 3:</u> The deformation (surface) invariants λ_ξ, $\varkappa_\xi$, χ_ξ and consequently the function $\zeta(\Theta^\alpha, \xi)$ do not depend on the components E_α and K_α of the strain measures $\underset{\sim}{E}_\alpha$ and $\underset{\sim}{K}_\alpha$, respectively.

The expressions for the components of metric tensor at any point $\bar{P} \in \bar{\mathbb{B}}$ can now be obtained in a similar manner. Introducing $(5.3)_1$ and $(5.4)_2$ into (3.4) the base vectors $\bar{\underset{\sim}{g}}_i$ read

$$\begin{aligned} &\bar{\underset{\sim}{g}}_\alpha = [a_{\alpha\beta} + E_{\alpha\beta} - \zeta(b_{\alpha\beta} + K_{\alpha\beta})]\underset{\sim}{A}^{\beta} + (\zeta,_\alpha + E_\alpha)\underset{\sim}{N} \;, \\ &\bar{\underset{\sim}{g}}_3 = \zeta,_\xi \underset{\sim}{N} \;. \end{aligned} \tag{5.9}$$

In view of (4.1) we also have

$$\zeta,_\xi = \lambda_\xi \frac{\mu}{\bar{\mu}} = \lambda_\xi \frac{1 - 2H\xi + K\xi^2}{1 - 2\tilde{H}\zeta + \tilde{K}\zeta^2} \tag{5.10}$$

It follows at once from (5.9) that

$$\begin{aligned} \bar{g}_{\alpha\beta} &= [\delta^\lambda_\alpha + E^{\cdot\lambda}_{\alpha\cdot} - \zeta(b^\lambda_\alpha + K^{\cdot\lambda}_{\alpha\cdot})][a_{\beta\lambda} + E_{\beta\lambda} - \zeta(b_{\beta\lambda} + K_{\beta\lambda})] \\ &\quad + (\zeta,_\alpha + E_\alpha)(\zeta,_\beta + E_\beta) , \\ \bar{g}_{\alpha 3} &= \zeta,_\xi(\zeta,_\alpha + E_\alpha) , \qquad \bar{g}_{33} = (\zeta,_\xi)^2 . \end{aligned} \tag{5.11}$$

The same form of $\bar{g}_{ij}$ can be obtained from (3.14) with the use of (5.6), $(5.3)_1$ and the identities (5.7). We note that by the definition $(5.3)_1$ and (3.7) $\underset{\sim}{E}_\alpha.\underset{\sim}{N} = \underset{\sim}{a}_\alpha.\underset{\sim}{N} = E_\alpha = \bar{\varphi}_\alpha$. The corresponding expressions for the components $\bar{g}^{ij}$ of the metric tensor in $\bar{\mathbb{B}}$ follow from (3.15), (5.6) and the aforementioned relations.

Since the right Cauchy - Green deformation tensor $\underset{\sim}{C} = \underset{\sim}{F}^T\underset{\sim}{F} = \bar{g}_{ij}\underset{\sim}{g}^i \otimes \underset{\sim}{g}^j$ is the appropriate measure of strains in the shell by virtue of (5.11), (5.8) and (4.4) we state the following

Remark 4: The strains in the shell are entirely determined by the surface strain measures $E_{\alpha\beta}$, E_α, $K_{\alpha\beta}$ and the derivatives of $E_{\alpha\beta}$, $K_{\alpha\beta}$ through the derivatives $\lambda_{\xi,\alpha}$, $\varkappa_{\xi,\alpha}$ and $\chi_{\xi,\alpha}$ of the invariants (4.3).

The discussion of the significance of this result we postpone to the next chapters. Here we note that the independence of strains in the shell of $K_\alpha = \underset{\sim}{K}_\alpha.\underset{\sim}{N}$ is natural in the following sense. To introduce strain measures (5.3) we have ascribed to each point $\bar{M} \in \bar{\Pi}$ the triad $(\underset{\sim}{A}_\alpha, \underset{\sim}{N})$. Physically it may be interpreted as requiring the material point of the reference surface to have six degrees of freedom characterized by $\bar{\underset{\sim}{r}}(\bar{M})$ and $\underset{\sim}{Q}(\bar{M})$ for fixed $\bar{M} \in \bar{\Pi}$. However, in this way we have endowed the reference surface with the kinematical structure richer than it follows from our assumptions I and II. Indeed, these assumptions implay that the only independent kinematical variables are $\bar{\underset{\sim}{r}}$ and $\underset{\sim}{N}$ with $\underset{\sim}{N}.\underset{\sim}{N} = 1$ (see our Remark 1 and 2). Thus for fixed $\bar{M} \in \bar{\Pi}$ only five components of $\bar{\underset{\sim}{r}}(\bar{M})$ and $\underset{\sim}{N}(\bar{M})$ are independent. In view of this and to provide better physical interpretation of the strain measures (5.3) we now present alternative description of the strains in the shell.

Let us define the non-singular tensor

$$\underset{\sim}{G} = \underset{\sim}{G}(\Theta^\alpha) = \bar{\underset{\sim}{a}}_\alpha \otimes \underset{\sim}{a}^\alpha + \bar{\underset{\sim}{n}} \otimes \underset{\sim}{n} , \qquad \det \underset{\sim}{G} = j = \sqrt{\frac{\bar{a}}{a}} . \tag{5.12}$$

By the polar decomposition theorem $\underset{\sim}{G} = \underset{\sim}{V}\underset{\sim}{R}$, where $\underset{\sim}{V}$ is a symmetric positive definite tensor and $\underset{\sim}{R}$ is a proper orthogonal tensor (rotation tensor). The tensor $\underset{\sim}{G}$ and its polar decomposition play a crucial role in the classical Kirchhoff-Love's types shell theories [1,11]. Here this tensor is adopted to show that whenever the deformation of the shell is consistent with the assumptions I and II the resulting strains can be decomposed into three parts; 1) strains due to stretching of the reference surface, 2) strains due to pure bending of the reference surface, 3) strains induced by additional rotations of the initially normal fibres due to shear.

Making use of the polar decomposition of $\underset{\sim}{G}$ we define the triad $(\underset{\sim}{r}_i)$ and its dual $(\underset{\sim}{r}^i)$ by [1]

$$\underset{\sim}{r}_i = \underset{\sim}{R}\underset{\sim}{a}_i \ , \qquad \underset{\sim}{r}^i = \underset{\sim}{R}\underset{\sim}{a}^i \ , \qquad \underset{\sim}{r}^i.\underset{\sim}{r}_j = \delta^i_j \ , \qquad (\underset{\sim}{a}_i) = (\underset{\sim}{a}_\alpha, \underset{\sim}{n}) . \tag{5.13}$$

It should be noted that $\underset{\sim}{r}_3 = \underset{\sim}{r}^3 = \bar{\underset{\sim}{n}}$, $\underset{\sim}{r}_\alpha . \underset{\sim}{r}_\beta = a_{\alpha\beta}$, etc. From (5.13) and (5.1) we obtain

$$\underset{\sim}{A}_i = \underset{\sim}{Q}\underset{\sim}{a}_i = \underset{\sim}{Q}\underset{\sim}{R}^T\underset{\sim}{r}_i = \underset{\sim}{\Gamma}\underset{\sim}{r}_i \ , \qquad \underset{\sim}{\Gamma} = \underset{\sim}{Q}\underset{\sim}{R}^T . \tag{5.14}$$

Obviously, $\underset{\sim}{\Gamma}$ is a proper orthogonal tensor. Let $\underset{\sim}{e}_\gamma$ be a unit vector along the axis of rotation of $\underset{\sim}{\Gamma}$, $\underset{\sim}{\Gamma}\underset{\sim}{e}_\gamma = \underset{\sim}{e}_\gamma$, and let γ denotes the angle of rotation of $\underset{\sim}{\Gamma}$, $\cos\gamma = \frac{1}{2}(\mathrm{tr}\underset{\sim}{\Gamma} - 1)$. We claim that

$$\bar{\underset{\sim}{n}}.\underset{\sim}{N} = \cos\gamma \ , \qquad \bar{\underset{\sim}{n}}\times\underset{\sim}{N} = \sin\gamma \underset{\sim}{e}_\gamma \ , \qquad |\gamma| < \frac{\pi}{2} . \tag{5.15}$$

Since $\underset{\sim}{R}$ is unique, the requirements (5.15) ensure that the rotation tensor $\underset{\sim}{Q} = \underset{\sim}{\Gamma}\underset{\sim}{R}$ is consistent with the assumptions I and II. We now define the finite rotation vector corresponding to $\underset{\sim}{\Gamma}$

$$\underset{\sim}{\gamma} = \underset{\sim}{\gamma}(\theta^\alpha) = 2\mathrm{tg}\frac{\gamma}{2}\underset{\sim}{e}_\gamma = \gamma_\alpha \underset{\sim}{r}^\alpha = \gamma^\alpha \underset{\sim}{r}_\alpha . \tag{5.16}$$

With the aid of general formulae given in [10] we find that

$$\begin{aligned} \underset{\sim}{A}_\alpha &= \underset{\sim}{\Gamma}\underset{\sim}{r}_\alpha = \underset{\sim}{r}_\alpha + P_\gamma[\tfrac{1}{2}(\gamma_\alpha\gamma^\beta - \delta^\beta_\alpha\gamma_\lambda\gamma^\lambda)\underset{\sim}{r}_\beta + \varepsilon_{\beta\alpha}\gamma^\beta\bar{\underset{\sim}{n}}] \ , \\ \underset{\sim}{N} &= \underset{\sim}{\Gamma}\bar{\underset{\sim}{n}} = \bar{\underset{\sim}{n}} + P_\gamma(\varepsilon^{\alpha\beta}\gamma_\beta\underset{\sim}{r}_\alpha - \tfrac{1}{2}\gamma_\lambda\gamma^\lambda\bar{\underset{\sim}{n}}) \ , \end{aligned} \tag{5.17}$$

$$P_\gamma = (1 + \tfrac{1}{4}\underset{\sim}{\gamma}.\underset{\sim}{\gamma})^{-1} = \cos^2\frac{\gamma}{2} \ , \tag{5.18}$$

where by virtue of (5.16)

$$\underset{\sim}{\gamma}.\underset{\sim}{\gamma} = \gamma_\lambda\gamma^\lambda = 4\mathrm{tg}^2\frac{\gamma}{2} . \tag{5.19}$$

We note that $\underset{\sim}{\Gamma}$ or, equivalently, $\underset{\sim}{\gamma}$ represent the additional rotations of the initially normal fibres due to shear.

Following [1,11] we next define the measure of stretching of the reference surface

$$\mathbf{h}_\alpha = \bar{\mathbf{a}}_\alpha - \mathbf{r}_\alpha = (\mathbf{V} - \mathbf{1})\mathbf{r}_\alpha = h_{\alpha\beta}\mathbf{r}^\beta , \qquad h_{\alpha\beta} = h_{\beta\alpha} . \tag{5.20}$$

Then from $(5.3)_1$, (5.20) and (5.17) we obtain

$$\mathbf{E}_\alpha = \bar{\mathbf{a}}_\alpha - \boldsymbol{\Gamma}\mathbf{r}_\alpha = \mathbf{h}_\alpha - P_\gamma \boldsymbol{\gamma}\times(\mathbf{r}_\alpha + \tfrac{1}{2}\boldsymbol{\gamma}\times\mathbf{r}_\alpha) , \tag{5.21}$$

or in the component form (it should be noted that $\mathbf{E}_\alpha$ and $\mathbf{h}_\alpha$ are resolved with respect to the different bases)

$$\begin{aligned} E_{\alpha\beta} &= h_{\alpha\beta} + \tfrac{1}{2}P_\gamma[\gamma_\alpha\gamma_\beta - (a_{\alpha\beta} + h_{\alpha\beta})\gamma_\lambda\gamma^\lambda + h_{\alpha\lambda}\gamma_\beta\gamma^\lambda] , \\ E_\alpha &= \varepsilon_{\lambda\beta}P_\gamma(\delta^\lambda_\alpha + h^\lambda_\alpha)\gamma^\beta . \end{aligned} \tag{5.22}$$

Let further $\mathbf{k}_\beta = \mathbf{k}_\beta(\Theta^\alpha)$ denotes the axial vector of the skew-symmetric tensor $\mathbf{R},_\beta\mathbf{R}^T$, i.e.

$$\mathbf{k}_\beta\times\mathbf{r}_\alpha = \mathbf{r}_{\alpha|\beta} - b_{\alpha\beta}\bar{\mathbf{n}} , \qquad \mathbf{k}_\beta\times\bar{\mathbf{n}} = \bar{\mathbf{n}},_\beta + b^\alpha_\beta\mathbf{r}_\alpha , \tag{5.23}$$

$$\mathbf{k}_\alpha = \varepsilon^{\lambda\beta}k_{\alpha\beta}\mathbf{r}_\lambda + k_\alpha\bar{\mathbf{n}} , \qquad k_{\alpha\beta} \neq k_{\beta\alpha} . \tag{5.24}$$

This vector, introduced in [1,11] in a slightly different manner, provides a suitable measure of pure bending of the reference surface. Differentiating now the relations (5.14) one gets

$$\mathbf{A}_{i|\beta} = \boldsymbol{\Gamma},_\beta\mathbf{r}_i + \boldsymbol{\Gamma}\mathbf{r}_{i|\beta} = \boldsymbol{\Gamma},_\beta\boldsymbol{\Gamma}^T\mathbf{A}_i + \boldsymbol{\Gamma}\mathbf{r}_{i|\beta} , \tag{5.25}$$

Substituting next (5.4) and (5.23) into (5.25) and recalling that $\mathbf{Q} = \boldsymbol{\Gamma}\mathbf{R}$ we obtain

$$\mathbf{K}_\beta\times\mathbf{A}_i = \boldsymbol{\eta}_\beta\times\mathbf{A}_i + \boldsymbol{\Gamma}(\mathbf{k}_\beta\times\mathbf{r}_i) . \tag{5.26}$$

Here $\boldsymbol{\eta}_\beta = \boldsymbol{\eta}_\beta(\Theta^\alpha)$ is defined as the axial vector of $\boldsymbol{\Gamma},_\beta\boldsymbol{\Gamma}^T$, i.e.

$$\boldsymbol{\Gamma},_\beta\boldsymbol{\Gamma}^T\mathbf{A}_i = \boldsymbol{\eta}_\beta\times\mathbf{A}_i , \qquad \boldsymbol{\eta}_\alpha = \varepsilon^{\lambda\beta}\eta_{\alpha\beta}\mathbf{A}_\lambda + \eta_\alpha\mathbf{N} . \tag{5.27}$$

Finally, keeping in mind that $\mathbf{Ta}\times\mathbf{Tb} = \mathbf{T}(\mathbf{a}\times\mathbf{b})$ for any proper orthogonal tensor $\mathbf{T}$ and any vectors $\mathbf{a}$, $\mathbf{b}$, we conclude from (5.26) that

$$\mathbf{K}_\beta - \boldsymbol{\eta}_\beta = \boldsymbol{\Gamma}\mathbf{k}_\beta , \qquad K_{\alpha\beta} - \eta_{\alpha\beta} = k_{\alpha\beta} , \qquad K_\alpha - \eta_\alpha = k_\alpha , \tag{5.28}$$

$$\boldsymbol{\eta}_\alpha = P_\gamma(\boldsymbol{\gamma},_\alpha + \tfrac{1}{2}\boldsymbol{\gamma}\times\boldsymbol{\gamma},_\alpha) . \tag{5.29}$$

To obtain (5.29) we have followed the line presented in [1,9]. Introducing now (5.22) and (5.28) into (5.11) and (5.8) we can obtain the expressions for $\bar{g}_{ij}$ in terms of $\mathbf{h}_\alpha$, $\mathbf{k}_\alpha$, $\boldsymbol{\gamma}$ and the derivatives of $\boldsymbol{\gamma}$. It is interesting to observe that $\lambda_\xi = j^{-1}\cos^{-1}\gamma$.

6. Equilibrium equations

The local equilibrium equations for the shell can be obtained as a straightforward consequence of the balance laws of linear momentum and moment of momentum of the classical continuum mechanics. In the case under consideration, keeping in mind that the position vector to any point $\bar{P} \in \bar{\mathbb{B}}$ is of the form (3.1), these equations are (for the lack of space we do not present here the detailed derivation, it follows the line shown in [9])

$$\underset{\sim}{N}^{\beta}{}_{|\beta} + \underset{\sim}{p} = \underset{\sim}{0} , \qquad \underset{\sim}{M}^{\beta}{}_{|\beta} + \underset{\sim}{\bar{r}},_{\beta} \times \underset{\sim}{N}^{\beta} + \underset{\sim}{l} = \underset{\sim}{0} . \tag{6.1}$$

Here the stress resultant $\underset{\sim}{N}^{\beta}$ and the stress couple $\underset{\sim}{M}^{\beta}$ are defined by

$$\underset{\sim}{N}^{\beta} = \int_{-}^{+} \underset{\sim}{T}^{\beta} \mu d\xi = N^{\beta\alpha} \underset{\sim}{A}_{\alpha} + Q^{\beta} \underset{\sim}{N} ,$$
$$\underset{\sim}{M}^{\beta} = \underset{\sim}{N} \times \int_{-}^{+} \zeta \underset{\sim}{T}^{\beta} \mu d\xi = \varepsilon_{\lambda\alpha} M^{\beta\alpha} \underset{\sim}{A}^{\lambda} , \tag{6.2}$$

and the external force $\underset{\sim}{p}$ and the external couple $\underset{\sim}{l}$ (both measured per unit area of the undeformed reference surface Π) incorporate the body force and the external loading acting on the faces Π^{+} and Π^{-}. In (6.2) $\underset{\sim}{T}^{\beta}$ denotes the nominal stress vector calculated from the first Piola-Kirchhoff stress tensor $\underset{\sim}{T} = T^{ij} \underset{\sim}{\bar{g}}_i \otimes \underset{\sim}{g}_j$ by $\underset{\sim}{T}^{\beta} = \underset{\sim}{T} \underset{\sim}{g}^{\beta}$, $(\underset{\sim}{A}_{\alpha}, \underset{\sim}{N})$ is the base defined by (5.1) and $\int_{-}^{+}$ stands for the integral in the range $[-h_o^{-}, +h_o^{+}]$.

We emphasize that the equilibrium equations (6.1) are obtained within the single assumption I. Consequently they are common to all shell theories that can be formulated in consistence with this assumption. In other words there do exist neither constitutive equations nor boundary conditions that would be uniquely associated with the equilibrium equations (6.1). Henceforth, to obtain a complete set of equations we confine ourselves to a theory of elastic shells.

7. Strain energy function

In consistence with the assumption II, let us consider a shell made of a hyperelastic incompressible material. For simplicity we assume that the material is isotropic and homogeneous. Then a strain energy function (per unit volume of $\mathbb{B}$) is of the form $W = W(I_1, I_2)$, where $I_1 = g^{ij}\bar{g}_{ij}$ and $I_2 = g_{ij}\bar{g}^{ij}$ are the first and second principal invariants of the Cauchy-Green deformation tensor $\underset{\sim}{C} = \underset{\sim}{F}^{T}\underset{\sim}{F} = \bar{g}_{ij}\underset{\sim}{g}^{i} \otimes \underset{\sim}{g}^{j}$. The strain energy U of the shell and the strain energy density Φ per unit area of the undeformed reference surface Π are defined as usual

$$U = \int_{\Pi} \Phi dA \ , \qquad \Phi = \int_{-}^{+} W(I_1, I_2)\mu d\xi \ , \tag{7.1}$$

where μ is given by $(2.4)_3$. By virtue of (2.4), (3.14), (3.15) and (5.10) we have

$$I_1 = g^{\alpha\beta}\bar{g}_{\alpha\beta} + \bar{g}_{33} \ , \qquad I_2 = g_{\alpha\beta}\bar{g}^{\alpha\beta} + \bar{g}^{33} \ , \tag{7.2}$$

$$g_{\alpha\beta} = \mu^{\lambda}_{\alpha}\mu^{\varkappa}_{\beta}a_{\lambda\varkappa} \ , \qquad g^{\alpha\beta} = (\mu^{-1})^{\alpha}_{\lambda}(\mu^{-1})^{\beta}_{\varkappa}a^{\lambda\varkappa} \ , \tag{7.3}$$

$$\begin{aligned} \bar{g}_{\alpha\beta} &= \bar{\mu}^{\lambda}_{\alpha}\bar{\mu}^{\varkappa}_{\beta}y_{\lambda\varkappa} + y_{\alpha}y_{\beta} \ , & \bar{g}_{33} &= \lambda^2_{\xi}\mu^2\bar{\mu}^{-2} \ , \\ \bar{g}^{\alpha\beta} &= (\bar{\mu}^{-1})^{\alpha}_{\lambda}(\bar{\mu}^{-1})^{\beta}_{\varkappa}y^{\lambda\varkappa} \ , & \bar{g}^{33} &= \bar{\mu}^2\lambda^{-2}_{\xi}\mu^{-2}(y_{\alpha}y_{\beta}\bar{g}^{\alpha\beta} + 1) \ , \end{aligned} \tag{7.4}$$

where for, convenience,the following notations have been introduced

$$y_{\alpha} = \zeta,_{\alpha} + \bar{\varphi}_{\alpha} \ , \qquad y_{\alpha\beta} = \bar{a}_{\alpha\beta} - \bar{\varphi}_{\alpha}\bar{\varphi}_{\beta} \ , \qquad y^{\alpha\beta} = \bar{a}^{\alpha\beta} - \bar{\varphi}_n^{-2}\bar{\varphi}^{\alpha}\bar{\varphi}^{\beta} \tag{7.5}$$

The above formulae with ζ being a solution (perhaps approximated) of the equation (4.2) give the invariants I_1, I_2 as the known function of the normal coordinate ξ. As W is a known function of I_1, I_2 we can evaluate the integral in $(7.1)_2$ to obtain the explicit form of the two-dimensional strain energy density Φ in terms of choosen surface strain measures. Since the expressions (7.4) contain the derivatives of the function ζ with respect to the surface coordinates Θ^{α}, $(7.1)_2$, (7.2) and (4.4) imply that

Remark 5: If the shell made of an incompressible hyperelastic (simple) material deforms such that initially normal fibres remain straight, then the two-dimensional strain energy density is a function of the corresponding surface strain measures as well as their derivatives.

Adopting the strain measures $\underset{\sim}{E}_{\alpha}$, $\underset{\sim}{K}_{\alpha}$ defined by (5.3) we note that the quantities (7.5) can be expressed in the form

$$\begin{aligned} y_{\alpha} &= \zeta,_{\alpha} + E_{\alpha} \ , \qquad y_{\alpha\beta} = a_{\alpha\beta} + E_{\alpha\beta} + E_{\beta\alpha} + E^{\cdot\lambda}_{\alpha\cdot}E_{\beta\lambda} \ , \\ y^{\alpha\beta} &= \lambda^2_{\xi}a_{\lambda\varkappa}(a^{\alpha\lambda} + \bar{\bar{E}}^{\alpha\lambda})(a^{\beta\varkappa} + \bar{\bar{E}}^{\beta\varkappa}) \ . \end{aligned} \tag{7.6}$$

Indeed, the above expressions for y_{α} and $y_{\alpha\beta}$ are straightforward consequences of the definition of the stretching vector $\underset{\sim}{E}_{\alpha}$. The expression for $y^{\alpha\beta}$ follows from (5.5) and (3.6). Furthermore, from (3.9), (5.6), (5.7) and $(7.6)_1$ we have

$$\begin{aligned} \bar{\mu}^{\lambda}_{\alpha}\bar{\mu}^{\varkappa}_{\beta}y_{\lambda\varkappa} = y_{\alpha\beta} &- \zeta[(a_{\alpha\lambda} + E_{\alpha\lambda})(b^{\lambda}_{\beta} + K^{\cdot\lambda}_{\beta\cdot}) + \\ &(a_{\beta\lambda} + E_{\beta\lambda})(b^{\lambda}_{\alpha} + K^{\cdot\lambda}_{\alpha\cdot})] + \zeta^2(b^{\lambda}_{\alpha} + K^{\cdot\lambda}_{\alpha\cdot})(b_{\beta\lambda} + K_{\beta\lambda}) \ , \end{aligned} \tag{7.7}$$

which may be verified with the aid of (5.11). Now, by virtue of (3.10) and $(3.15)_1$ the components $\bar{g}^{\alpha\beta}$ of the metric tensor can be expressed in the form

$$\bar{g}^{\alpha\beta} = \frac{\lambda_\xi^2}{\bar{\mu}^2}\varepsilon^{\alpha\lambda}\varepsilon^{\beta\varkappa}\bar{\mu}^\rho_\lambda\bar{\mu}^\eta_\varkappa Y_{\rho\eta} \ . \tag{7.8}$$

Combining next (7.2) - (7.8) and (4.4), (5.8) we find out that the two-dimensional strain energy density Φ is of the form

$$\Phi = \Phi(E_{\alpha\beta}, E_\alpha, K_{\alpha\beta}, \lambda_{\xi,\beta}, \varkappa_{\xi,\beta}, \chi_{\xi,\beta}; b_{\alpha\beta}, H_{,\beta}, K_{,\beta}) . \tag{7.9}$$

It should be pointed out that (7.9) is a mathematically exact implication of the assumptions I and II.

8. Principle of virtual work

In this chapter we summarize our previous results in the form of a complete set of equations for the shells undergoing the deformation consistent with the assumptions I and II. The most straightforward way to gain this aim is to start with the three-dimensional principle of virtual work.

Like in the preceding chapter we shall consider the shell made of an incompressible hyperelastic material. The deformation of the shell from the initial configuration $\mathbb{B}$ into the deformed one $\bar{\mathbb{B}}$ is defined by the mapping $\underset{\sim}{x} = \chi(\underset{\sim}{X})$. Let the shell be in an equilibrium under the body forces $\underset{\sim}{f}$ and the external surface loads $\underset{\sim}{t}^+$, $\underset{\sim}{t}^-$ and $\underset{\sim}{t}^*$ acting on the faces Π^+, Π^- and the lateral surface $\partial\mathbb{B}^*$, respectively. The equilibrium conditions we express in the weak form (which is just the principle of virtual work)

$$\int_{\mathbb{B}} \delta W dV = \int_{\mathbb{B}} \underset{\sim}{f}.\delta\underset{\sim}{x} dV + \int_{\Pi^+} \underset{\sim}{t}^+.\delta\underset{\sim}{x}^+ dA^+ + \int_{\Pi^-} \underset{\sim}{t}^-.\delta\underset{\sim}{x}^- dA^- + \int_{\partial\mathbb{B}^*} \underset{\sim}{t}^*.\delta\underset{\sim}{x} dA^* . \tag{8.1}$$

The left-hand side of (8.1) represents the virtual change in the strain energy and, by virtue of (7.1), can be rewritten as

$$\int_{\mathbb{B}} \delta W dV = \int_{\Pi} \delta\Phi dA \ . \tag{8.2}$$

The right-hand side of (8.1) is the work done by the body forces and the surface loads in a virtual displacement $\delta\underset{\sim}{x}$. Since the deformation of the shell is subject to the constraints resulting from the assumptions I and II, the virtual displacement $\delta\underset{\sim}{x}$ must be consistent with (3.1) and (4.4). We have

$$\delta \underset{\sim}{x} = \delta \bar{\underset{\sim}{r}} + \zeta \delta \underset{\sim}{N} + \underset{\sim}{N} \delta \zeta \ , \tag{8.3}$$

$$\delta \zeta = \frac{\partial \zeta}{\partial \lambda_\xi} \delta \lambda_\xi + \frac{\partial \zeta}{\partial \varkappa_\xi} \delta \varkappa_\xi + \frac{\partial \zeta}{\partial \chi_\xi} \delta \chi_\xi = \zeta_{(1)} \delta \lambda_\xi + \zeta_{(2)} \delta \varkappa_\xi + \zeta_{(3)} \delta \chi_\xi \ . \tag{8.4}$$

To proceed further we adopt the position vector $\bar{\underset{\sim}{r}} = \bar{\underset{\sim}{r}}(\Theta^\alpha)$ and the proper orhtogonal tensor $\underset{\sim}{Q} = \underset{\sim}{Q}(\Theta^\alpha)$ as the basic independent kinematical variables. Then, in view of (5.1) we obtain

$$\delta \underset{\sim}{N} = \delta \underset{\sim}{Q} \underset{\sim}{n} = \delta \underset{\sim}{Q} \underset{\sim}{Q}^T \underset{\sim}{N} = \underset{\sim}{w} \times \underset{\sim}{N} \ . \tag{8.5}$$

Clearly $\underset{\sim}{w} = \underset{\sim}{w}(\Theta^\alpha)$ is the axial vector of the skew-symmetric tensor $\delta \underset{\sim}{Q} \underset{\sim}{Q}^T$. Using now (8.3) and (8.5) the last term on the right-hand side of (8.1) can be expressed in the form

$$\int_{\partial \mathbb{B}^*} \underset{\sim}{t}^* . \delta \underset{\sim}{x} dA^* = \int_{\partial \Pi} (\underset{\sim}{N}^* . \delta \bar{\underset{\sim}{r}} + \underset{\sim}{M}^* . \underset{\sim}{w} + q^* \delta \lambda_\xi + n^* \delta \varkappa_\xi + m^* \delta \chi_\xi) ds \ , \tag{8.6}$$

where the resultant boundary loads are defined by

$$\underset{\sim}{N}^* = \int_-^+ \underset{\sim}{t}^* \mu d\xi \ , \qquad \underset{\sim}{M}^* = \underset{\sim}{N} \times \int_-^+ \zeta \underset{\sim}{t}^* \mu d\xi \ , \tag{8.7}$$

$$q^* = \int_-^+ \zeta_{(1)} t_N^* \mu d\xi \ , \quad n^* = \int_-^+ \zeta_{(2)} t_N^* \mu d\xi \ , \quad m^* = \int_-^+ \zeta_{(3)} t_N^* \mu d\xi \ , \tag{8.8}$$

with $t_N^*(s,\xi) = \underset{\sim}{N} . \underset{\sim}{t}^*$. In a similar manner we can reduce the body forces and the loads acting on the shell faces to statically equivalent loads defined on the reference surface Π. Without giving the details of the derivation, we only note that $dV = \mu d\xi dA$ and $\underset{\sim}{t}^{\pm} = \underset{\sim}{T} \underset{\sim}{n}^{\pm}$, where $\underset{\sim}{T}$ denotes the first Piola-Kirchhoff stress tensor and $\underset{\sim}{n}^{\pm}$ are the outward unit normal vectors to the faces Π^+ and Π^-, respectively. Then, with the aid of (8.2) - (8.6) the principle of virtual work (8.1) can be rewritten as

$$\int_\Pi \delta \Phi dA = \int_\Pi (\underset{\sim}{p}' . \delta \bar{\underset{\sim}{r}} + \underset{\sim}{l}' . \underset{\sim}{w} + p_{(1)} \delta \lambda_\xi + p_{(2)} \delta \varkappa_\xi + p_{(3)} \delta \chi_\xi) dA + \int_{\partial \Pi} (\underset{\sim}{N}^* . \delta \bar{\underset{\sim}{r}} + \underset{\sim}{M}^* . \underset{\sim}{w} + q^* \delta \lambda_\xi + n^* \delta \varkappa_\xi + m^* \delta \chi_\xi) ds \ , \tag{8.9}$$

where the resultant surface loads are defined by

$$\begin{aligned} \underset{\sim}{p}' &= \int_-^+ \underset{\sim}{f} \mu d\xi + \left[\mu (\underset{\sim}{T}^3 - h_{\circ,\alpha} \underset{\sim}{T}^\alpha) \right]_-^+ \ , \\ \underset{\sim}{l}' &= \underset{\sim}{N} \times \int_-^+ \zeta \underset{\sim}{f} \mu d\xi + \left[\mu \zeta \underset{\sim}{N} \times (\underset{\sim}{T}^3 - h_{\circ,\alpha} \underset{\sim}{T}^\alpha) \right]_-^+ \ , \end{aligned} \tag{8.10}$$

$$p_{(k)} = \int_{-}^{+} \zeta_{(k)} \underset{\sim}{N} \cdot \underset{\sim}{f} \mu d\xi + \left[\mu \zeta_{(k)} \underset{\sim}{N} \cdot (\underset{\sim}{T}^3 - h_{\circ,\alpha} \underset{\sim}{T}^{\alpha}) \right]_{-}^{+}, \quad k = 1,2,3 \tag{8.11}$$

with $\underset{\sim}{T}^i = T\underset{\sim}{g}^i$ and $\zeta_{(k)}$ defined by (8.4). Here it was taken into account that $h_\circ^+$ and $h_\circ^-$ may be variable with respect to Θ^α. The terms containing $p_{(k)}$, $k=1,2,3$, in (8.9) represent a correction to the classical shell theories. They can be transformed further as follows. First, using the definition (4.3) and the relations (5.8) we calculate the variations of λ_ξ, $\varkappa_\xi$ and χ_ξ

$$\begin{aligned} \delta\lambda_\xi &= \frac{\partial\lambda_\xi}{\partial E_{\alpha\beta}} \delta E_{\alpha\beta} = G^{\alpha\beta}_{(1)} \delta E_{\alpha\beta}, \\ \delta\varkappa_\xi &= \frac{\partial\varkappa_\xi}{\partial E_{\alpha\beta}} \delta E_{\alpha\beta} + \frac{\partial\varkappa_\xi}{\partial K_{\alpha\beta}} \delta K_{\alpha\beta} = G^{\alpha\beta}_{(2)} \delta E_{\alpha\beta} + H^{\alpha\beta}_{(1)} \delta K_{\alpha\beta}, \\ \delta\chi_\xi &= \frac{\partial\chi_\xi}{\partial E_{\alpha\beta}} \delta E_{\alpha\beta} + \frac{\partial\chi_\xi}{\partial K_{\alpha\beta}} \delta K_{\alpha\beta} = G^{\alpha\beta}_{(3)} \delta E_{\alpha\beta} + H^{\alpha\beta}_{(2)} \delta K_{\alpha\beta}. \end{aligned} \tag{8.12}$$

Next, from (5.3) and the definition of $\underset{\sim}{w}$ we find that (compare [8,9])

$$\begin{aligned} \delta E_{\alpha\beta} \underset{\sim}{A}^\beta + \delta E_\alpha \underset{\sim}{N} &= \delta \underset{\sim}{E}_\alpha - \underset{\sim}{w} \times \underset{\sim}{E}_\alpha = \delta \bar{\underset{\sim}{r}}_{,\alpha} - \underset{\sim}{w} \times \bar{\underset{\sim}{r}}_{,\alpha}, \\ \varepsilon^{\lambda\beta} \delta K_{\alpha\beta} \underset{\sim}{A}_\lambda + \delta K_\alpha \underset{\sim}{N} &= \delta \underset{\sim}{K}_\alpha - \underset{\sim}{w} \times \underset{\sim}{K}_\alpha = \underset{\sim}{w}_{,\alpha}. \end{aligned} \tag{8.13}$$

Finally, introducing (8.12) and (8.13) into (8.9) and applying the divergence theorem we obtain

$$\begin{aligned} \int_\Pi \delta\Phi dA = \int_\Pi (\underset{\sim}{p} \cdot \delta\bar{\underset{\sim}{r}} + \underset{\sim}{l} \cdot \underset{\sim}{w}) dA + \int_{\partial\Pi} [(\underset{\sim}{N}^* + \underset{\sim}{P}^\beta \nu_\beta) \cdot \delta\bar{\underset{\sim}{r}} + \\ (\underset{\sim}{M}^* + \underset{\sim}{L}^\beta \nu_\beta) \cdot \underset{\sim}{w} + q^* \delta\lambda_\xi + n^* \delta\varkappa_\xi + m^* \delta\chi_\xi] ds, \end{aligned} \tag{8.14}$$

wherein

$$\underset{\sim}{p} = \underset{\sim}{p}' - \underset{\sim}{P}^\beta{}_{|\beta}, \qquad \underset{\sim}{l} = \underset{\sim}{l}' - (\underset{\sim}{L}^\beta{}_{|\beta} + \bar{\underset{\sim}{r}}_{,\beta} \times \underset{\sim}{P}^\beta), \tag{8.15}$$

$$\begin{aligned} \underset{\sim}{P}^\alpha &= (p_{(1)} G^{\alpha\beta}_{(1)} + p_{(2)} G^{\alpha\beta}_{(2)} + p_{(3)} G^{\alpha\beta}_{(3)}) \underset{\sim}{A}_\beta = P^{\alpha\beta} \underset{\sim}{A}_\beta, \\ \underset{\sim}{L}^\alpha &= \varepsilon_{\lambda\beta} (p_{(2)} H^{\alpha\beta}_{(1)} + p_{(3)} H^{\alpha\beta}_{(2)}) \underset{\sim}{A}^\lambda = \varepsilon_{\lambda\beta} L^{\alpha\beta} \underset{\sim}{A}^\lambda. \end{aligned} \tag{8.16}$$

To finish the analysis it remains to calculate the virtual change in the strain energy density Φ. According to (7.9) we have

$$\begin{aligned} \delta\Phi = \tilde{N}^{\alpha\beta} \delta E_{\alpha\beta} + Q^\alpha \delta E_\alpha + \tilde{M}^{\alpha\beta} \delta K_{\alpha\beta} - q^\beta{}_{|\beta} \delta\lambda_\xi - n^\beta{}_{|\beta} \delta\varkappa_\xi - \\ m^\beta{}_{|\beta} \delta\chi_\xi + (q^\beta \delta\lambda_\xi + n^\beta \delta\varkappa_\xi + m^\beta \delta\chi_\xi)_{|\beta}, \end{aligned} \tag{8.17}$$

wherein

$$\tilde{N}^{\alpha\beta} = \frac{\partial\Phi}{\partial E_{\alpha\beta}} , \qquad Q^{\alpha} = \frac{\partial\Phi}{\partial E_{\alpha}} , \qquad \tilde{M}^{\alpha\beta} = \frac{\partial\Phi}{\partial K_{\alpha\beta}} ,$$
$$q^{\alpha} = \frac{\partial\Phi}{\partial\lambda_{\xi'\alpha}} , \qquad n^{\alpha} = \frac{\partial\Phi}{\partial\kappa_{\xi'\alpha}} , \qquad m^{\alpha} = \frac{\partial\Phi}{\partial\chi_{\xi'\alpha}} . \tag{8.18}$$

Further, with the help of (8.12), (8.13) and the divergence theorem we find that the virtual change in the strain energy (8.17) takes the form

$$\int_{\Pi} \delta\Phi dA = - \int_{\Pi} [\underset{\sim}{N}^{\beta}{}_{|\beta}\cdot\delta\underset{\sim}{\bar{r}} + (\underset{\sim}{M}^{\beta}{}_{|\beta} + \underset{\sim}{\bar{r}}_{,\beta}\times\underset{\sim}{N}^{\beta})\cdot\underset{\sim}{w}]dA + \int_{\partial\Pi} (\underset{\sim}{N}^{\beta}\cdot\delta\underset{\sim}{\bar{r}} +$$
$$\underset{\sim}{M}^{\beta}\cdot\underset{\sim}{w} + q^{\beta}\delta\lambda_{\xi} + n^{\beta}\delta\kappa_{\xi} + m^{\beta}\delta\chi_{\xi})\nu_{\beta} ds , \tag{8.19}$$

where the resultant stress vector $\underset{\sim}{N}^{\alpha}$ and the stress couple $\underset{\sim}{M}^{\alpha}$ are defined by

$$\underset{\sim}{N}^{\alpha} = N^{\alpha\beta}\underset{\sim}{A}_{\beta} + Q^{\alpha}\underset{\sim}{N} ,$$
$$N^{\alpha\beta} = \tilde{N}^{\alpha\beta} - (q^{\lambda}{}_{|\lambda}G^{\alpha\beta}_{(1)} + n^{\lambda}{}_{|\lambda}G^{\alpha\beta}_{(2)} + m^{\lambda}{}_{|\lambda}G^{\alpha\beta}_{(3)}) ,$$
$$\underset{\sim}{M}^{\alpha} = \varepsilon_{\lambda\beta}M^{\alpha\beta}\underset{\sim}{A}^{\lambda},$$
$$M^{\alpha\beta} = \tilde{M}^{\alpha\beta} - (n^{\lambda}{}_{|\lambda}H^{\alpha\beta}_{(1)} + m^{\lambda}{}_{|\lambda}H^{\alpha\beta}_{(2)}) . \tag{8.20}$$

In view of (8.19) and (8.14) the final form of the principle of virtual work is

$$- \int_{\Pi} [(\underset{\sim}{N}^{\beta}{}_{|\beta} + \underset{\sim}{p})\cdot\delta\underset{\sim}{\bar{r}} + (\underset{\sim}{M}^{\beta}{}_{|\beta} + \underset{\sim}{\bar{r}}_{,\beta}\times\underset{\sim}{N}^{\beta} + \underset{\sim}{l})\cdot\underset{\sim}{w}]dA +$$
$$\int_{\partial\Pi} [(\underset{\sim}{N}^{\beta}\nu_{\beta} - \underset{\sim}{N}^{*} - \underset{\sim}{P}^{\beta}\nu_{\beta})\cdot\delta\underset{\sim}{\bar{r}} + (\underset{\sim}{M}^{\beta}\nu_{\beta} - \underset{\sim}{M}^{*} - \underset{\sim}{L}^{\beta}\nu_{\beta})\cdot\underset{\sim}{w} +$$
$$(q^{\beta}\nu_{\beta} - q^{*})\delta\lambda_{\xi} + (n^{\beta}\nu_{\beta} - n^{*})\delta\kappa_{\xi} + (m^{\beta}\nu_{\beta} - m^{*})\delta\chi_{\xi}]ds = 0 . \tag{8.21}$$

From (8.21) we can read off the differential equilibrium equations and the corresponding boundary conditions. The equilibrium equations resulting from (8.21) coincide with those of (6.1), as ought to be the case. These equations, the constitutive equations given by (8.18) and (8.20) and the compatibility equations for $\underset{\sim}{E}_{\alpha}$, $\underset{\sim}{K}_{\alpha}$ obtained in [8,9] form the complete set of field equations for the shell theory consistent with the assumptions I and II. With the reference to the boundary conditions we note that the three-dimensional displacement boundary conditions require $\underset{\sim}{x}(s,\xi)$ to be prescribed on the lateral surface $\partial\mathbb{B}^{*}$. Since $\underset{\sim}{x}(s,\xi) = \underset{\sim}{\bar{r}}(s) + \zeta(s,\xi)\underset{\sim}{N}(s)$ with $\zeta(s,\xi)$ being of the form (4.4), this requires specification of $\underset{\sim}{\bar{r}}$, $\underset{\sim}{N} = \underset{\sim}{Q}\underset{\sim}{n}$, λ_{ξ}, κ_{ξ} and χ_{ξ} along the boundary curve $\partial\Pi$. Then the corresponding static boundary conditions follow from (8.21). It is apparent that prescribing $\bar{\underset{\sim}{r}}$, $\underset{\sim}{N}$, λ_ξ, $\varkappa_\xi$ and χ_ξ along $\partial\Pi$ we can exactly satisfy the three-dimensional displacement boundary conditions on $\partial\mathbb{B}^*$ that are consistent with the assumptions I and II. As an example, consider the case $\lambda_\xi = 1$, $\varkappa_\xi = \chi_\xi = 0$. Then $\zeta(s,\xi) = \xi$ what corresponds to the shell deformation without the change in the thickness at the boundary. In the case $\varkappa_\xi = \chi_\xi = 0$ we have $\zeta(s,\xi) = \lambda_\xi(s)\xi$. This admits an uniform transverse normal deformation of the shell along $\partial\Pi$. Obviously, these types of boundary conditions have no their counterparts in classical theories. We also note that by virtue of (5.8) the boundary conditions on λ_ξ, $\varkappa_\xi$ and χ_ξ can be expressed in terms of appropriate components of $\underset{\sim}{E}_\alpha \nu^\alpha$ and $\underset{\sim}{K}_\alpha \nu^\alpha$.

Furthermore, an alternate form of the field equations and boundary conditions can be obtained by choosing as the basic unknows the position vector $\bar{\underset{\sim}{r}}$ and the finite rotation vector $\underset{\sim}{\gamma}$ defined by (5.16).

Finally, we emphasize that throughout the paper no assumptions have been made as to the magnitude of strains, rotations and the initial shell thickness. In this sense the derived field equations and boundary conditions are exact.

The work of this paper represents a substantial extension of the earlier results [3-6] insofar as the earlier theories of shells did not include of the transverse shear deformation. Moreover, in [4-6] the transverse normal deformation was included in a very restricted form (being the particular cases of our general approach).

Perhaps, the theory presented in this paper is far more general then necessary for practical purposes. However, our work provides the bases for the formulations of simplified theories without inconsistencies that may arise if unnecessary assumptions are introduced at too early stage of considerations.

References

1. PIETRASZKIEWICZ W., Finite Rotations and Lagrangean Description in the Non-Linear Theory of Shells, Polish Sci. Publ., Warszawa 1979.
2. NAGHDI P. M., The Theory of Plates and Shells, in: Handbuch der Physik, vol. VIa/2, 425 - 640, Springer - Verlag, New York 1972.
3. STUMPF H., MAKOWSKI J., On large strain deformation of shells. Acta Mechanica (submitted for publication).
4. BIRICIKOGLU V., KALNINS A., Large elastic deformations of shells with inclusion of transverse normal strain, Int. J. Solids Struct., vol. 7 (1971), 431 - 444.
5. CHERNYKH K., F., Nonlinear theory of isotropically elastic thin shells (in Russian), Mekh. Tverdogo Tela, vol. 15 (1980), 148 - 159.

6. CHERNYKH K., F., The theory of thin shells of elastomers (in Russian), Advances in Mechanics, vol. 6 (1983), 111 - 147.

7. STUMPF H., MAKOWSKI J., Large strain deformation of shell - like bodies, GAMM Congress'85, Dubrownik, 1-4 April 1985 (to be published: ZAMM, vol. 66, 1986).

8. REISSNER E., Linear and nonlinear theory of shells, in: Thin - Shell Structures (Sechler Anniversory Volume), 29 - 44, Prentice Hall, Englewood Cliffs, New Jersey 1974.

9. LIBAI A., SIMMONDS J., G., Nonlinear elastic shell theory, in: Advances in Applied Mechanics, vol. 23, 271 - 371, Academic Press, 1983.

10. PIETRASZKIEWICZ W., BADUR J., Finite rotations in the description of continuum deformation, Int. J. Engng Sci., vol. 21 (1983), 1097 - 1115.

11. SIMMONDS J. G., DANIELSON D. A., Nonlinear shell theory with a finite rotation vector, Proc. Kon. Ned. Ak. Wet., vol. 73 (1970), 460 - 478.

THEORY OF THIN-WALLED ELASTIC BEAMS WITH FINITE DISPLACEMENTS

H. MØLLMANN
Department of Structural Engineering
Technical University of Denmark
2800 Lyngby, Denmark

1. Introduction

The well known linear theory of thin-walled elastic beams of open cross-section (Vlasov, 1940) is a useful tool which can be used to treat a wide range of problems involving torsional-flexural interaction in thin-walled beams. However, several problems of practical importance (such as the postbuckling behaviour of thin-walled beams) involve the effects of finite displacements and are therefore outside the range of a purely linear theory. It follows that there is a need for a nonlinear theory of thin-walled beams capable of accounting for the effects of finite displacements. In recent years, several attempts have been made to develop such a nonlinear theory (see e.g. [1], [2], [3], [4], [9], [10]).

It will be recalled that the linear theory of thin-walled beams of open cross-section is based on the following fundamental geometrical assumptions (Vlasov's constraints):

1) The shearing strains in the middle surface of the beam are negligible.

2) Any cross-section of the beam is not deformable in its own plane.

In the following, a consistent nonlinear theory of thin-walled elastic beams of open cross-section is presented. The theory is based on Vlasov's constraints and is valid for large displacements and rotations, but the strains are assumed to be small throughout the beam. A detailed derivation of the theory is given in [7].

2. The strains of the thin-walled beam

The thin-walled beam will be regarded as a thin shell, and we shall use the results of classical nonlinear shell theory in the derivations. We assume that the beam is straight and of constant open cross-section in the undeformed state. The middle surface is therefore a cylinder in the undeformed state. We introduce a fixed Cartesian coordinate system xyz with the x-axis parallel to the generators of the cylinder, and we define convected curvilinear coordinates (θ_1, θ_2) of a particle on the middle surface as follows:

$$\theta_1 = x \; , \qquad \theta_2 = s \; , \tag{1}$$

where x is the x-coordinate and s the arc length measured along a section curve x = const. of the particle in the undeformed state.

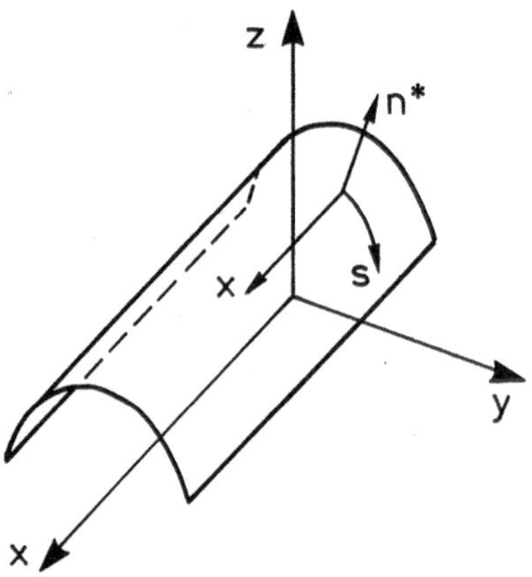

Fig. 1

According to classical nonlinear shell theory, the strains throughout the shell are determined by the middle surface strain and bending tensors:

$$\varepsilon_{\alpha\beta} = \frac{1}{2}(a_{\alpha\beta} - a^*_{\alpha\beta}) \quad , \quad k_{\alpha\beta} = b_{\alpha\beta} - b^*_{\alpha\beta} \quad , \tag{2}$$

which are proportional to the increments of the first and second fundamental tensors of the middle surface in the transition from the undeformed to the deformed state (Greek indices have the range 1, 2 and refer to the convected curvilinear coordinates θ_α, see (1)).

The required generalization of Vlasov's constraints for the thin-walled beam with finite displacements is now expressed by the following three geometrical constraints:

1. The shearing strains of the middle surface may be neglected, i.e.

$$\varepsilon_{12} = 0. \tag{3a}$$

2. The extension of the middle surface in the transverse direction may be neglected, i.e.

$$\varepsilon_{22} = 0. \tag{3b}$$

3. The bending deformation in the transverse direction may be neglected, i.e.

$$k_{22} = 0. \tag{3c}$$

In 1966, Koiter [5] derived a set of exact compatibility equations for shells with finite displacements. These exact equations are complicated nonlinear tensor equations. However, Koiter also derived a set of simplified compatibility equations which can be used when the strains are small throughout the shell. Introducing Vlasov's constraints, we find that the simplified compatibility equations assume the form:

$$\left.\begin{aligned} &\varepsilon_{11,22} - \frac{1}{\rho} k_{11} - k_{12}^2 = 0 , \\ &k_{12,2} = 0 , \\ &k_{12,1} - k_{11,2} - \frac{1}{\rho} \varepsilon_{11,2} = 0 , \end{aligned}\right\} \tag{4}$$

where $\rho(s)$ is the radius of curvature of the section curve, see Fig. 2. From $(4)_2$ we obtain $k_{12} = f(x)$, where $f(x)$ is an arbitrary function of x, and the remaining two equations may then be solved as a system of two linear, ordinary differential equations (for a fixed value of x). In this manner, the following general solution of (4) is obtained:

$$\left.\begin{aligned} \varepsilon_{11} &= \alpha(x) + \beta(x)\, y(s) + \gamma(x)\, z(s) + \\ &\quad + f'(x)\, \omega(s) + f^2(x)\, \frac{1}{2} r^2(s) , \\ k_{11} &= \beta(x) \sin\varphi - \gamma(x) \cos\varphi + f'(x)\, b(s) - \\ &\quad - f^2(x)\, h(s) , \\ k_{12} &= f(x) , \end{aligned}\right\} \tag{5}$$

where $\alpha(x)$, $\beta(x)$, $\gamma(x)$, and $f(x)$ are arbitrary functions of x, the meaning of the quantities $y(s)$, $z(s)$, $\varphi(s)$, $b(s)$ and $h(s)$ is shown in Fig. 2,

$$r(s) = \sqrt{y^2(s) + z^2(s)} ,$$

and

$$\omega(s) = \int_{0}^{s} h(t)\, dt$$

is the usual sector coordinate.

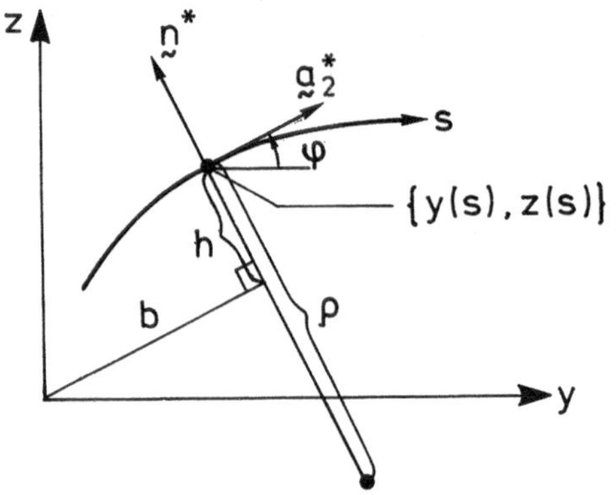

Fig. 2

Note that k_{12} depends only on x, and the sum of the first four terms in the formula for ε_{11} has the same form as the strain expression used in the linear theory. It will be seen that we have determined the general form of the strain and bending tensors of the thin-walled beam solely with the help of the compatibility equations and Vlasov's constraints, but without any reference to the displacements.

3. The associated Bernoulli beam

We now seek a geometric interpretation of the arbitrary functions appearing in (5). For this purpose we introduce an associated Bernoulli beam (i.e. a beam for which cross-sections that are plane and perpendicular to the centre-line before deformation remain so after deformation and undergo no strain in their planes), and we write the displacements of the middle surface of the thin-walled beam as the sum of two contributions:

1) The displacements of the Bernoulli beam, plus

2) certain additional displacements, denoted by $\underset{\sim}{u}$.

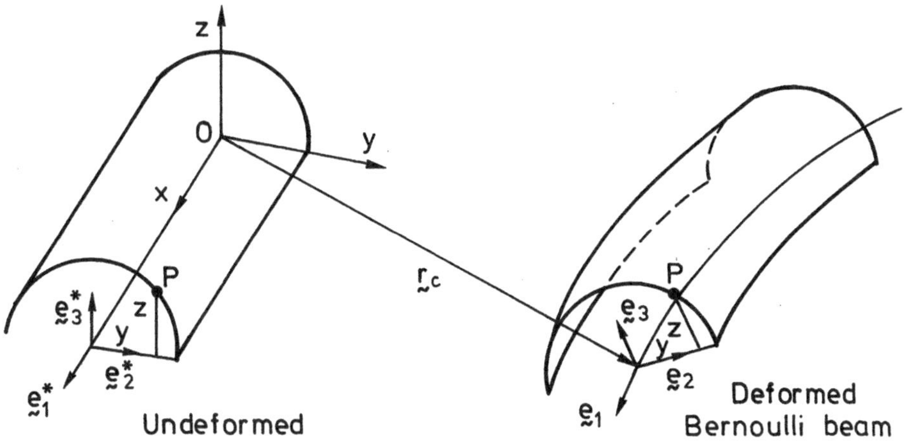

Fig. 3

In the undeformed state, we shall regard the x-axis as the centre-line of the beam. When a displacement takes place, the plane cross-sections of the Bernoulli beam undergo rigid-body displacements, and the centre-line is generally deformed into a space curve. Let $\underset{\sim}{e}_i^*$ (i = 1, 2, 3) be unit vectors in the directions of the x-, y-, and z-axes, respectively. The rigid-body displacement of a cross-section carries the unit vectors $\underset{\sim}{e}_2^*$, $\underset{\sim}{e}_3^*$ into unit vectors $\underset{\sim}{e}_2$, $\underset{\sim}{e}_3$ in the deformed state, see Fig. 3. The unit vector $\underset{\sim}{e}_1$ is tangent to the centre-line in the deformed state and is perpendicular to $\underset{\sim}{e}_2$ and $\underset{\sim}{e}_3$.

The deformation of the Bernoulli beam may be characterized by the elongation ε and the curvatures κ_i (i = 1, 2, 3). These quantities are defined in the following manner: ε is the elongation per unit length of the centre-line, and the curvatures are defined by

$$\kappa_1 = \underset{\sim}{e}_2' \cdot \underset{\sim}{e}_3, \quad \kappa_2 = \underset{\sim}{e}_3' \cdot \underset{\sim}{e}_1, \quad \kappa_3 = \underset{\sim}{e}_1' \cdot \underset{\sim}{e}_2, \tag{6}$$

where $(\ )' = \frac{d}{dx}(\ )$. κ_2 and κ_3 determine the bending about the 2- and 3-directions (see Fig. 3), and κ_1 determines the twist.

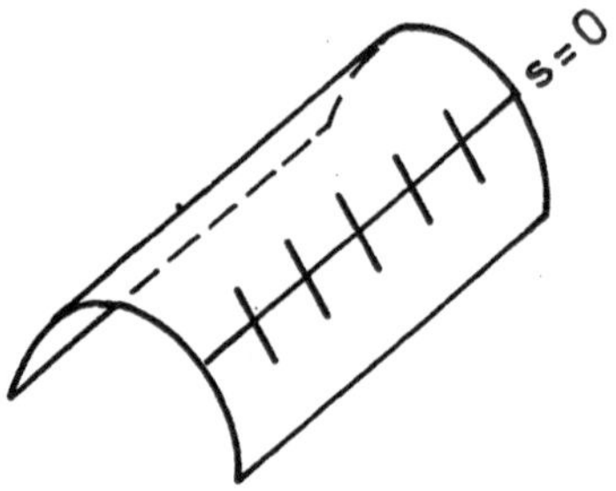

Fig. 4

We now demand that the additional displacement $\underset{\sim}{u}$ should vanish along the generator s = 0 on the middle surface, see Fig. 4, and that the tangent planes to the thin-walled beam and the Bernoulli beam should coincide along this line. It can then be shown (see [7], Part 1) that the arbitrary functions in (5) may be identified with the elongation and the curvatures (within the accuracy of a small strain theory), and we obtain the following final formulae for the non-vanishing components of the strain and bending tensors of the thin-walled beam:

$$\left.\begin{aligned} \varepsilon_{11} &= \varepsilon(x) - \kappa_3(x)\, y(s) + \kappa_2(x)\, z(s) + \kappa_1'(x)\, \omega(s) + \\ &\qquad + \tfrac{1}{2}\kappa_1^2(x)\, r^2(s)\,, \\ k_{11} &= -\kappa_3(x)\sin\varphi - \kappa_2(x)\cos\varphi + \kappa_1'(x)\, b(s) - \\ &\qquad - \kappa_1^2(x)\, h(s)\,, \\ k_{12} &= \kappa_1(x)\,. \end{aligned}\right\} \tag{7}$$

Moreover, it is found that the additional displacement may be taken to be

$$\underset{\sim}{u}(x,s) = \omega(s)\,\kappa_1(x)\,\underset{\sim}{e}_1(x)\,. \tag{8}$$

The orders of magnitude of the quantities ε and κ are given by

$$\begin{aligned} &\varepsilon = O(\bar{\varepsilon})\,, \quad \kappa_2\bar{b} = O(\bar{\varepsilon})\,, \quad \kappa_3\bar{b} = O(\bar{\varepsilon})\,, \\ &\kappa_1'\bar{b}^2 = O(\bar{\varepsilon})\,, \quad \kappa_1\bar{b} = O(\bar{\varepsilon}^{\frac{1}{2}})\,, \end{aligned} \tag{9}$$

where $\bar{\varepsilon}$ is the maximum principal strain in the beam, and $\bar{b}$ is a characteristic cross-sectional dimension. It will be seen that κ_1 may be considerably greater than κ_2 and κ_3, which confirms the well known fact that thin-walled beams of open cross-section are relatively weak in torsion.

4. The strain energy of the beam

We now make use of the known expression for the strain energy of a thin shell consisting of a homogeneous and isotropic elastic material. The strain energy measured per unit area of undeformed middle surface is therefore given by

$$\left.\begin{aligned} W = \frac{1}{2}\,\frac{E\delta}{(1-\nu^2)}\,[(1-\nu)\,\varepsilon^{\alpha\beta}\,\varepsilon_{\alpha\beta} + \nu\,\varepsilon^{\alpha}_{\alpha}\,\varepsilon^{\beta}_{\beta}] &+ \\ + \frac{1}{24}\,\frac{E\delta^3}{(1-\nu^2)}\,[(1-\nu)\,k^{\alpha\beta}\,k_{\alpha\beta} + \nu\,k^{\alpha}_{\alpha}\,k^{\beta}_{\beta}]&\,, \end{aligned}\right\} \tag{10}$$

where δ is the shell thickness, E is Young's modulus, ν is Poisson's ratio, and we have used the summation convention for Greek indices. Introducing Vlasov's constraints (3a) and (3c) in (10), but replacing the constraint $\varepsilon_{22} = 0$ by $\sigma_{22} = 0$, since the transverse stresses σ_{22} will be small compared with the longitudinal stresses σ_{11}, we obtain

$$W = \frac{1}{2}E\delta\,\varepsilon_{11}^2 + \frac{1}{24}\,\frac{E\delta^3}{(1-\nu^2)}\,[k_{11}^2 + 2(1-\nu)\,k_{12}^2]\,. \tag{11}$$

An estimate of the order of magnitude of the terms in this equation reveals that the term k_{11}^2 in the bracket may be omitted. Inserting the expressions for ε_{11} and k_{12} given by (7), and integrating with respect to s, we obtain the following expression for <u>the strain energy $\mathcal{W}$ measured per unit length of undeformed centre-line</u>:

$$\left.\begin{aligned} \mathcal{W} = \frac{1}{2}E\Big\{A\,\varepsilon_o^2 + I_2\,\kappa_2^2 + I_3\,\kappa_3^2 + I_\omega\,\kappa_1'^2 &+ \\ + I_p\,\kappa_1^2\,\varepsilon_o - I_{\eta r}\,\kappa_1^2\,\kappa_3 + I_{\zeta r}\,\kappa_1^2\,\kappa_2 + I_{\omega r}\,\kappa_1^2\,\kappa_1' &+ \\ + \frac{1}{4}I_4\,\kappa_1^4\Big\} + \frac{1}{2}GI_t\,\kappa_1^2&\,, \end{aligned}\right\} \tag{12}$$

where the section constants are given by

$$\left.\begin{aligned} A &= \int_c \delta\, ds\ , \quad I_2 = \int_c \zeta^2 \delta\, ds\ , \quad I_3 = \int_c \eta^2\, \delta\, ds\ , \\ I_\omega &= \int_c \omega^2\, \delta\, ds\ , \quad I_p = \int_c r^2\, \delta\, ds\ , \quad I_{\eta r} = \int_c \eta\, r^2\, \delta\, ds\ , \\ I_{\zeta r} &= \int_c \zeta\, r^2\, \delta\, ds\ , \quad I_{\omega r} = \int_c \omega\, r^2\, \delta\, ds\ , \\ I_4 &= \int_c r^4\, \delta\, ds\ , \quad I_t = \frac{1}{3} \int_c \delta^3\, ds\ , \end{aligned}\right\} \qquad (12a)$$

and the integrations are performed along the section curve. It is assumed that the coordinate system in the cross-section of the undeformed beam is located as indicated in Fig. 5, η and ζ are measured from the centroid, the quantity ε_o is defined by

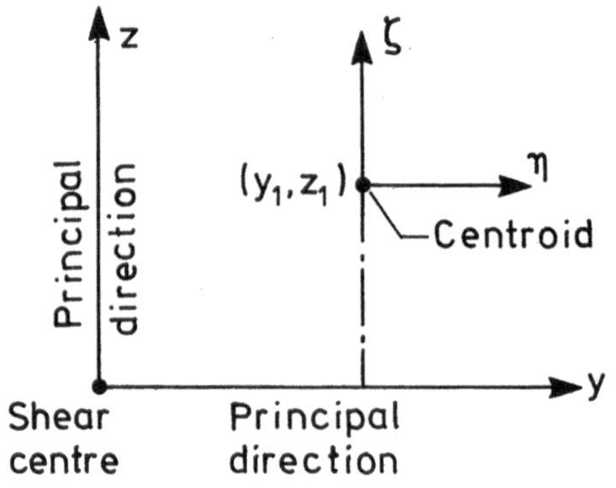

Fig. 5

$$\varepsilon_o = \varepsilon - \kappa_3 y_1 + \kappa_2 z_1\ ,$$

and

$$G = \frac{E}{2(1+\nu)}\ .$$

The generalized stress resultants are now determined as the partial derivatives of W with respect to the generalized strains ε, κ_1, κ_2, κ_3, and κ_1'. In this way we obtain the constitutive equations:

$$\left.\begin{aligned}
N_1 &= \frac{\partial W}{\partial \varepsilon} = E\,A\,\varepsilon_o + \frac{1}{2}\,E\,I_p\,\kappa_1^2 \,, \\
M_2 &= \frac{\partial W}{\partial \kappa_2} = E\,I_2\,\kappa_2 + \frac{1}{2}\,E\,I_{\zeta r}\,\kappa_1^2 + N_1\,z_1 \,, \\
M_3 &= \frac{\partial W}{\partial \kappa_3} = E\,I_3\,\kappa_3 - \frac{1}{2}\,E\,I_{\eta r}\,\kappa_1^2 - N_1\,y_1 \,, \\
M_t &= \frac{\partial W}{\partial \kappa_1} = G\,I_t\,\kappa_1 + E\,I_p\,\varepsilon_o\,\kappa_1 - E\,I_{\eta r}\,\kappa_1\,\kappa_3 + \\
&\quad + E\,I_{\zeta r}\,\kappa_1\,\kappa_2 + E\,I_{\omega r}\,\kappa_1\,\kappa_1' + \frac{1}{2}\,E\,I_4\,\kappa_1^3 \,, \\
B &= \frac{\partial W}{\partial \kappa_1'} = E\,I_\omega\,\kappa_1' + \frac{1}{2}\,E\,I_{\omega r}\,\kappa_1^2 \,,
\end{aligned}\right\} \qquad (13)$$

where

N_1 = axial force.

M_2, M_3 = bending moments about axes through the shear centre parallel with $\underset{\sim}{e}_2$ and $\underset{\sim}{e}_3$.

M_t = contribution to torsional moment.

B = bimoment, as defined by Vlasov.

5. Finite rotation vector

Because of the finite displacements, the cross-sections of the Bernoulli beam may undergo finite rotations. These will be described by means of finite rotation vectors. Suppose that a finite rotation (angle θ) takes place about an axis 1 - 1 in space. The corresponding finite rotation vector $\underset{\sim}{q}$ is defined by

$$\underset{\sim}{q} = \underset{\sim}{e} \tan \frac{\theta}{2} \,, \qquad (14)$$

where $\underset{\sim}{e}$ is a unit vector parallel to the axis of rotation, and the direction of $\underset{\sim}{e}$ is determined by the right-hand rule.

It can be shown that the following results concerning finite rotation vectors are valid: Consider a rotation $\underset{\sim}{q}_1$ followed by a rotation $\underset{\sim}{q}_2$, and let the resulting rotation be $\underset{\sim}{q}_3$. We then have

$$\underset{\sim}{q}_3 = (\underset{\sim}{q}_1 + \underset{\sim}{q}_2 + \underset{\sim}{q}_2 \times \underset{\sim}{q}_1)/(1 - \underset{\sim}{q}_1 \circ \underset{\sim}{q}_2) \,. \qquad (15)$$

The presence of the term $\underset{\sim}{q}_2 \times \underset{\sim}{q}_1$ shows that finite rotations do not commute.

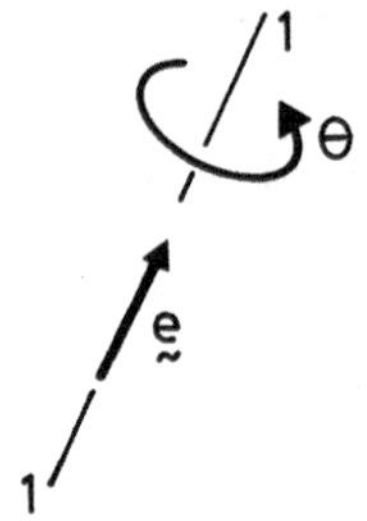

Fig. 6

The curvatures are given by

$$\kappa_i(x) = 2\,\frac{\underset{\sim}{q}'(x) + \underset{\sim}{q}'(x) \times \underset{\sim}{q}(x)}{1 + \underset{\sim}{q}(x) \cdot \underset{\sim}{q}(x)} \quad \underset{\sim}{e}_i^* \tag{16}$$

and the unit vectors of the deformed state $\underset{\sim}{e}_i(x)$ can be expressed as follows:

$$\underset{\sim}{e}_i(x) = \underset{\sim}{e}_i^* + \frac{2}{1 + \underset{\sim}{q}(x) \cdot \underset{\sim}{q}(x)} \left\{ \underset{\sim}{q}(x) \times \underset{\sim}{e}_i^* + \underset{\sim}{q}(x) \times (\underset{\sim}{q}(x) \times \underset{\sim}{e}_i^*) \right\} . \tag{17}$$

The configuration of the deformed Bernoulli beam will be described by means of the elongation $\varepsilon(x)$ and the finite rotation vector $\underset{\sim}{q}(x)$. Since $\underset{\sim}{q}$ determines the vectors $\underset{\sim}{e}_i$ of the deformed state, the position vector of the deformed centre-line $\underset{\sim}{r}_c(x)$ can be found by integrating the equation

$$\underset{\sim}{r}_c' = (1 + \varepsilon)\ \underset{\sim}{e}_1 \ , \tag{18}$$

which follows from the definition of ε as the elongation per unit length.

6. Variational principle

It follows from the geometrical investigation in Sections 2 and 3 that the displacement of an arbitrary point of the middle surface, measured from the undeformed state, can be written in the form

$$\underset{\sim}{v}(x,s) = \underset{\sim}{v}_c(x) + \underset{\sim}{d}(x,s) - \underset{\sim}{d}^*(s) + \underset{\sim}{u}(x,s) \ , \tag{19}$$

where

$$\left.\begin{aligned} \underset{\sim}{v}_c(x) &= \underset{\sim}{r}_c(x) - x\, \underset{\sim}{e}_1^* , \\ \underset{\sim}{d}(x,s) &= y(s)\, \underset{\sim}{e}_2(x) + z(s)\, \underset{\sim}{e}_3(x) , \\ \underset{\sim}{d}^*(s) &= y(s)\, \underset{\sim}{e}_2^* + z(s)\, \underset{\sim}{e}_3^* , \\ \underset{\sim}{u}(x,s) &= \omega(s)\, \kappa_1(x)\, \underset{\sim}{e}_1(x) . \end{aligned}\right\} \qquad (19a)$$

$\underset{\sim}{d}$ and $\underset{\sim}{d}^*$ are vectors in the cross-section of the Bernoulli beam in the deformed and the undeformed state, respectively.

Assuming that the prescribed loads are "dead loads", the total potential energy of the thin-walled beam can be written in the form

$$V = \int_{x_1}^{x_2} W(\varepsilon, \kappa_i, \kappa_1') \, dx - \int_S \underset{\sim}{p} \cdot \underset{\sim}{v} \, dS - \left[\int_C \underset{\sim}{t} \cdot \underset{\sim}{v} \, ds \right]_{x_1}^{x_2} , \qquad (20)$$

where W is given by (12), $x_1 = 0$ and $x_2 = \ell$ (the end sections of the beam), the surface integral is extended over the area S of the undeformed middle surface, $\underset{\sim}{p}$ = prescribed surface load measured per unit area of undeformed middle surface,

$-\underset{\sim}{t}(x_1,s)$ = prescribed load on the end section $x_1 = 0$ (measured per unit length of section curve), and

$\underset{\sim}{t}(x_2,s)$ = prescribed load on the end section $x_2 = \ell$, see Fig. 7.

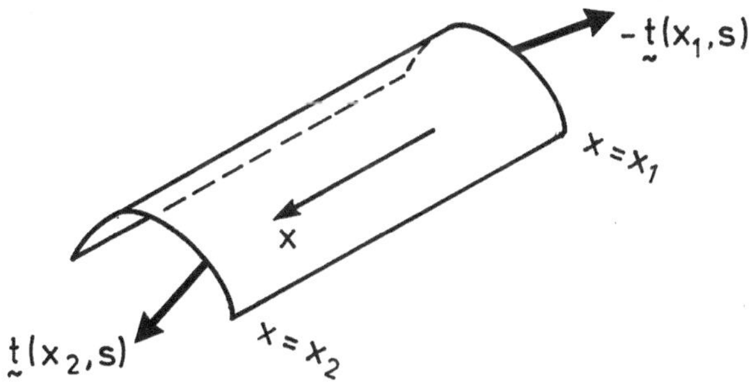

Fig. 7

The vector function $\underset{\sim}{t}(x_\alpha,s)$, $\alpha = 1,2$ is thus defined for all values of s along the section curve, but only for two discrete x-values (x_1 and x_2). The term in brackets in (20) has the usual meaning, i.e.

$$[\int_C \underset{\sim}{t} \cdot \underset{\sim}{v}\, ds]_{x_1}^{x_2} =$$

$$= \int_C \left\{ \underset{\sim}{t}(x_2,s) \cdot \underset{\sim}{v}(x_2,s) - \underset{\sim}{t}(x_1,s) \cdot \underset{\sim}{v}(x_1,s) \right\} ds \ , \tag{21}$$

and equals the work of the prescribed loads on the end sections.

We now put

$$\underset{\sim}{p}(x) = \int_C \underset{\sim}{p}(x,s)\, ds \ , \tag{22}$$

this is the load measured per unit length of undeformed centre-line. We then have

$$\int_S \underset{\sim}{p}(x,s) \cdot \underset{\sim}{v}_c(x)\, dS = \int_{x_1}^{x_2} \underset{\sim}{p}(x) \cdot \underset{\sim}{v}_c(x)\, dx \ . \tag{23}$$

We recall that the geometrical quantities $\varepsilon(x)$ and $\underset{\sim}{q}(x)$ are used to determine the configuration of the beam. It will be seen that ε, $\underset{\sim}{q}$, and $\underset{\sim}{v}_c(x_\alpha)$ are subjected to the following condition

$$\underset{\sim}{v}_c(x_2) - \underset{\sim}{v}_c(x_1) - \int_{x_1}^{x_2} \left\{ (1+\varepsilon)\, \underset{\sim}{e}_1 - \underset{\sim}{e}_1^* \right\} dx = \underset{\sim}{0} \ , \tag{24}$$

which is deduced by integration of (18). The three scalar conditions obtained by expressing the vectors in (24) in the global xyz-system are now multiplied by corresponding Lagrange multipliers and added to the potential energy functional (20). The resulting <u>modified functional</u> can be written in the form:

$$\Psi = \int_{x_1}^{x_2} \left\{ W(\varepsilon,\kappa_i,\kappa_1') - \underset{\sim}{N} \cdot [(1+\varepsilon)\underset{\sim}{e}_1 - \underset{\sim}{e}_1^*] \right\} dx -$$

$$- \int_S \underset{\sim}{p} \cdot (\underset{\sim}{d} - \underset{\sim}{d}^* + \underset{\sim}{u})\, dS - [\int_C \underset{\sim}{t} \cdot (\underset{\sim}{d} - \underset{\sim}{d}^* + \underset{\sim}{u})\, ds]_{x_1}^{x_2} +$$

$$+ [(\underset{\sim}{N} - \underset{\sim}{N}) \cdot \underset{\sim}{v}_c]_{x_1}^{x_2} \ . \tag{25}$$

In this equation,

$$\underset{\sim}{N}(x) = \underset{\sim}{N}_o - \int_{x_1}^{x} \underset{\sim}{p}(\xi)\, d\xi \tag{25a}$$

is the resultant internal force in the beam, this is known except possibly for the constant term $\underset{\sim}{N}_o$ (the latter may be known or unknown depending on the types of support), and

$$\underset{\sim}{N}(x_\alpha) = \int_C \underset{\sim}{t}(x_\alpha, s)\, ds \tag{25b}$$

is the total prescribed force on the end section $x = x_\alpha$ (apart, possibly, from a change of sign). The global components of the vector $\underset{\sim}{N}_o$ are the Lagrange multipliers.

If we now impose the condition of stationarity for the modified functional, we obtain the governing equations of the thin-walled beam (including equilibrium conditions) as the Euler equations of the variational problem (see [7]). However, a more important property of the variational formulation is that it can be used to obtain numerical solutions to beam problems by means of variational methods such as the finite element method.

7. Applications

We now briefly review some of the applications that have been made of the above general, nonlinear theory of thin-walled beams of open cross-section.

a. Nonlinear torsion of thin-walled cantilever beams

The beam is built-in without warping at one end, and it is subjected to a torsional moment at the other (free) end. In a state of pure torsion, the cross-sections rotate about a straight axis of rotation which is parallel to the x-axis.

It is found that pure torsion is a possible state of equilibrium for a cantilever beam of doubly symmetrical cross-section. However, for a monosymmetrical channel section, pure torsion is not a possible state of equilibrium, since the twist will be accompanied by small bending curvatures (see [8]).

b. Determination of critical loads, comparison with Vlasov's equations

The linearized differential equations derived from our general, nonlinear theory agree substantially with Vlasov's stability equations. Our equations contain all the terms that appear in Vlasov's equations, but two of the three equations also contain a few additional terms that do not appear in Vlasov's equations (see [8]).

c. Influence of prebuckling deformations on critical load

For a simply supported beam loaded by equal and opposite bending moments at the ends, it is found that the critical moment is increased as a result of the prebuckling deformations. This is a well known result, but the quantitative agreement between our results and previous results tends to confirm the validity of the present theory (see [8]).

d. Postbuckling behaviour of thin-walled beams

The general nonlinear theory was used in a comprehensive investigation by Carl Pedersen [8] on the postbuckling behaviour of thin-walled beams of open cross-section.

Cantilever beams and simply supported beams with several different types of thin-walled open cross-section were studied. Various cases of lateral stability and column stability were investigated, and imperfection sensivity was detected on several occasions.

The imperfection sensitivity of the beams was found to be moderate (typically, the reduction in load carrying capacity was found to be about 25% for imperfections of a realistic magnitude).

8. Summary and conclusions

A general, nonlinear theory of thin-walled elastic beams of open cross-section has been presented, account being taken of finite displacements and warping rigidity. The beam is regarded as a thin shell, and appropriate geometrical constraints are introduced which constitute a generalization of those employed in Vlasov's linear theory. Using Koiter's general equations of compatibility for shells, expressions for the strains and the strain energy of the thin-walled beam are derived. The rotations of the beam are described by means of finite rotation vectors which are closely associated with W.R. Hamilton's quaternions. A new variational principle is formulated which may be used as a basis of numerical methods for the solution of beam problems. Some applications of the theory are also discussed.

The present nonlinear theory of thin-walled beams constitutes a consistent generalization of the well known linear theory into the range of finite displacements. The theory is capable of describing nonlinear phenomena such as postbuckling behaviour and nonlinear torsion of thin-walled beams of open cross-section.

References

1. EPSTEIN M., MURRAY D.W., Three-dimensional large deformation analysis of thin walled beams, Int. Journ. Solids Structures 12 (1976), 867-876.
2. EPSTEIN M., Thin-walled beams as directed curves, Acta Mechanica 33 (1979), 229-242.
3. GHOBARAH A.A., TSO W.K., A non-linear thin-walled beam theory, Int. Journ. Mech. Sciences 13 (1971), 12, 1025-1038.
4. GRIMALDI A., PIGNATARO M., Postbuckling behaviour of thin-walled open cross-section compression members, Journ. Structural Mech. 7 (1979), 2, 143-159.
5. KOITER W.T., On the nonlinear theory of thin elastic shells, Parts I to III, Koninkl. Nederl. Akademie van Wetenschappen, Proc., Series B 69 (1966), 1, 1-54.
6. MØLLMANN H., Thin-walled elastic beams with finite displacements, Dept. of Struct. Engineering, Techn. Univ. of Denmark, Report No. R 142, 1981.
7. MØLLMANN H., Finite displacements of thin-walled beams, Parts 1 and 2, Danish Center for Appl. Math. and Mech., Techn. Univ. of Denmark, Reports Nos. 252 and 253, 1982.
8. PEDERSEN C., Stability properties and non-linear behaviour of thin-walled elastic beams of open cross-section, Parts 1 and 2, Dept. of Struct. Engineering, Techn. Univ. of Denmark, Reports Nos. R 149 and R 150, 1982.
9. SZYMCZAK C., Buckling and initial post-buckling behaviour of thin-walled I columns, Computers & Structures 11 (1980), 481-487.
10. WOOLCOCK S.T., TRAHAIR N.S., Post-buckling behaviour of determinate beams, Journ. Eng. Mech. Div., ASCE 101 (1974), EM2, 151-171.

ONE-DIMENSIONAL FINITE ROTATION SHELL PROBLEMS IN DISPLACEMENT FORMULATION

L.-P. NOLTE
Institute of Mechanics
Ruhr-Universität Bochum
D-4630 Bochum
Fed. Rep. of Germany

1. Introduction

The present report deals with certain consequences for the displacement formulation of nonlinear first approximation shell theories if shell problems are concerned in which the partial differential equations reduce to ordinary ones. In these cases a necessary geometrical constraint is that the reference middle surface admit one-dimensional strain fields. Accordingly the shell strains, depending on differences of the metric and curvature tensors in the deformed and undeformed configuration are functions of one independent variable only. It has been shown by Simmonds [1] that then during the deformation process the shell middle surface must be a general helicoid, which additionally implies special boundary conditions, material properties and type of loadings. There are several reasons for the analysis of one-dimensional reduced shell problems. By that general nonlinear shell equations remain rather complicated and require approximate solutions by using large computer codes such that the associated one-dimensional equations may considerably reduce the cost and/or time of the nonlinear solution. Besides, the derivation of general and simplified one-dimensional shell theories gains a good insight into similar investigations of the general theory of shells.
General helicoidal shells have been studied in [2-4]. Extensive literature, however, may be found for three special cases (see e.g. [5-11] and literature cited therein): the cylindrical deformation of shells, the torsionless axisymmetric deformation of shells of revolution, we only mention the famous Reissner equations and the pure bending of curved tubes. The associated governing equations may be formulated either in terms of displacements or in terms of stress resultants (or stress functions) and rotations as basic unknowns. In spite of relative simplicity the latter approach processes certain drawbacks from the point of view of numerical applications. These and other shortcomings

discussed in [8] make the displacement formulation in conjunction with powerful approximation procedures as the finite element method (FEM) competitive within the analyses of one-dimensional shell problems. However, existing finite rotation displacement formulations [12-14] remain quite complicated even when reduced to one-dimensional shell problems.

Therefore in the present paper the construction of one-dimensional shell equations in terms of displacements is reconsidered for the torsionless axisymmetric deformation of shells of revolution and the cylindrical deformation of shells, where the latter may be treated as a limiting case of the former one. For these problems the axis of rotation does not change its direction during the deformation process, which is a basic simplifying fact. First, we establish pertinent kinematical relations in terms of displacements on the one hand and in terms of mid-surface stretches and the rotation angle on the other hand. Then fundamental estimates [15-17] allow to derive the complete set of equations in as simple form as is consistent within the first approximation shell theory.
Further attention is devoted to the formulation of simplified versions undergoing restricted rotations. In particular, it is shown that for the class of shell problems under consideration there exists no consistent improvement of the well-known Sanders-Koiter strain measures [12] that would be simpler than our general ones valid for finite rotations. The theoretical considerations are illustrated by various comparative numerical examples.

2. Notations and Basic Kinematical Relations

In order to make the present paper self-contained we first recall some basic relations of the axisymmetric deformation of shells of revolution.
Let **M** be the shell middle surface in the undeformed reference configuration. The position vector measured from a fixed cylindrical coordinate system with an orthonormal base $\{\mathbf{e}_z, \mathbf{e}_r, \mathbf{e}_\theta = \mathbf{e}_r \times \mathbf{e}_z\}$ according to fig. 2.1 reads

$$\mathbf{r}(s,\theta) = z(s)\mathbf{e}_z + r(s)\mathbf{e}_r(\theta) , \tag{2.1}$$

where s is taken to be the arc length along a meridian of **M**, $s \in [s_1, s_2]$, $\theta \in [0, 2\pi]$. The notation $(\ldots)'$ indicates derivatives on **M** with respect to s and we make use of the fact that any axisymmetric vector-valued function **g** on **M** depends on the circumferential coordinate

θ only by the base vectors, which implies the following relation for derivates of **g** with respect to θ

$$\partial \mathbf{g}/\partial\theta = \mathbf{g} \times \mathbf{e}_z \ . \tag{2.2}$$

With a cross we denote the usual vector product. For this choice of the (material) surface coordinates the orthogonal base vector triad at any point $M \in \mathbf{M}$ is given by

$$\begin{aligned} &\mathbf{a}_1 \hat{=} \mathbf{a}_s = \mathbf{r}' = \cos\Phi\mathbf{e}_z + \sin\Phi\mathbf{e}_r \ , \quad \mathbf{a}_2 \hat{=} \mathbf{a}_\theta = - \mathbf{e}_z \times \mathbf{r} = r\mathbf{e}_\theta \ , \\ &\mathbf{n} = r^{-1}\mathbf{a}_s \times \mathbf{a}_\theta = - \sin\Phi\mathbf{e}_z + \cos\Phi\mathbf{e}_r \ , \end{aligned} \tag{2.3}$$

from which the components of the surface metric and curvature tensors can be determined as

$$\begin{aligned} &a_{11} \hat{=} a_{ss} = 1 \ , \quad a_{22} \hat{=} a_{\theta\theta} = r^2 \ , \\ &b_{11} \hat{=} b_{ss} = \sigma_s = \Phi' \ , \quad b_{22} \hat{=} b_{\theta\theta} = r^2\sigma_\theta = - r\cos\Phi \ . \end{aligned} \tag{2.4}$$

Here σ_s, σ_θ stand for the principle curvatures and Φ is the angle between the tangent to the meridian of **M** and the z-axis (see fig. 2.1).

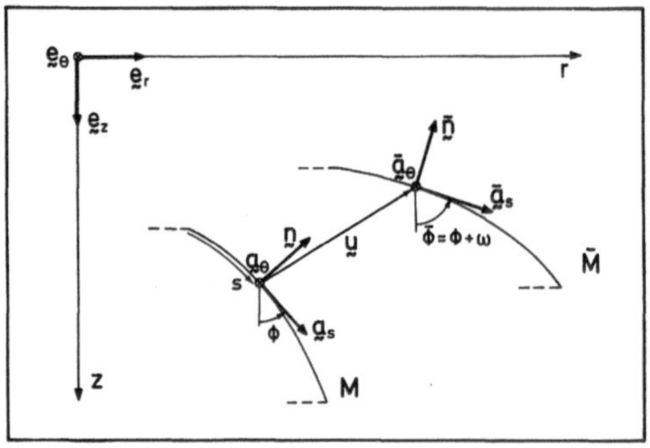

Fig. 2.1 Notations

Let us consider an axisymmetric deformation χ of the shell from the reference configuration to the actual configuration. Under Kirchhoff-Love constraints ($\bar{a}_\alpha \cdot \bar{n} = 0$, $\bar{n} \cdot \bar{n} = 1$) complete information of the local shell deformation is provided by the deformation gradient tensor **G** of the middle surface

$$\mathbf{G} = \bar{\mathbf{a}}_\alpha \otimes \mathbf{a}^\alpha + \bar{\mathbf{n}} \otimes \mathbf{n} \ , \tag{2.5}$$

where a bar is used to indicate corresponding quantities on **M** and a $\otimes$ denotes the tensor product of two vectors. According to the polar decomposition theorem the tensor **G** can be represented in the form

$$\mathbf{G} = \mathbf{R}\mathbf{U} \ , \tag{2.6}$$

where $\mathbf{U}$ and $\mathbf{R}$ are the right stretch and finite rotation tensor, which for an axisymmetric deformation can be given by

$$\mathbf{R} = \cos\omega\ \mathbf{1} - r^{-1}\sin\omega\ \mathbf{a}_\theta \times \mathbf{1} + r^{-2}(1-\cos\omega)\ \mathbf{a}_\theta \otimes \mathbf{a}_\theta\ ,$$
$$\mathbf{U} = \lambda_s\ \mathbf{a}_s \otimes \mathbf{a}_s + r^{-2}\lambda_\theta\ \mathbf{a}_\theta \otimes \mathbf{a}_\theta\ . \tag{2.7}$$

In (2.7) λ_s, λ_θ and $\omega = \bar{\Phi} - \Phi$ are the stretches along $\mathbf{a}_s$, $\mathbf{a}_\theta$ and the finite rotation angle, respectively.

For the purpose of the subsequent analysis it is essential to give an equivalent kinematical formulation on the one hand in terms of the stretches and the rotation angle and on the other hand in terms of the components of the axisymmetric displacement vector

$$\mathbf{u}(s,\theta) = u(s)\mathbf{a}_s(s,\theta) + w(s)\mathbf{n}(s,\theta)\ . \tag{2.8}$$

Accordingly the orthogonal base triad at any point $M \in \bar{\mathbf{M}}$ can be determined by

$$\bar{\mathbf{a}}_s = \lambda_s(\cos\omega\mathbf{a}_s + \sin\omega\mathbf{n}) = (1+\theta_{ss})\mathbf{a}_s + \varphi_s\mathbf{n}\ ,$$
$$\bar{\mathbf{a}}_\theta = \lambda_\theta\mathbf{a}_\theta = (1+\theta_{\theta\theta})\mathbf{a}_\theta\ , \tag{2.9}$$
$$\bar{\mathbf{n}} = -\sin\omega\mathbf{a}_s + \cos\omega\mathbf{n} = [(1+\theta_{ss})^2 + \varphi_s^2]^{-1/2}\ [-\varphi_s\mathbf{a}_s + (1+\theta_{ss})\mathbf{n}]\ ,$$

from which we may find the components of the metric and curvature tensors of the deformed middle surface $\bar{\mathbf{M}}$

$$\bar{a}_{ss} = \lambda_s^2 = (1+\theta_{ss})^2 + \varphi_s^2\ , \quad \bar{a}_{\theta\theta} = r^2\lambda_\theta^2 = r^2(1+\theta_{\theta\theta})^2\ ,$$
$$\bar{b}_{ss} = \ldots\ , \quad \bar{b}_{\theta\theta} = \ldots\ , \tag{2.10}$$

where the explicit form of $\bar{b}_{ss}$ and $\bar{b}_{\theta\theta}$ is dropped for the sake of brevity. In (2.9-10) we have introduced the linearized in-surface strains θ_{ss}, $\theta_{\theta\theta}$ and the linearized rotation φ_s

$$\theta_{ss} = \lambda_s\cos\omega - 1 = u' - \sigma_s w\ , \quad \theta_{\theta\theta} = \lambda_\theta - 1 = (\sin\Phi/r)u - \sigma_\theta w\ ,$$
$$\varphi_s = \lambda_s\sin\omega = w' + \sigma_s u\ , \tag{2.11}$$

Hence, the shell stretches and displacements are related via linearized quantities by

$$\lambda_s = [(1+\theta_{ss})^2 + \varphi_s^2]^{1/2}\ , \quad \lambda_\theta = 1 + \theta_{\theta\theta}\ . \tag{2.12}$$

Note that the expression for the circumferential direction λ_θ becomes linear.

In order to formulate later boundary conditions at the shell edges s_1, s_2 (which are either circles or points) we have to express the rotation angle ω in terms of the displacements

$$\omega(\mathbf{u}) = \arcsin\{\varphi_s\ [(1+\theta_{ss})^2 + \varphi_s^2]^{-1/2}\}\ . \tag{2.13}$$

It is interesting to note that the argument of the arcsin-function in

(2.13) is used as the fourth independent geometric boundary parameter in the general nonlinear first approximation shell theory [14].

3. Shell Strain Measures

The geometry of the deformed middle surface $\bar{M}$ is described by the components of the metric and curvature tensors $\bar{a}_{\alpha\beta}$ and $\bar{b}_{\alpha\beta}$, which should satisfy the Codazzi-Gauss equations. To describe the surface deformation in the general theory of shells the conventional strain measures are defined by [12,14]

$$\gamma_{\alpha\beta} = 1/2(\bar{a}_{\alpha\beta}-a_{\alpha\beta}) , \quad \varkappa_{\alpha\beta} = -(\bar{b}_{\alpha\beta}-b_{\alpha\beta}) . \tag{3.1}$$

Hence, using the relations (2.4) and (2.10) they can be expressed in terms of displacements as follows

$$\begin{aligned} \gamma_{ss} &= \vartheta_{ss} + (1/2)\vartheta_{ss}^2 + (1/2)\varphi_s^2 , \\ \gamma_{\theta\theta} &= \vartheta_{\theta\theta} + (1/2)\vartheta_{\theta\theta}^2 , \\ \varkappa_{ss} &= -\lambda_s^{-1}[\varphi_s'(1+\vartheta_{ss}) - \varphi_s\vartheta_{ss}' + \sigma_s(1+2\gamma_{ss})] + \sigma_s , \\ \varkappa_{\theta\theta} &= -\lambda_s^{-1}(1+\vartheta_{\theta\theta})[(\sin\Phi/r)\varphi_s + \sigma_\theta\vartheta_{ss} + \sigma_\theta] + \sigma_\theta , \end{aligned} \tag{3.2}$$

with λ_s according to $(2.12)_1$. It should be underlined that even for the axisymmetric deformation considered here the strain-displacement relations remain rather complicated and virtually useless from the point of view of applications. In what follows we will show that they may considerably be simplified within the accuracy of the first approximation shell theory [16]. The fundamental estimation of the two-dimensional strain energy function will serve as a basis. It implies that terms $O(\eta^2/h, \eta/R)$ in the shell strain measures do not effect the accuracy of Love's first approximation to the strain energy [12,15]. Here h, R and η denote the shell thickness, the minimum principle radius of curvature and the maximum strain in the shell space, respectively.
Recalling further that [17]

$$\gamma_{\alpha\beta} = O(\eta) , \quad \varkappa_{\alpha\beta} = O(\eta/h) , \tag{3.3}$$

the following estimates for the particular case of axisymmetric deformation of shells of revolution can be given

$$\omega' = O(\eta/h) , \quad [\cos(\Phi+\omega) - \cos\Phi]/r = O(\eta/h) , \tag{3.4}$$

where we have used the relations (2.4) and (2.10) and kept in mind that

$$\max\{|\lambda_s(s)-1|,|\lambda_\theta(s)-1|\} \lessdot \eta , \quad \max\{|\sigma_s(s)|,|\sigma_\theta(s)|\} \lessdot 1/R \tag{3.5}$$

In consideration of (3.3-5) various simplifying transformations and estimates for terms in (3.2) can be given applying the previously shown relations

$$\begin{aligned}
&\varphi_s'(1+\vartheta_{ss}) - \varphi_s\vartheta_{ss}' = \lambda_s^2\omega' = O(\eta/h) \ , \\
&(\sin\Phi/r)\varphi_s + \sigma_\theta\vartheta_{ss} = \lambda_s[\cos(\Phi+\omega) - \cos\Phi]/r - \sigma_\theta(\lambda_s-1) = O(\eta/h) \ , \\
&\lambda_s^{-1} = (1+2\gamma_{ss})^{-1/2} = 1 + O(\eta) \qquad (3.6) \\
&|\vartheta_{\theta\theta}| \leq \eta \ , \quad \gamma_{\theta\theta} = \vartheta_{\theta\theta} + O(\eta^2) \ .
\end{aligned}$$

Hence, within the accuracy of the first approximation shell theory which implies a relative error margin of $O(\eta)$ the consistent set of strain-displacement relations for the axisymmetric deformation of shells of revolution undergoing finite rotations are obtained in the form

$$\begin{aligned}
&\gamma_{ss} = \vartheta_{ss} + (1/2)\varphi_s^2 + (1/2)\vartheta_{ss}^2 \ , \quad \gamma_{\theta\theta} = \vartheta_{\theta\theta} \ , \\
&\varkappa_{ss} = -\varphi_s' - \varphi_s'\vartheta_{ss} + \varphi_s\vartheta_{ss}' \ , \quad \varkappa_{\theta\theta} = -(\sin\Phi/r)\varphi_s - \sigma_\theta\vartheta_{ss} \ .
\end{aligned} \qquad (3.7)$$

Note that the meridional strains γ_{ss} and $\varkappa_{ss}$ contain only two quadratic terms whereas the circumferential strains $\gamma_{\theta\theta}$ and $\varkappa_{\theta\theta}$ both are linear with respect to the displacements and their derivatives. Moreover, the expression (3.7) would differ from any set of strain-displacement relations obtained from general shell equations by formal linearization and reduction to the one-dimensional problems under consideration. This is due to the second term in $(3.7)_4$. However, within the constrained versions of small (linear theory) and moderate rotations this contribution may be neglected (see chapter 6). The equivalence of (3.7) to the Reissner strains as well as an interesting simplified transformation of (3.7) to the global cylindrical coordinates can be given [20].

4. Principle of Virtual Work

Based on the consistently simplified strain-displacement relations (3.7) the associated equations of equilibrium and boundary conditions can now be derived with the help of the principle of virtual work.

Let us consider an axisymmetric equilibrium state of a shell of revolution under an axisymmetric surface load $\mathbf{p}(s)$, defined per unit area of the undeformed middle surface, a boundary force $\mathbf{T}^*(s_q)$ and a boundary couple $M^*(s_q)$, $q = 1,2$, both measured per unit length of the boundary circles. All loads may in general be deformation-dependent. Let us note that in the case of mixed boundary conditions the kinematical and associated force variables prescribed at the same edge

s_q should be mutually complementary. Then within a fully Lagrangian description the principle of virtual work reads

$$2\pi\{\int_{s_1}^{s_2}(N_{ss}\delta\gamma_{ss} + N_{\theta\theta}\delta\gamma_{\theta\theta} + M_{ss}\delta\varkappa_{ss} + M_{\theta\theta}\delta\varkappa_{\theta\theta} - \mathbf{p}\cdot\delta\mathbf{u})\, r ds - $$
$$- [(r\mathbf{T}^*)\cdot\delta\mathbf{u} - (rM^*)\delta\omega]_{s_1}^{s_2}\} = 0 \, , \tag{4.1}$$

where $\delta\gamma_{ss}, \ldots, \delta\varkappa_{\theta\theta}$ denote the virtual changes of the strain measures (3.7) and $N_{ss}, \ldots, M_{\theta\theta}$ are as the mechanical variables of the axisymmetric shell theory, the stress resultants and stress couples, respectively. With (3.7) and (2.13) the virtual work expression (4.1) can be transformed further by partial integration and application of Gauss' divergence theorem, leading to the weak form of the equilibrium conditions of the shell

$$-\int_{s_1}^{s_2}[(r\mathbf{T}_s)' - \mathbf{e}_z \times \mathbf{T}_\theta + r\mathbf{p}]\cdot\delta\mathbf{u}\, ds +$$
$$+ [r(\mathbf{T}_s - \mathbf{T}^*)\cdot\delta\mathbf{u} - r\{[(1+\vartheta_{ss})^2 + \varphi_s^2]M_{ss} - M^*\}\delta\omega]_{s_1}^{s_2} = 0 \, . \tag{4.2}$$

In order to derive the boundary terms in (4.2) properly an identity has been used

$$\varphi_s\delta\vartheta_{ss} - (1+\vartheta_{ss})\delta\varphi_s = -[(1+\vartheta_{ss})^2 + \varphi_s^2]\delta\omega \, , \tag{4.3}$$

which can be derived from (2.13).
Whenever $\mathbf{T}_s$, $\mathbf{T}_\theta$ and $\mathbf{p}$ are continuous functions, (4.2) yields the local equations of equilibrium

$$(r\mathbf{T}_s)' - \mathbf{e}_z \times \mathbf{T}_\theta + r\mathbf{p} = \mathbf{0} \, , \tag{4.4}$$

and the corresponding boundary conditions

$$\mathbf{T}_s = \mathbf{T}^* \text{ or } \mathbf{u} = \mathbf{u}^* \, , \tag{4.5}$$

at $s = s_1, s_2$

$$[(1+\vartheta_{ss})^2 + \varphi_s^2]M_{ss} = M^* \text{ or } \omega(\mathbf{u}) = \omega^* \, , \tag{4.6}$$

where an asterisk indicates prescribed quantities at the boundary. In (4.2) and (4.4-5) the following internal stress resultant vectors have been introduced

$$\mathbf{T}_s = [(1+\vartheta_{ss})N_{ss} - \varphi_s' M_{ss} - \sigma_\theta M_{\theta\theta} - (1/r)(r\varphi_s M_{ss})']\mathbf{a}_s +$$
$$+ \{\varphi_s N_{ss} + \vartheta_{ss}' M_{ss} - (\sin\Phi/r)M_{\theta\theta} + (1/r)[r(1+\vartheta_{ss})M_{ss}]'\}\mathbf{n}, \tag{4.7}$$
$$\mathbf{T}_\theta = (1/r)N_{\theta\theta}\mathbf{a}_\theta \, .$$

Let us note that the third geometrical boundary condition (4.6) being linear with respect to the rotation angle, becomes nonlinear in terms of displacements according to (2.13). A discussion of various correspondingly modified rotational boundary conditions will be given in [20].

5. Variational Formulation

For a thin isotropic and elastic shell of revolution undergoing axisymmetric deformations the strain energy function Σ, defined per unit area of the undeformed middle surface, can be given by [16]

$$\Sigma = 1/2\ \{C[(\gamma_{ss}+\nu\gamma_{\theta\theta})^2 + (\gamma_{\theta\theta}+\nu\gamma_{ss})^2] + D[(\varkappa_{ss}+\nu\varkappa_{\theta\theta})^2 + (\varkappa_{\theta\theta}+\nu\varkappa_{ss})^2]\}\ , \tag{5.1}$$

where $C = Eh/(1-\nu^2)$ and $D = Eh^3/[12(1-\nu^2)]$ denote the extensional and flexural rigidity of the shell.

The linear constitutive equations corresponding to (5.1) take the form

$$\begin{aligned} N_{ss} &= C(\gamma_{ss}+\nu\gamma_{\theta\theta})\ , \quad N_{\theta\theta} = C(\gamma_{\theta\theta}+\nu\gamma_{ss})\ , \\ M_{ss} &= D(\varkappa_{ss}+\nu\varkappa_{\theta\theta})\ , \quad M_{\theta\theta} = D(\varkappa_{\theta\theta}+\nu\varkappa_{ss})\ , \end{aligned} \tag{5.2}$$

which complete the set of differential equations for the finite axisymmetric deformation of shells of revolution.

However, from the point of view of numerical applications it is more convenient to start from a corresponding variational formulation. In the case of a thin elastic shell the internal virtual work can be expressed as a variation of the shell strain energy $J_i(\mathbf{u})$

$$J_i(\mathbf{u}) = 2\pi \int_{s_1}^{s_2} \Sigma\ [\gamma_{ss}(\mathbf{u}),\ \ldots,\ \varkappa_{\theta\theta}(\mathbf{u})]\ r ds\ . \tag{5.3}$$

Under the assumption of conservative external loads with potentials $J_{es}(\mathbf{u})$ and $J_{eb}(\mathbf{u},\omega(\mathbf{u}))$ of the surface and boundary loads, respectively, the principle of virtual displacements can be transformed into a variational principle $\delta J(\mathbf{u}) = 0$ for the functional

$$J(\mathbf{u}) = J_i(\mathbf{u}) + J_{es}(\mathbf{u}) + [J_{eb}(\mathbf{u},\omega(\mathbf{u}))]_{s_1}^{s_2}\ . \tag{5.4}$$

Then the set of statical equations (4.4-6) can be derived as the corresponding Euler-Lagrange equations for the principle of stationary total potential energy. The potential functionals $J_{es}(\mathbf{u})$ and $J_{eb}(\mathbf{u},\omega(\mathbf{u}))$ will not be specified here. Let us only mention that examples of conservative forces are provided by dead loadings, pressure loadings with certain constrained geometrical boundary conditions [19] and loadings due to a steady shell rotation about the z-axis.

6. Simplified Strain Measures

The shell relations given in the previous chapters are valid for small strain axisymmetric deformation of shells of revolution, but no restrictions have been imposed on the rotations of the shell material elements so far. But within a geometrically nonlinear analysis of many axisymmetric shell structures it is hardly necessary to allow rotations of any magnitude. Hence, it is worth to consider possible simplifications of the displacement formulation under restricted rotations. To this end we need various estimates of nonlinear terms and coupled terms in (3.7) which follow from (2.11), (3.4) and (3.5)

$$\begin{aligned} \vartheta_{ss} &= (1+\varepsilon_s)\cos\omega - 1 = \varepsilon_s - (1/2)\omega^2 + \dots , \\ \varphi_s &= (1+\varepsilon_s)\sin\omega = \omega - (1/6)\omega^3 + \dots , \\ -\varphi_s'\vartheta_{ss} + \varphi_s\vartheta_{ss}' &= -\omega'(1+\varepsilon_s)[\varepsilon_s + 2\sin(\omega/2)] + \varepsilon_s'\sin\omega = \\ &= -\omega'(\varepsilon_s + (1/2)\omega^2 - \dots) + \varepsilon_s'(\omega - (1/6)\omega^3 + \dots) , \\ \varepsilon_s &= \lambda_s - 1. \end{aligned} \tag{6.1}$$

With wavelengths of deformation pattern bounded below a "natural" wavelength of $O(\sqrt{Rh})$ it can be shown that $\max|\varepsilon_s'(s)| < \eta h/hR$ [15,16]. From this fact together with (3.4-5) and (6.1) it follows that we have to assume the magnitude of rotation $|\omega| \lessdot \eta$, classified as moderate rotations in the general first approximation shell theory, to derive any consistent simplification of (3.7). Then we may have the following estimates

$$\begin{aligned} (1/2)\vartheta_{ss}^2 &= O(\eta^2), \quad -\varphi_s'\vartheta_{ss} + \varphi_s\vartheta_{ss}' = O(\eta^2/h) , \\ \sigma_s\vartheta_{ss} &= O(\eta/R) . \end{aligned} \tag{6.2}$$

With (6.2) the strain-displacement relations reduce to the Sanders-Koiter form [12]

$$\begin{aligned} \gamma_{ss} &= \vartheta_{ss} + (1/2)\varphi_s^2 , \quad \gamma_{\theta\theta} = \vartheta_{\theta\theta} \\ \varkappa_{ss} &= -\varphi_s' , \quad \varkappa_{\theta\theta} = -(\sin\Phi/r)\varphi_s , \end{aligned} \tag{6.3}$$

when the latter ones are specialized to the one-dimensional shell problems under consideration. They contain as a special case the axisymmetric shallow shell theory for which in (6.3) φ_s should be replaced by $w_{,s}$ and are identical with those in the best linear theory after formal linearization.

Let us note that within the shell literature various axisymmetric shell models have been presented with nonlinear membran and nonlinear bending strains. A detailed discussion of corresponding shell equations may be found in [20].

7. Cylindrical Deformation of Shells

Let us consider now the cylindrical deformation of infinite cylindrical shells (beamshells). In this case, replacing the cylindrical coordinate system (see fig. 2.1) by a Cartesian one $\{x,y,z\}$ with a unit base triad $\{\mathbf{i},\mathbf{j},\mathbf{k}\}$, the position vector of a beamshell middle surface $\mathbf{M}$ reads

$$\mathbf{r}(s,z) = x(s)\mathbf{i} + y(s)\mathbf{j} + z\mathbf{k} , \tag{7.1}$$

with $s \in [s_1, s_2]$ and $z \in]-\infty,+\infty[$. If we assume that all material properties, external disturbances and boundary conditions are independent of z such that every cross section z = const. undergoes the same motion in its initial plane we call this a cylindrical deformation. Then following the procedure outlined in chapter 2 and 3 the strain-displacement relations for beamshells can be given as

$$\begin{aligned} \gamma_{ss} &= \vartheta_{ss} + (1/2)\varphi_s^2 + \underline{(1/2)\vartheta_{ss}^2} , \\ \varkappa_{ss} &= -\varphi_s' - \underline{\varphi_s'\vartheta_{ss}} + \underline{\varphi_s\vartheta_{ss}'} , \end{aligned} \tag{7.2}$$

while the rotation angle ω may be determined according to (2.13). Note that within a constrained theory undergoing moderate rotations terms in (7.2) marked by a solid line may be omitted.

The corresponding equilibrium equations may then be obtained from the principle of virtual work or by specialization of the general equations (4.4) in the form

$$\mathbf{T}_s' + \mathbf{p} = \mathbf{0} , \tag{7.3}$$

while the boundary conditions (4.5-6) remain unchanged. In (7.3) we have introduced the internal stress resultant vector

$$\begin{aligned} \mathbf{T}_s &= [(1 + \vartheta_{ss})N_{ss} - \varphi_s'M_{ss} - (\varphi_s M_{ss})']\mathbf{a}_s + \\ &\quad +[\varphi_s N_{ss} + \vartheta_{ss}'M_{ss} + M_{ss}' + (M_{ss}\vartheta_{ss})']\mathbf{n} . \end{aligned} \tag{7.4}$$

The strain energy function Σ for a beamshell defined per unit length of the undeformed reference line is

$$\Sigma = 1/2(C\gamma_{ss}^2 + D\varkappa_{ss}^2) , \tag{7.5}$$

where $C = Eh$ and $D = Eh^3/12$ denote the extensional and flexural rigidity. An associated variational formulation may then be constructed following the expositions in chapter 5. For further information we refer to [24].

8. Comparative Numerical Investigation

To integrate our results into a more general frame of consideration let us carry out a comparative numerical investigation in which additionally three axisymmetric shell theories with nonlinear membran and bending strains [21-23] will be included. A FE-code has been developed for the analysis of cylindrical deformation of shells and the axisymmetric deformation of shells of revolution using a high precision element with cubic shape functions for both displacement components. Details of the numerical procedures employed are presented in [27]. These algorithms shall now be applied to highly nonlinear shell problems which in fact are indicators for inconsistent simplifications of the general shell equations. To provide a reference solution the complicated one-dimensional reduced equations of a general first approximation shell theory [14] are also implemented.

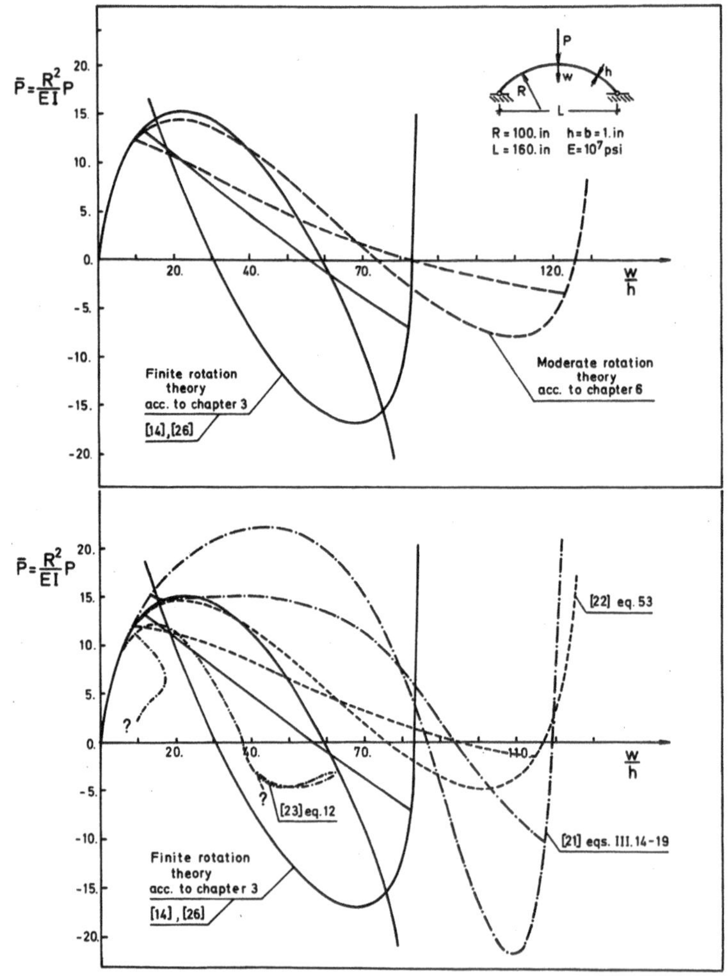

Fig. 8.1 Point loaded infinite cylindrical panel. Load vs deflection

As an introductory example let us consider the point loaded infinite cylindrical panel as shown in fig. 8.1. A FE-discretization with 20 elements was performed. The real behaviour of this structure indicates the so-called "looping-effects", i.e. the equlibrium path contains "loops" with many local extrema (in fig. 8.1 only parts of the equlibrium paths are shown). It is interesting to note that the solutions based on our finite rotation theory, the general shell theory [14] and an analytical solution given in [26] are in full agreement. This is of considerable importance as it supports the supposition that equivalent shell theories (in the sense of [16]) should lead to negligible differences in the solutions. The consistent moderate rotation theory of chapter 6 leads to accurate solutions within its range of validity. However, it is rather difficult to establish precisely this range as it depends on the problem under consideration. Nevertheless we have observed in all analysed examples that as long as maximum rotations do not exceed about 20° the maximum differences in the displacements and stresses remain in an acceptable range of a few % as compared to the solutions using strain-displacement relations valid for finite rotations. As outlined before, the refined variants [21-23] cannot be treated as consistent improvements of the moderate rotation theory. In particular, strong deviation from the real critical loads in case of [21] and [23] and convergence problems during the iteration process for [23] (indicated by a question-mark) occur. In the latter case it seems that the corresponding equilibrium equations are ill-conditioned.

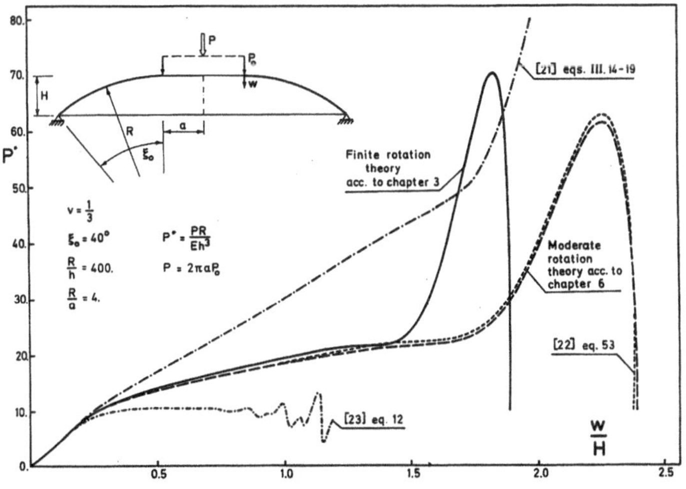

Fig. 8.2 Torodial segment under ring load. Load vs deflection

The example of a toroidal segment under ring load according to fig. 8.2 displays highly nonlinear shell behaviour. The FE-discretization was again performed with 20 elements to provide also accurate solutions for the shell stresses. Both finite rotation variants fully harmonize in the prediction of the solution. Again the modified variants [21] and [23] give inconsiderable results, showing structural stiffening or loss of stiffening, whereas in accordance to the foregoing example the version [22] differs only slightly from our moderate rotation theory.

Both numerical examples outline the significance regarding the omission or retention of various terms in the strain-displacement relations. The results presented there clearly show that inconsistent strain-displacement relations may lead to an unrealistic prediction of the structural behaviour. This problem is extensively analysed in [24-25]. Further examples concerning the equivalence of our equations to Reissner ones we have given in [20].

Acknowledgements - The author gratefully acknowledges continuous support of the Deutsche Forschungsgemeinschaft and the Polish Academy of Sciences in Gdansk.

References

1. SIMMONDS J.G., Surfaces with metric and curvature tensors that depend on one coordinate only are general helicoids, Q. Appl. Math. 37 (1979), 82.
2. SIMMONDS J.G., General helicoidal shells undergoing large, one-dimensional strains or large inextensional deformations, Int. J. Solids Structures Vol. 20 (1984), 1, 13.
3. REISSNER E., WAN F., On axial extension and torsion of helicoidal shells, J. Math. Phys. 47 (1968), 1.
4. REISSNER E., On finite bending and twisting of circular ring sector plates and shallow helicoidal shells, Q. Appl. Math. 11 (1953), 473.
5. REISSNER E., On axisymmetric deformations of thin shells of revolution, Proc. Symp. Appl. Math. 27, New York (1950).
6. EMMERLING, F.A., Nonlinear bending of tubes, Proc. EUROMECH Colloquium Nr. 165: Flexible Shells, Theory and Applications, München (1983).
7. MESCALL J., Numerical solutions of nonlinear equations for shells of revolution, Trans. ASCE J. Appl. Mech. 4 (1966), 2041.
8. GRIGOLYUK E.I., MAMAI V.I., FROLOV A.N., Investigation of the stability of nonshallow spherical shells for the case of finite displacements based on various equations of the theory of shells, Izv. AN SSSR MTT 7 (1972), 154.
9. SHILKRUT D.I., VYRLAN P.M., Stability of nonlinear shells, Izd-Vo "Shtiintsa", Kishinev (1977), English translation by FTD (1979).

10. VALISHVILI N.V., Methods of computer calculations of shells of revolution, "Mashinostroenije", Moscow (1976) (in Russian).

11. BUSHNELL D., Computerized analysis of shells - governing equations, Comp. & Struct. 18 (1984), 471.

12. KOITER W.T., On the nonlinear theory of thin elastic shells, Proc. Kon. Ned. Akad. Wet. B69, 1 (1966).

13. BUDIANSKY B., Notes on nonlinear shell theory, Trans. ASCE, J. Appl. Mech. 5 (1968), 393.

14. PIETRASZKIEWICZ W., SZWABOWICZ M., Entirely Lagrangian non-linear theory of thin shells, Arch. Mech. 33 (1981), 273.

15. KOITER W.T., SIMMONDS J.G., Foundations of shell theory, Proc. 12th Int. Congr. Theor. Appl. Mech., Moscow 1972, Springer-Verlag Berlin (1973).

16. KOITER W.T., A consistent first approximation in the general theory of thin elastic shells, Proc. IUTAM Symp. Theory of Thin Shells, North-Holland P.Co., Amsterdam (1960).

17. JOHN F., Refined interior shell equations, Proc. IUTAM Symp. Theory of Thin Shells, Springer-Verlag Berlin (1969).

18. BUDIANSKY, B., SANDERS J.L., On the "best" first-order linear shell theory, Progr. in Appl. Mech., Macmillian, New York (1963).

19. NOLTE L.-P., MAKOWSKI J., Pressure loaded shells undergoing different levels of nonlinearity, Mech. Res. Comm. (1986), in print.

20. MAKOWSKI J., NOLTE L.-P., Simple equations in terms of displacements for finite axisymmetric deflections of shells of revolution, Int. J. Non-Linear Mech. (1986) in print.

21. YAGHMAI S., Incremental analysis of large deformations in mechanics of solids with applications to axisymmetric shells of revolution, Ph.D. Dissertation, University of California, Berkeley (1968).

22. YAGHMAI S., POPOV E.P., Incremental analysis of large deflections of shells of revolution, Int. J. Solids Struct. 7 (1971), 1375.

23. VARPASUO P., Incremental analysis of axisymmetric shallow shells with varying strain-displacement relations, Comp. Math. Appl. Mech. Engng. 21 (1980), 153.

24. MAKOWSKI J., NOLTE L.-P., STUMPF H., Finite in-plane deformations of flexible rods - insight into nonlinear shell problems, IfM Mitt. 46, Ruhr-Univ. Bochum (1985).

25. NOLTE L.-P., MAKOWSKI J., STUMPF H., On the derivation and comparative analysis of large rotation shell theories, Ing. Arch. (1985) in print.

26. HUDDLESTON J.V., Finite deflections and snap through of high circular arches, Trans. ASME, J. Appl. Mech. (1968), 763.

27. CHROSCIELEWSKI J., NOLTE L.-P., Strategies for the solution of nonlinear problems in structural mechanics and their implementation into the MESY-concept (in German), IFM Mitt. 48, Ruhr-Univ. Bochum (1985).

ON THE DERIVATION AND EFFICIENT COMPUTATION OF LARGE ROTATION SHELL MODELS

L.-P. NOLTE
Institute of Mechanics
Ruhr-Universität Bochum
D-4630 Bochum
Fed. Rep. of Germany

1. Introduction

Within the past few years remarkable progress has been achieved on the nonlinear analysis of so termed flexible shells, mainly influenced by the fundamental results of Reissner, John and Koiter [1-4]. A novel approach to the displacement formulation of general and constrained geometrically nonlinear shell equations has been given in recent years by Pietraszkiewicz [5-7]. In particular, based on polar decomposition of shell strains and rotations it was suggested to classify small strain shell models according to the magnitude of the rotation angle of the material elements. It is known from the literature that most of the engineering shell problems may be tackled accurately enough with the help of moderate rotation shell theories [5-12]. Numerical applications of corresponding shell equations may be found e.g. in [12-14]. There are, however, various one- and two-dimensional shell problems to which large rotation shell models with nonlinear membran and bending strains should be applied. In these situations the rotational part of deformation dominates. Several shell theories with nonlinear change of curvature expressions have been already given (see [15-17] and literature cited therein). It turned out, however, that most of the corresponding field equations are very complicated and virtually useless for numerical applications. On the other hand we have shown recently [16] that various approaches used in engineering practice may lead to inconsistent shell equations. As a lack of precise order estimates an inaccurate prediction of the shell response may then occur.

Moreover, additional problems arise, when the nonlinear shell equations are approximately solved using the Finite Element Method (FEM). The limit capacity of even large scalar computer installations may then be exceeded for highly nonlinear calculations of complex shell structures. So certain trend can be observed to reduce the cost and/or time of solving nonlinear shell problems on different levels. Most effort has been devoted to the modeling of elements with reduced number of kinematical variables. However, within those concepts one has to put up with certain drawbacks as certain digression from the two-dimensional shell model.

In view of the aforementioned shortcomings we first give a brief derivation of efficient and numerically applicable large rotation shell models. Most simple kinematical relations are presented on which a system of variationally derivable first approximation shell theories can be based. The used refined classifications assure an optimal adaption to the nonlinear shell problem under consideration. Next, a detailed comparative analysis of 14 geometrically nonlinear shell theories is carried out within well choosen one-dimensional shell problems. A general finite rotation shell theory [6] is used to provide a reference solution. Thereby a distinct advantage of shell models based on the well founded decomposition of strains and rotations over comparable variants can be illustrated.
For the analysis of two-dimensional shell problems we make use of modernst hardware. An algorithm for the vectorisation of higher order elements [18] is applied to a doubly curved triangular high precision shell element with 54 degrees of freedom recently given in [14]. The element algorithms are then adapted to the two-pipe CDC Cyber 205 vector computer installation at Bochum university.
Various numerical examples are solved to illustrate the efficiency of the presented theoretical and numerical contributions.

2. Notations and Lagrangian Shell Equations

By $\mathbf{M}$ we denote the shell middle surface in the undeformed configuration with material coordinates θ^α, $\alpha = 1,2$. With each point of $\mathbf{M}$ we associate a position vector $\mathbf{r}(\theta^\alpha)$, standard surface base vectors $\mathbf{a}_\alpha = \mathbf{r}_{,\alpha}$, a unit normal vector $\mathbf{n} = (1/2)\varepsilon^{\alpha\beta}\mathbf{a}_\alpha \times \mathbf{a}_\beta$, the metric tensor components $a_{\alpha\beta} = \mathbf{a}_\alpha \cdot \mathbf{a}_\beta$ with $\det(a_{\alpha\beta}) = a$ and the curvature tensor components $b_{\alpha\beta} = -\mathbf{a}_\alpha \cdot \mathbf{n}_{,\beta}$. Here, a cross, a comma and a vertical stroke preceding a subscript will stand for the usual vector product, the partial and covariant differentiation on $\mathbf{M}$, $\varepsilon^{\alpha\beta}$ denotes the surface permutation tensor and δ^α_β is the Kronecker symbol. Let $\mathbf{C}$ be the boundary contour of $\mathbf{M}$ with the arc length s. The orthonormal triad $\{\boldsymbol{\nu}, \mathbf{t}, \mathbf{n}\}$ at any regular point $M \in \mathbf{C}$ is defined by $\mathbf{t} = \mathbf{r}_{,s} = t^\alpha \mathbf{a}_\alpha$, $\boldsymbol{\nu} = \mathbf{t} \times \mathbf{n}$.

We shall consider a smooth deformation of the middle surface $\chi : \mathbf{M} \rightarrow \bar{\mathbf{M}}$ caused by Lagrangian surface loads $\mathbf{p} = p^\alpha \mathbf{a}_\alpha + p\mathbf{n}$ and/or Lagrangian boundary forces and static moments $\mathbf{T}^*_\nu$, $\mathbf{H}^*_\nu$ [6], represented by a displacement field $\mathbf{u} = u^\alpha \mathbf{a}_\alpha + w\mathbf{n}$. Then analogous quantities on $\bar{\mathbf{M}}$ are marked by a bar. The base triad on $\bar{\mathbf{M}}$ can be determined by

$$\bar{\mathbf{a}}_\alpha = 1^\lambda_{\cdot\alpha}\mathbf{a}_\lambda + \varphi_\alpha \mathbf{n} \;,\quad \bar{\mathbf{n}} = n^\lambda \mathbf{a}_\lambda + n\mathbf{n} \;, \tag{2.1}$$

$$\sqrt{\bar{a}/a}\; n^\lambda = \varepsilon^{\alpha\beta}\varepsilon^{\mu\lambda}\varphi_\alpha 1_{\mu\beta} \equiv m^\lambda \;,\quad \sqrt{\bar{a}/a}\; n = (1/2)\varepsilon^{\alpha\beta}\varepsilon^{\lambda\mu}1_{\lambda\alpha}1_{\mu\beta} \equiv m \;,$$

where the usual linearized quantities have been introduced

$$1_{\alpha\beta} = a_{\alpha\beta} + \theta_{\alpha\beta} - \omega_{\alpha\beta} \;,\quad \theta_{\alpha\beta} = 1/2(u_{\alpha|\beta} + u_{\beta|\alpha}) - b_{\alpha\beta}w \;,$$
$$\omega_{\alpha\beta} = 1/2(u_{\beta|\alpha} - u_{\alpha|\beta}) \;,\quad \varphi_\alpha = w_{,\alpha} + b^\lambda_\alpha u_\lambda \;,\quad \varphi = (1/2)\varepsilon^{\alpha\beta}u_{\beta|a} \;. \tag{2.2}$$

The deformation of the shell boundary surface may be described by the vectors $\mathbf{u}$ and $\bar{\mathbf{n}}$, with their decomposition according to $\mathbf{u} = u_\nu \boldsymbol{\nu} + u_t \mathbf{t} + w\mathbf{n}$ and $\bar{\mathbf{n}} = n_\nu \boldsymbol{\nu} + n_t \mathbf{t} + n\mathbf{n}$. Under Kirchhoff-Love (K-L) constraints

$$\bar{\mathbf{a}}_\alpha \cdot \bar{\mathbf{n}} = 0 \;,\quad \bar{\mathbf{n}}\cdot\bar{\mathbf{n}} = 1 \tag{2.3}$$

and rotations not exceeding $\pm\pi/2$ we have the unique representation

$$\begin{Bmatrix} n_t \\ n \end{Bmatrix} = \frac{1}{c_\nu^2 - |\bar{a}_t|^2}\left[c_\nu n_\nu \begin{Bmatrix} c_t \\ c_n \end{Bmatrix} + \sqrt{|\bar{a}_t|^2(1-n_\nu^2) - c_\nu^2}\begin{Bmatrix} c_n \\ -c_t \end{Bmatrix}\right] , \tag{2.4}$$

where $\bar{\mathbf{a}}_t$ is the deformed boundary vector

$$\bar{\mathbf{a}}_t = c_\nu \boldsymbol{\nu} + c_t \mathbf{t} + c_n \mathbf{n} \;,\quad c_\nu = u_{\nu,s} + \tau_t w - \varkappa_t u_t \;,$$
$$c_t = 1 + u_{t,s} + \varkappa_t u_\nu - \sigma_t w \;,\quad c_n = w_{,s} + \sigma_t u_t - \tau_t u_\nu \;. \tag{2.5}$$

Hence it follows from (2.4-5) that the six variables describing the boundary surface deformation can be reduced to four u_ν, u_t, w and $n_\nu = \sqrt{a/\bar{a}}\,(\mathbf{u}_{,s}\times\boldsymbol{\nu}-\mathbf{n})\cdot\mathbf{u}_{,\nu}$ as the independent rotational representation [6]. Within the geometrically nonlinear first approximation theory of thin isotropic and elastic shells the strain energy function Σ is given by

$$\Sigma = (h/2)H^{\alpha\beta\lambda\mu}[\gamma_{\alpha\beta}\gamma_{\lambda\mu} + (h^2/12)\varkappa_{\alpha\beta}\varkappa_{\lambda\mu}][1+O(\theta^2)] \;, \tag{2.6}$$

involving a relative error E0 of $O(\theta^2)$, where θ is a common small parameter [3,5]

$$\theta = \max\{h/d,\; h/L,\; h/L^*,\; \sqrt{h/R},\; \sqrt{\eta}\} \;,\quad \theta^2 \ll 1 . \tag{2.7}$$

Here, $H^{\alpha\beta\lambda\mu}$ are the components of the modified elasticity tensor while R, h, L, L^*, η and d denote the minimum principle radius of curvature, the shell thickness, the wave length of deformation and curvature pattern, the maximum principle strain in the shell space and the distance of any point under consideration from the shell boundary, respectively. In (2.6) $\gamma_{\alpha\beta}$ and $\varkappa_{\alpha\beta}$ are the Langrangian mid-surface strain and change of curvature tensors defined by

$$\gamma_{\alpha\beta} = 1/2(\bar{a}_{\alpha\beta} - a_{\alpha\beta}) = 1/2(l^{\lambda}_{\cdot\alpha} l_{\lambda\beta} + \varphi_\alpha \varphi_\beta - a_{\alpha\beta}) \,, \tag{2.8}$$
$$\varkappa_{\alpha\beta} = -(\bar{b}_{\alpha\beta} - b_{\alpha\beta}) = (n^{\lambda}|_{\beta} - b^{\lambda}_{\beta} n) l_{\lambda\alpha} + (b_{\lambda\beta} n^{\lambda} + n_{,\beta})\varphi_\alpha + b_{\alpha\beta} \,.$$

Partial derivatives of (2.6) yield the linear constitutive equations

$$N^{\alpha\beta} = \partial\Sigma/\partial\gamma_{\alpha\beta} = hH^{\alpha\beta\lambda\mu}\gamma_{\lambda\mu} \,, \quad M^{\alpha\beta} = \partial\Sigma/\partial\varkappa_{\alpha\beta} = (h^3/12)H^{\alpha\beta\lambda\mu}\varkappa_{\lambda\mu} \,. \tag{2.9}$$

If we assume the external boundary loads $\mathbf{T}^*_\nu$, $\mathbf{H}^*_\nu$ to be of dead load type and the surface load $\mathbf{p}$ either to be a uniform pressure load with the intensity q or a dead load, too, then the functional of total potential energy is given by

$$\begin{aligned} J(\mathbf{u}) = & \iint_{\mathbf{M}} [\Sigma(\mathbf{u}) - \mathbf{p}\cdot\mathbf{u}] \, dA - \int_{C_f} [\mathbf{T}^*_\nu \cdot \mathbf{u} + \mathbf{H}^*_\nu \cdot (\bar{\mathbf{n}}(\mathbf{u}) - \mathbf{n})] \, ds \\ & + \iint_{\mathbf{M}} [q\{\mathbf{n} + (1/2)\varepsilon^{\alpha\beta}\mathbf{a}_\alpha \times \mathbf{u}|_\beta + (1/6)\varepsilon^{\alpha\beta}\mathbf{u}|_\alpha \times \mathbf{u}|_\beta\}\cdot\mathbf{n}] dA \,. \end{aligned} \tag{2.10}$$

It should be noted, however, that for the case of a uniform pressure load certain geometrical constraints have to be fulfilled to assure the load operator to be potential [19]. The derivation of the statical equations as Euler-Lagrange equations of the principle of stationary total potential energy $\delta J(\mathbf{u}) = 0$ may follow now conventional line. For the sake of brevity we refer to [15-17].

3. Simplified Large Rotation Shell Models

An exact description of the polar decomposition of the shell gradient deformation in terms of engineering strains and a finite rotation vector $\boldsymbol{\Omega} = \sin\omega\,\mathbf{i}$ has been given by Pietraszkiewicz [5]. Here $\mathbf{i}$ and ω denote a unit vector along the axis of rigid-body rotation and the rotation angle. Under K-L constraints (2.3) and assumed small strains $\gamma_{\alpha\beta} = O(\theta^2)$ $\boldsymbol{\Omega}$ may be approximated as

$$\boldsymbol{\Omega} = (1/2)\varepsilon^{\beta\alpha}[(2 + \theta^{\lambda}_{\lambda})\varphi_\alpha - \varphi^{\lambda}(\theta_{\lambda\alpha} - \omega_{\lambda\alpha})]\mathbf{a}_\beta + \varphi\mathbf{n} \,. \tag{3.1}$$

In what follows we restrict our considerations to the class of shell problems in which rotations about tangents to the middle surface may be large, $\boldsymbol{\Omega}\cdot\mathbf{a}_\alpha = O(\sqrt{\theta})$, while rotations about the normals remain small, $\boldsymbol{\Omega}\cdot\mathbf{n} = O(\theta^2)$. It should be pointed out that those assumptions of large/small rotations (l/s) are valid for most flexible shell problems of engineering interest. Then (3.1) with $(2.8)_1$ yields the estimates $\varphi_\alpha = O(\sqrt{\theta})$, $\varphi = O(\theta^2)$, $\theta_{\alpha\beta} = O(\theta)$. However, as already pointed out in [15-17,20] correspondingly simplified shell strain measures using term by term estimates do not allow for a proper variational formulation of an associated shell theory. Therefore let us expand the non-rational K-L constraint $(2.3)_1$ into series with respect to the linearized parameters (2.2) using (2.1), (2.8) and the expression $\bar{a}/a = 1 + 2\gamma^\alpha_\alpha + 2(\gamma^\alpha_\alpha\gamma^\lambda_\lambda + \gamma^\alpha_\lambda\gamma^\lambda_\alpha)$ [15]. After truncation within the desired accuracy E0 we get the simplifying relation

$$\theta_{\alpha\beta} = -(1/2)\varphi_\alpha\varphi_\beta + O(\theta^2) \ , \tag{3.2}$$

from which the following estimates of coupled terms can be obtained

$$\varphi_\lambda\theta^\varkappa_\varkappa - \varphi^\mu\theta_{\mu\lambda} = O(\theta^2\sqrt{\theta}) \ , \qquad \theta^\varkappa_\varkappa\theta^\rho_\rho - \theta^\rho_\varkappa\theta^\varkappa_\rho = O(\theta^3) \ . \tag{3.3}$$

Furthermore let us admit a slightly greater relative error of $O(\theta\sqrt{\theta})$ in the strain energy function (2.6) denoted by E1. A detailed discussion of 1/s rotation shell equations valid within the margin E0 will be published elsewhere [16]. Then the shell strain measures (2.8) may be approximated as

$$\begin{aligned}
\gamma_{\alpha\beta} &= \theta_{\alpha\beta} + (1/2)\varphi_\alpha\varphi_\beta - \underline{1/2(\theta^\lambda_\alpha\omega_{\lambda\beta} + \theta^\lambda_\beta\omega_{\lambda\alpha})} + \underline{(1/2)\theta^\lambda_\alpha\theta_{\lambda\beta}} \ , \\
\varkappa_{\alpha\beta} &= 1/2[(\delta^\lambda_\alpha + \underline{\theta^\lambda_\alpha})\tilde{m}_\lambda|_\beta + (\delta^\lambda_\beta + \underline{\theta^\lambda_\beta})\tilde{m}_\lambda|_\alpha] + \underline{1/2(\varphi_\alpha\tilde{m}|_\beta + \varphi_\beta\tilde{m}|_\alpha)} - \\
&\quad - \underline{1/2(b^\lambda_\alpha\theta_{\lambda\beta} + b^\lambda_\beta\theta_{\lambda\alpha})} - \underline{1/2(b^\lambda_\alpha\varphi_\beta + b^\lambda_\beta\varphi_\alpha)\varphi_\lambda} + \underline{(1/2)b_{\alpha\beta}\varphi^\lambda\varphi_\lambda} \ , \\
\tilde{m}_\lambda &= -\varphi_\lambda \ , \qquad \tilde{m} = 1 + \underline{\theta^\varkappa_\varkappa} \ .
\end{aligned} \tag{3.4}$$

To derive the dependent geometrical boundary parameters of this 1/s rotation shell theory a similar expansion and truncation procedure can be used

$$\begin{aligned}
n_t &= -\varphi_t = -w_{,t} + \tau_t u_\nu - \sigma_t u_t \ , \\
n &= 1 - \underline{(1/2)n^2_\nu} - \underline{(1/2)\varphi^2_t} - \underline{(1/8)(n^2_\nu + \varphi^2_t)^2} \ .
\end{aligned} \tag{3.5}$$

Note, that in (3.5) the fourth independent (rotational) boundary parameter n_ν appears in a linear form

$$n_\nu = -\varphi_\nu \ , \tag{3.6}$$

which is of overwhelming importance, because it enables the formulation of linear geometrical boundary conditions, such that powerful approximation procedures as the FEM can be applied.

Within the assumptions of 1/s rotations the load operator of a uniform pressure field may be approximated as [19]

$$q\,[-1/2\, u^\lambda\varphi_\lambda + w(1 + (1/2)\theta^\varkappa_\varkappa)]. \tag{3.7}$$

When (3.4-7) are introduced into the functional of total potential energy (2.10) an immediate consequence of its vanishing first Gâteaux variation at an equilibrium state **u** are the Euler-Lagrange equations

$$\begin{aligned}
&\mathbf{T}^\beta(\mathbf{u})|_\beta + \mathbf{p} + \mathbf{Q} = \mathbf{0} && \text{in } \mathbf{M} \ , \\
&\mathbf{P}(\mathbf{u}) - \mathbf{P}^*(\mathbf{u}) = \mathbf{0} \ , \quad M(\mathbf{u}) - M^*(\mathbf{u}) = 0 && \text{on } \mathbf{C}_f \ , \\
&F_j(\mathbf{u}) - F^*_j(\mathbf{u}) = 0 && \text{at } s = s_{fj} \in \mathbf{C}_f, \ j = 1, \dots, n \ ,
\end{aligned} \tag{3.8}$$

where $(3.8)_1$ are the equilibrium equations, $(3.8)_{2,3}$ the static boundary and $(3.8)_4$ the static corner conditions. In addition we have the consistently simplified geometric boundary and corner conditions

$$u_\nu = u^*_\nu, \quad u_t = u^*_t, \quad w = w^*, \quad \varphi_\nu = \varphi^*_\nu \quad \text{on } C_u ,$$
$$w_i = w^*_i \qquad \text{at } s=s_{ui} \in C_u, \ i=1,\dots,m . \tag{3.9}$$

In (3.8) we have introduced the following internal and external stress resultants

$$\mathbf{T}^\beta = T^{\lambda\beta}\mathbf{a}_\lambda + T^\beta \mathbf{n}, \qquad \mathbf{Q} = Q^\lambda \mathbf{a}_\lambda + Q\mathbf{n} ,$$

$$T^{\lambda\beta} = (1/2)qa^{\lambda\beta}w + N^{\lambda\beta} + 1/2[(\underline{\theta^{\lambda\alpha}} + \underline{\varphi^{\lambda\alpha}})N^\beta_\alpha + (\underline{\theta^{\alpha\beta}} - \underline{\varphi^{\alpha\beta}})N^\lambda_\alpha] -$$
$$- 1/2[(\underline{\varphi^\lambda|_\alpha} + \underline{b^\lambda_\alpha})M^{\alpha\beta} + (\underline{\varphi^\beta|_\alpha} + \underline{b^\beta_\alpha})M^{\alpha\lambda}] - \underline{a^{\lambda\beta}(M^{\varkappa\rho}\varphi_\varkappa)|_\rho} ,$$

$$T^\beta = -(1/2)qu^\beta + \varphi_\alpha N^{\alpha\beta} + M^{\alpha\beta}|_\alpha + \underline{(\theta^\beta_\varkappa M^{\varkappa\rho})|_\rho} + \underline{M^{\alpha\beta}\theta^\varkappa_\varkappa|_\alpha} -$$
$$- \underline{(M^{\beta\rho}b^\lambda_\rho + M^{\lambda\rho}b^\beta_\rho)\varphi_\lambda} + \underline{b_{\varkappa\rho}M^{\varkappa\rho}\varphi^\beta} ,$$

$$Q^\lambda = -(1/2)q\varphi^\lambda , \quad Q = q\,[1 + (1/2)\theta^\varkappa_\varkappa] ,$$

$$\mathbf{P} = P_\nu \boldsymbol{\nu} + P_t \mathbf{t} + P_n \mathbf{n} , \quad F_j = F(s_{fj}+0) - F(s_{fj}-0) ,$$
$$P_\nu = T_{\nu\nu} + \tau_t F , \quad P_t = T_{t\nu} - \sigma_t F , \quad P_n = T_{n\nu} + d/ds\, F , \tag{3.10}$$
$$F = M^{\lambda\beta}(\delta^\alpha_\lambda + \underline{\theta^\alpha_\lambda} + \underline{\varphi_\lambda \varphi^\alpha})t_\alpha \nu_\beta , \quad M = M^{\lambda\beta}(\delta^\alpha_\lambda + \underline{\theta^\alpha_\lambda} + \underline{\varphi_\lambda\varphi^\alpha})\nu_\alpha\nu_\beta ,$$

$$\mathbf{P}^* = P^*_\nu \boldsymbol{\nu} + P^*_t \mathbf{t} + P^*_n \mathbf{n} , \quad F^*_j = F^*(s_{fj}+0) - F^*(s_{fj}-0) ,$$
$$P^*_\nu = T^*_{\nu\nu} + \tau_t F^* , \quad P^*_t = T^*_{t\nu} - \sigma_t F^* , \quad P^*_n = T^*_{n\nu} + d/ds\, F^* ,$$
$$F^* = H^*_{t\nu} + \varphi_t[\underline{1} + \underline{1/2(\varphi^2_\nu + \varphi^2_t)}]\, H^*_{n\nu} ,$$
$$M^* = H^*_{\nu\nu} + \varphi_\nu[\underline{1} + \underline{1/2(\varphi^2_\nu + \varphi^2_t)}]\, H^*_{n\nu} .$$

The set of partial differential equations for this simplified 1/s rotation shell theory based on (3.4-6) take now the form according to (3.8-9). However, for a uniform pressure loaded shell certain set of geometric boundary conditions have to be fulfilled [19].

An interesting and numerically efficient variant of the above given theory may be derived if we admit a greater relative error E2 of $O(\theta)$ in the energy expression to concentrate on main significant terms. In this case terms in (3.4-5) and (3.10) marked by a broken line may be omitted. The equations of the well known simplified moderate rotation shell theory [4,5,11] may then be obtained neglecting additionally all terms underlined by a solid line.

Let us note, that transformations of the type (3.2-3) should be used carefully within the relaxed errors E1 and E2. Likewise, modified 1/s rotation shell variants recently derived in [15,18] turn out to be less efficient and show a restricted domain of applicability.

4. Comparative Investigation

In order to integrate and to judge the results of the present approach let us recall various shell models undergoing different levels of non-linearity as summarized in fig. 4.1. By M1-M3 and L1-L3 we denote such moderate and large rotation theories which can be derived via decomposition of strains and rotations, while N1-N7 stand for variants based on a priori different assumptions.

Theory	Graph	Denotation	References
Moderate rotations			
M1	--- -·-·-·--	"shallow shell theory"	Donnell, Vlasov [9, 10]
M2	-----------	"small finite deflections" "moderately small rotations"	Koiter [4] Sanders [11]
M3	----·-----	"moderate rotations"	Pietraszkiewicz [5]
Various nonlinear variants			
N1	·-··-··-··	"moderate deflections"	Koiter [4]
N2	··-··-··-	"moderately large rotations"	Basar, Krätzig, Harte [22, 12, 14]
N3	-----·-----	"cubic approximation"	Chuyko ([24] eq. 3.6)
N4	----·-----	"quadratic approximation"	Shapovalov ([23] eq. 1.9, 2.12)
N5	--·--·----	"large deformations and rotations"	Yaghmai ([25] eq. III. 20-24)*
N6	-----·-·--	"large displacements"	Varpasuo ([26] eq. 12-13)*
N7	------·---	"large rotations"	Harte [14]
Large rotations			
L1	----------	large/small {E0, E1, E2}	acc. to chapter 3 Nolte, Stumpf [20]
L2	··········	large/small E1	Pietraszkiewicz ([15] eq. 5.1$_1$, 5.3-4)
L3	-·-·-·-·-	large/small E1	Schmidt ([21] eq. 4.2-3)
Finite rotations			
F1	———	"finite rotations"	Pietraszkiewicz, Szwabowicz [6]

*) theory of axisymmetric shells of revolution

Fig. 4.1 Collection of general and various simplified shell theories

A detailed critical discussion of the above mentioned theories has been given recently in [15,16]. There, possible errors are pointed out which may lead to inconsistent approximations of the shell strain measures and geometrical boundary parameters. Moreover, after reduction of the nonlinear shell equations for the cylindrical bending of shells certain a priori estimates for relative errors of associated strain measures can be constructed with a graphical representation according to fig. 4.2 [16]. It is remarkable that only our l/s rotation variants denoted by L1 do not exhibit errors in the mid-surface and change of curvature expressions $e_{\tilde{\gamma}}$ and $e_{\tilde{\varkappa}}$, respectively. Their one-dimensional reduced form turns out to be exact in the sense of the first approximation to the shell strain energy.

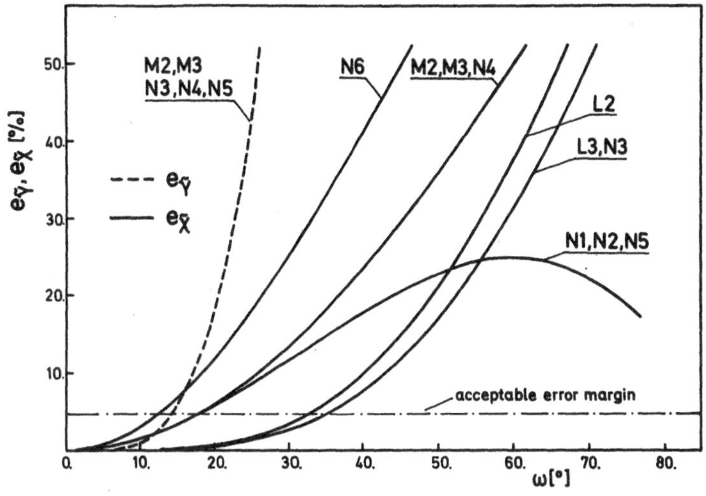

Fig. 4.2 Relative errors for various strain measures

In what follows we want to support the foregoing theoretical considerations by a detailed comparative numerical investigation. The FE-code has been developed for the analysis of cylindrical bending of shells and the axisymmetric deformation of shells of revolution. A conforming two-node high precision element with 8 degress of freedom, exact geometrical description, cubic shape functions for both displacement components and three point Gauss integration is used. The relative simplicity of one-dimensional reduced shell equations allows the implementation of all constrained (M1-M3, L1-L3, N1-N6) and a general geometrically nonlinear shell theory F1 [6] to assure an accurate reference solution. These FE-algorithms shall now be applied to nonlinear examples which are indicators for the quality of a considered shell theory. In this context the well founded decision about the consistency of a shell model and the clear definition of its range of applicability are of special importance.

As an introductory example let us consider the pure stress problem of a nonsymmetrically loaded infinite deep circular cylindrical panel as shown in fig. 4.3-4. This shell structure exhibits highly nonlinear behaviour and may serve as a good illustration for various levels of approximation in the theory of shells. For all examples within the class of cylindrical bending of shells, due to equivalence in the sense of the first approximation, the l/s rotation solutions L1 are in full agreement with the reference solution F1' as long as classical boundary conditions are concerned. Likewise, the consistent moderate rotation shell theories M2, M3 can predict the structural response properly only within their range of applicability, while with increasing rotations ($\omega > 25$-30°) they fail completely. Compared to this the modified l/s rotation variants L2, L3 lead to an inconsiderable improvement and predict, as the theories M2, N4, qualitative different structural behaviour. The stress problem is changed into a stability problem with the

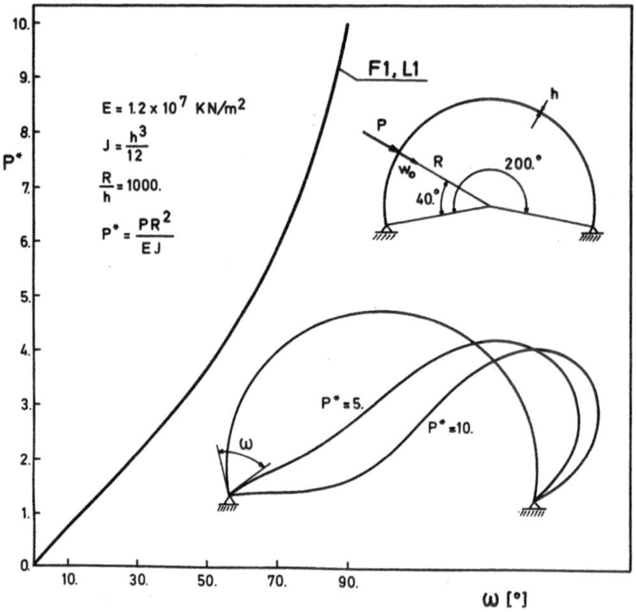

Fig. 4.3 Infinite deep circular cylindrical panel. Load vs rotation and deformed shapes

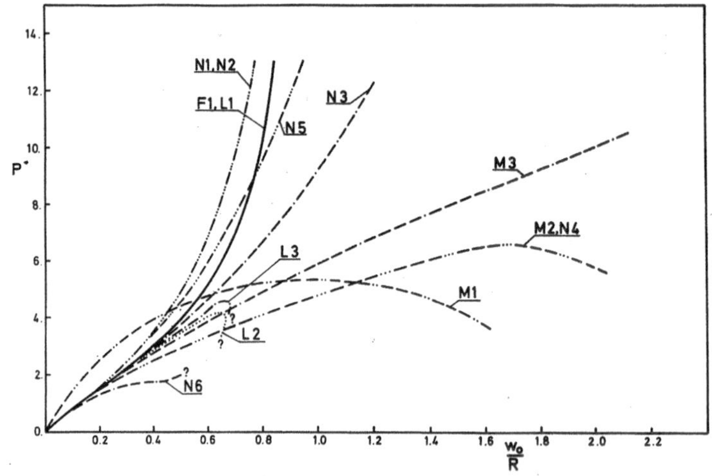

Fig. 4.4 Infinite deep circular cylindrical panel. Load vs deflection predicted by various shell theories

occurrence of snap-through buckling. Note that the shallow shell theory M1 begins the solution with an incorrect tangent and thus cannot be used even to describe the linear shell problem. The good correspondence of the solutions given in fig. 4.4 to the estimates in fig. 4.2 is strongly underlined by lower deviations of the load-displacement paths based on the variants N1, N2, N3, N5. In the case of N3, N5 this results from an interaction of different errors in the membran and bending part of the strain energy. Remarkable is, that only for those shell theories L2, L3 and N6, whose change of curvature expressions do not depend on second derivatives of the tangential displacement, convergence problems arised during the iteration process. This behaviour is indicated by a questionmark.

Tori under a ring load with varying distance L_i to the center of rotation figs. 4.5-6 have been calculated to demonstrate that it is possible to generate various phenomena of shell structural behaviour by changing only one geometrical parameter. Besides, Tori with $L_i \rightarrow \infty$ like complete cylinders under diametrical concentrated forces [14] are reliable indicators for different levels of geometrically nonlinear shell theories.

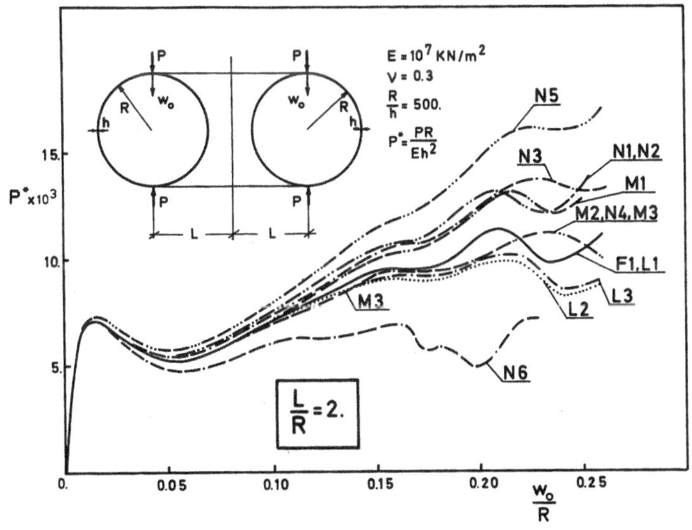

Fig. 4.5 Torus under ring-load L/R = 2. Load vs deflection predicted by various shell theories

Hence for L/R = 2000 (fig. 4.6), a pure stress problem, the structure displays a typical "one-dimensional" load carrying behaviour as in the foregoing example. With the exception of the shallow shell theory M1, which again fails in the prediction of the linear solution and the inconsistent variant N6 with non-satisfying results all other theories deviate in almost the same manner from the reference solution F1, L1 showing certain structural stiffening or loss of stiffening. The modified variants L2, L3, N1-N6 do not essentially improve the range of applicability as compared to the simple consistent moderate rotation theories M2, M3.

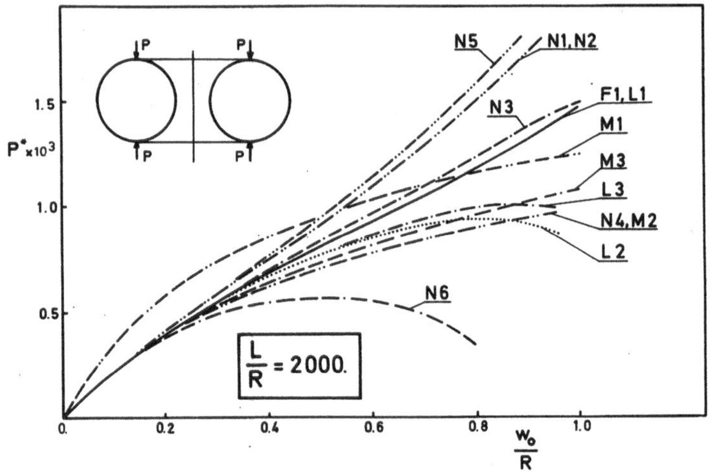

Fig. 4.6 Torus under ring-load L/R = 2000. Load vs deflection predicted by various shell theories

Changing the center-distance to L/R = 2. (fig. 4.5) typical "two-dimensional" load carrying behaviour appears generated by increasing circumferential strains. This slightly nonlinear shell problem may be analysed correctly within the prebuckling range by using any simplified variant except N5, which predicts a too high snap load. Without detailed interpretation of certain differences in the post-buckling range we only mention that dominating circumferential strains lead to similar load-displacement paths with the occurrence of various local extrema.

5. Two-dimensional FE - Shell Analysis

A critical review of various double curved shell elements has been presented recently in [14]. Thereon, for the purpose of the present analysis we have choosen the condensed high precision triangular finite element with 54 degrees of freedom, which allows a rather direct realization of the choosen nonlinear shell model, here the l/s E2 version denoted by L1. Quintic polynomial shape functions for all displacement components and an exact geometrical description in the 21 Gauss integration points provide excellent convergence properties.

However, as already pointed out in the introduction even in the case of simplest shell theories the limit capacity of large scalar computers may be exceeded if an extensive shell analysis is performed.
This crucial situation might change with respect to the actual and future market for vector computers, as it is illustrated by the growing number of installations of various producers. For many applications the computer power of these equipements effecting nearly 1 GFLOPS reach one or even two orders of magnitude beyond the power of conventional scalar computers. Accordingly an efficient algorithm for the vectorisation of higher order elements recently derived by the author [18] is used for the choosen high precision element. It is based on a combination of the so called "natural" vector length, here e. g. given by 54 degress of freedom, and "calculating parallel", which stands for the simultaneous application of algorithms at several "points" , e. g. the numerical integration points and/or a group of elements, here the 21 Gauss points. A detailed description of the vectorized shell element is given in [18]. Let us only mention that the computation of one element routine on the CDC Cyber 205 two-pipe installation at Bochum university requires 0.14 sec CPU time. Paging is avoided by a special manner of storage. A speed up factor of 27 in relation to the CDC Cyber 205 scalar unit could be achieved. The element routine is implemented into a program package which allows the choice between various efficient trace-algorithms, which is important for the analysis of complex non-linear problems in BATCH-mode without any interactive facilities.

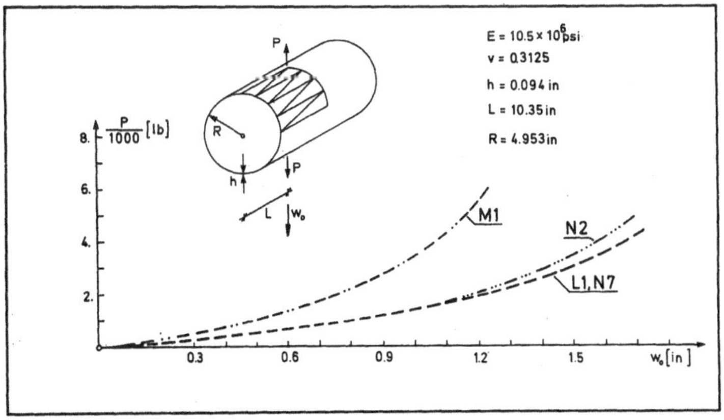

Fig. 5.1 Diametrical point loaded cylindrical tube. Load vs deflection predicted by various shell theories

The highly geometrically nonlinear problem of a diametrical point loaded free cylindrical tube is shown in fig. 5.1. A FE-discretisation with 8 vectorized elements for one eighth of the shell was performed using symmetry properties. In a former investigation a comparative analysis using the shell variants M1, N2 and N7 has been carried out [14]. Our results based on the simplest 1/s rotation theory L1 are in full agreement with those predicted by the complicated version N7, in which all cubic contributions of the displacement gradients to the shell strain measures are taken into account. Deviations in the case of M1 and N2 confirm the results of the comparative investigation given in chapter 4.

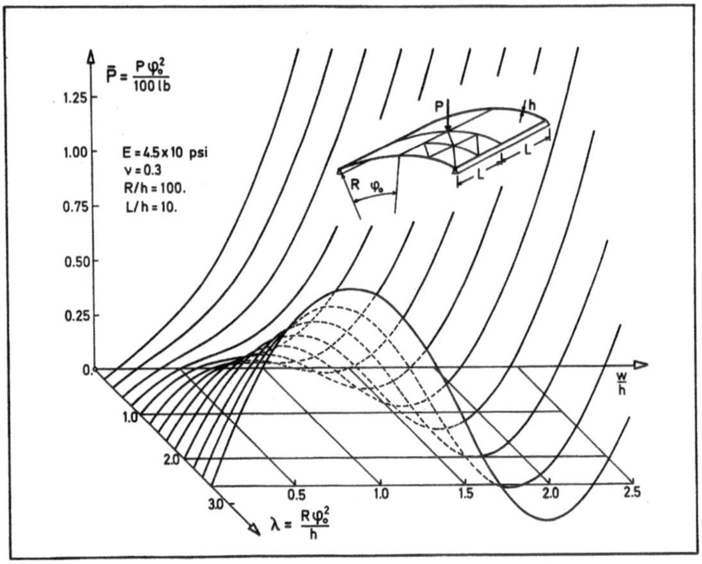

Fig. 5.2 Point loaded cylindrical panel. Equilibrium surface I

The point loaded cylindrical panel in fig. 5.2-3 was analysed to demonstrate the effectiveness of the applied vectorized element algorithm. The shell boundary consisted of two hinged and two free edges at opposite sides, respectively. 8 elements were used for one quarter of the shell with different meshes according to fig. 5.2-3. Fig. 5.2 shows the catastrophe map of a two-parameter system with λ describing the shell flexibility. Projections into the $\bar{P}$-λ- plane yields the graph of the well known cusp catastrophe. For increasing λ a very complicated self-penetrating equilibrium surface is derived in fig. 5.3, characterized by instable "loops" of the load-displacement paths. It was obtained with 679 incremental steps and averagely 4 to 5 Standard-Newton-Raphson iterations per step. The total CPU time was 62 min.

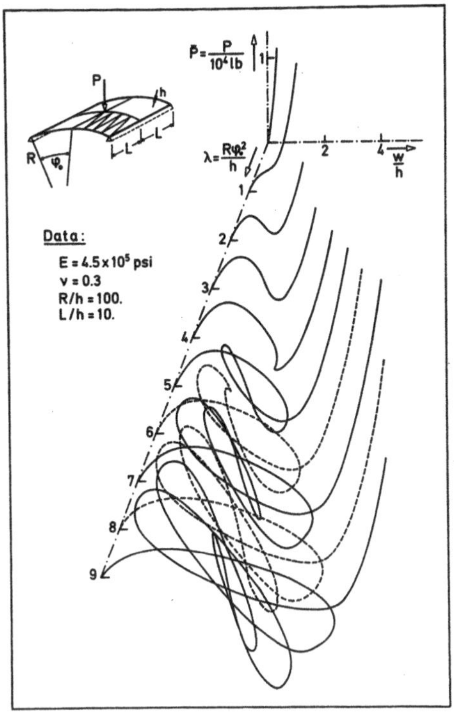

Fig. 5.3

Point loaded cylindrical panel. Equilibrium surface II

Acknowledgements - The author gratefully acknowledges continuous support of the Deutsche Forschungsgemeinschaft and the Polish Academy of Sciences in Gdansk.

6. References

1. REISSNER E., On axisymmetric deformations of thin shells of revolution, Proc. Symp. Appl. Math. 3, New York (1950) 27-52.

2. JOHN F., Estimates for the derivatives of stresses in a thin shell and interior shell equations, Comm. Pure and Appl. Math. 18 (1965) 235-267.

3. KOITER W.T., A consistent first approximation in the general theory of thin elastic shells, Proc. IUTAM Symp. Delft 1959, North-Holland Publ. Co., Amsterdam (1960) 12-33.

4. KOITER W.T., On the nonlinear theory of thin elastic shells, Proc. Kon. Ned. Ak. Wet., Ser B, 69 (1966) 1-54.

5. PIETRASZKIEWICZ W., Finite rotations in the non-linear theory of thin shells, in: Thin Shell Theory, New Trends and Applications, Ed. by W. Olszak, Springer-Verlag, Wien (1980) 153-208.

6. PIETRASZKIEWICZ W., SZWABOWICZ M.L., Entirely Lagrangean nonlinear theory of thin shells, Arch. Mech. 33, 2 (1981) 273-288.

7. PIETRASZKIEWICZ W., On consistent approximations in the geometrically non-linear theory of thin shells, IFM-Mitt. 26, Ruhr-Univ. Bochum, (1981).

8. MARGUERRE K., Zur Theorie der gekrümmten Platte großer Formänderungen, Proc. 5th Int. Congr. of Appl. Mech., Cambridge/Mass. (1938) 93-101, Wiley and Sons (1939).

9. DONNELL L.H., Stability of thin walled tubes under torsion, NACA TR 479 (1933).

10. VLASOV V.Z., General theory of shells and its application in engineering, NASA TT F-99 (1964).

11. SANDERS J.L., Nonlinear theories for thin shells, Quart. Appl. Math., Vol. 21 (1963) 21-36.

12. KRÄTZIG W.B., Basar Y., Wittek U., Nonlinear behaviour and elastic stability of shells, Buckling of Shells, Proc. of a state-of-the-Art Colloquium. Springer Verlag, Berlin-Heidelberg-New York (1982).

13. STEIN E., BERG A., WAGNER W., Different levels of non-linear shell theory in finite element stability analysis, Buckling of Shells, Proc. of a state-of-the-Art-Colloquium. Springer Verlag, Berlin-Heidelberg-New York (1982).

14. HARTE R., Doubly curved triangular finite elements for the linear and geometrically nonlinear analysis of general shell structures (in German), KIB-TWM 82-10, Ruhr-Univ. Bochum (1982).

15. PIETRASZKIEWICZ W., Lagrangian description and incremental formulation in the non-linear theory of thin shells, Int. J. Non-Linear Mech. 19 (1984) 115-140.

16. NOLTE L-P., MAKOWSKI J., STUMPF H., On the derivation and comparative analysis of large rotation shell theories, Ing.-Arch. (1985) in print.

17. SCHMIDT R., A current trend in shell theory: Constrained geometrically nonlinear Kirchhoff-Love type theories based on polar decomposition of strains and rotations, Comp. & Struct. Vol. 20, No. 1-3 (1985) 265-275.

18. NOLTE L.-P., SCHIECK B., Vectorized high precision finite elements with applications to nonlinear shell problems, Parallel Computing 85, North-Holland, Amsterdam (1986) in print.

19. NOLTE L.-P., MAKOWSKI J., Pressure loaded shells undergoing different levels of nonlinearity, Mech. Res. Comm. (1986) in print.

20. NOLTE L.-P., STUMPF H., Energy-consistent large rotation shell theories in Lagrangean description, Mech. Res. Comm. Vol. 10 (4), (1983) 213-221.

21. SCHMIDT R., On geometrically non-linear theories for thin elastic shells, Proc. EUROMECH Coll. "Flexible Shells, Theory and Applications", München (1983).

22. BASAR Y., A geometrically nonlinear shell theory (in German), KIB-Berichte 38/39, Vulkan-Verlag, Essen (1981).

23. SHAPOVALOV L.A., On a simplest variant of equations of geometrically non-linear theory of thin shells (in Russian), Mekh. Tv. Tela 3, No. 1, 1968, 56-62.

24. CHUYKO A.N., On nonlinear equations for the theory of thin shells, Soviet Applied Mechanics, Vol. 19, No. 11 (1983) 66-71.

25. YAGHMAI S., Incremental analysis of large deformations in mechanics of solids with applications to axisymmetric shells of revolution, Dissertation, University of California, Berkeley (1969).

26. VARPASUO P., Incremental analysis of axisymmetric shallow shells with varying strain-displacement equations, Comp. Meth. in Appl. Mech. and Eng. 21 (1980) 153-169.

ROTATIONS AS PRIMARY UNKNOWNS IN THE NONLINEAR THEORY OF SHELLS AND CORRESPONDING FINITE ELEMENT MODELS

L. RECKE, W. WUNDERLICH
Ruhr-Universität Bochum, F.R. Germany

Abstract

A consistent geometrically nonlinear theory of shells is derived based on the formulation of a generalized variational principle. In addition to displacements and stresses, the components of a finite rotation vector are introduced as primary unknowns. This is accomplished by applying the polar decomposition of the deformation gradient. The basic equations are described in an incremental Lagrangian frame and are given in clear operator form to emphasize the additional equations of the angular momentum balance and the rigid body kinematics as a special feature of the present theory. Mixed variant tensorial components render advantages in the reduction to the shell equations in which shear effects are included. The two-dimensional principle serves then as the adequate basis for the formulation of different mixed-type finite element models. Numerical results are presented.

1. Introduction

In the analysis of flexible structures realistic results can only be achieved by inclusion of nonlinear effects. Especially large deflections and large rotations have to be considered and require an adequate formulation of the underlying theory as well as effective numerical solution procedures and finite element models. The behaviour of curved surfaces is of special interest, and it follows that nonlinear theories of shells and shell models have to be developed further. Most of the existing theories are restricted to moderate rotations or by other assumptions. On the other hand it is well known in the finite element approximation using shell models that the proper description of rigid body rotations plays an important role. Large rotations no longer have vector properties and the deformation cannot be described by displacement and rotation vectors additively, but a sequence of rotations may be connected in a multiplicative manner. For this reason it seems to be consequent to use the rotations as primary unknowns in the formulation and solution of shells problems.

Based on similar considerations for bars /1/, polar decomposition of the deformation gradient into a rigid rotation and a pure stretch part is employed in this paper. This derivation of the nonlinear shell theory is based on a generalized variational principle of the Hellinger-Reissner type which in the context of continuum formulations was also used by de Veubecke /2/ and other authors as the starting point of their work /3,4/. The application to shells renders a theory with displacements, rotations and stress-resultants as basic unknowns. In the framework of a consistent theory stresses and stress-resultants conjugate to the "engineering" strain quantities are used. In an incremental formulation their symmetric parts are connected by a semilinear material law while the nonsymmetric part of the stresses and stress-resultants give additional imformation with respect to rigid body rotations as dual counterparts. For the nonlinear solution an incremental form of the theory in Lagrangian description is given. It is possible to employ nonlinear increments of the rotations as they may be expressed in terms of the linearized solution vector. This is of special importance for an effective iteration.

In the formulation of the generalized variational principle for thin shells also shear effects are included, which seems to lead to a more compact form compared to theories under the Kirchhoff-Love hypothesis. In the derivation assumptions for the distribution across the thickness are made for the stresses as well as for the displacements and rotations. In addition, the theory is written with mixed-variant tensor components following the lines of /5/ leading to conceptional advantages in the reduction from the continuum to the shell surface, especially with respect to the constitutive equations.

The numerical analysis for shells is most commonly performed by the finite element method. Rotations are included into finite element models described in the literature in different manner, e. g. tangential to the deformed surface /6/, or as semitangential rotations /7/, or as difference of the displacements of the inner and outer surface of the shell /8/. In this paper the generalized variational principle is consequently employed also as the basis of a mixed-type finite element approach, in which the primary unknowns of the theory serve as nodal parameters of the finite element models. Before assembling the elements to the whole shell structure the generalized element matrices are condensed to a displacement model to be easily included into a standard finite element code.

2. Basic relations for the continuum

2.1 Description of deformation

The configurations of a deformable body may be referred to the undeformed initial state with coordinates X^A and base vectors $\mathbf{G}_A$, or to a current, deformed state with coordinates $\overset{\circ}{x}{}^i$ and base vectors $\overset{\circ}{\mathbf{g}}_i$. Denote the position vector of a particle in the undeformed body by $\mathbf{R}$, then the position vector $\overset{\circ}{\mathbf{r}}$ in the deformed state is $\overset{\circ}{\mathbf{r}} = \mathbf{R} + \overset{\circ}{\mathbf{u}}$ with the displacement vector $\overset{\circ}{\mathbf{u}}$, and corresponding strains and stresses, which have to be in equilibrium for this deformed state. The use of capital and lower case indices helps to differentiate whether a tensor is associated with initial or deformed base vectors, or both, in which case it is a two-point tensor. The superscript $^\circ$ marks the deformed (fundamental) state while corresponding increments from that state remain unlabelled.

As a result of the deformation the neighborhood of a particle is transformed according to

$$d\overset{\circ}{x}{}^i = \frac{\partial \overset{\circ}{x}{}^i}{\partial X^A}\, dX^A = \overset{\circ}{F}{}^i_{\cdot A}\, dX^A, \tag{1}$$

or with reference to curvilinear, convective coordinates θ^j, the symbol $|$ denoting the covariant derivative

$$d\overset{\circ}{x}{}^i = \overset{\circ}{y}{}^i|_j \, dX^j = (\delta^i_j + \overset{\circ}{u}{}^i|_j)\, dX^j. \tag{2}$$

The deformation gradient $\overset{\circ}{F}{}^i_{\cdot A}$ (or $\overset{\circ}{F}{}^i_{\cdot j}$) may be split up into a rigid body rotation and a stretch by polar decomposition

$$\overset{\circ}{F}{}^i_{\cdot A} = \overset{\circ}{R}{}^i_{\cdot A}\, \overset{\circ}{U}{}^A_B = \overset{\circ}{R}{}^i_{\cdot C}(\delta^C_A + \overset{\circ}{h}{}^C_A). \tag{3}$$

In (3), $\overset{\circ}{R}{}^i_{\cdot C}$ are the components of a twofield orthonormal rigid rotation tensor, and $\overset{\circ}{U}{}^A_B$ those of the "right" stretch tensor containing the engineering strains $\overset{\circ}{h}{}^C_A$ which for principal axis give the stretch $(d\overset{\circ}{s}_i - dS_i)/dS_i$ of the corresponding fibers.

An incremental form of these quantities and of the basic equations is necessary for the solution of the nonlinear problem. For this purpose a neighboring state to the current configuration is considered. The corresponding quantities are marked with a superscript ~, whereas the increments between the two states remain unlabelled:

$$\tilde{\mathbf{F}} = \overset{\circ}{\mathbf{F}} + \mathbf{F} \; ; \quad \tilde{\mathbf{R}} = \overset{\circ}{\mathbf{R}} + \mathbf{R} \; ; \quad \tilde{\mathbf{h}} = \overset{\circ}{\mathbf{h}} + \mathbf{h} \; , \text{ etc.} \tag{4}$$

The increments as well as all the other quantities shall be referred to the same reference configuration, e. g. according to the Total-Lagrange formulation to the initial state.

Special attention has to be given to an adequate formulation of the rotation tensor $\overset{\circ}{\mathbf{R}}$ and its increment $\mathbf{R}$ when - as in this paper - large rotations shall be described properly. For this purpose the sequence of an orthogonal triad undergoing rigid rotation is considered, see Fig. 1.

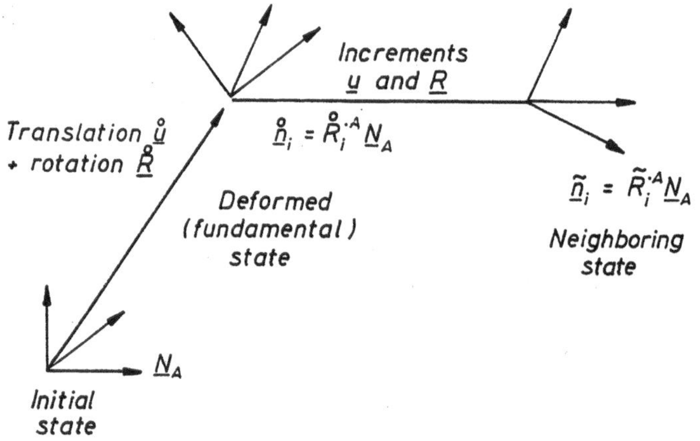

Fig. 1: Rigid rotations of triads

The difference between the rotated triad $\overset{\circ}{\mathbf{n}}_i$ and its neighboring state $\tilde{\mathbf{n}}_i$ gives the increment $\mathbf{R}$ of the rotation. Its components referred to the initial triad $\mathbf{N}_A$ are $R_i^{\cdot A}$, and its components with reference to $\overset{\circ}{\mathbf{n}}_j$ shall be denoted as $W_i^{\cdot j}$. Then, with the relations given in Fig. 1 and the orthonormal properties of $\overset{\circ}{\mathbf{R}}$ the incremental quantity

$$\tilde{\mathbf{n}}_i - \overset{\circ}{\mathbf{n}}_i = R_i^{\cdot A} \, \overset{\circ}{R}{}^{j}_{\cdot A} \, \overset{\circ}{\mathbf{n}}_j = W_i^{\cdot j} \, \overset{\circ}{\mathbf{n}}_j \tag{5}$$

is obtained which connects the two triads referred to the configuration $\overset{\circ}{\mathbf{n}}$. For a total Lagrangian approach, reference to the initial triad is adequate, and the difference between the two quantities may be expressed through

$$\tilde{\mathbf{n}}_i - \overset{\circ}{\mathbf{n}}_i = R_i^{\cdot A}\mathbf{N}_A = W_i^{\cdot j}\,\overset{\circ}{R}{}_j^{\cdot A}\,\mathbf{N}_A$$
$$= \overset{\circ}{R}{}_i^{\cdot C}\,W_C^{\cdot A}\,\mathbf{N}_A \quad \text{with } W_C^{\cdot A} = \overset{\circ}{R}{}^k_{\cdot C}\,W_k^{\cdot j}\,\overset{\circ}{R}{}_j^{\cdot A}. \tag{6}$$

As the increment of the rotation tensor describes the rigid rotation of a corotating base triad during deformation the nine components of this tensor are not independent. They can be written in terms of three independent parameters ω_A of a pseudo-vector $\boldsymbol{\omega}$:

$$\boldsymbol{\omega} = \left[\ \omega_1\ \omega_2\ \omega_3\ \right]. \tag{7}$$

The components $W_i^{\cdot j}$ of the increment of the updated rotation tensor are the linearized coefficients contained in the skew-symmetric rotation matrix

$${}_1\mathbf{W} = \begin{bmatrix} 0 & -\omega_3 & \omega_2 \\ \omega_3 & 0 & -\omega_1 \\ -\omega_2 & \omega_1 & 0 \end{bmatrix}. \tag{8}$$

With respect to the initial frame the increment

$$R_i^{\cdot A} = \overset{\circ}{R}{}_i^{\cdot C}\,W_C^{\cdot A} \tag{9}$$

is a nonlinear quantity that may be expressed in terms of ${}_1\mathbf{W}$ as /9/

$$\mathbf{R}({}_1\mathbf{W}) = \overset{\circ}{\mathbf{R}}\left(\frac{1}{1!}\,{}_1\mathbf{W} + \frac{1}{2!}\,{}_1\mathbf{W}\,{}_1\mathbf{W} + \ \ldots + \frac{1}{n!}\,{}_1\mathbf{W}^{(n)}\right) \tag{10}$$

$$= \overset{\circ}{\mathbf{R}}\,(e^{{}_1\mathbf{W}} - \mathbf{I}). \tag{11}$$

In the nonlinear solution process this relation makes it possible to use nonlinear increments for the rigid rotations and to calculate it as an exponential of the linearized matrix ${}_1\mathbf{W}$. The matrix exponential may either be evaluated by the series expansion (10) directly, or by the corresponding minimal polynomial

$$\mathbf{R} = \overset{\circ}{\mathbf{R}}\,\left(\mathbf{I} + \frac{\sin(\|\boldsymbol{\omega}\|)}{\|\boldsymbol{\omega}\|}\,{}_1\mathbf{W} + \frac{1}{2}\,\left(\frac{\sin(\|\boldsymbol{\omega}\|/2)}{\|\boldsymbol{\omega}\|/2}\right)^2{}_1\mathbf{W}\,{}_1\mathbf{W}\right). \tag{12}$$

2.2 Conjugate stress measures and constitutive relations

From the consideration of the strain energy density written in terms of the strain tensor $\overset{\circ}{\mathbf{h}}$ one'arrives quite naturally at the definition of the "engineering" or "Jaumann" stress tensor $\overset{\circ}{\mathbf{r}}$ see e. g. /2/. It is nonsymmetric and can be expressed in terms of the equally nonsymmetric, Lagrangian, or first Piola-Kirchhoff, stress tensor $\mathbf{t}$ via

$$\overset{\circ}{\mathbf{r}} = \overset{\circ}{\mathbf{t}}\,\overset{\circ}{\mathbf{R}}, \tag{13}$$

where $\overset{\circ}{\mathbf{R}}$ is the finite rotation tensor. Thus, both $\overset{\circ}{\mathbf{t}}$ and $\overset{\circ}{\mathbf{r}}$ are Lagrangian quantities, and the components of $\overset{\circ}{\mathbf{r}}$, referred to the set of rigidly rotated base vectors $\hat{\mathbf{G}}_i$, follow from those of $\overset{\circ}{\mathbf{t}}$, which are referred to the initial base triad $\mathbf{G}_i$, through an orthogonal transformation.

It is shown in /2/, that the symmetric part

$${}_s\overset{\circ}{\mathbf{r}} = \frac{1}{2}\,(\overset{\circ}{\mathbf{r}} + \overset{\circ}{\mathbf{r}}^T) \tag{14}$$

of $\overset{\circ}{\mathbf{r}}$ is conjugate to the symmetric engineering stress tensor $\overset{\circ}{\mathbf{h}}$, and that the Legendre transformation

$$W(\overset{\circ}{\mathbf{h}}) + W^*({}_s\overset{\circ}{\mathbf{r}}) = {}_s\overset{\circ}{\mathbf{r}}^T\overset{\circ}{\mathbf{h}} \tag{15}$$

holds. In (15) $W(\overset{\circ}{\mathbf{h}})$ and $W^*({}_s\overset{\circ}{\mathbf{r}})$ are the strain energy and the complementary energy densities, respectively. Differentiation of W and W^* with respect to $\overset{\circ}{\mathbf{h}}$ and ${}_s\overset{\circ}{\mathbf{r}}$, respectively, gives stress-strain relations, as well as their inverses, which are valid for a so-called semilinear material /1/. In the following we will assume that equivalent relationships hold for the increments ${}_s\mathbf{r}$ and $\mathbf{h}$ as well. The anti-symmetric part ${}_a\overset{\circ}{\mathbf{r}} = \frac{1}{2}\,(\overset{\circ}{\mathbf{r}} - \overset{\circ}{\mathbf{r}}^T)$ which does not figure in the constitutive equations and is neglected by most authors will be considered in chapter 3.2 in connection with the local description of the incremental rigid rotations.

2.3 Formulation of the variational principle

Using the deformation gradient $\overset{\circ}{\mathbf{F}}$, the finite rotation tensor $\overset{\circ}{\mathbf{R}}$ and the Jaumann stress tensor $\overset{\circ}{\mathbf{r}}$ as field variables a variational principle can be given in the form /1/

$$\delta \overset{\circ}{J}_{CB} = \delta\{\int_{V_o} [-W^*({}_s\overset{\circ}{r}) + \overset{\circ}{r}{}^A_B \, \overset{\circ}{h}{}^B_A - \rho_o \overset{\circ}{f}{}^A \, \overset{\circ}{u}_A] dV$$

$$- \int_{S_p} \overset{\circ}{\bar{t}}{}^A \, \overset{\circ}{u}_A \, dS - \int_{S_u} \overset{\circ}{t}{}^A (\overset{\circ}{u}_A - \overset{\circ}{\bar{u}}_A) dS\} = 0 \qquad (16)$$

which ensures satisfaction of the kinematic relations, the force equilibrium equations, the angular momentum balance and all displacement and traction boundary conditions in a weak sense. Application of Gauss' theorem in the usual manner leads to the respective Euler equations. From (16) a related principle governing the increments of the respective field quantities may be derived which reads

$$\delta J_{CB} = \delta \{\int_{V_o} [-W^*({}_s\mathbf{r}) + \mathbf{r} \, \overset{\circ}{\mathbf{R}}{}^T \, \mathbf{F} + \mathbf{r} \, \mathbf{R}^T \, \overset{\circ}{\mathbf{F}}$$

$$+ \overset{\circ}{\mathbf{r}} \, (\mathbf{R}^T \, \mathbf{F} - \frac{1}{2} \, \mathbf{R}^T \, \mathbf{R} \, \overset{\circ}{\mathbf{R}}{}^T \, \overset{\circ}{\mathbf{F}})$$

$$+ \underline{\overset{\circ}{\mathbf{r}} \, \mathbf{R}^T \, \overset{\circ}{\mathbf{F}}} + \rho_o \, \mathbf{b} \, \mathbf{u}] \, d V_o \qquad (17)$$

$$- \int_{S_p} [\bar{\mathbf{p}}^T \, \mathbf{u}] \, dS - \int_{S_u} [\bar{\mathbf{p}}^T (\mathbf{u} - \bar{\mathbf{u}})] dS\} = 0 \, .$$

Carrying out the variations with respect to the independent incremental variables appearing in (17) again gives the appropriate incremental Euler equations. In the derivation, the treatment of the underlined term in (17) requires some special consideration. Due to the strongly nonlinear nature of the rotation tensor as well as its finite increment (see equation (11)) the following terms appear in the angular momentum balance equations:

$$\delta \mathbf{R}^T \overset{\circ}{\mathbf{F}} \, \overset{\circ}{\mathbf{r}} = - \, \delta_1 \mathbf{W} \, \tilde{\mathbf{R}}^T \, [\overset{\circ}{\mathbf{F}} \, \overset{\circ}{\mathbf{r}}]$$

$$= - \, \delta_1 \mathbf{W} \, [\overset{\circ}{\mathbf{U}} \, \overset{\circ}{\mathbf{r}} + \mathbf{W} \, \overset{\circ}{\mathbf{U}} \, \overset{\circ}{\mathbf{r}}] \, . \qquad (18)$$

While the first term in the last line of (18) expresses the angular momentum balance condition of the reference state, which is presumed to be satisfied identically, the second term represents an additional nonlinear influence which needs to be included in the incremental form of the angular momentum equations.

3. A nonlinear theory of shells including shear effects

3.1 Assumptions

The reduction from the general three-dimensional equations and variational principle (17) to the two-dimensional theory of thin shells is achieved along the lines given in /5/ for a linear theory using mixed-type components. First, the equations are transformed without any approximation to curved but still three-dimensional coordinates of the shell space. For details of this step the reader is referred to /5/. Then, by introducing assumptions for the variation across the shell, two-dimensional equations of a theory for thin shells may be derived.

In the derivation of a shell theory including shear effects the stresses r^3_3 across the thickness are neglected, and the corresponding strains arising from Poisson's ratio's influences may be assumed to be small as a general theory with small strains but large rotations is considered. On the kinematic side this is consistent with the assumption that a normal to the shell surface remains straight but not normal after deformation. Due to the use of a semilinear material law the assumptions for the shear stresses result in similar distributions for the shear strains.

As primary hypothesis the change of the stress resultants through the shell thickness is assumed to be constant and linear in the coordinate θ^3. Through the definition in mixed tensor components this assumption holds also for the components of the strains. For the engineering stresses, related to the shell middle surface, the distributions across the thickness are chosen as

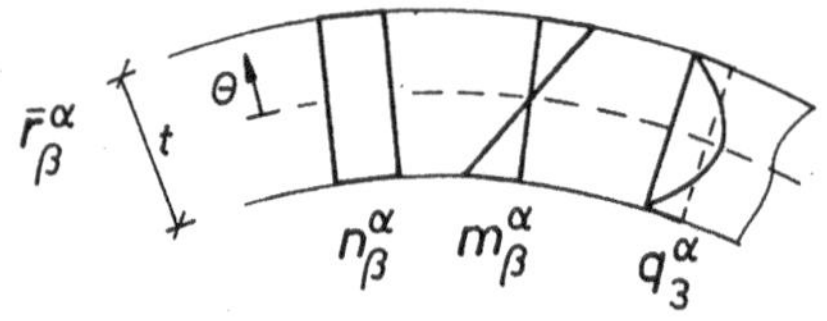

$$_s\bar{r}^\alpha_\beta = \frac{1}{t}\, n^\alpha_\beta + \frac{6}{t^2}\, m^\alpha_\beta\, \frac{\theta}{t/2}\,,$$

$$_s\bar{r}^\alpha_3 = \frac{3}{2t}\, q^\alpha_3\, \left(1 - \theta^2/(t/2)\right)\,, \qquad (19)$$

$$_s\bar{r}^3_3 = 0\,.$$

The parameters in the definitions of equ. (19) may be identified as symmetric stress resultants, but keeping in mind the definition of stress resultants resulting from the integration of the stress measures including the shell shifter $(\mu^{-1})^\alpha_\beta$, the non-symmetric resultants are:

$$n^{\alpha}_{\cdot\beta} = n^{\alpha}_{\beta} + (b^{\alpha}_{\lambda} - 2H\, \delta^{\alpha}_{\lambda})\, m^{\lambda}_{\beta} \,,$$

$$m^{\alpha}_{\cdot\beta} = m^{\alpha}_{\beta} + \underline{\frac{B}{D} (b^{\alpha}_{\lambda} - 2H\, \delta^{\alpha}_{\lambda}) \overset{(2)}{m}{}^{\lambda}_{\beta}} \,. \tag{20}$$

The underlined term may be neglected in the framework of a first approximation for thin shells. A comparative study of the different definition of stress resultants is given in /5/.

To describe the deformed position of a point with reference to the shell surface supplementary kinematic assumptions are necessary which are admissible to the stress assumptions for the first approximation of a shell theory. According to the characterization of the normal after the deformation, the displacements of the shell space may be split up into:

$$\bar{u}^{B} = {}_{o}u^{B} + \Theta^{3} F^{B}_{\cdot 3} \,. \tag{21}$$

With the deformation gradient $F^{B}_{\cdot 3}$ the position of the rotated normal is described exactly including shear effects. But equation (21) is valid for the three-dimensional continuum only and has to be modified for the utilization in a two-dimensional shell theory. First the linear part of equation (21) is written for the incremental form of equation (3):

$$F^{B}_{\cdot 3} = R^{B}_{\cdot A} (\delta^{A}_{3} + \overset{o}{h}{}^{A}_{3}) + \tilde{R}^{B}_{\cdot\beta}\, h^{\beta}_{3} + \tilde{R}^{B}_{\cdot 3}\, h^{3}_{3} . \tag{22}$$

By neglecting the shear deformations of the reference state with respect to unity (small strains) and of the last term of equation (22) according to the assumed rigidity of the normal we obtain

$$F^{B}_{\cdot 3} = R^{B}_{\cdot 3} + \tilde{R}^{B}{}_{\beta}\, h^{\beta}_{3} , \tag{23}$$

in which the shear deformation part can be interpreted as a rotation in addition to the rigid body rotation. The relation (23) then shows clearly the multiplicative nature of subsequent rotations. As the change of the length of the normal is not considered a point of the outer surface $\Theta^{3} = \pm t/2$ of the shell continuum moves on a sphere with radius t/2 measured from the local base triad. Then, the rotation of the normal - consisting of two parts - may be described with a rotation angle ϕ and its components with respect to the fundamental (updated) state:

$$\mathbf{r} = (u^B + \theta^3 \; {}_{\phi}R^B{}_3)\, \mathbf{A}_B \,, \tag{24}$$

$$\text{with} \quad {}_{\phi}R^B{}_3 = R^B{}_3(\phi_\lambda) \tag{25}$$

$$\varepsilon^{\beta\lambda} \, \phi_\lambda = \varepsilon^{\beta\lambda} \, \omega_\lambda + h^\beta_3 \,, \tag{26}$$

in which the components of the pure rigid part of the rotation are denoted by ω_λ. With respect to a vector description of the momentum equations a variant to equation (24) will be given in terms of the (global) rotation parameters ${}_g\phi^B$, measured with respect to the base vectors $\mathbf{A}_B$ of the undeformed shell surface:

$$\mathbf{r} = (\, u^B + \theta^3 \; {}_g\phi^B \,)\, \mathbf{A}_B \,, \tag{27}$$

$$\text{with} \quad \phi_\lambda = \tilde{R}^B_{\cdot\,\lambda} \; {}_g\phi_B + R^B_{\cdot\,\lambda} \; {}_g\overset{\circ}{\phi}_B \tag{28}$$

ϕ_λ - local components, defined by equation (26), $\lambda=1,2$
${}_g\phi^B$ - global components, B=1,2,3.

The rotations ${}_g\phi^B$ referred to the initial state can be identified as conjugate to the components of the momentum vector along the edge of a surface element of the shell. The increment of the components of the rotation ${}_g\boldsymbol{\phi}$ is written in terms of the local components ϕ^λ, as well as equation (24).

3.2 Variational principle

The further reduction to shell equations is achieved by inserting the assumptions for the displacements and rotations (24,27) and those for the stresses (19) into the variational principle (17).

In the formulation of the specific complementary energy $W^*(n^\alpha_\beta, m^\alpha_\beta, q^\alpha)$ only the symmetric parts of stress resultants (20) are included. By writing W^* in terms of the unsymmetric components and applying the Legendre-Transformation we obtain the inverse constitutive equations

$$\phi^{\alpha}_{\cdot\beta} = \frac{\partial W^*}{\partial n^{\cdot\beta}_{\alpha}} = \frac{1}{D}\,[(1+\nu)\,n^{\alpha}_{\cdot\beta} - \nu\,\delta^{\alpha}_{\beta}\,n^{\cdot\lambda}_{\lambda}]$$
$$- \frac{1}{D}\,[(1+\nu)\,b^{\lambda}_{\beta}\,m^{\alpha}_{\lambda} - \nu\,\delta^{\alpha}_{\beta}\,b^{\lambda}_{\nu}\,m^{\nu}_{\lambda}]\ ,$$

$$\chi^{\alpha}_{\cdot\beta} = \frac{\partial W^*}{\partial m^{\beta}_{\alpha}} = \frac{1}{B}\,[(1+\nu)\,m^{\alpha}_{\beta} - \nu\,\delta^{\alpha}_{\beta}\,m^{\lambda}_{\lambda}] \qquad (29)$$
$$- \frac{1}{D}\,b^{\lambda}_{\beta}\,[(1+\nu)\,n^{\alpha}_{\cdot\lambda} - \nu\,\delta^{\alpha}_{\lambda}\,n^{\cdot\nu}_{\nu}]\ ,$$

$$\gamma_{\alpha} = \frac{\partial W^*}{\partial q_{\alpha}} = \frac{1}{Gt}\,\frac{6}{5}\,q_{\alpha}\ .$$

It is characteristic for the present approach that the components of the rotations as well as their rigid parts are used as primary unknowns in the basic relations. Thus, additional local equations corresponding to the rotations appear in the derivation. Furthermore, from the study of the structural scheme of a shell theory the duality of static and kinematic variables is well known /5/. To identify a similar correspondence also for the components of the rigid rotation, conjugate stress resultants have to be defined, for which again a set of related equations has to exist. For this purpose the engineering stress tensor is split up into its symmetric and antimetric part. This results in the symmetric definition of equation (19) and in corresponding antisymmmtric stress resultants $\Delta\mathbf{s} = \{{}_ar^{12}\ \ {}_ar^{\gamma 3}\}$:

$$ {}_ar^{12} = \frac{1}{t}\,{}_an^{12}\ ,$$
$$ {}_ar^{\gamma 3} = \frac{3}{2t}\,{}_aq^{\gamma 3}\,(1-\theta^2/(t/2))\ . \qquad (30)$$

Introducing equation (30) into the variational principle, equation (17), the antisymmetric stress resultants may be identified as the static variables which are conjugate to the components of the rigid rotation $\boldsymbol{\omega}$, and as Lagrangian multipliers for the definition of the rigid body kinematics as subsidiary conditions:

$$ {}_a\phi_{12} = \phi_{21} - \phi_{12} = 0\ ,$$
$$ {}_ah_{\gamma 3} = h_{\gamma 3} - h_{3\gamma} = 0\ . \qquad (31)$$

These kinematical symmetry relations serve as nonlinear equations for the iterative calculation of the components of the rigid body rotation. The antisymmetric stress resultants of equation (30) appear in the equations for the angular momentum balance.

Finally, the system of kinematical relations and equilibrium, contained in the variational principle in incremental form is written in operator form. It reveals clearly the structure of a consistent theory of shell, especially with respect to the two sets of additional Euler equations arising from the path of derivation chosen in this paper:

$$\begin{array}{l} \text{Equilibrium} \\ \text{Angular momentum balance} \\ \text{Kinematics} \\ \text{Rigid body motion} \end{array} \quad \int_F \begin{bmatrix} \delta \mathbf{u}^T \\ \delta \boldsymbol{\omega}^T \\ \delta \mathbf{s}^T \\ \delta \Delta \mathbf{s}^T \end{bmatrix} \begin{bmatrix} \mathbf{G} & \mathbf{M}_1^T & {}_u\mathbf{D}^T & {}^r_u\mathbf{D}^T \\ \mathbf{M}_1 & \mathbf{M}_2 & \mathbf{M}_3 & \mathbf{M}_4 \\ \mathbf{D}_u & \mathbf{M}_3^T & -\mathbf{E}^{-1} & \mathbf{0} \\ {}^r\mathbf{D}_u & \mathbf{M}_4^T & \mathbf{0} & \mathbf{0} \end{bmatrix} \begin{bmatrix} \mathbf{u} \\ \boldsymbol{\omega} \\ \mathbf{s} \\ \Delta \mathbf{s} \end{bmatrix} dF \qquad (32)$$

By variation of the functional J_{CB} and the application of Gauß divergence theorem the complete nonlinear local Euler equations for the description of the shell problem are obtained. The equations are given here without further derivation:

$$\phi^{\alpha}_{\cdot\gamma} = \tilde{R}^{\cdot\alpha}_{B}\, u^{B}|_{\gamma} + R^{\cdot\alpha}_{B}\, \overset{\circ}{y}^{B}|_{\gamma} - b^{\beta}_{\gamma}\,(\tilde{R}^{\cdot\alpha}_{\beta}\, u^{3} + R^{\cdot\alpha}_{\beta}\, \overset{\circ}{y}^{3}) + b_{\gamma\beta}(\tilde{R}^{\cdot\alpha}_{3}\, u^{\beta} + R^{\cdot\alpha}_{3}\, \overset{\circ}{y}^{\beta}) \;, \qquad (33)$$

$$\chi^{\alpha}_{\cdot\gamma} = \varepsilon^{\alpha\lambda}(\tilde{R}_{B\lambda}\, g^{\phi^{B}}|_{\gamma} + R_{B\lambda}\, g^{\overset{\circ}{\phi}^{B}}|_{\gamma} - b^{\beta}_{\gamma}(\tilde{R}_{\beta\lambda}\, g^{\phi^{3}} + R_{\beta\lambda}\, g^{\overset{\circ}{\phi}^{3}}) + b_{\gamma\beta}(\tilde{R}_{3\lambda}\, g^{\phi^{\beta}} + R_{3\lambda}\, g^{\overset{\circ}{\phi}^{\beta}})) \;, \qquad (34)$$

$$h^{3}_{\gamma} = \tilde{R}^{\cdot 3}_{B}\, u^{B}|_{\gamma} + R^{\cdot 3}_{B}\, \overset{\circ}{y}^{B}|_{\gamma} - b^{\beta}_{\gamma}(\tilde{R}^{\cdot 3}_{\beta}\, u^{3} + R^{\cdot 3}_{\beta}\, \overset{\circ}{y}^{3}) + b_{\gamma\beta}(\tilde{R}^{\cdot 3}_{3}\, u^{\beta} + R^{\cdot 3}_{3}\, \overset{\circ}{y}^{\beta}) \;, \qquad (35)$$

$$h_{3\gamma} = \tilde{R}_{B\gamma}\, {}_{\phi}R^{B}_{\cdot 3} + R_{B\gamma}\, {}_{\phi}\overset{\circ}{R}{}^{B}_{\cdot 3} \;. \qquad (36)$$

Equilibrium equations:

$$\begin{aligned} s^{\gamma}_{\cdot\beta}|_{\gamma} - b_{\gamma\beta}\, s^{\gamma}_{\cdot 3} + p_{\beta} &= 0 \;, \\ s^{\gamma}_{\cdot 3}|_{\gamma} + b^{\beta}_{\gamma}\, s^{\gamma}_{\cdot\beta} + p_{3} &= 0 \;, \end{aligned} \qquad (37)$$

$$B^{\cdot\gamma}_{\cdot\beta}|_{\gamma} - b_{\gamma\beta}\, B^{\cdot\gamma}_{\cdot 3}$$
$$- \epsilon_{\gamma\beta}[(\tilde{R}_{33}\, \tilde{R}^{\gamma\alpha} - \tilde{R}^{\cdot\gamma}_{\cdot 3}\, \tilde{R}^{\cdot\alpha}_{3})q^{\cdot 3}_{\cdot\alpha} + (\tilde{R}_{33}\, R^{\gamma\alpha} - \tilde{R}^{\cdot\gamma}_{\cdot 3} R^{\cdot\alpha}_{3})\overset{\circ}{q}{}^{\cdot 3}_{\cdot a}] = 0\,, \tag{38}$$

$$B^{\cdot\gamma}_{\cdot 3}|_{\gamma} + b^{\beta}_{\gamma}\, B^{\cdot\gamma}_{\cdot\beta}$$
$$- \epsilon_{\gamma\beta}[\tilde{R}^{\cdot\gamma}_{\cdot 3}(\tilde{R}^{\beta\alpha}\, q^{\cdot 3}_{\cdot\alpha} + R^{\beta\alpha}\, \overset{\circ}{q}{}^{\cdot 3}_{\cdot\alpha})] = 0\,. \tag{39}$$

Static boundary condition:

$$\underbrace{(n^{\cdot\gamma}_{\cdot\alpha}\, \tilde{R}^{\cdot\alpha}_{B} + q^{\cdot\gamma}_{\cdot 3}\, \tilde{R}^{\cdot 3}_{B} + \overset{\circ}{n}{}^{\cdot\gamma}_{\cdot\alpha}\, R^{\cdot\alpha}_{B} + \overset{\circ}{q}{}^{\cdot\gamma}_{\cdot 3}\, R^{\cdot 3}_{B})}_{s^{\cdot\gamma}_{\cdot B}}\, N_{\gamma} - \bar{p}_{B} = 0\,, \tag{40}$$

$$[\underbrace{\epsilon^{\alpha\lambda}(\tilde{R}_{B\lambda}\, m^{\gamma}_{\alpha} + R_{B\lambda}\, \overset{\circ}{m}{}^{\gamma}_{\alpha})}_{B^{\cdot\gamma}_{\cdot B}}\, N_{\gamma} - \bar{m}_{B}] = 0 \rightarrow (\epsilon^{\alpha\lambda} m^{\gamma}_{\alpha}\, N_{\gamma} - \bar{m}^{\lambda}) = 0\,. \tag{41}$$

Geometric boundary condition:

$$u^{B} - \bar{u}^{B} = 0\,, \tag{42}$$

$$_{g}\phi^{B} - {}_{g}\bar{\phi}^{B} = 0 \rightarrow \phi^{\lambda} - \bar{\phi}^{\lambda} = 0\,. \tag{43}$$

It should be remembered that in contrast to usual formulations of non-linear shell equations we obtain this set under the given assumptions without further series expansions and order-of-magnitude considerations. Due to reference to the initial configuration (Total-Lagrange-Description) the resulting equations are rather simple in form. This holds also for the equations of momentum balance which arise by variation of the rotation components $_{g}\phi^{B}$, referred to the initial reference frame. It is to notice, however, that these three components may be expressed by two independent components ϕ^{λ}, referred to the fundamental (updated) state of the shell surface. By introduction of equation (28) one equation may be eliminated to obtain the system of differential equations for this reference configuration. Also the geometric boundary conditions can be formulated in terms of two independent rotation parameters of the shell surface, see equation (43).

While the equations of the rigid body kinematics may be described easily using equation (31) and the equations (35,36), the angular momentum balance equations are not given here in detail due to limited space. For details the reader is referred to /10/.

4. Formulation of element models

Starting from (32), discretization of the governing incremental field equations has been performed in two ways. First, a plane rectangular finite element with orthogonal local coordinates has been developed employing low order polynomial interpolation for all variables appearing in (32), see also /10/. Secondly, a ring finite element applicable to the analysis of shells of revolution has been derived using a semi-analytical procedure whereby Fourier decomposition of the Euler equations associated with (32) is used in the circumferential direction and the remaining set of ordinary differential equations is integrated numerically along a finite interval of the meridian in the manner described in detail e. g. in /11/. In both cases the stress resultants have been eliminated on the element level so that the remaining degrees of freedom are analogous to those which would be obtained in a pure displacement formulation, and the resulting stiffness matrices may be incorporated in a standard displacement finite element program.

4.1 Plane quadrilateral element

Because the angle of rotation of the normal to the middle surface and the normal component of the displacement vector may be treated independently, their approximating functions need only satisfy C^0-continuity. Accordingly, the unknown quantities at the nodes are the increments of the three components of the displacement vector, referred to the undeformed base triad, as well as the two components of the incremental rotation vector which are referred to the current, updated reference triad, see Fig. 2.

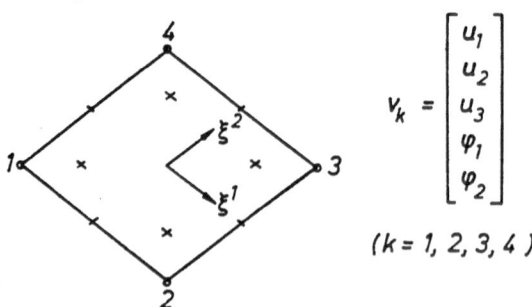

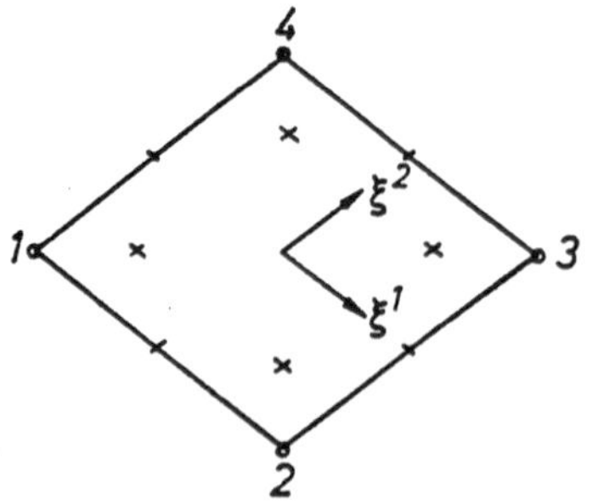

$$v_k = \begin{bmatrix} u_1 \\ u_2 \\ u_3 \\ \varphi_1 \\ \varphi_2 \end{bmatrix}$$

$(k = 1, 2, 3, 4)$

Fig. 2: Element and nodal degrees-of-freedom

Because the underlying variational principle is of mixed type the displacements, rotations and stress resultants may be approximated independently. Nevertheless, the various interpolation functions should be chosen with sufficient care so as to avoid the occurence of locking phenomena, zero energy modes or other features which may have adverse effects on the resulting element's performance. Thus, we have chosen a balanced approximation with bilinear interpolation functions for the displacements and matching functions for the stress resultants and the components of the rigid rotation given below. Their choice has been guided by considerations presented in detail in /10, 13/.

Interpolation functions:

Displacements $$\mathbf{u} = \sum_{i=1}^{4} \Omega_i \mathbf{v}_i \tag{44a}$$

with
$$\Omega_1 = 1/4\ (1-\xi^1)\ (1-\xi^2) \qquad \Omega_3 = 1/4\ (1+\xi^1)\ (1+\xi^2)$$
$$\Omega_2 = 1/4\ (1+\xi^1)\ (1-\xi^2) \qquad \Omega_4 = 1/4\ (1-\xi^1)\ (1+\xi^2)$$

Stress resultants

$$\begin{aligned}
n^1{}_1 &= \bar{n}^1{}_1 + \xi^2\, \hat{n}^1{}_1 & m^1_1 &= \bar{m}^1_1 + \xi^2\, \hat{m}^1_1 & q^1{}_3 &= \bar{q}^1{}_3 + \xi^2\, \hat{q}^1{}_3 \\
n^2{}_2 &= \bar{n}^2{}_2 + \xi^1\, \hat{n}^2{}_2 & m^2_2 &= \bar{m}^2_2 + \xi^1\, \hat{m}^2_2 & q^2{}_3 &= \bar{q}^2{}_3 + \xi^1\, \hat{q}^2{}_3 \\
n^1{}_2 &= \bar{n}^1{}_2 & m^1_2 &= \bar{m}^1_2 & q^3{}_1 &= \bar{q}^3{}_1 + \xi^2\, \hat{q}^3{}_1 \\
n^2{}_1 &= \bar{n}^2{}_1 & m^2_1 &= \bar{m}^2_1 & q^3{}_2 &= \bar{q}^3{}_2 + \xi^1\, \hat{q}^3{}_2
\end{aligned} \tag{44b}$$

Components of rigid rotation

$$\omega^1 = \bar{\omega}^1 + \xi^1\, \hat{\omega}^1, \qquad \omega^2 = \bar{\omega}^2 + \xi^2\, \hat{\omega}^2, \qquad \omega^3 = \bar{\omega}^3. \tag{44c}$$

Introducing (44a-c) into (32) and carrying out the integration over the area of the element leads to the discrete form of (32) and to the associated mixed finite element matrix. At this stage then the discrete degrees of freedom of the stress resultants and of the rigid rotation vector may be eliminated to ultimately give a standard stiffness matrix connecting the remaining displacement degrees of freedom. As they are the only discrete unknowns appearing at the global level, all other quantities must be evaluated at the element level. Having been eliminated beforehand, this also holds for the components of the rotation vector which must be computed using relationship (31). As

(31) is nonlinear this analysis must be carried out iteratively in such a way that the symmetry of the engineering strain is maintained.

4.2 Ring element

The derivation of the ring finite element starts from the Euler equations associated with (42). They consist of two coupled sets of ordinary and partial differential equations in the variables

$$\mathbf{z}^T = [u_{(1)}\; u_{(2)}\; u_{(3)}\; \phi_{(1)}\; \phi_{(2)}\; n^{(2)}_{(1)}\; n^{(2)}_{(2)}\; q^{(2)}\; m^{(2)}_{(1)}\; m^{(2)}_{(2)}] ,$$
$$\mathbf{z}_E^T = [n^{(1)}_{(1)}\; n^{(1)}_{(2)}\; q^{(1)}\; m^{(1)}_{(1)}\; m^{(1)}_{(2)}] , \tag{45}$$

and may be combined to a system of differential equations containing only first order derivatives with respect to the meridional coordinate θ^2. Written in matrix notation it takes the form

$$\frac{\partial}{\partial\theta^2}\left((I + \mathbf{B}^G(\overset{\circ}{\mathbf{R}}))\;\mathbf{z}\right) = (\mathbf{A}^L + \mathbf{A}^G(\overset{\circ}{\mathbf{z}}))\;\mathbf{z} + \bar{\mathbf{z}} \;. \tag{46}$$

Using now a Fourier representation of all independent variables in $\mathbf{z}$ with respect to the circumferential coordinate θ^1, i. e.

$$\mathbf{z} = \sum_{n=0}^{N} \overset{(n)}{\mathbf{z}}(\theta^2) \begin{bmatrix} \sin(n\theta^1) \\ \cos(n\theta^1) \end{bmatrix}$$

one obtains a coupled set of ordinary first-order differential equations with respect to θ^2:

$$\frac{d}{d\theta^2}\overset{(n)}{\mathbf{z}} = \overset{(n)}{\mathbf{A}}{}^L\,\overset{(n)}{\mathbf{z}} + \overset{(\underline{n})}{\bar{\mathbf{z}}} + \overset{(n)}{\mathbf{z}}{}^G + \overset{(n)}{\mathbf{z}}{}^R \tag{47}$$

$$\text{with}\quad \overset{(n)}{\mathbf{z}}{}^G = \overset{(n)}{\mathbf{A}}{}^G(\overset{\circ}{\mathbf{z}})\,\overset{(n)}{\mathbf{z}}\,, \qquad \overset{(n)}{\mathbf{z}}{}^R = -\frac{d}{d\theta^2}\left(\overset{(n)}{\mathbf{B}}{}^G(\overset{(n)}{\overset{\circ}{\mathbf{R}}})\,\overset{(n)}{\mathbf{z}}\right).$$

Matrices **A** and **B** appearing in (47) consist of linear parts $\mathbf{A}_L$ and $\mathbf{B}_L$ and nonlinear matrices $\mathbf{A}_G$ and $\mathbf{B}_G$. Whereas the former are block-diagonal, the latter are fully populated and thus lead to a strong coupling of all Fourier terms. Because **A** and **B** are the sums of $\mathbf{A}_L$ and $\mathbf{A}_G$ and $\mathbf{B}_L$ and $\mathbf{B}_G$, respectively, the vectors $(\mathbf{A}^G + \mathbf{B}^G)\mathbf{z}$ may be considered as pseudo-load vectors. The coefficient matrix on the right-hand side of

(47) is the same as for the linear problem and does not change during the loading history. Equations (47) may now be integrated numerically in the manner described in /11,14/, and the resulting transfer matrix may finally be transformed to the associated standard stiffness matrix.

As before, special attention has to be given to the treatment of the angular momentum balance equation as well as to the evaluation of the finite increment of components of the rotation tensor. Thus, equations (31) and the equations of angular momentum may be viewed as geometrically nonlinear subsidiary conditions to be satisfied at discrete points along the nodal circles of the ring elements. This is equivalent to evaluating the components of the rotation tensor (11) as well as its derivatives at these discrete stations around the circumference. Their number depends on the number of Fourier harmonics chosen in any particular case. Likewise all other quantities appearing in the pseudo-load vectors in (47) are evaluated at these points, and finally, a Fourier analysis with respect to the circumferential coordinate θ^1 of the discrete values of the latter's components is performed. Thus, (47) is reduced to an uncoupled set of ordinary differential equations which may be integrated numerically as mentioned above. This procedure ultimately results in an uncoupled set of N global linear algebraic equations of the form

$$\overset{(n)}{\bar{\mathbf{P}}} = \overset{(n)}{\mathbf{K}}_L \overset{(n)}{\mathbf{V}} + \overset{(n)}{\mathbf{P}}{}^G + \overset{(n)}{\mathbf{P}}{}^R \qquad (48)$$

which must be solved iteratively due to the occurence of the pseudo-load vectors $\overset{(n)}{\mathbf{P}}{}^G$ and $\overset{(n)}{\mathbf{P}}{}^R$ which depend on $\overset{(n)}{\mathbf{z}}$ and account for all nonlinear influences. The iterative solution of (48) is performed very efficiently by using a preconditioned conjugate direction method given in /14/.

5. Example

To demonstrate the capability of the approach presented in this paper results of the numerical analysis for a special construction undergoing large rotations are given. The system consists of an assembly of short tubes which is used in automobiles as shockabsorbing device. Although most of the kinematic energy is changed by dissipation into plastic deformation also the geometrically nonlinear elastic part is of influence as the behaviour of the assembly is characterized by geometric locking with large deflections and large rotations. As a preliminary study such a nonlinear elastic analysis was performed without taking into account contact and plasticity. The tubes are assumed to be rigidly fastened at the connection nodes and the whole assembly is subjected to loads at the top with an angle of $\alpha = 15^{\circ}$ from the vertical symmetry line, see Fig. 3. The structure was modelled with 120 two-dimensional elements of the type described in chapter 4.1, and the loads were assumed to remain unchanged in direction during deformation.

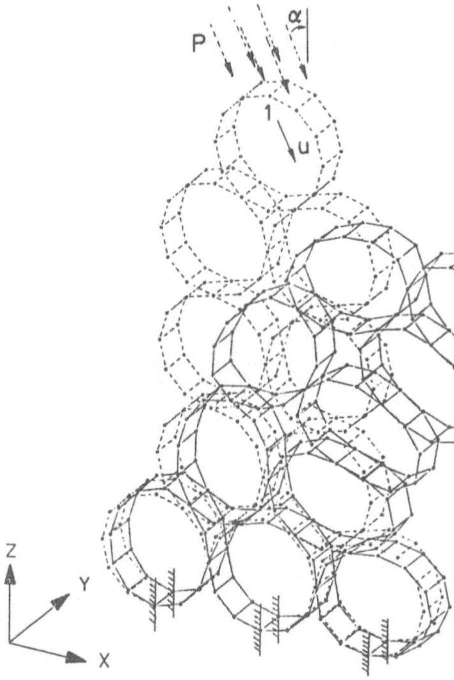

Fig. 3: Tube-structure, undeformed and deformed configuration

In Fig. 3 also the deformed structure is shown for the case of maximal load carrying capacity, and it is obvious that the structure has been undergoing large rotations with respect to the undeformed state. A specific load-deflection diagram for the point 1 is given in Fig. 4. The interconnections between the tubes give the possibility of unload-

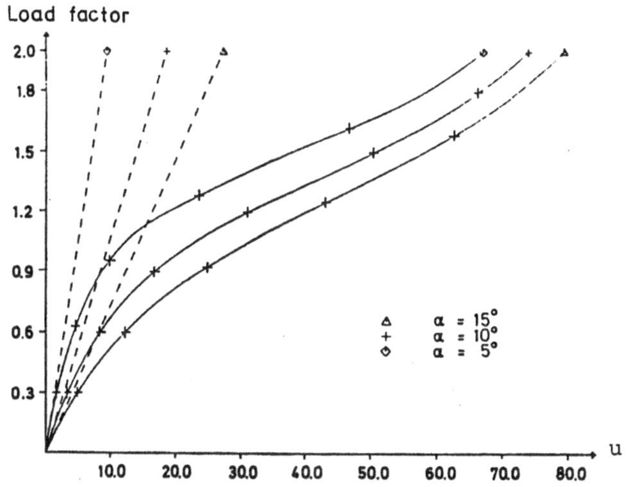

Fig. 4: Load-deflection diagram for point 1, system of tubes

ing of a single member and redistribution of the stresses resulting in a stiffening effect of the whole system, see Fig. 4. Different curves are given for various directions of the load.

6. References

/1/ WUNDERLICH, W., OBRECHT, H.: Large spatial deformations of rods using generalized varitional principles. In: Nonlinear finite element analysis in structural mechanics. (Wunderlich, W., Stein, E., Bathe, K.-J., Eds.), Springer-Verlag Berlin, 185-216 (1981).

/2/ FRAEIJS DE VEUBEKE, B.: A new variational principle for finite elastic displacements. Int. J. Eng. Sci. **10**, 745-763 (1972).

/3/ MURAKAWA, H.,ATLURI, S.N.: Finite elasticity solutions using hybrid finite elements based on a complementary energy principle. J. of Appl. Mech. **45**, 539-547 (1978).

/4/ CARNOY, E.G.: Mixed finite elements based on Marguerre theory for the study of geometrically nonlinear behavior of thin shells. Comp. Meth. in Appl. Mech. and Engng. **29**, 121-146 (1981).

/5/ WUNDERLICH, W.: On a consistent shell theory in mixed tensor formulation. In: Theory of Shells. (Koiter, W. T., Mikhailov, G. K., Eds.), North Holland Publ. Co, 607-633 (1980).

/6/ RAMM, E.: A plate/shell element for large deflections and rotations. In: Formulations and computational algorithms in finite element analysis. (Bathe, K.-J., Oden, T. J., Wunderlich, W., Eds.), M.I.T. Press, Boston, 264-293 (1977).

/7/ ARGYRIS, J.H., DUNNE, P.C., MALEJANNAKIS, G., SCHARPF, D.W.: On large displacement - small strain analysis of structures with rotational degrees of freedom. Comp. Meth. in Appl. Mech. and Engng. **14**, 401-451; **15**, 99-135 (1978).

/8/ KANOK-NUKULCHAI, W., TAYLOR, R.L., HUGHES, T.J.R.: A large deformation formulation for shell analysis by the finite element method. Computers and Structures **13**, 19-27 (1981).

/9/ PIETRASZKIEWICZ, W., BADUR, I.: Finite rotations in the description of continuum deformation. Int. J. Engng. Sci. **21**, 1097-1115 (1983).

/10/ RECKE, L.: Behandlung großer Rotationen elastischer Flächentragwerke auf der Basis einer Theorie schubweicher Schalen und eines gemischten Finite Element Konzepts. Technisch-wissenschaftliche Mitteilungen, Ruhr-Universität Bochum (1986).

/11/ WUNDERLICH, W.: Differentialsystem und Übertragungsmatrizen der Biegetheorie allgemeiner Rotationsschalen. Dissertation, TH Hannover (1966), or: Ing.-Archiv **36**, 267-279, (1967).

/12/ ARGYRIS, J.: An excursion into large rotations. Comp. Meth. in Appl. Mech. and Engng. **32**, 85-155 (1982).

/13/ WEMPNER, G., TALASLIDIS, D., HWANG, C.-M.: A simple and efficient approximation of shells via finite quadrilateral elements. Trans. ASME. J. Appl. Mech. **49**, 115-120 (1982).

/14/ WUNDERLICH, W., CRAMER, H., OBRECHT, H.: Application of Ring Elements in the Nonlinear Analysis of Shells of Revolution under Nonaxisymmetric Loading. Comp. Meth. in Appl. Mech. and Engng. **51**, 259-275 (1985).

POLAR DECOMPOSITION AND FINITE ROTATION VECTOR IN FIRST-ORDER FINITE ELASTIC STRAIN SHELL THEORY

R. SCHMIDT
University of Wuppertal
Institute of Civil Engineering Mechanics
Pauluskirchstrasse 7
D-5600 Wuppertal 2, Fed. Rep. of Germany

1. Introduction

While large strain membrane theories are well established in literature since more than three decades, there exist considerably less papers which deal with large elastic strain shell theory incorporating also the bending effects into the nonlinear analysis. Recently, however, this topic has gained considerable interest. Important contributions have been given by CHERNYKH [1,2] , LIBAI and SIMMONDS [5,6] , and SIMMONDS [20] , where also additional references on related works may be found. The aforementioned authors agree that such a theory should be based on a refined Kirchhoff-Love type model which admits at least changes in shell thickness. Due to bending this thickness change is in general asymmetric about the undeformed midsurface so that its deformed configuration is no longer the geometrical midsurface of the deformed shell. This requires a representation of the position vector of the deformed shell space which incorporates at least quadratic terms with respect to the thickness coordinate.

In the framework of the aforementioned modified Kirchhoff-Love hypothesis it is the aim of the present paper to develop a basis for the derivation of consistently simplified strain-displacement relations for problems in which the shell material elements undergo strains and rotations which are finite, but restricted (small, moderate, large) in the sense of a clear and physically meaningful classification scheme. First, assuming arbitrarily large strains and straight but extensible normal fibres we shall investigate the deformation of the shell by means of the polar decomposition theorem. The rotational part of deformation will be described by a suitable finite rotation vector which can be related easily to basic kinematical parameters of the shell middle surface. This makes it possible to restrict the order of magnitude of the rotations and to determine then the order of magnitude of displacements and their gradients, respectively. As an application of these results we shall derive consistently simplified strain-displacement relations for a shell undergoing moderate elastic strains and moderate rotations in the framework of CHERNYKH's theory [1,2] . Our developments parallel and generalize those given by PIETRASZKIEWICZ [9-11]for small strain shell theory (see also [13] for additional references to related works).

2. Notations and Basic Relations

Let M be the shell middle surface in the undeformed (reference) configuration with convected surface coordinates $\theta^\alpha, \alpha = 1,2$, and position vectors $\underset{\sim}{r}(\theta^\alpha) = x^k(\theta^\alpha)\underset{\sim}{i}_k$. Here x^k denotes the spatial components of $\underset{\sim}{r}$ with respect to a fixed Cartesian frame. Throughout the paper the Einsteinian summation convention will be used with Latin indices ranging from 1 to 3 and Greek indices ranging from 1 to 2. With each point of M we associate standard covariant surface base vectors $\underset{\sim}{a}_\alpha = \underset{\sim}{r}_{,\alpha}$, a unit normal vector $\underset{\sim}{n} = (1/2)\,\varepsilon^{\alpha\beta}\underset{\sim}{a}_\alpha \times \underset{\sim}{a}_\beta$, covariant components of the surface metric tensor $a_{\alpha\beta} = \underset{\sim}{a}_\alpha \cdot \underset{\sim}{a}_\beta$ (with $\det(a_{\alpha\beta}) \equiv a$), and of the curvature tensor $b_{\alpha\beta} = -\underset{\sim}{a}_\alpha \cdot \underset{\sim}{n}_{,\beta}$. Here $\varepsilon^{\alpha\beta}$ denotes the skew-symmetric surface permutation tensor, a comma denotes partial differentiation, $(\ldots)_{,k} = \partial(\ldots)/\partial\theta^k$, and the notation $(\ldots)|_\alpha$ will be used for covariant differentiation on M. The displacement field mapping each point of M into a point on its deformed configuration $\bar{M}$ is denoted by $\underset{\sim}{u} = u^\alpha \underset{\sim}{a}_\alpha + w\underset{\sim}{n}$. With $\bar{M}$ we associate position vectors $\bar{\underset{\sim}{r}}(\theta^\alpha) = \bar{x}^k(\theta^\alpha)\underset{\sim}{i}_k = \underset{\sim}{r}(\theta^\alpha) + \underset{\sim}{u}(\theta^\alpha)$, covariant surface base vectors $\bar{\underset{\sim}{a}}_\alpha = \bar{\underset{\sim}{r}}_{,\alpha}$, unit normal vectors $\bar{\underset{\sim}{n}} = (1/2)\,\bar{\varepsilon}^{\alpha\beta}\bar{\underset{\sim}{a}}_\alpha \times \bar{\underset{\sim}{a}}_\beta$, covariant components of the metric tensor $\bar{a}_{\alpha\beta} = \bar{\underset{\sim}{a}}_\alpha \cdot \bar{\underset{\sim}{a}}_\beta$ (with $\det(\bar{a}_{\alpha\beta}) = \bar{a}$) and of the curvature tensor $\bar{b}_{\alpha\beta} = -\bar{\underset{\sim}{a}}_\alpha \cdot \bar{\underset{\sim}{n}}_{,\beta}$. This yields the well known relations

$$\bar{\underset{\sim}{a}}_\alpha = \ell^{\lambda}_{\cdot\alpha}\underset{\sim}{a}_\lambda + \varphi_\alpha \underset{\sim}{n} \quad , \quad \bar{\underset{\sim}{n}} = n^\alpha \underset{\sim}{a}_\alpha + n\,\underset{\sim}{n} \tag{2.1}$$

where we have introduced the notations

$$m_\lambda = -(1+\theta^\mu_\mu)\varphi_\lambda + \varphi^\mu \varphi_{\mu\lambda} \quad , \quad n_\lambda = J\,m_\lambda \quad , \quad J = \sqrt{\frac{a}{\bar{a}}} \quad , \tag{2.2}$$

$$m = 1 + \theta^\lambda_\lambda + \frac{1}{2}(\theta^\lambda_\lambda\theta^\mu_\mu - \theta^\lambda_\mu\theta^\mu_\lambda) + \varphi^2 \quad , \quad n = J\,m \quad , \tag{2.3}$$

$$\ell_{\alpha\beta} = a_{\alpha\beta} + \theta_{\alpha\beta} - \omega_{\alpha\beta} \quad , \quad \varphi_{\alpha\beta} = u_{\alpha|\beta} - b_{\alpha\beta}w \quad , \tag{2.4}$$

$$\theta_{\alpha\beta} = \frac{1}{2}(u_{\alpha|\beta} + u_{\beta|\alpha}) - b_{\alpha\beta}w \quad , \quad \omega_{\alpha\beta} = \frac{1}{2}(u_{\beta|\alpha} - u_{\alpha|\beta}) \quad , \tag{2.5}$$

$$\varphi_\alpha = w_{,\alpha} + b^\beta_\alpha u_\beta \quad , \quad \varphi = \frac{1}{2}\varepsilon^{\alpha\beta}u_{\beta|\alpha} \quad . \tag{2.6}$$

By θ^3, $-\frac{h}{2} < \theta^3 < \frac{h}{2}$, we denote the coordinate normal to M, h being the shell thickness. Let $\underset{\sim}{p} = \underset{\sim}{r}(\theta^\alpha) + \theta^3\underset{\sim}{n}$ and $\bar{\underset{\sim}{p}}$ be the position vectors of a point in the undeformed and deformed shell space, respectively. In finite strain shell theory the change of shell thickness is a major effect which has to be taken into account. Therefore, we shall assume that normal fibres to M remain straight and normal also to $\bar{M}$, but can change their length. Due to bending this change of length is generally asymmetric about M and leads to a shift of the original midsurface, so that $\bar{M}$ is no longer the geometrical midsurface of the deformed shell. This asymmetry requires at least a quadratic representation of the position vector $\bar{\underset{\sim}{p}}$ given here in the form used by CHERNYKH [1,2]

$$\bar{\underset{\sim}{p}} = \bar{\underset{\sim}{r}} + (\theta^3 + \tfrac{1}{2}(\theta^3)^2\, k)\, \lambda \bar{\underset{\sim}{n}} \quad . \tag{2.7}$$

The theory based on the representation (2.7) will be called first-order transverse normal strain theory, since it is the simplest modification of the classical Kirchhoff-Love theory which is able to describe the problem under consideration. With (2.7) we obtain the deformed shell thickness as $\bar{h} = \lambda h$ and the shift of the shell mid-surface as $h^2\lambda k/8$. For $k=0$ this shift vanishes and (2.7) describes a symmetric change of the shell thickness, as it is the case for pure finite membrane strains. For $k=0$ and $\lambda=1$ (2.7) yields the classical Kirchhoff-Love assumption pertinent to small strain first-approximation shell theory.

Let $\underset{\sim}{g}_i = \underset{\sim}{p}_{,i}$ and $\bar{\underset{\sim}{g}}_i = \bar{\underset{\sim}{p}}_{,i}$ denote the covariant base vectors in the undeformed and deformed shell space, respectively, and let $g_{ij} = \underset{\sim}{g}_i \cdot \underset{\sim}{g}_j$ (with det $(g_{ij}) = g$) and $\bar{g}_{ij} = \bar{\underset{\sim}{g}}_i \cdot \bar{\underset{\sim}{g}}_j$ (with det $(\bar{g}_{ij}) = \bar{g}$) be the associated covariant metric tensor components. Under the refined kinematical hypothesis (2.7) we have

$$\bar{\underset{\sim}{g}}_\alpha = \bar{\mu}^\beta_\alpha\, \bar{\underset{\sim}{a}}_\beta + \bar{\mu}^3_\alpha\, \bar{\underset{\sim}{n}} \quad , \quad \bar{\underset{\sim}{g}}_3 = \bar{\mu}^3_3\, \bar{\underset{\sim}{n}} \quad ; \tag{2.8}$$

$$\bar{g}_{\alpha\beta} = \bar{\mu}^\lambda_\alpha\, \bar{\mu}^\varkappa_\beta\, \bar{a}_{\lambda\varkappa} + \bar{\mu}^3_\alpha\, \bar{\mu}^3_\beta \quad , \quad \bar{g}_{\alpha 3} = \bar{\mu}^3_\alpha\, \bar{\mu}^3_3 \quad , \quad \bar{g}_{33} = (\bar{\mu}^3_3)^2 \quad ; \tag{2.9}$$

$$\bar{\mu}^\beta_\alpha = \delta^\beta_\alpha - (\theta^3 + \tfrac{1}{2}(\theta^3)^2\, k)\, \lambda \bar{b}^\beta_\alpha \ , \quad \bar{\mu}^3_\alpha = \theta^3 \lambda_{,\alpha} + \tfrac{1}{2}(\theta^3)^2\, (\lambda k)_{,\alpha} \ , \quad \bar{\mu}^3_3 = \lambda(1 + \theta^3 k) \ ; \tag{2.10}$$

$$\frac{\bar{g}}{g} = \frac{\bar{a}}{a}\, \lambda^2\, (\, 1 + \theta^3\, 2(k - \lambda \bar{b}^\alpha_\alpha + b^\alpha_\alpha) + \ldots\ldots)\ . \tag{2.11}$$

This yields the following representation of the Green strain tensor across the shell thickness :

$$E_{\alpha\beta} = \sum_{n=0}^{4} (\theta^3)^n \overset{(n)}{e}_{\alpha\beta} \quad , \quad E_{\alpha 3} = \sum_{n=0}^{3} (\theta^3)^n \overset{(n)}{e}_{\alpha 3} \quad , \quad E_{33} = \sum_{n=0}^{2} (\theta^3)^n \overset{(n)}{e}_{33} \quad ; \tag{2.12}$$

$$\overset{(o)}{e}_{\alpha\beta} = \tfrac{1}{2}\,(\, \bar{a}_{\alpha\beta} - a_{\alpha\beta}\,) \quad , \qquad \overset{(1)}{e}_{\alpha\beta} = -\,(\, \lambda \bar{b}_{\alpha\beta} - b_{\alpha\beta}\,) \quad ,$$

$$\overset{(2)}{e}_{\alpha\beta} = \tfrac{1}{2}\,(\, \lambda^2 \bar{b}^\lambda_\alpha\, \bar{b}_{\lambda\beta} - \lambda k\, \bar{b}_{\alpha\beta} + \lambda_{,\alpha}\, \lambda_{,\beta} - b^\lambda_\alpha\, b_{\lambda\beta}\,) \quad ,$$

$$\overset{(3)}{e}_{\alpha\beta} = \tfrac{1}{2}\,(\, \lambda^2\, k\, \bar{b}^\lambda_\alpha\, \bar{b}_{\lambda\beta} + \tfrac{1}{2}\, \lambda_{,\alpha}\, (\lambda k)_{,\beta} + \tfrac{1}{2}\, \lambda_{,\beta}\, (\lambda k)_{,\alpha}\,) \quad ,$$

$$\overset{(4)}{e}_{\alpha\beta} = \tfrac{1}{8}\,(\, \lambda^2 k^2\, \bar{b}^\lambda_\alpha\, \bar{b}_{\lambda\beta} + (\lambda k)_{,\alpha}\, (\lambda k)_{,\beta} \quad ; \tag{2.13}$$

$$\overset{(o)}{e}_{\alpha 3} = 0 \quad , \quad \overset{(1)}{e}_{\alpha 3} = \tfrac{1}{2}\, \lambda \lambda_{,\alpha} \quad , \quad \overset{(2)}{e}_{\alpha 3} = \tfrac{1}{4}\, (3\lambda\lambda_{,\alpha} k + \lambda^2 k_{,\alpha}) \quad , \quad \overset{(3)}{e}_{\alpha 3} = \tfrac{1}{8}\, (\lambda^2 k^2)_{,\alpha} \quad ; \tag{2.14}$$

$$\overset{(o)}{e}_{33} = \tfrac{1}{2}\,(\, \lambda^2 - 1\,) \quad , \quad \overset{(1)}{e}_{33} = \lambda^2 k \quad , \quad \overset{(2)}{e}_{33} = \tfrac{1}{2}\, \lambda^2 k^2 \quad . \tag{2.15}$$

In CHERNYKH's theory [1,2] further simplifications are introduced by truncating all series expressions for the base vectors (2.8) and metric tensor components (2.9) (and subsequently for the strains, strain invariants, strain energy function, and stresses) after the respective term which is linearly dependent on the normal coordinate θ^3, thus retaining only the expressions

$$\bar{g}_{\alpha\beta} = \bar{a}_{\alpha\beta} - \theta^3\, 2\lambda\bar{b}_{\alpha\beta} \quad , \quad \bar{g}_{\alpha 3} = \theta^3 \lambda\lambda,_{\alpha} \quad , \quad \bar{g}_{33} = \lambda^2\,(1 + \theta^3\, 2k) \quad , \tag{2.16}$$

and, in consequence, only the Green strain tensor components $(2.13)_{1,2}$, $(2.14)_{1,2}$, and $(2.15)_{1,2}$. In contrast to $(2.13)_2$, however, in [1] (eq.(2.3)) a modified bending strain tensor $\varkappa^{\beta}_{\alpha} = -\,(\,\lambda\bar{b}^{\beta}_{\alpha} - b^{\beta}_{\alpha}\,)$ was introduced, while in [2] (eq.(2.7)) another modified bending strain tensor $\varkappa_{\alpha\beta} = -\lambda\bar{b}_{\alpha\beta} + (1/2)\,(\bar{a}_{\alpha\lambda}b^{\lambda}_{\beta} + \bar{a}_{\beta\lambda}b^{\lambda}_{\alpha})$ was preferred. In terms of these strain measures $(2.16)_1$ is further simplified in [1] and [2] by using in both papers the reduced relation $-\lambda\bar{b}_{\alpha\beta} = \varkappa_{\alpha\beta}$ yielding $\bar{g}_{\alpha\beta} = \bar{a}_{\alpha\beta} + \theta^3\, 2\bar{a}_{\lambda\beta}\varkappa^{\lambda}_{\alpha}$ in [1] and $\bar{g}_{\alpha\beta} = \bar{a}_{\alpha\beta} + \theta^3\, 2\varkappa_{\alpha\beta}$ in [2] , respectively. Observing further that with the simplified relations (2.16) the strain invariants $I_1 = g^{ij}\bar{g}_{ij}$, $I_2 = I_3 g_{ij}\bar{g}^{ij}$, $I_3 = \bar{g}/g$ and, in consequence, also the associated approximate strain energy function do not depend on the transverse shear strains, pertinent constitutive equations for compressible and incompressible material behaviour as well as the 2-D virtual work principle and consistent static equations are given in [1,2]. The latter theory is an intrinsic theory, not referring to the displacements at all. A related displacement form of shell equations is given in [14], where we adopted some of the basic assumptions of [1,2] , omitting, however, certain simplifications used in [1,2] concerning the geometrical relations of both the undeformed and deformed shell space.

In the following developments we shall restrict our considerations to the case of incompressible material behaviour. Then the incompressibility $I_3 = \bar{g}/g \equiv 1$, enforced throughout the shell thickness, yields with the aid of (2.11)

$$\lambda = \sqrt{\frac{a}{\bar{a}}} \quad , \quad k = \sqrt{\frac{a}{\bar{a}}}\; \bar{a}^{\alpha\beta}\bar{b}_{\alpha\beta} - b^{\alpha}_{\alpha} \quad . \tag{2.17}$$

These relations, along with the equations

$$\frac{\bar{a}}{a} = 1 + 2\,(\,\overset{(o)}{e}{}^{\lambda}_{\lambda} + \overset{(o)}{e}{}^{\lambda}_{\lambda}\overset{(o)}{e}{}^{\mu}_{\mu} - \overset{(o)}{e}{}^{\lambda}_{\mu}\overset{(o)}{e}{}^{\mu}_{\lambda}) \quad ,$$

$$\bar{a}^{\alpha\beta} = \frac{a}{\bar{a}}\,\{\, a^{\alpha\beta}(1 + 2\,\overset{(o)}{e}{}^{\lambda}_{\lambda}) - 2\,\overset{(o)}{e}{}^{\alpha\beta}\,\} \tag{2.18}$$

can be used to eliminate λ and k from the strain-displacement relations (2.13) - (2.15). The displacemental form of these strain-displacement relations (and, in consequence, of associated variationally consistent static equations) is of course very complicated, since they are full nonlinear expressions and up to now no assumptions as to the magnitude of strains or rotations have been used in order to simplify them. On the other hand, for most applications it is hardly necessary to use a theory which admits the occurence of arbitrarily large strains and rotations. As to the strains it is obvious in which way a restriction imposed on their order of magnitude can yield a simplification of the strain-displacement relations. Since the strains appear explicitly in (2.13) - (2.15) (e.g. by (2.17) - (2.18)) it is easily possible to omit all such product terms whose order of magnitude - under the respective constraints imposed on the strains - is small, if compared with the leading terms. It is more difficult, however, to identify those terms in the strain-displacement relations which are negligibly small under the assumption of restricted rotations. To achieve this a detailed investigation of the shell deformation by means of the polar decomposition theorem is necessary.

3. Polar Decomposition of Shell Deformation

In this chapter we shall investigate the finite deformation of a shell based on the refined kinematical model (2.7) by using the polar decomposition theorem. Our developments parallel those given by PIETRASZKIEWICZ [9 - 11] for small strain first-order transverse shear approximation theory and Kirchhoff-Love type theory.

The deformation gradient tensor $\mathbf{G}$ at the shell middle surface is defined by

$$\bar{\mathbf{a}}_i = \mathbf{G}\cdot\mathbf{a}_i \quad ; \quad \mathbf{G} = \bar{\mathbf{a}}_i\,\mathbf{a}^i \quad , \tag{3.1}$$

where for the shell theory based on the kinematical hypothis (2.7) $\bar{\mathbf{a}}_3 = \lambda\bar{\mathbf{n}}$ (see $(2.8)_2$ and $(2.1o)_3$). Hence, for an incompressible material $\mathbf{G}$ can be expressed in terms of displacements by using (2.1) along with $(2.17)_1$. The shell deformation gradient tensor (3.1) contains complete information as to the deformation of the neighbourhood of a material point of the midsurface. According to the polar decomposition theorem we shall decompose the local shell deformation into a pure stretch along the principal directions of strain transforming the base vectors $(\mathbf{a}_\alpha,\mathbf{n})$ into an intermediate basis $(\check{\mathbf{a}}_\alpha,\check{\mathbf{a}}_3)$ and a subsequent rigid body rotation transforming the intermediate basis into the final basis $(\bar{\mathbf{a}}_\alpha,\bar{\mathbf{a}}_3)$ of the deformed configuration. For the intermediate base vector triad we find with the aid of $(2.9)_{2,3}$ and $(2.1o)_{2,3}$ for the shell theory under consideration

$$\check{\mathbf{a}}_3 = \lambda\mathbf{n} \quad , \quad \check{\mathbf{a}}^3 = \frac{1}{\lambda}\,\mathbf{n} \quad , \quad \check{\mathbf{a}}^i = \bar{a}^{ij}\,\check{\mathbf{a}}_j \quad , \tag{3.2}$$

$$\check{a}_{\alpha\beta} = \bar{a}_{\alpha\beta} \quad , \quad \check{a}_{\alpha 3} = \bar{a}_{\alpha 3} = 0 \quad , \quad \check{a}_{33} = \bar{a}_{33} = \lambda^2 \quad . \tag{3.3}$$

By means of the polar decomposition theorem the deformation gradient tensor $\mathbf{G}$ is decomposed as follows

$$\mathbf{G} = \mathbf{R}\cdot\mathbf{U} \quad , \tag{3.4}$$

where $\mathbf{U}$ is the (symmetric) right stretch tensor and $\mathbf{R}$ is the (orthogonal) finite rotation tensor

$$\mathbf{U} = \check{\mathbf{a}}_\alpha \mathbf{a}^\alpha + \check{\mathbf{a}}_3 \mathbf{n} \quad , \qquad \mathbf{R} = \bar{\mathbf{a}}_\alpha \check{\mathbf{a}}^\alpha + \bar{\mathbf{a}}_3 \check{\mathbf{a}}^3 \quad , \tag{3.5}$$

so that

$$\mathbf{U} \cdot \mathbf{a}_\alpha = \check{\mathbf{a}}_\alpha \quad , \qquad \mathbf{U} \cdot \mathbf{n} = \check{\mathbf{a}}_3 = \lambda \mathbf{n} \quad , \tag{3.6}$$

$$\mathbf{R} \cdot \check{\mathbf{a}}_\alpha = \bar{\mathbf{a}}_\alpha \quad , \qquad \mathbf{R} \cdot \check{\mathbf{a}}_3 = \bar{\mathbf{a}}_3 = \lambda \bar{\mathbf{n}} \quad , \tag{3.7}$$

where by virtue of $(3.2)_3$

$$\mathbf{R}^T = \mathbf{R}^{-1} \qquad\qquad \mathbf{R} = \mathbf{R}^{-1T} \quad , \tag{3.8}$$

i.e. $\mathbf{R} \cdot \mathbf{R}^T = \mathbf{1}$ and $\mathbf{R}^{-1} \cdot \mathbf{R}^{-1T} = \mathbf{1}$, $\mathbf{1}$ being the identity tensor. From (3.5) and (3.8) we obtain the following relations for the contravariant bașe vectors :

$$\mathbf{U}^{-1T} \cdot \mathbf{a}^\alpha = \check{\mathbf{a}}^\alpha \quad , \qquad \mathbf{U}^{-1T} \cdot \mathbf{n} = \check{\mathbf{a}}^3 = \frac{1}{\lambda} \mathbf{n} \quad ; \tag{3.9}$$

$$\mathbf{R} \cdot \check{\mathbf{a}}^\alpha = \bar{\mathbf{a}}^\alpha \quad , \qquad \mathbf{R} \cdot \check{\mathbf{a}}^3 = \bar{\mathbf{a}}^3 = \frac{1}{\lambda} \bar{\mathbf{n}} \quad . \tag{3.1o}$$

By means of $(2.13)_1$, $(2.15)_1$ and $(3.1)_2$ the Green strain tensor of the shell middle surface can be expressed as

$$\mathbf{e} = \overset{(o)}{e}_{\alpha\beta} \mathbf{a}^\alpha \mathbf{a}^\beta + \overset{(o)}{e}_{33} \mathbf{n}\mathbf{n} = \frac{1}{2} (\mathbf{G}^T \cdot \mathbf{G} - \mathbf{1}) = \frac{1}{2} (\mathbf{U} \cdot \mathbf{U} - \mathbf{1}) \quad . \tag{3.11}$$

For the following developments it proves to be convenient to introduce also the engineering strain tensor

$$\check{\mathbf{e}} = \mathbf{U} - \mathbf{1} = (\mathbf{1} + 2\mathbf{e})^{1/2} - \mathbf{1} \quad , \tag{3.12}$$

where by (3.2) and $(3.6)_1$ for the shell theory under consideration

$$\check{\mathbf{e}} = \check{e}_{\alpha\beta} \mathbf{a}^\alpha \mathbf{a}^\beta + \check{e}_{33} \mathbf{n}\mathbf{n} = (\check{\mathbf{a}}_\alpha - \mathbf{a}_\alpha) \mathbf{a}^\alpha + (\lambda - 1) \mathbf{n}\mathbf{n} \quad . \tag{3.13}$$

From $(3.6)_1$, (3.12) and (3.13) it follows that

$$\check{\mathbf{a}}_\alpha = (\delta^\beta_\alpha + \check{e}^\beta_\alpha) \mathbf{a}_\beta \quad , \tag{3.14}$$

which, by means of (3.3), yields

$$\bar{a}_{\alpha\beta} = a_{\alpha\beta} + 2\overset{\vee}{e}_{\alpha\beta} + \overset{\vee}{e}{}^{\lambda}_{\alpha}\overset{\vee}{e}_{\lambda\beta} \quad . \tag{3.15}$$

Now, from (3.11), (3.13) and (3.15), we obtain the following relations between the Green and engineering strain tensor components :

$$\overset{(o)}{e}_{\alpha\beta} = \overset{\vee}{e}_{\alpha\beta} + \frac{1}{2}\overset{\vee}{e}{}^{\lambda}_{\alpha}\overset{\vee}{e}_{\lambda\beta} \quad , \qquad \overset{(o)}{e}_{33} = \overset{\vee}{e}_{33} + \frac{1}{2}\overset{\vee}{e}{}^{2}_{33} \quad . \tag{3.16}$$

Starting from the relation $\bar{a}^{\lambda\mu} = \bar{\varepsilon}^{\alpha\lambda}\bar{\varepsilon}^{\beta\mu}\bar{a}_{\alpha\beta}$ and making use of (3.14), (3.15) and $(3.2)_3$ we can derive successively the following set of equations given in [9,11]

$$\begin{aligned}
\bar{a}^{\alpha\beta} &= \frac{a}{\bar{a}}\,\varepsilon^{\alpha\lambda}\varepsilon^{\beta\mu}\,(\,\delta^{\varkappa}_{\lambda} + \overset{\vee}{e}{}^{\varkappa}_{\lambda}\,)(\,\delta^{\eta}_{\mu} + \overset{\vee}{e}{}^{\eta}_{\mu}\,)\,a_{\varkappa\eta} \quad , \\
\overset{\vee}{\mathbf{a}}{}^{\alpha} &= \sqrt{\frac{a}{\bar{a}}}\,\varepsilon^{\alpha\beta}\varepsilon_{\lambda\mu}\,(\,\delta^{\mu}_{\beta} + \overset{\vee}{e}{}^{\mu}_{\beta}\,)\,\mathbf{a}^{\lambda} \quad , \\
\sqrt{\frac{\bar{a}}{a}} &= \frac{1}{2}\,\varepsilon^{\alpha\beta}\varepsilon_{\lambda\mu}\,(\,\delta^{\lambda}_{\alpha} + \overset{\vee}{e}{}^{\lambda}_{\alpha}\,)(\,\delta^{\mu}_{\beta} + \overset{\vee}{e}{}^{\mu}_{\beta}\,) \quad , \\
\bar{\varepsilon}_{\alpha\beta} &= (\,\delta^{\lambda}_{\alpha} + \overset{\vee}{e}{}^{\lambda}_{\alpha}\,)(\,\delta^{\mu}_{\beta} + \overset{\vee}{e}{}^{\mu}_{\beta}\,)\,\varepsilon_{\lambda\mu} \quad , \qquad \bar{\varepsilon}^{\alpha\beta} = \frac{a}{\bar{a}}\,(\,\delta^{\alpha}_{\varkappa} + \overset{\vee}{e}{}^{\alpha}_{\varkappa}\,)(\,\delta^{\beta}_{\gamma} + \overset{\vee}{e}{}^{\beta}_{\gamma}\,)\,\varepsilon^{\varkappa\gamma} \quad .
\end{aligned} \tag{3.17}$$

For the ensuing developments it is convenient to describe the rotational part of the shell deformation by means of a finite rotation vector instead of the finite rotation tensor $(3.5)_2$. In literature various different definitions for finite rotation vectors are commonly used to describe the rigid-body rotation of a vector about an arbitrarily orientated axis of rotation (see e.g. CHERNYKH [3] , KORN and KORN [4] , LUR'E [7] , PARS [8] , PIETRASZKIEWICZ [9-11] , PIETRASZKIEWICZ and BADUR [12], SCHROEDER [15,16] , SHAMINA [17] , SIMMONDS and DANIELSON [18,19]). The respective choice of a suitably defined definition depends on the particular problem under consideration. We point out here that unlike some of the papers mentioned above we do not want to present a shell theory with a rotation vector (or rotation angles) as independent kinematical variables. Our developments are intended for the purpose to create a basis for the derivation of appropriate simplified geometric shell relations for problems in which only restricted rotations occur. For this purpose a finite rotation vector has to be introduced which can be related easily to basic kinematical parameters of shell deformation, in particular to the displacements and displacement gradients of the midsurface. If then constraints are imposed on the order of magnitude of the rotations, it is possible to infer the order of magnitude of midsurface displacements and displacement gradients, respectively, pertinent to the particular case of restricted rotational part of shell deformation. This, in turn, makes it possible to simplify consistently the general strain-displacement relations by omitting those terms which have been proved to be negligibly small for the respective case of restricted rotations. In the framework of small strain shell theory it has been shown in [9-11] that for the purpose described above it is particular convenient to introduce a finite rotation vector in the form

$$\underset{\sim}{\Omega} = \sin\omega \; \underset{\sim}{e}_1 \quad , \tag{3.18}$$

where $\underset{\sim}{e}_1$ is a unit vector in direction of the rotation axis and ω denotes the rotation angle about this axis. In order to make the present paper self-contained we shall record briefly in the following formulae (3.19) - (3.21) some results available in [9-12] concerning the description of a rigid body rotation of a vector by means of the finite rotation vector (3.18).

According to Fig.1 any vector $\underset{\sim}{w}$ can be decomposed into a vector $\underset{\sim}{w}_1$ along the rotation axis and a vector $\underset{\sim}{w}_p$ in a plane orthogonal to $\underset{\sim}{e}_1$ yielding the relations

$$\underset{\sim}{w}_p = \underset{\sim}{w} - \underset{\sim}{w}_1 = \underset{\sim}{w} - \frac{\underset{\sim}{\Omega} \cdot \underset{\sim}{w}}{\sin^2\omega}\,\underset{\sim}{\Omega} = - \frac{\underset{\sim}{\Omega} \times (\underset{\sim}{\Omega} \times \underset{\sim}{w})}{\sin^2\omega} \quad . \tag{3.19}$$

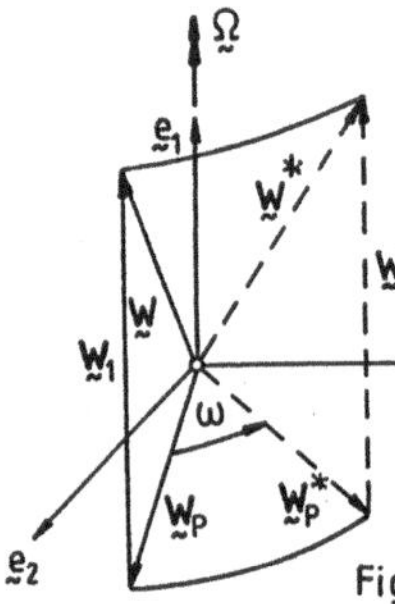

Fig. 1

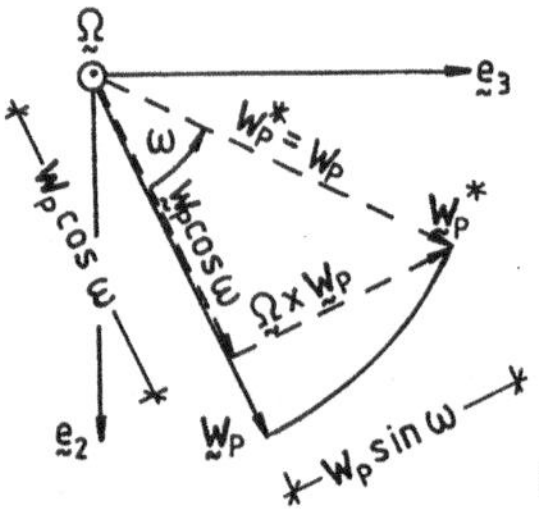

Fig. 2

Let us rotate now $\underset{\sim}{w}$ about $\underset{\sim}{e}_1$ with a rotation angle ω yielding the rotated vector $\underset{\sim}{w}^*$. It follows from Fig. 2 that

$$\underset{\sim}{w}^* = \underset{\sim}{w}_1 + \cos\omega\,\underset{\sim}{w}_p + \underset{\sim}{\Omega} \times \underset{\sim}{w}_p \quad . \tag{3.2o}$$

This yields with the aid of (3.19)

$$\underset{\sim}{w}^* = \underset{\sim}{w} + \underset{\sim}{\Omega} \times \underset{\sim}{w} + \frac{1}{2\cos^2(\omega/2)}\,\underset{\sim}{\Omega} \times (\underset{\sim}{\Omega} \times \underset{\sim}{w}) \quad . \tag{3.21}$$

If (3.21) is applied to describe the rotational part of shell deformation by means of the finite rotation vector $\underset{\sim}{\Omega}$ one obtains

$$\begin{aligned} \bar{\underset{\sim}{a}}_\alpha &= \check{\underset{\sim}{a}}_\alpha + \underset{\sim}{\Omega} \times \check{\underset{\sim}{a}}_\alpha + \frac{1}{2\cos^2(\omega/2)}\,\underset{\sim}{\Omega} \times (\underset{\sim}{\Omega} \times \check{\underset{\sim}{a}}_\alpha) \quad , \\ \bar{\underset{\sim}{a}}_3 &= \check{\underset{\sim}{a}}_3 + \underset{\sim}{\Omega} \times \check{\underset{\sim}{a}}_3 + \frac{1}{2\cos^2(\omega/2)}\,\underset{\sim}{\Omega} \times (\underset{\sim}{\Omega} \times \check{\underset{\sim}{a}}_3) \quad . \end{aligned} \tag{3.22}$$

Starting from (3.22), applying repeatedly standard rules of vector calculus and making use of the relations

$$\frac{1}{2}\bar{\varepsilon}^{\alpha\beta}\,\bar{\underset{\sim}{a}}_\alpha \times \bar{\underset{\sim}{a}}_\beta = \lambda\bar{\underset{\sim}{a}}^3 \quad , \qquad \check{\underset{\sim}{a}}_3 \times \check{\underset{\sim}{a}}_\alpha = \lambda\,\bar{\varepsilon}_{\alpha\beta}\check{\underset{\sim}{a}}^\beta \quad , \tag{3.23}$$

we obtain after some rather involved transformations

$$\bar{\varepsilon}^{\alpha\beta}\,\bar{\underset{\sim}{a}}_\alpha\cdot\check{\underset{\sim}{a}}_\beta = 2\lambda\,\underset{\sim}{\Omega}\cdot\check{\underset{\sim}{a}}^3 \quad , \qquad \bar{\varepsilon}^{\alpha\beta}\,(\,\bar{\underset{\sim}{a}}_3\cdot\check{\underset{\sim}{a}}_\alpha - \bar{\underset{\sim}{a}}_\alpha\cdot\check{\underset{\sim}{a}}_3\,) = 2\lambda\,\underset{\sim}{\Omega}\cdot\check{\underset{\sim}{a}}^\beta \;. \tag{3.24}$$

Thus, for the shell theory under consideration the finite rotation vector of the shell middle surface can be presented in the form

$$\underset{\sim}{\Omega} = \frac{1}{2}\bar{\varepsilon}^{\alpha\beta}\left\{\frac{1}{\lambda}\,(\,\bar{\underset{\sim}{a}}_3\cdot\check{\underset{\sim}{a}}_\alpha - \bar{\underset{\sim}{a}}_\alpha\cdot\check{\underset{\sim}{a}}_3\,)\,\check{\underset{\sim}{a}}_\beta + (\,\bar{\underset{\sim}{a}}_\alpha\cdot\check{\underset{\sim}{a}}_\beta\,)\,\frac{1}{\lambda}\,\check{\underset{\sim}{a}}_3\right\} \;. \tag{3.25}$$

With the aid of (2.1), (3.14) and $(3.17)_{2,4}$ the finite rotation vector can also be given as

$$\underset{\sim}{\Omega} = \frac{1}{2}\varepsilon^{\alpha\beta}\left\{\left(\,n_\alpha - \bar{a}^{\lambda\mu}\,(\,a_{\alpha\lambda} + \check{e}_{\alpha\lambda}\,)\,\varphi_\mu\,\right)\underset{\sim}{a}_\beta + \bar{a}^{\lambda\mu}\,(\,a_{\alpha\lambda} + \check{e}_{\alpha\lambda}\,)\,\ell_{\beta\mu}\,\underset{\sim}{n}\right\}. \tag{3.26}$$

It may be observed that the latter form of the finite rotation vector obtained on the basis of the refined kinematical hypothesis (2.7) is the same as for the Kirchhoff-Love model, as long as no explicit restriction is imposed on the order of magnitude of the strains (compare also [9-11]). In contrast to the limitations imposed by the Kirchhoff-Love hypothesis, however, we shall assume now that finite elastic strains may occur in the shell.

Problems in which finite elastic strains have to be taken into account will be classified according to the order of magnitude of the largest principal strain (denoted by η) as follows :

small strains :	$\eta \ll 1$,	large strains :	$\eta^4 \ll 1$,
moderate strains :	$\eta^2 \ll 1$,	unrestricted strains :	$\eta \geq O(1)$.

In the sense of this classification scheme the range of moderate strains permits the occurence of elongations up to 2o% (yielding engineering and Green strains of 2o% and 22%, respectively) while the range of large strains reaches approximately up to elongations of 4o% (yielding engineering and Green strains of 4o% and 48%, respectively).

Since small strain bending shell problems can be described accurately enough by the Kirchhoff-Love model and unrestricted elastic strain problems are nearly pure membrane problems for most of the relevant cases, the ranges of applicability of finite strain bending shell theory are mainly the cases in which moderate or large strains (in the sense of the above classification scheme) occur.

We shall restrict ourselves here to the case of moderate membrane strains and, in view of the refined kinematical hypothesis (2.7), to thin-walled structures. In this case the finite rotation vector (3.26) may be simplified further by using (2.2), (2.18), and (3.16) and estimating carefully all terms under the assumption $\eta^2 \ll 1$. Then all those terms are omitted whose contribution to (3.26) is negligibly small, if compared with the respective leading terms. This yields the following consistently simplified formula for the finite rotation vector describing the rotational part of the deformation in the nonlinear theory of shells undergoing moderately large elastic membrane strains:

$$\underset{\sim}{\Omega} = \Omega^{\beta} \underset{\sim}{a}_{\beta} + \Omega^{3} \underset{\sim}{n} \quad , \tag{3.27}$$

$$\Omega^{\beta} = \frac{1}{2}\varepsilon^{\beta\alpha}\{2\varphi_{\alpha} + (\varphi_{\alpha}\theta^{\lambda}_{\lambda} - \varphi^{\lambda}(\theta_{\lambda\alpha} - \omega_{\lambda\alpha}))(1 - \overset{(o)}{e}{}^{\varkappa}_{\varkappa}) - \varphi_{\alpha}\overset{(o)}{e}{}^{\lambda}_{\lambda} - \varphi^{\lambda}\overset{(o)}{e}_{\lambda\alpha}\} + O(\varphi_{\alpha}\eta^{2}) \quad ,$$

$$\Omega^{3} = \varphi + \frac{1}{2}\varepsilon^{\beta\alpha}\{\overset{(o)}{e}{}^{\mu}_{\alpha}(\theta_{\beta\mu} - \omega_{\beta\mu}) + \frac{5}{2}\overset{(o)}{e}{}^{\lambda}_{\alpha}\overset{(o)}{e}{}^{\varkappa}_{\lambda}\theta_{\varkappa\beta} - 4\overset{(o)}{e}{}^{\varkappa}_{\varkappa}\overset{(o)}{e}{}^{\lambda}_{\alpha}\theta_{\lambda\beta}\} \tag{3.28}$$

$$+ O(\varphi\eta^{2}, \theta_{\alpha\beta}\eta^{3}, \eta^{4}) \quad . \tag{3.29}$$

For the case of small strains eqs.(3.27) - (3.29) yield exactly the finite rotation vector of small strain Kirchhoff-Love type shell theory given by PIETRASZKIEWICZ [9-11].

4. Moderate Strains Accompanied by Restricted Rotations

Now, with the help of the finite rotation vector (3.27) we may reach our goal to determine the order of magnitude of displacements and displacement gradients for cases in which moderately large elastic strains are accompanied by restricted rotations. Following [9-11] we shall classify the rotations according to the order of magnitude of the largest rotation angle ω according to (3.18) as follows:

small rotations :	$\omega \ll 1$,	large rotations :	$\omega^4 \ll 1$,
moderate rotations :	$\omega^2 \ll 1$,	unrestricted rotations :	$\omega \geq O(1)$.

The range of moderate rotations permits the occurence of rotation angles up to 10° approximately, while the limits of applicability of large rotation shell theory in the sense of the above classification scheme is reached when the rotations of the shell material elements are 25° at most. Note in this context also that for the aforementioned values of rotation angles by virtue of (3.18) we have $O(|\underset{\sim}{\Omega}|) = O(\sin\omega) = O(\omega)$.

If the above restrictions are imposed on the rotational part of deformation represented by the finite rotation vector (3.27) and additionally the exact relation

$$\overset{(o)}{e}_{\alpha\beta} = \theta_{\alpha\beta} + \frac{1}{2}\varphi_{\alpha}\varphi_{\beta} + \frac{1}{2}a_{\alpha\beta}\varphi^{2} - \frac{1}{2}(\theta^{\lambda}_{\alpha}\omega_{\lambda\beta} + \theta^{\lambda}_{\beta}\omega_{\lambda\alpha}) + \frac{1}{2}\theta^{\lambda}_{\alpha}\theta_{\lambda\beta} \tag{4.1}$$

is used to estimate $\theta_{\alpha\beta}$ intermediately in terms of $\overset{(o)}{e}_{\alpha\beta}$, φ_α and φ, eq.(3.27) yields after rather involved order of magnitude considerations the following set of estimates pertinent to moderate elastic strain theory

Rotations of the normal	Rotations about the normal	φ_α	φ	$\theta_{\alpha\beta}$
small	small	θ^2	θ^2	θ
moderate	small	θ	θ^2	θ
moderate	moderate	θ	θ	θ
large	small	$\sqrt{\theta}$	θ^2	θ
large	moderate	$\sqrt{\theta}$	θ	θ
large	large	$\sqrt{\theta}$	$\sqrt{\theta}$	θ
unrestricted	small	1	θ	1
unrestricted	moderate	1	θ	1
unrestricted	large	1	$\sqrt{\theta}$	1
unrestricted	unrestricted	1	1	1

where $\theta^2 \ll 1$. Now, with the help of these estimates it is possible to derive from (2.13)-(2.15) consistently simplified strain-displacement relations for each of the cases indicated above.

As an example let us derive appropriate shell strain-displacement relations for the case of moderate strains and moderate rotations in the framework of a theory which may be regarded as a counterpart of CHERNYKH's theory [1,2] in displacement form . As a consequence of his simplifications described in chapter 2, the only strain measures appearing in his internal virtual work expression (eq.(6.3) of [2]) are $\overset{(o)}{e}_{\alpha\beta}$, a modification of our $\overset{(1)}{e}_{\alpha\beta}$, $\overset{(o)}{e}_{33}$, and $\overset{(1)}{e}_{33}$. Conjugate stress measures and associated constitutive equations are given in eqs.(6.6)-(6.7) of [2]. With $(2.13)_{1,2}$, $(2.15)_{1,2}$, (2.17) , and (2.1) the pertinent full nonlinear strain-displacement relations of our approach read

$$\overset{(o)}{e}_{\alpha\beta} = \frac{1}{2}\,(\ell^{\lambda}_{.\alpha}\ell_{\lambda\beta} + \varphi_{\alpha}\varphi_{\beta} - a_{\alpha\beta}) \;,$$

$$\overset{(1)}{e}_{\alpha\beta} = (\,(Jn^{\mu})|_{\beta} - b^{\mu}_{\beta}Jn\,)\ell_{\mu\alpha} + (\,b_{\mu\beta}Jn^{\mu} + (Jn)_{,\beta}\,)\varphi_{\alpha} + b_{\alpha\beta} \;,$$

$$\overset{(o)}{e}_{33} = \frac{1}{2}\,(\,J^{2} - 1\,) \;, \qquad \overset{(1)}{e}_{33} = J^{3}\bar{a}^{\alpha\beta}\bar{b}_{\alpha\beta} - b^{\alpha}_{\alpha} \;, \tag{4.2}$$

where (2.2) - (2.6) along with (2.17) - (2.18) have to be used in order to express them explicitly in terms of displacements. These equations can be considerably simplified for problems in which only restricted strains and rotations occur, if each term is estimated using the above order of magnitude considerations. For the case of moderate strains and moderate rotations we obtain after a rather lengthy estimation procedure:

$$\overset{(o)}{e}_{\alpha\beta} = \theta_{\alpha\beta} + \frac{1}{2}\theta^{\lambda}_{\alpha}\theta_{\lambda\beta} + \frac{1}{2}\varphi_{\alpha}\varphi_{\beta} + \underline{\frac{1}{2}a_{\alpha\beta}\varphi^{2}} - \underline{\frac{1}{2}(\theta^{\lambda}_{\alpha}\omega_{\lambda\beta} + \theta^{\lambda}_{\beta}\omega_{\lambda\alpha})} \;,$$

$$\begin{aligned}\overset{(1)}{e}_{\alpha\beta} = -\frac{1}{2}\,\{ & ((1+\theta^{\lambda}_{\lambda})\varphi_{\alpha} - \varphi^{\lambda}(\theta_{\lambda\alpha} - \underline{\omega_{\lambda\alpha}}))|_{\beta} + ((1+\theta^{\lambda}_{\lambda})\varphi_{\beta} - \varphi^{\lambda}(\theta_{\lambda\beta} - \underline{\omega_{\lambda\beta}}))|_{\alpha} \\ & - \varphi_{\alpha}\theta^{\lambda}_{\lambda}|_{\beta} - \varphi_{\beta}\theta^{\lambda}_{\lambda}|_{\alpha} + \varphi^{\lambda}|_{\beta}(\theta_{\lambda\alpha} - \underline{\omega_{\lambda\alpha}}) + \varphi^{\lambda}|_{\alpha}(\theta_{\lambda\beta} - \underline{\omega_{\lambda\beta}}) \\ & + b^{\lambda}_{\beta}(\theta_{\lambda\alpha} - \underline{\omega_{\lambda\alpha}}) + b^{\lambda}_{\alpha}(\theta_{\lambda\beta} - \underline{\omega_{\lambda\beta}}) - 2b_{\alpha\beta}\theta^{\lambda}_{\lambda}\}\;\{1+O(\theta^{2})\} \;,\end{aligned}$$

$$\overset{(o)}{e}_{33} = \{-\theta^{\lambda}_{\lambda} + \theta^{\lambda}_{\lambda}\theta^{\mu}_{\mu} + \frac{1}{2}\theta^{\lambda}_{\mu}\theta^{\mu}_{\lambda} - \frac{1}{2}\varphi_{\lambda}\varphi^{\lambda} - \underline{\varphi^{2}}\}\;\{1+O(\theta^{2})\} \;,$$

$$\overset{(1)}{e}_{33} = \{\varphi^{\lambda}|_{\lambda} - \theta^{\lambda}_{\mu}(b^{\mu}_{\lambda} + 2\varphi^{\mu}|_{\lambda}) - 3\theta^{\lambda}_{\lambda}(b^{\mu}_{\mu} + \varphi^{\mu}|_{\mu}) - \varphi_{\lambda}(\theta^{\lambda\mu} - \underline{\omega^{\lambda\mu}})|_{\mu}\}\;\{1+O(\theta^{2})\} \;. \tag{4.3}$$

The terms marked in (4.3) by a solid line may be dropped additionally, if only the rotations of the normal are moderate while the in-surface rotations remain small.

In the same way, by using the order estimates given in this chapter it is possible to derive consistently simplified geometric shell equations for problems in which moderate strains occur accompanied by large or unrestricted rotations in the sense of the above classification scheme. Such extended relations (and also analogous ones for the ranges of large and unrestricted strains) will be given along with associated variationally derived static equations in a companion paper.

Acknowledgement

The author wishes to thank the Deutsche Forschungsgemeinschaft and the Polish Academy of Sciences for partial support of this research.

References

1. CHERNYKH K.F., Mechanics of Solids 15 (198o), No.2, 118-127, Transl. of Mekhanika Tverdogo Tela 15 (198o), No.2, 148-159.
2. CHERNYKH K.F., Advances in Mechanics 6 (1983), 111-147 (in Russian).
3. CHERNYKH K.F., In: Mechanics of Deformable Continuum, 9-72 (in Russian), Kuybyshev State University 1976.
4. KORN G.A., KORN T.M., Mathematical Handbook for Scientists and Engineers, 2nd ed., Mc Graw-Hill, New York 1968.
5. LIBAI A., SIMMONDS J.G., Int.J.Non-Linear Mechanics 16 (1981), 91-1o3.
6. LIBAI A., SIMMONDS J.G., Int.J.Non-Linear Mechanics 18 (1983), 181-197.
7. LUR'E A.I., Analytical Mechanics (in Russian), Nauka, Moscow 1961.
8. PARS L.A., A Treatise on Analytical Dynamics, Heinemann, London 1965.
9. PIETRASZKIEWICZ W.,Finite Rotations and Langrangean Description in the Non-Linear Theory of Shells, Polish Scientific Publishers, Warszawa-Poznań 1979.
1o. PIETRASZKIEWICZ W.,In: Theory of Shells, eds. W.T. Koiter; G.K. Mikhailov, 445-471, North-Holland Publ. Co., Amsterdam-New York-Oxford 198o.
11. PIETRASZKIEWICZ W.,Mitteilungen aus dem Institut für Mechanik, Nr. 1o, Ruhr-Universität Bochum 1977.
12. PIETRASZKIEWICZ W.,BADUR J., Int.J.Engrg.Sci. 21 (1983), 1o97-1115.
13. SCHMIDT R., Proc. of the Symp. on Advances and Trends in Structures and Dynamics, Washington 1984, eds. A.K. Noor; R.J. Hayduk, 265-275, Pergamon Press, New York, reprinted as Computers and Structures 2o (1985), 265-275.
14. SCHMIDT R., Contributions to the Nonlinear Theory of Thin Elastic Shells, Part II, Lecture XVI IUTAM Int.Congr. of Theoretical and Applied Mechanics, Lyngby, 19.-25.8.1984, to be published.
15. SCHROEDER F.-H., Ingenieur-Archiv 39 (197o), 87-1o3.
16. SCHROEDER F.-H., Cosserat Theory of Shells with Large Rotations and Displacements, Lecture EUROMECH-Colloquium Nr. 165 "Flexible Shells, Theory and Applications", München, May 17-2o, 1983.
17. SHAMINA V.A., Mechanics of Solids 9 (1974), No.1, 9-16, Transl. of Mekhanika Tverd. Tela 9 (1974), No.1, 14-22.
18. SIMMONDS J.G., DANIELSON D.A., Proc.Kon.Ned.Ak.Wet.Ser.B 73 (197o), 46o-478.
19. SIMMONDS J.G., DANIELSON D.A., J.Appl.Mech. 39 (1972), 1o85-1o9o.
2o. SIMMONDS J.G., Int.J.Solids Structures 21 (1985), 67-77.

NONLINEAR MODELS OF DEFORMED THIN BODIES WITH SEPARATION OF THE FINITE ROTATION FIELD

L.I. SHKUTIN
Computer Center, the Siberian Branch
of the USSR Academy of Sciences
660036 Krasnoyarsk, USSR

In scientific literature there are many works concerning the formulation of the two-dimensional shell and one-dimensional rod deformation models on the basis of directed material surface and curve conception [1-4]. Such formalism is mathematically strong. However this is not satisfactory from the physical point of view because it breaks the natural connections with spatial formulation. As a result the problem of the construction of reological equations and of the reconstruction of the displacement, strain and stress spatial fields has been outside the scope of such formal approach.

The procedure realized in the present work is free from these shortcomings. Its fundamental conception treats a thin body as a three-dimensional continuum with internal kinematic constraints. The rigid-body rotation of coordinate basis is separated from its complete transformation in the process of deformation.

The first section of the paper contains the nontraditional three-dimensional formulation of the nonlinear deformation problem for an arbitrary solid body. A free rotation field is introduced together with the displacement field.

In the second section a two-dimensional model of a deformed shell is formulated. The internal kinematic constraint transforms the shell into the deformed material surface. Two variable vectors (the radius-vector and the normal one) are connected with its every point. The model contains six kinematic degrees of freedom.

In the third section a one-dimensional model of a deformed rod is formulated. The rod is transformed into the deformed material curve. Three variable vectors (the radius-vector and two normal vectors) are connected with its every point. The model contains nine kinematic degrees of freedom.

In the notations the capital indices take the values 1;2;3 and small indices - 1;2. The time dependence is implicit.

1. Three-dimensional model of a deformed body

At the initial moment of time a body occupies a spatial region $\mathcal{A}$ with a boundary surface $\mathcal{B}_V$. The region is parametrized by the material coordinates s^N. $\underline{S} = (s^1; s^2; s^3)$ is an arbitrary point of the body, $\underline{A}(\underline{S})$ is its initial radius-vector, $\underline{A}_{(N)}(\underline{S}) = \partial_N \underline{A}$ is the natural initial coordinate basis.

The body deformation is generated by the mechanical and/or thermal influences. They are the surface temperature fields. The kinematic constraints, surface and volume force fields refer to the mechanical influences. According to d'Alamber's principle, the inertial force field is included in the volume force field. This is denoted by $\underline{P}(\underline{S})$. The surface mechanical influences are distributed as follows: the force field $\underline{P}^V(\underline{S})$ is given on the section $\mathcal{B}_V^+$, the displacement field $\underline{F}_V(\underline{S})$ is given on the section $\mathcal{B}_V^-$ of the boundary surface.

The body deformation transforms the initial radius-vector $\underline{A}$ into the current one $\underline{A} + \underline{U}$, $\underline{U}(\underline{S})$ is the displacement vector field. The initial basis $\underline{A}_{(N)}$ transforms into the current one

$$\underline{A}_{\{N\}} = \partial_N (\underline{A} + \underline{U}) = \underline{A}_{(N)} + \partial_N \underline{U} \tag{1.1}$$

where ∂_N is the partial derivative operator.

The total transformation $\underline{A}_{(N)} \to \underline{A}_{\{N\}}$ decomposes into two consequent transformations $\underline{A}_{(N)} \to \underline{A}_N \to \underline{A}_{\{N\}}$. The first one is a finite rigid-body rotation of the initial basis into the turned basis $\underline{A}_N(\underline{S})$. The second transformation is a finite deformation of the turned basis into the current basis $\underline{A}_{\{N\}}(\underline{S})$. The turned and initial bases have the same metrical tensor.

The rotational transformation is expressed in terms of a finite rotation vector field $\underline{V}(\underline{S})$ by the formula

$$\begin{gathered} \underline{A}_N = \underline{A}_{(N)} + R_1 \underline{V} \times \underline{A}_{(N)} + R_2 \underline{V} \times (\underline{V} \times \underline{A}_{(N)}), \\ R_1 = (V)^{-1} \sin V, \qquad R_2 = (V)^{-2}(1 - \cos V). \end{gathered} \tag{1.2}$$

Vector $\underline{V}$ is the natural rotation vector because its length coincides with the rotation angle $V = (\underline{V} \cdot \underline{V})^{1/2}$. Such condition eliminates the singularities which the rotational transformation acquires when the length of the rotation vector is defined as a trigonometrical function of the rotation angle [3,5,6,10-12].

The introduced rotation field retains space metrics. It produces pure bending of curves and surfaces in the body space. The measure of the pure bending is the bending tensor field

$$\underline{W}_N = R_1 \partial_N \underline{V} + R_2 \underline{V} \times \partial_N \underline{V} + (R_3 \partial_N V)\underline{V}, \quad R_3 = (V)^{-1}(1 - R_1). \tag{1.3}$$

The measure of the body deformation is the strain tensor field

$$\underline{U}_N = \underline{A}_{\{N\}} - \underline{A}_N = \partial_N \underline{U} - R_1 \underline{V} \times \underline{A}_N + R_2 \underline{V} \times (\underline{V} \times \underline{A}_N). \tag{1.4}$$

Its principal values are the elongations of the linear elements. Such definition of the strain tensor is not traditional.

The body deformation generates the stress tensor field $\underline{X}^N(\underline{S})$. It is defined on the current coordinate areas and referred to the initial metrics. The current state of the body is subjected to the global equation of virtual work

$$\int_{\mathcal{B}_v^+} \underline{P}^v \cdot \delta \underline{U}\, dB_v + \int_{\mathcal{A}} \underline{P} \cdot \delta \underline{U}\, dA - \int_{\mathcal{A}} \Delta Z\, dA = 0. \tag{1.5}$$

Besides, the local equation of moments is satisfied

$$\underline{A}_{\{N\}} \times \underline{X}^N = 0. \tag{1.6}$$

With the help of (1.6) the virtual strain energy ΔZ is expressed by the equalities

$$\Delta Z = \underline{X}^N \cdot \delta \partial_N \underline{U} = \underline{X}^N \cdot (\delta \partial_N \underline{U} - \delta_0 \underline{V} \times \underline{A}_{\{N\}}) = \underline{X}^N \cdot \delta_0 \underline{U}_N,$$
$$\Delta Z = X^{NM} \delta U_{NM}, \quad X^{NM} = \underline{X}^N \cdot \underline{A}^M, \quad U_{NM} = \underline{U}_N \cdot \underline{A}_M \tag{1.7}$$

In formulae (1.5), (1.7) δ is the operator of the total variation; δ_0 is the operator of the variation with respect to the turned basis; $\delta_0 \underline{V}$ is the angular velocity vector.

The formulation of the three-dimensional nonlinear model of a deformed body is completed by the reological equations of the continuum. They connect kinematic and dynamic parameters of deformation.

When isothermal deformation processes are studied, the strain energy density Z is given as a function of the strain tensor U_{NM}. As a consequence one has the reological equations

$$Z = E(U_{NM}), \qquad X^{NM} = \partial E / \partial U_{NM}. \tag{1.8}$$

They complete the formulation of the three-dimensional nonlinear model of a deformed body.

The formulation discussed above is specific in that it separates the finite rotation field. For the considered momentless continuum this field is free according to its definition. As long as it remains free, the formulated equation system is not closed, the strain tensor U_{NM} is not symmetrical. To eliminate the freedom, the rotation field is to be subjected to some additional conditions, for example, to the ones which symmetrize or triangularize the matrix of the strain tensor.

For the moment continuum the rotation field is not free. It generates the moment stress field and is connected with the latter by the reological equations. The moment continuum models with the finite rotations are described in papers [5,6].

2. Two-dimensional model of a deformed shell

When a spatial body has a shape of a shell the coordinate system is connected with a fixed basic surface $\mathcal{B}$. The parameters s^1, s^2 are defined as its internal coordinates, s^3 is the normal coordinate. The shell boundary surface consists of two external surfaces $\mathcal{B}_1$, $\mathcal{B}_2$ and the edging surface $\mathcal{B}_3$. The latter is orthogonal to the basic surface along its boundary contour $\mathcal{C}$. Thus

$$\mathcal{B}_\nu = \mathcal{B}_1 \cup \mathcal{B}_2 \cup \mathcal{B}_3, \qquad \mathcal{B}_3 \perp \mathcal{B}, \qquad \mathcal{C} = \mathcal{B}_3 \cap \mathcal{B}.$$

Each point $\underline{S} = (s^1; s^2; s^3)$ of the shell (original-point) is imaged into the point $\underline{s} = (s^1; s^2)$ of the basic surface (image-point).

If $a(\underline{s})$ is the initial radius-vector of the image-point and $a_{(3)}(\underline{s})$ is the initial normal unit vector, the radius-vector $\underline{A}(\underline{S})$ of the original-point is defined by the equality

$$\underline{A} = \underline{a} + s^3 \underline{a}_{(3)} . \tag{2.1}$$

As a consequence one obtains the equations

$$\underline{A}_{(3)} = \underline{a}_{(3)} , \quad \underline{A}_{(n)} = \underline{a}_{(n)} + s^3 \underline{b}_{(n)} ,$$
$$\underline{a}_{(n)} = \partial_n \underline{a} , \quad \underline{b}_{(n)} \equiv \partial_n \underline{a}_{(3)} \tag{2.2}$$

which connect the initial spatial basis $\underline{A}_{(N)}(\underline{S})$ with the initial surface basis $\underline{a}_{(N)}(\underline{s})$. The Jacobians of these bases are denoted by $J(\underline{S})$ and $j(\underline{s})$, respectively. The surfaces $\mathcal{B}$ and $\mathcal{B}_m$ are given by the equations

$$s^3 = 0, \qquad s^3 = b_m(\underline{s}),$$

in particular, b_m may be constant.

The surface mechanical influences are distributed as follows: the force fields $\underline{P}^m(\underline{S})$ are given on the external surfaces $\mathcal{B}_m$; the force field $\underline{P}^3(\underline{S})$ is given on the surface section $\mathcal{B}_3^+$ and the displacement field $\underline{F}_3(\underline{S})$ is given on the surface section $\mathcal{B}_3^-$.

The shell deformation generates spatial and surface displacement fields $\underline{U}(\underline{S})$ and $\underline{u}(\underline{s})$, respectively. The initial tangential vectors $\underline{a}_{(n)}$ are transformed into the current ones

$$\underline{a}_{\{n\}} \equiv \partial_n (\underline{a} + \underline{u}) = \underline{a}_{(n)} + \partial_n \underline{u} . \tag{2.3}$$

The initial normal vector $\underline{a}_{(3)}$ is transformed into the current vector $\underline{a}_{\{3\}}$ which is not normal to the deformed basic surface.

To construct a two-dimensional model of the shell, its deformation has to be subjected to the internal kinematic constraint

$$\underline{A}_{\{3\}} = \underline{a}_{\{3\}} . \tag{2.4}$$

As a consequence there appears a linear distribution of the spatial displacement field along the normal coordinate

$$\underline{U} = \underline{u} + s^3 \underline{u}_{(3)} , \quad \underline{u}_{(3)} \equiv \underline{a}_{\{3\}} - \underline{a}_{(3)} \tag{2.5}$$

With the help of (2.4) and (2.5) the equalities (1.1) take the form

$$\underline{A}_{\{3\}} = \underline{a}_{\{3\}}, \quad \underline{A}_{\{n\}} = \underline{a}_{\{n\}} + s^3 \underline{b}_{\{n\}}, \quad \underline{b}_{\{n\}} \equiv \partial_n \underline{a}_{\{3\}}. \tag{2.6}$$

To realize rotational transformation, a two-dimensional rotation field $\underline{v}(\underline{s})$ is introduced. The initial bases $\underline{A}_{(N)}(\underline{S})$ and $\underline{a}_{(N)}(\underline{s})$ are transformed into the turned bases $\underline{A}_N(\underline{S})$ and $\underline{a}_N(\underline{s})$ according to (1.2). They are connected by the equalities

$$\underline{A}_3 = \underline{a}_3, \quad \underline{A}_n = \underline{a}_n + s^3 \underline{b}_n, \quad \underline{b}_n \cdot \underline{a}_M = \underline{b}_{(n)} \cdot \underline{a}_{(M)} \tag{2.7}$$

Local rotations generate a two-dimensional tensor field of pure bendings of the basic surface

$$\underline{w}_n = \tau_1 \partial_n \underline{v} + \tau_2 \underline{v} \times \partial_n \underline{v} + (\tau_3 \partial_n v)\underline{v}, \quad v \equiv (\underline{v} \cdot \underline{v})^{1/2},$$
$$\tau_1 \equiv (v)^{-1} \sin v, \quad \tau_2 \equiv (v)^{-2}(1-\cos v), \quad \tau_3 \equiv (v)^{-1}(1-\tau_1). \tag{2.8}$$

The equations

$$\underline{U}_3 = \underline{u}_3, \quad \underline{U}_n = \underline{u}_n + s^3 \underline{v}_n, \tag{2.9}$$

following from (1.4),(2.6) and (2.7), express the spatial strain tensor in terms of two surface tensors: the metrical strain tensor

$$\underline{u}_N \equiv \underline{a}_{\{N\}} - \underline{a}_N = \underline{u}_{(N)} - \tau_1 \underline{v} \times \underline{a}_N + \tau_2 \underline{v} \times (\underline{v} \times \underline{a}_N),$$
$$\underline{u}_{(n)} \equiv \partial_n \underline{u}, \quad u_{NM} \equiv \underline{u}_N \cdot \underline{a}_M \tag{2.10}$$

and the bending strain tensor

$$\underline{v}_n \equiv \underline{b}_{\{n\}} - \underline{b}_n = \underline{w}_n \times \underline{a}_3 + \partial_n \underline{u}_3, \quad v_{nM} \equiv \underline{v}_n \cdot \underline{a}_M. \tag{2.11}$$

The virtual work equation (1.5), being subjected to the kinematic constraint (2.5) and integrated over s^3, transforms into the equation

$$\int_{\mathcal{B}} (\underline{p} \cdot \delta \underline{u} + \underline{q} \cdot \delta \underline{u}_{(3)} - \Delta z)\, dB +$$
$$+ \int_{\mathcal{C}} (\underline{p}^3 \cdot \delta \underline{u} + \underline{q}^3 \cdot \delta \underline{u}_{(3)})\, dC = 0 \tag{2.12}$$

defined in the two-dimensional space of the basic surface.

Formulae

$$
\begin{aligned}
j\underline{p} &= \underline{P}^m J_m + \int_{b_1}^{b_2} \underline{P} J\, ds^3, \qquad p^3 = \int_{b_1}^{b_2} \underline{P}^3 J_3\, ds^3, \\
j\underline{q} &= \underline{P}^m J_m b_m + \int_{b_1}^{b_2} \underline{P} J s^3 ds^3, \qquad \underline{q}^3 = \int_{b_1}^{b_2} \underline{P}^3 J_3 s^3 ds^3
\end{aligned}
$$

define the surface and contour fields of the resultant forces and moments generated by the external and inertial forces.

The quantity

$$
\begin{aligned}
\Delta z &= (j)^{-1} \int_{b_1}^{b_2} (\Delta Z) J\, ds^3 = \\
&= \underline{x}^N \cdot \delta_0 \underline{u}_N + \underline{y}^n \cdot \delta_0 \underline{v}_n = x^{NM} \delta u_{NM} + y^{nM} \delta v_{nM}
\end{aligned}
\tag{2.13}
$$

has the meaning of the virtual strain energy surface density. Formula (2.13) introduces the surface tensor fields of the resultant internal forces and moments

$$
\begin{aligned}
j\underline{x}^N &= \int_{b_1}^{b_2} \underline{X}^N J\, ds^3, \qquad j\underline{y}^n = \int_{b_1}^{b_2} \underline{X}^n J s^3 ds^3, \\
x^{NM} &= \underline{x}^N \cdot \underline{a}^M, \qquad y^{nM} = \underline{y}^n \cdot \underline{a}^M.
\end{aligned}
\tag{2.14}
$$

As a consequence of (2.12) one obtains the local dynamic equations

$$
\begin{aligned}
&\partial_n (j\underline{x}^n) + j\underline{p} = 0, \qquad \partial_n (j\underline{y}^n) - j\underline{x}^3 + j\underline{q} = 0, \\
&\underline{a}_{\{N\}} \times \underline{x}^N + \underline{b}_{\{n\}} \times \underline{y}^n = 0
\end{aligned}
\tag{2.15}
$$

and the boundary dynamic conditions

$$
(\underline{e}^3 \cdot \underline{a}_{(n)}) \underline{x}^n = \underline{p}^3, \quad (\underline{e}^3 \cdot \underline{a}_{(n)}) \underline{y}^n = \underline{q}^3, \quad \forall \underline{s} \in \mathcal{C}^+
\tag{2.16}
$$

where $\underline{e}^3$ is the unit vector normal to $\mathcal{B}_3$.

To formulate the two-dimensional reological equations in the case of isothermal elastic deformation, the strain energy surface density z is defined as a function of both surface strain tensors u_{NM} and v_{nM}. As a consequence one has the reological equations

$$z = (j)^{-1} \int_{b_1}^{b_2} ZJ\, d\varsigma^3 = e(u_{NM}, v_{nM}),$$
$$x^{NM} = \partial e / \partial u_{NM}, \qquad y^{nM} = \partial e / \partial v_{nM}. \tag{2.17}$$

The system of two-dimensional equations (2.8),(2.10),(2.11), (2.15)-(2.17) describes the deformed shell as the two-dimensional moment continuum and contains three unknown independent kinematic vectors $\underline{u}$, $\underline{u}_3$, $\underline{v}$. The finite rotation vector $\underline{v}$ is free according to its definition. The freedom should be eliminated by some additional conditions. First of all, this could be the trivial variant with zero rotation vector $\underline{v}=0$. The gradiental formulation without separating the rotation field corresponds to this variant. Three nontrivial physically visual variants could be suggested.

In the first variant basis rotation is performed in such a manner that the rotation field is subjected to the conditions

$$u_{21} = u_{12}, \quad u_{31} = u_{13}, \quad u_{32} = u_{23}. \tag{2.18}$$

Hence the metrical strain tensor is represented by the symmetrical matrix. The reological equations (2.17) define the symmetrized components of the tensor x^{NM} only. Its antisymmetrized components are defined by the last of equations (2.15).

In the second variant the rotation field is subjected to the conditions

$$u_{21} = 0, \quad u_{31} = 0, \quad u_{32} = 0. \tag{2.19}$$

The metrical strain tensor is represented by the upper triangular matrix. The third vector of the turned basis is collinear to that of the current basis. The force parameters x^{21}, x^{31}, x^{32} are determined by the last of equations (2.15) and not by the reological ones (2.17).

In the third variant the rotation field is subjected to the conditions

$$u_{12} = 0, \quad u_{13} = 0, \quad u_{23} = 0. \tag{2.20}$$

The metrical strain tensor is represented by the lower triangular matrix. The third vector of the turned basis is collinear to the momentary normal to the deformed basic surface. The force parameters x^{12}, x^{13}, x^{23} are determined by the last of equations (2.15).

In a certain sense the second and third variants of the deformed shell model are the mirror images of each other.

The conditions of shape (2.18), or (2.19), or (2.20) connect the rotation vector $\underline{v}$ with two other kinematic vectors $\underline{u}$ and $\underline{u}_3$. The former is the measure of the radius-vector variation, the latter is the measure of the normal vector variation due to the shell deformation. Hence the model proposed treats a deformed shell as the two-dimensional moment continuum with six scalar kinematic degrees of freedom. The model contains also three-dimensional equations (2.5), (2.9) and (1.8) for the reconstruction of spatial kinematic and dynamic fields in the shell. As a result it defines the total strain and stress tensors.

The turned basis was used in the monograph [7] for an approximate description of shell nonlinear kinematics in the scope of Kirchhoff-Love constraints. A consistent and detailed development of this approach was given by ALUMYAE [8]. Under the same constraints he introduced the strain tensors (2.10) and (2.11) (with $\underline{u}_3=0$) and the resultant force and moment tensors (2.14) (with $x^3=0$). However the finite rotation vector was not yet introduced by him. The correct and complete formulation of the Kirchhoff-Love type nonlinear model was given in the papers [9,10]. The generalized nonlinear models of a deformed shell can be found in the papers [11,12]. The formulation given in [12] is identical to the present one with the additional conditions (2.19) (the second variant).

3. One-dimensional model of a deformed rod

When a spatial body has a shape of a rod, the coordinate system is connected with a fixed basic curve $\mathcal{C}_3$. The coordinate s^3 is the natural parameter of the curve, coordinates s^1 and s^2 form a plane system in any rod cross-section orthogonal to the curve. The rod boundary surface consists of two edging cross-sections $\mathcal{B}_1$, $\mathcal{B}_2$ and a tubelike surface $\mathcal{B}_3$. Thus

$$\mathcal{B}_\nu = \mathcal{B}_1 \cup \mathcal{B}_2 \cup \mathcal{B}_3\,, \qquad \mathcal{B}_m \perp \mathcal{C}_3.$$

Each point $\underline{s} \equiv (s^1; s^2; s^3)$ of the rod (original-point) is imaged in the point s^3 of the basic curve (image-point). An arbitrary cross-section of the rod and its contour are denoted by $\mathcal{B}$ and $\mathcal{C}$, respectively.

If $\underline{a}(s^3)$ is the initial radius-vector of the image-point and $\underline{a}_{(n)}$ is the orthonormal basis of the plane coordinate system, the radius-vector $\underline{A}(\underline{S})$ of the original-point is defined by the equality

$$\underline{A} = \underline{a} + s^n \underline{a}_{(n)} . \quad (3.1)$$

As a consequence one has the following equations

$$\underline{A}_{(n)} = \underline{a}_{(n)}, \quad \underline{A}_{(3)} = \underline{a}_{(3)} + s^n \underline{b}_{(n)}, \quad \underline{a}_{(3)} \equiv \partial_3 \underline{a}, \quad \underline{b}_{(n)} \equiv \partial_3 \underline{a}_{(n)} \quad (3.2)$$

which connect the initial spatial basis $\underline{A}_{(N)}(\underline{S})$ with the initial contour basis $\underline{a}_{(N)}(s^3)$.

The edging cross-sections $\mathcal{B}_m$ are given by the equations

$$s^3 = c_m , \qquad c_m = const .$$

The surface mechanical influences are distributed as follows: the force field $\underline{P}^3(\underline{S})$ is given on the tubelike surfaces $\mathcal{B}_3$, the force field $\underline{P}^m(\underline{S})$ or the displacement field $\underline{F}_m(\underline{S})$ is given on every edging cross-section $\mathcal{B}_m$.

The shell deformation generates spatial and contour displacement fields $\underline{U}(\underline{S})$ and $\underline{u}(s^3)$, respectively.

The initial tangential vector $\underline{a}_{(3)}$ is transformed into the current one

$$\underline{a}_{\{3\}} \equiv \partial_3 (\underline{a} + \underline{u}) = \underline{a}_{(3)} + \partial_3 \underline{u} . \quad (3.3)$$

The initial normal vectors $\underline{a}_{(n)}$ are transformed into the current vectors $\underline{a}_{\{n\}}$ which are not normal to the deformed basic curve.

To construct a one-dimensional model of the rod, its deformation is subjected to the internal kinematic constraint

$$\underline{A}_{\{n\}} = \underline{a}_{\{n\}} . \quad (3.4)$$

As a consequence there appears a linear distribution of spatial displacement field along the normal coordinates

$$\underline{U} = \underline{u} + s^n \underline{u}_{(n)}, \qquad \underline{u}_{(n)} \equiv \underline{a}_{\{n\}} - \underline{a}_{(n)} . \quad (3.5)$$

With the help of (3.4) and (3.5) equalities (1.1) take the form

$$\underline{A}_{\{n\}} = \underline{a}_{\{n\}}, \quad \underline{A}_{\{3\}} = \underline{a}_{\{3\}} + s^n \underline{b}_{\{n\}}, \quad \underline{b}_{\{n\}} \equiv \partial_3 \underline{a}_{\{n\}} . \quad (3.6)$$

To realize rotational transformation a contour rotation field $\underline{v}(s^3)$ is introduced. The initial bases $\underline{A}_{(N)}(\underline{S})$ and $\underline{a}_{(N)}(s^3)$ are transformed into the turned bases $\underline{A}_N(\underline{S})$ and $\underline{a}_N(s^3)$ according to (1.2). They are connected by the equalities

$$\underline{A}_n = \underline{a}_n, \quad \underline{A}_3 = \underline{a}_3 + s^n \underline{b}_n, \quad \underline{b}_n \cdot \underline{a}_M = \underline{b}_{(n)} \cdot \underline{a}_{(M)}. \tag{3.7}$$

Local rotations generate the contour tensor field of pure bendings of the basic curve

$$\underline{w}_n = \tau_1 \partial_n \underline{v} + \tau_2 \underline{v} \times \partial_n \underline{v} + (\tau_3 \partial_n v) \underline{v}. \tag{3.8}$$

The equations

$$\underline{U}_n = \underline{u}_n, \quad \underline{U}_3 = \underline{u}_3 + s^n \underline{v}_n, \tag{3.9}$$

following from (1.4),(3.6) and (3.7), express the spatial strain tensor in terms of two contour tensors: the metrical strain tensor

$$\begin{aligned} &\underline{u}_N \equiv \underline{a}_{\{N\}} - \underline{a}_N = \underline{u}_{(N)} - \tau_1 \underline{v} \times \underline{a}_N + \tau_2 \underline{v} \times (\underline{v} \times \underline{a}_N), \\ &\underline{u}_{(3)} \equiv \partial_3 \underline{u}, \qquad u_{NM} \equiv \underline{u}_N \cdot \underline{a}_M \end{aligned} \tag{3.10}$$

and the bending strain tensor

$$\underline{v}_n \equiv \underline{b}_{\{n\}} - \underline{b}_n = \underline{w}_3 \times \underline{a}_n + \partial_3 \underline{u}_n, \qquad v_{nM} \equiv \underline{v}_n \cdot \underline{a}_M. \tag{3.11}$$

The virtual work equation (1.5), being subjected to kinematic constraint (3.5) and integrated over s^n, transforms into the equation

$$\begin{aligned} &\int_{c_1}^{c_2} (\underline{p} \cdot \delta \underline{u} + \underline{q}^n \cdot \delta \underline{u}_{(n)} - \Delta z)\, ds^3 + \\ &+ \underline{p}^m \cdot \delta \underline{u}\big|_{s^3 = c_m} + \underline{q}^{mn} \cdot \delta \underline{u}_{(n)}\big|_{s^3 = c_m} = 0 \end{aligned} \tag{3.12}$$

defined in the one-dimensional space of the basic curve.

Formulae

$$\underline{p} = \int_{\mathcal{B}} \underline{P} J \, dB + \int_{\mathcal{C}} \underline{P}^3 J_3 \, dC, \qquad \underline{p}^m = \int_{\mathcal{B}_m} \underline{P}^m J_m \, dB_m,$$

$$\underline{q}^n = \int_{\mathcal{B}} \underline{P} J s^n dB + \int_{\mathcal{C}} \underline{P}^3 J_3 s^n dC, \quad \underline{q}^{mn} = \int_{\mathcal{B}_m} \underline{P}^m J_m s^n dB_m$$

define the contour and one-point fields of the resultant forces and moments generated by the external and inertial forces.

The quantity

$$\Delta z \equiv \int_{\mathcal{B}} (\Delta Z) J \, dB =$$

$$= \underline{x}^{N} \cdot \delta_0 \underline{u}_N + \underline{y}^{n} \cdot \delta_0 \underline{v}_n = x^{NM} \delta u_{NM} + y^{nM} \delta v_{nM} \tag{3.13}$$

has the meaning of the virtual energy contour density.

Formula (3.13) introduces the contour tensor fields of the internal resultant forces and moments

$$\begin{aligned} \underline{x}^{N} &= \int_{\mathcal{B}} \underline{X}^{N} J \, dB, & \underline{y}^{n} &= \int_{\mathcal{B}} \underline{X}^{3} J s^{n} \, dB, \\ x^{NM} &= \underline{x}^{N} \cdot \underline{a}^{M}, & y^{nM} &= \underline{y}^{n} \cdot \underline{a}^{M}. \end{aligned} \tag{3.14}$$

As a consequence of (3.12) one obtains the local dynamic equations

$$\begin{aligned} & \partial_3 \underline{x}^{3} + \underline{p} = 0, \qquad \partial_3 \underline{y}^{n} - \underline{x}^{n} + \underline{q}^{n} = 0, \\ & \underline{a}_{\{N\}} \times \underline{x}^{N} + \underline{b}_{\{n\}} \times \underline{y}^{n} = 0 \end{aligned} \tag{3.15}$$

and the boundary dynamic conditions

$$(\underline{e}^{m} \cdot \underline{a}_{(3)}) \underline{x}^{3} = \underline{p}^{m}, \qquad (\underline{e}^{m} \cdot \underline{a}_{(3)}) \underline{y}^{n} = \underline{q}^{mn} \tag{3.16}$$

where $\underline{e}^{m}$ is the unit vector normal to $\mathcal{B}_m$.

To formulate the one-dimensional reological equations in the case of isothermal elastic deformation, the strain energy contour density is defined as a function of both contour strain tensors u_{NM} and v_{nM}. As a consequence, one has the reological equations

$$\begin{aligned} & z = \int_{\mathcal{B}} Z J \, dB = e(u_{NM}, v_{nM}), \\ & x^{NM} = \partial e / \partial u_{NM}, \qquad y^{nM} = \partial e / \partial v_{nM}. \end{aligned} \tag{3.17}$$

The system of one-dimensional equations (3.8),(3.10),(3.11), (3.15)-(3.17) describes the deformed rod as the one-dimensional moment continuum and contains four unknown independent kinematic vectors $\underline{u}, \underline{u}_n, \underline{v}$. The finite rotation vector $\underline{v}$ is free according to its definition. The freedom should be eliminated by some additional conditions. Just as for the shell, three nontrivial physically visual variants could be suggested.

In the first variant basis rotation is performed in such a manner that the rotation field is subjected to the conditions (2.18). Hence the metrical strain tensor is represented by the symmetrical matrix.

The reological equations (3.17) define the symmetrized components of the tensor x^{NM} only. Its antisymmetrized components are determined by the last of equations (3.15).

In the second variant the rotation field is subjected to the conditions (2.19). The metrical strain tensor is represented by the upper triangular matrix. The third vector of the turned basis is collinear to the momentary tangential vector. The force parameters x^{21}, x^{31}, x^{32} are defined by the last of equations (3.15) and not by the reological equations (3.17).

In the third variant the rotation field is subjected to the conditions (2.20). The metrical strain tensor is represented by the lower triangular matrix. The third vector of the turned basis is normal to the rod deformed cross-section. The force parameters x^{12}, x^{13}, x^{23} are defined by the last of equations (3.15).

The conditions of shape (2.18), or (2.19), or (2.20) connect the rotation vector $\underline{v}$ with three other kinematic vectors: $\underline{u}$ and $\underline{u}_n$. The former is the measure of the radius-vector variation, the latter are the measures of variations of the normal vectors due to the rod deformation. Hence the model proposed treats a deformed rod as a one-dimensional moment continuum with nine scalar kinematic degrees of freedom. The model contains also three-dimensional equations (3.5),(3.9) and (1.8) for reconstruction of spatial kinematic and dynamic fields in the rod. As a result it defines the total spatial strain and stress tensors.

References

1. ERICKSEN J.L., TRUESDELL C., Exact theory of stress and strain in rods and shells, Arch.Rat.Mech.Anal. 1 (1958), 4, 295-322.
2. GREEN A.E., NAGHDI P.M., Non-isothermal theory of rods, plates and shells, Int.J.Solids & Struct. 6 (1970), 209-244.
3. SIMMONDS J.G., DANIELSON D.A., Nonlinear shell theory with finite rotation and stress-function vectors, Trans.ASME E39 (1972), 4, 1085-1090.
4. DE SILVA C.N., TSAI P.J., A general theory of directed surfaces, Acta Mech. 18 (1973), 1-2, 89-101.
5. REISSNER E., On kinematics and statics of finite-strain force and moment stress elasticity, Stud.Appl.Math. 52 (1973), 2, 97-101.
6. SHKUTIN L.I., The nonlinear models of deformed moment media, J.Appl.Mech. & Tech.Phys. 6 (1980), 111-117 (in Russian).
7. LOVE A., A treatise on the mathematical theory of elasticity, 4-th ed., Cambridge University Press, 1927.

8. ALUMYAE N.A., Differential equations of equilibrium states of thin-walled elastic shells in the post-critical states, Appl. Math. & Mech. 13 (1949), 1, 95-106 (in Russian).

9. SHKUTIN L.I., The correct formulation of thin shell nonlinear deformation equations, in: Appl.Problems of Strength & Plasticity, Gorki University Press 7 (1977), 8 (1978), 9 (1978) (in Russian).

10. PIETRASZKIEWICZ W., Finite rotation and Lagrangean description in the nonlinear theory of shells, Polish Sci.Publ., Warszawa, 1979.

11. SHKUTIN L.I. The nonlinear model of a shell with nontensile cross-fibres, J.Appl.Mech. & Tech.Phys. 1 (1982), 163-167 (in Russian).

12. SHKUTIN L.I., The nonlinear model of a shell with tensile cross-fibres, J.Appl.Mech. & Tech.Phys. 1 (1984), 168-174 (in Russian).

ULTIMATE LOAD ANALYSIS OF THIN WALLED STEEL STRUCTURES WITH ELASTOPLASTIC DEFORMATION PROPERTIES USING FEM - THEORETICAL, ALGORITHMIC AND NUMERICAL INVESTIGATIONS -

E. STEIN, K.H. LAMBERTZ and L. PLANK
Institut für Baumechanik und Numerische Mechanik
der Universität Hannover
Callinstr. 32, D-3ooo Hannover 1

1. Introduction, technical problems and suitable theoretical frame

For the prediction of static ultimate loads of thin walled stiffened structures it is necessary to take into account geometrical and material non-linearities. An adequate method for the calculation of different kinds of structures, e.g. bridges, vessels, tall buildings and masts, is the finite element method (FEM). The geometrical non-linearity is treated by using an updated Lagrangian incremental formulation [1] assuming small strains but moderately large displacements and rotations in the increments [2] .

The starting point for the material non-linearity is the von MISES yield condition. Alternatively, elastic-perfectly plastic behaviour or elastic deformations with isotropic hardening according to HILL is considered [3] .

2. The treatment of the geometrical non-linearity

2.1 Incremental principle of virtual work in the general 3-D-case

The principle of virtual displacements is used as the weak equilibrium condition of the body in the current configuration "2", see fig. 2.1. (The upper left index and the lower left index denote the state concerned and the reference configuration, resp.).

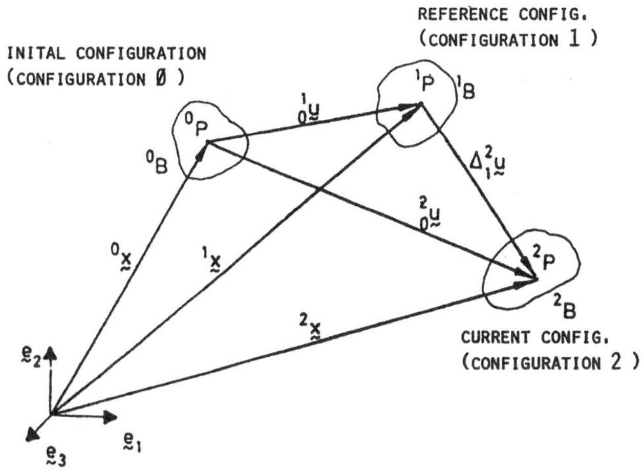

Fig. 2.1: Three configurations of a solid body

In the updated formulation all variables are referred to the reference configuration "1" which has been reached during the foregoing load increments.

The principle of virtual work then requires

$$\int_{^1B} {}^2_1\underset{\sim}{S} \cdot \delta\Delta^2_1\underset{\sim}{E}\, dB = {}^2R \tag{2.1}$$

with

$$^2R = \int_{^0B} {}^2\underset{\sim}{f} \cdot \delta\underset{\sim}{u}\, dB + \int_{\partial^0B} {}^2\bar{\underset{\sim}{t}} \cdot \delta\underset{\sim}{u}\, dA \tag{2.2}$$

where the boundary forces ${}^2\bar{\underset{\sim}{t}}$ and the body forces ${}^2\underset{\sim}{f}$ are assumed to be independent of the deformation of the body. In equ. (2.1), ${}^2_1\underset{\sim}{S}$ represents the 2. PIOLA-KIRCHHOFF stress tensor, and $\Delta^2_1\underset{\sim}{E}$ is the increment of the GREEN-LAGRANGE strain tensor from the configuration "1" to configuration "2" with

$$\begin{aligned}\Delta^2_1\underset{\sim}{E} = \tfrac{1}{2}[& {}_1\mathrm{Grad}\,\Delta^2_1\underset{\sim}{u} + ({}_1\mathrm{Grad}\,\Delta^2_1\underset{\sim}{u})^T \\ & + ({}_1\mathrm{Grad}\,\Delta^2_1\underset{\sim}{u})({}_1\mathrm{Grad}\,\Delta^2_1\underset{\sim}{u})^T].\end{aligned} \tag{2.3}$$

The additive incremental stress decomposition yields

$${}^2_1\underset{\sim}{S} = {}^1\underset{\sim}{T} + \Delta^2_1\underset{\sim}{S}\,. \tag{2.4}$$

with ${}^1\underset{\sim}{T}$ representing the CAUCHY stress tensor of configuration "1", and so equ. (2.1) can be written as

$$\begin{aligned}&{}_1\int_B \Delta_1^2 \underset{\sim}{S}\cdot \delta\Delta_1^2 \underset{\sim}{E}^{lin}\, dB\\&+{}_1\int_B {}^1\underset{\sim}{T}\cdot({}_1\text{Grad}\ \delta\Delta_1^2 \underset{\sim}{u})^T({}_1\text{Grad}\ \Delta_1^2\underset{\sim}{u})\, dB\\&={}^2R-{}_1\int_B {}^1\underset{\sim}{T}\cdot({}_1\text{Grad}\ \delta\Delta_1^2\underset{\sim}{u})^T dB\end{aligned} \tag{2.5}$$

where $\Delta_1^2\underset{\sim}{E}$ is linearized in the strain energy density as

$$\Delta_1^2\underset{\sim}{E}^{lin}=\tfrac{1}{2}[{}_1\text{Grad}\ \Delta_1^2\underset{\sim}{u}+{}_1\text{Grad}\ \Delta_1^2\underset{\sim}{u}^T]\, dB\,. \tag{2.6}$$

This geometrical linearization causes a residual work

$$\Delta R={}^2R-{}_2\int_B {}^2\underset{\sim}{T}\cdot(\text{Grad}\ \Delta_1^2\underset{\sim}{u})^T dB\,. \tag{2.7}$$

2.2 Incremental principle of folded plate structures

In the non-linear numerical treatment of folded structures the following assumptions are usual:

- small strains
- plane stress state
- the cross section remains plane during the deformation (Kirchhoff or Reissner Mindlin theory, resp.).

These assumptions allow the following simplifications in the finite element formulation:

- the elements remain plane during the deformation (Facet-elements), see chapt. 6)
- the shape of the elements does not change (small strain assumption).
- the configurations are described by an orthogonal transformation from the global to the local coordinate system of the element.

Equ. (2.5) and (2.7) then yield for a single element

$$^1\underline{K}^e = {}^1\underline{R}^e\ {}^l\underline{K}^e_{lin}\ {}^1\underline{R}^{eT}, \qquad {}^l\underline{K}^e_{lin} = \int_{^oB} \underline{B}^T\,\underline{C}\,\underline{B}\; dV, \tag{2.8}$$

and

$$\Delta\underline{P}^e = {}^2\underline{\bar{P}}^e - {}^1\underline{R}^e\,{}^l\underline{P}^e, \qquad {}^l\underline{P}^e = \int_{^oB} \underline{B}^T\,\underline{T}\, dV\,. \tag{2.9}$$

In equ. (2.8) and (2.9) the upper right index e denotes the element, the upper left index ℓ the configuration, where ℓ means local coordinate systems. $\underline{K}_{\ell in}$ is the linear elastic and $\underline{K}_\sigma$ the initial stress element matrix [4], $\underline{\bar{p}}$ the external load vector, and $\underline{p}$ results from the element stresses. The orthogonal transformation matrix $\underline{R}$ characterizes the current deformation state and includes possibly large rotations.

The element terms are assembled in the usual way to the incremental linear system. The residual load due to the linearization is added to the next load increment (algorithm 1) [2]. This procedure causes a deviation from the solution path, which depends on the increment size. An advantage of this method is the simple formulation because only the initial stress matrix $\underline{K}_\sigma$ is added to the linear formulation. It is also suitable for rate dependent problems (e.g. plastic analysis), because in this case the increment size has to be small.

In an alternative algorithm 2, the residual load is considered by equilibrium iterations. From equ. (2.1) it follows for each iteration step for an element e

$$\Delta \underline{P}^{e(i)} = \int_B \underline{B}^T \, {}^2\underline{T}^{(i)} dv + {}^l\underline{K}_\sigma^{(i)} \Delta \underline{u}^{e(i)} \tag{2.10}$$

where the first term results from the linear and the second one from the nonlinear strains. ${}^2\underline{T}^{(i)}$ depends on the stress resultants as well as $\underline{K}_\sigma^{(i)}$. $\underline{B}$ contains the derivatives of the shape functions.

This iteration method reduces the incremental error and saves computation time because larger increments are possible (see chapt. 8).

3. Material laws for elastoplastic deformations

The strain tensor for small strains is given by

$$\varepsilon_{ij} = \frac{1}{2}(u_{i,j} + u_{j,i}) \; , \tag{3.1}$$

and the elastic material law (HOOKE) is

$$\sigma_{ij} = 2G\,\varepsilon_{ij}^{el} + \lambda\,\varepsilon_{kk}^{el}\,\delta_{ij} \; . \tag{3.2}$$

The additive decomposition of the strain tensor (3.1) into elastic and plastic parts yields

$$\varepsilon_{ij} = \varepsilon_{ij}^{el} + \varepsilon_{ij}^{pl} \; . \tag{3.3}$$

The well known von MISES yield condition with isotropic hardening holds for the calculation of inealstic deformations, especially for metals:

$$F(\sigma, e_v) = \sigma_{ij}^D \sigma_{ij}^D - \frac{2}{3}\bar{\sigma}^2(e_v) \leq 0 \; , \tag{3.4}$$

with the effective strain rate

$$\dot{e}_v = \sqrt{\frac{2}{3}\,\dot{\varepsilon}_{ij}^{pl}\,\dot{\varepsilon}_{ij}^{pl}} \qquad (3.5)$$

and the yield stress $\bar{\sigma}$. σ_{ij}^D denotes the deviatoric part of σ_{ij}. The flow rule according to PRANDTL-REUSS for isotropic hardening using (3.4) can be given by the variational inequality with the trial state $(\tilde{\sigma}, \tilde{e}_v)$ [5], [6]

$$\dot{\varepsilon}_{ij}^{pl}(\tilde{\sigma}_{ij} - \sigma_{ij}) - \frac{d\bar{\sigma}}{de_v}\,\dot{e}_v(\tilde{e}_v - e_v) \leq 0 \qquad (3.6)$$

$$\forall (\tilde{\sigma}, \tilde{e}_v) \quad \text{with } F(\tilde{\sigma}, \tilde{e}_v) \leq 0 \; .$$

Assuming $\tilde{e}_v = e_v$, equ. (3.6) yields

$$\Longrightarrow \dot{\varepsilon}_{ij}^{pl} = \mu\,\frac{\partial F}{\partial \sigma_{ij}} \; . \qquad (3.7)$$

The variational inequality (3.6) states that the plastic strain rates are orthogonal to the yield surface, see fig. 3.1.

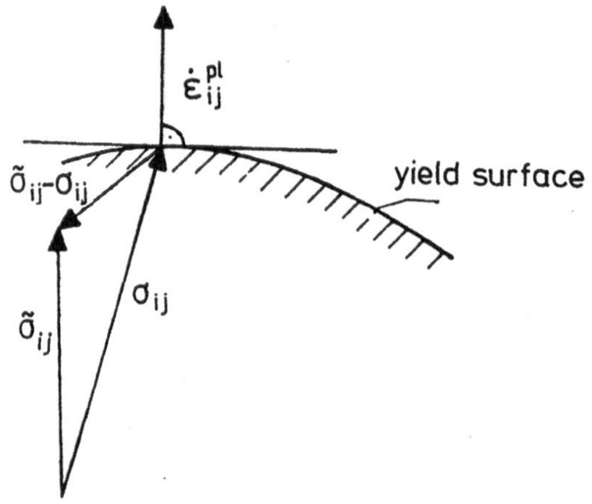

Fig. 3.1: The normality condition for the plastic strain rate

4. Stress projections onto the yield surface

4.1 Explicit integration scheme = tangential stiffness method

For $F(\underset{\sim}{\sigma}, e_v) = 0$ and $\frac{d}{dt} F = 0$, the PRANDTL-REUSS constitutive equation follows from equ. (3.7).

$$\dot{\underset{\sim}{\sigma}} = \underset{\sim}{\mathbb{C}}\,\dot{\underset{\sim}{\varepsilon}} - \underset{\sim}{\mathbb{C}}\,\underset{\sim}{\sigma}^D \frac{\underset{\sim}{\sigma}^D \cdot \underset{\sim}{\mathbb{C}}\,\dot{\underset{\sim}{\varepsilon}}}{\underset{\sim}{\sigma}^D \cdot \underset{\sim}{\mathbb{C}}\,\underset{\sim}{\sigma}^D + \frac{4}{9}\bar{\sigma}^2 \frac{\partial \bar{\sigma}}{\partial e_v} + \frac{4}{3}\gamma \bar{\sigma}^2} \tag{4.1a}$$

or

$$\dot{\underset{\sim}{\sigma}} = \underset{\sim}{\mathbb{C}}^{elpl}(\underset{\sim}{\sigma}, e_v)\,\dot{\underset{\sim}{\varepsilon}} \; . \tag{4.1b}$$

Numerical integration

$$\underset{\sim}{\sigma}(t^\circ + \Delta t^\circ) = \underset{\sim}{\sigma}(t^\circ) + \Delta t\,\dot{\underset{\sim}{\sigma}}(t^\circ) \tag{4.2}$$

yields the stress increments $\Delta_1^2 \underset{\approx}{S}$

$$\Rightarrow \Delta_1^2 \underset{\sim}{S} = \underset{\sim}{\mathbb{C}}^{elpl}({}^1\underset{\sim}{T}, {}^1E_v)\,\Delta_1^2 \underset{\approx}{E} \tag{4.3}$$

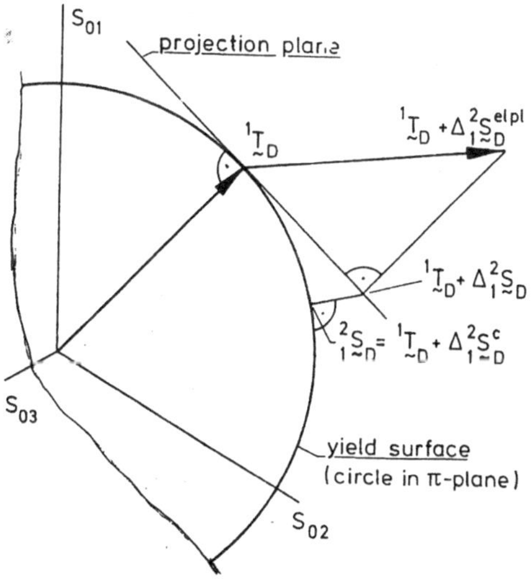

Fig. 4.1: Projection onto the tangential hyperplane

As shown in fig. 4.1, equ. (4.3) describes an orthogonal projection of $\Delta_1^2 \underset{\approx}{S}^{elpl}$ onto a hyperplane tangential to the yield surface. Shortcomings of the explicit methods are:

- $F({}^1\underset{\sim}{T} + \Delta_1^2 \underset{\approx}{S}) \neq 0$ in general

and therefore additional corrections are necessary:

- modifications for $F({}^1\underset{\sim}{T}) < 0$
- not stable for large increments.

4.2 Implicit integration schemes, projection methods (for ideal plastic material only) [7], [8]

In incremental notation, equ. (3.6) can be written as a variational inequality for the unknown stresses

$$ {}^2_1\underset{\sim}{S} = \underset{\sim}{T} + \Delta{}^2_1\underset{\sim}{S} \quad \text{WITH} \quad F({}^2_1\underset{\sim}{S}) = 0 \tag{4.4}$$

and

$$\Delta{}^2_1\underset{\sim}{E}^{pl} \cdot ({}^2_1\tilde{\underset{\sim}{S}} - {}^2_1\underset{\sim}{S}) \leq 0 \qquad \forall\, {}^2_1\tilde{\underset{\sim}{S}} \quad \text{WITH} \quad F({}^2_1\tilde{\underset{\sim}{S}}) \leq 0 \,. \tag{4.5}$$

Alternatively this can be written in the form

$$\Delta{}^2_1\underset{\sim}{E}^{pl} = \Delta{}^2_1\underset{\sim}{E} - \Delta{}^2_1\underset{\sim}{E}^{el} = \underset{\sim}{\mathbb{C}}^{-1}(\Delta{}^2_1\underset{\sim}{S}^{elpl} - \Delta{}^2_1\underset{\sim}{S}) \,, \tag{4.6}$$

and equ. (4.5) yields

$$(\Delta{}^2_1\underset{\sim}{S}^{elpl} - \Delta{}^2_1\underset{\sim}{S}) \cdot \underset{\sim}{\mathbb{C}}^{-1}(\Delta{}^2_1\tilde{\underset{\sim}{S}} - \Delta{}^2_1\underset{\sim}{S}) \leq 0 \qquad \forall\, \Delta{}^2_1\tilde{\underset{\sim}{S}} \quad \text{WITH} \quad F({}^1\underset{\sim}{T} + \Delta{}^2_1\tilde{\underset{\sim}{S}}) \leq 0 \,. \tag{4.7}$$

In equ. (4.7), ${}^2_1\underset{\sim}{S}$ (given by equ. (4.4)) is the projection of

$${}^1\underset{\sim}{T} + \Delta{}^2_1\underset{\sim}{S}^{elpl} \tag{4.8}$$

onto the yield surface, orthogonal with respect to the inner product (see fig. 4.2)

$$\langle \underset{\sim}{x}, \underset{\sim}{y} \rangle = \underset{\sim}{x} \cdot \underset{\sim}{\mathbb{C}}^{-1} \underset{\sim}{y} \,. \tag{4.9}$$

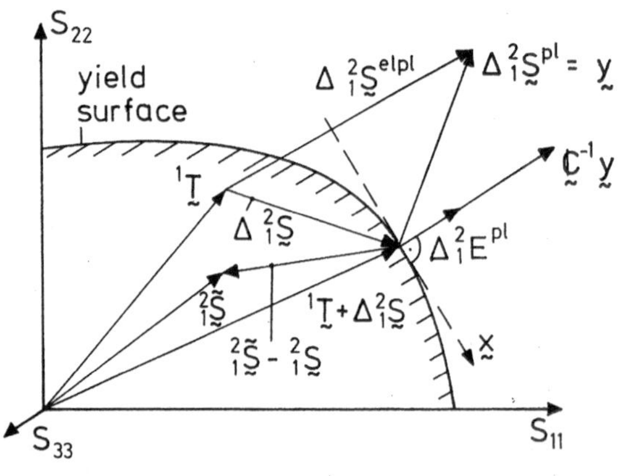

Fig. 4.2: Orthogonal projection onto the yield surface

Therefore, (4.7) is equivalent to the minimum problem

$$\min_{\Delta_1^2 S} J(\Delta_1^2 \underset{\sim}{S}) := \langle \Delta_1^2 \underset{\sim}{S}^{elpl} - \Delta_1^2 \underset{\sim}{S}, \Delta_1^2 \underset{\sim}{S}^{elpl} - \Delta_1^2 \underset{\sim}{S} \rangle \; . \tag{4.10}$$

4.3 The radial return method for plane stress

In this paper the radial return method is developed for the plane stress state [9], [10], which is usually assumed in thin walled structures [3], see fig. 4.3.

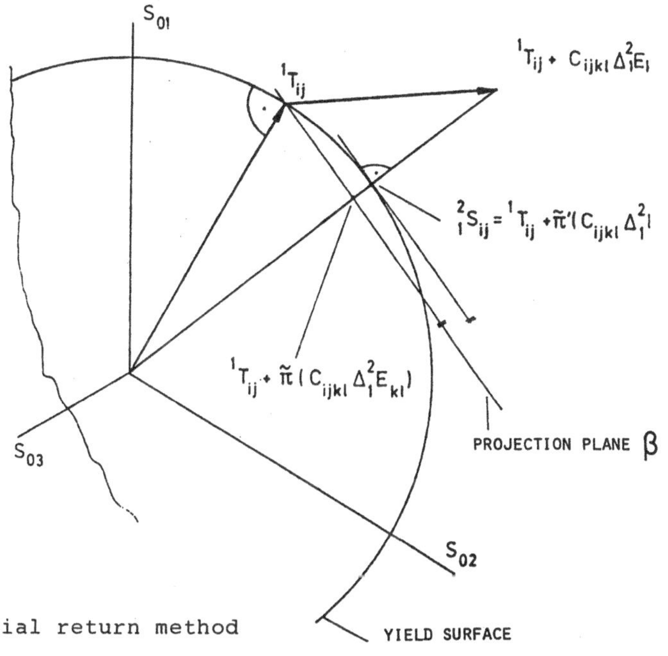

Fig. 4.3: The radial return method

From equ. (4.10) we define

$$\min_{\Delta_1^2 \underset{\sim}{S}} J(\Delta_1^2 \underset{\sim}{S}) := (\Delta_1^2 \underset{\sim}{S}^{elpl} - \Delta_1^2 \underset{\sim}{S}) \overset{*}{\mathbb{C}}^{-1} (\Delta_1^2 \underset{\sim}{S}^{elpl} - \Delta_1^2 \underset{\sim}{S}) \tag{4.11}$$

with the constraint

with
$$F({}^1\underset{\sim}{T} + \Delta_1^2 \underset{\sim}{S}, {}^2_1\underset{\sim}{E}_v) = 0 \quad . \tag{4.12}$$

The solution of (4.11) is the closest point projection $\Delta_1^2 \underset{\sim}{S}$ of $\Delta_1^2 \underset{\sim}{S}^{elpl}$ onto the yield surface with respect to the plane stress condition. In (4.11), $\overset{*}{\underset{\sim}{\mathbb{C}}}$ denotes the well known elasticity tensor for plane stress.

An equivalent formulation of equ. (4.11) is given by the determination of the saddle-point of

$$J^*(\Delta_1^2\underset{\sim}{S},\mu) := J(\Delta_1^2\underset{\sim}{S}) + \mu F({}^1\underset{\sim}{T} + \Delta_1^2\underset{\sim}{S}, {}^2\underset{\sim}{E}_v) . \tag{4.13}$$

After some algebra, one arrives at the three equations

$$\begin{pmatrix} ({}^1T + \Delta\, {}_1^2S)_{11} \\ ({}^1T + \Delta\, {}_1^2S)_{22} \end{pmatrix} = \frac{2}{D} \begin{pmatrix} 2/E - 2\mu & 2\nu/E - \mu \\ 2\nu/E - \mu & 2/E - 2\mu \end{pmatrix} \begin{pmatrix} E_{11} \\ E_{22} \end{pmatrix} \tag{4.14}$$

$$({}^1\underset{\sim}{T} + \Delta_1^2\underset{\sim}{S})_{12} = E/[2(1+\nu) - 3\mu E] \; E_{12}$$

with

$$D = (2/E - 2\mu)^2 - (2\nu/E - \mu)^2 \tag{4.15}$$

and μ as the solution of the biquadratic equation

$$\begin{aligned} 0 = {} & [\,2(1-\nu)/E - \mu\,]^2 \; (E_{11}^2 + E_{22}^2 + 3E_{12}^2 - E_{11}\,E_{22}) \\ & + 4/E^2 \; (1 - \mu E)(\nu - \mu E/2)(E_{11} + E_{22})^2 \\ & + 3/4\, D^2\, k^2(E_v^{pl}) \; . \end{aligned} \tag{4.16}$$

The solution of equ. (4.16) requires only few iteration steps with the secant method.

5. Correct tangential elastoplasticity tensor consistent with the projection [11], [12]

The yield condition is given by

$$F({}_1^2\underset{\sim}{S},\mu) = \frac{1}{2}({}_1^2\underset{\sim}{S}_D^T \cdot {}_1^2\underset{\sim}{S}) - Y^2(e_v) \tag{5.1}$$

with $Y(e_v) = \frac{\bar{\sigma}}{\sqrt{3}}$ from equ. (3.4) and ${}_1^2\underset{\sim}{S}_D$ denoting the deviatoric part of ${}_1^2\underset{\sim}{S}$.

Equ. (4.13) requires

$$\frac{\partial J^*(\Delta_1^2\underset{\sim}{S},\mu)}{\partial \Delta_1^2\underset{\sim}{S}} = -2\underset{\sim}{\overset{*}{C}}{}^{-1} \cdot (\Delta_1^2\underset{\sim}{S}^{elpl} - \Delta_1^2\underset{\sim}{S}) + \mu\Delta_1^2\underset{\sim}{S}_D \overset{!}{=} 0 \; . \tag{5.2}$$

With

$$\Delta_1^2\underset{\sim}{S}_D = \underset{\sim}{A}\,\Delta\,{}_1^2\underset{\sim}{S} \tag{5.3}$$

$$\underset{\sim}{\overset{*}{\bar{C}}}(\mu) = (\underset{\sim}{\overset{*}{C}}{}^{-1} + \frac{\mu}{2}\underset{\sim}{A})^{-1} , \tag{5.4}$$

and

$${}_1^2\underset{\sim}{E} = \underset{\sim}{\overset{*}{C}}{}^{-1} \cdot {}^1\underset{\sim}{T} + \Delta\,{}_1^2\underset{\sim}{E} , \tag{5.5}$$

equ. (5.2) yields

$$\Delta_1^2 \underset{\sim}{S}({}_1^2\underset{\sim}{E},\mu) = \overset{*}{\underset{\sim}{\mathbb{C}}}(\mu)\cdot{}_1^2\underset{\sim}{E} . \tag{5.6}$$

Equ. (4.13) also requires

$$\frac{\partial J^*(\Delta_1^2 \underset{\sim}{S},\mu)}{\partial\mu} = \frac{1}{2}\underset{\sim}{S}_D^T\cdot\underset{\approx}{S}_D - Y^2(e_v) \overset{!}{=} 0 . \tag{5.7}$$

From the projection method, see chapter 4, we obtain from equ. (4.16) the actual value of

$$\mu = \mu({}_1^2\underset{\sim}{E}) . \tag{5.8}$$

The effective plastic strain increment is given by

$$\Delta e_v = \mu\sqrt{\frac{1}{2}({}_1^2\underset{\approx}{S}_D^T\cdot{}_1^2\underset{\approx}{S}_D)}, \tag{5.9}$$

and therefore we get

$${}^2e_v = {}^1e_v + \Delta e_v . \tag{5.10}$$

The tangential elastoplasticity tensor is

$$\underset{\sim}{\bar{\mathbb{P}}} = \frac{d\,{}_1^2\underset{\sim}{S}(\mu,\,{}_1^2\underset{\sim}{E})}{d\Delta\,{}_1^2\underset{\sim}{E}} . \tag{5.11}$$

Using equ. (5.6) and $\frac{d_1^2\underset{\sim}{E}}{d\Delta_1^2\underset{\sim}{E}} = \underset{\sim}{1}$, we obtain

$$\underset{\sim}{\bar{\mathbb{P}}} = \overset{*}{\underset{\sim}{\mathbb{C}}} + \frac{\partial\Delta\,{}_1^2\underset{\approx}{S}}{\partial\mu}\otimes\frac{\partial\mu}{\partial\,{}_1^2\underset{\sim}{E}} . \tag{5.12}$$

From the yield condition $F(\mu,\ {}_1^2\underset{\sim}{E})$ and

$$\frac{\partial F}{\partial{}_1^2\underset{\sim}{E}} + \frac{\partial F}{\partial\mu}\,\frac{\partial\mu}{\partial{}_1^2\underset{\sim}{E}} = \underset{\sim}{0} , \tag{5.13}$$

$\frac{\partial\mu}{\partial_1^2\underset{\sim}{E}}$ is given by

$$\frac{\partial\mu}{\partial\,{}_1^2\underset{\sim}{E}} = -\frac{\dfrac{\partial F}{\partial\,{}_1^2\underset{\sim}{E}}}{\dfrac{\partial F}{\partial\mu}} , \tag{5.14}$$

and therefore

$$\frac{d\Delta_1^2\underset{\sim}{S}}{d\,{}_1^2\underset{\sim}{E}} = \overset{*}{\underset{\sim}{\mathbb{C}}} - \frac{\dfrac{\partial\Delta_1^2\underset{\approx}{S}}{\partial\mu}\otimes\dfrac{\partial F}{\partial{}_1^2\underset{\sim}{E}}}{\dfrac{\partial F}{\partial\mu}} . \tag{5.15}$$

With

$$\frac{\partial\Delta\,{}_1^2\underset{\sim}{S}(\mu,{}_1^2\underset{\sim}{E})}{\partial\mu} = -\frac{1}{2}\,\overset{*}{\underset{\sim}{\mathbb{C}}}\,\underset{\sim}{A}\,\overset{*}{\underset{\sim}{\mathbb{C}}}\,{}_1^2\underset{\sim}{E}, \tag{5.16}$$

and

$$\frac{\partial F}{\partial\,{}_1^2\underset{\sim}{E}} = \overset{*}{\underset{\sim}{\mathbb{C}}}\,{}_1^2\underset{\sim}{S}_D , \tag{5.17}$$

and $$\frac{\partial F}{\partial \mu} = -\frac{1}{2}({}^2_1\underset{\sim}{S}_D)^T \cdot \overset{*}{\bar{\underset{\sim}{C}}}\, {}^2_1\underset{\sim}{S}_D - 2Y^2\xi\left(\frac{1}{1-\mu\xi}\right) \tag{5.18}$$

where $\xi = \frac{\partial Y(e_v)}{\partial e_v}$, the tangential elastoplasticity tensor

$$\bar{\underset{\sim}{P}} = \overset{*}{\bar{\underset{\sim}{C}}} - \frac{\overset{*}{\bar{\underset{\sim}{C}}}\, {}^2_1\underset{\sim}{S}_D \otimes {}^2_1\underset{\sim}{S}_D^T\, \overset{*}{\bar{\underset{\sim}{C}}}^T}{\frac{4}{9}\xi Y^2(e_v)\left(\frac{1}{1-\mu\xi}\right) + {}^2\underset{\sim}{S}_D^T \cdot \overset{*}{\bar{\underset{\sim}{C}}}\, {}^2_1\underset{\sim}{S}_D} \; . \tag{5.19}$$

With $\mu = 0$ we obtain the well known PRANDTL-REUSS equation

$$\tilde{\underset{\sim}{P}} = \overset{*}{\underset{\sim}{C}} - \frac{\overset{*}{\underset{\sim}{C}}\, {}^2_1\underset{\sim}{S}_D \otimes {}^2_1\underset{\sim}{S}_D^T\, \overset{*}{\underset{\sim}{C}}^T}{\frac{4}{9}\xi Y^2(e_v) + \underset{\sim}{S}_D^T \cdot \overset{*}{\underset{\sim}{C}}\, {}^2_1\underset{\sim}{S}_D} \; . \tag{5.20}$$

6. FE-Discretization

6.1 The extended DKT-element

For the computations a plane triangular element with the 6 primary DOF's at each nodal point (see fig. 6.1) was applied [13], [4].

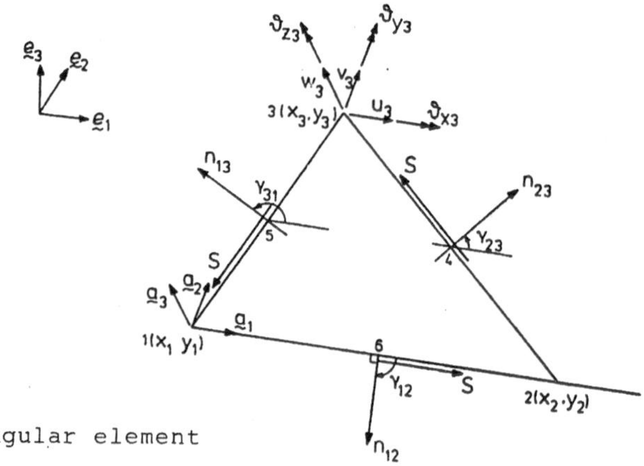

Fig. 6.1: Plane triangular element

The element stiffness is given as a linear superposition of the following independent parts (see fig. 6.2).

- membran stiffness matrix $\underline{K}_M$
- bending stiffness matrix $\underline{K}_B$
- formal rotational stiffness matrix $\underline{K}_{\Theta_z}$ for rotations with respect to the normal direction of the element.

$$\underline{K}_0 = \begin{bmatrix} \underline{K}_M & & \\ & \underline{K}_B & \\ & & \underline{K}_{\vartheta z} \end{bmatrix}$$

MEMBRANE STIFFNESS
- CONSTANT STRAIN/PLANE STRESS

BENDING STIFFNESS
- DISCRETE KIRCHHOFF THEORY

FORMAL ROTATIONAL STIFFNESS

Fig. 6.2: The complete element stiffness matrix

$\underline{K}_B$ is the stiffness matrix of the DKT-element (DKT means discrete KIRCHHOFF theory) investigated by BATHE et al. in [13]. $\underline{K}_M$ is the constant strain plane stress stiffness matrix of a three node element.

6.2 Layer model

For progressing plastic zones over the thickness a model is used, which is based on the introduction of (so called) layer points (fig. 6.3) distributed over the element thickness [3].

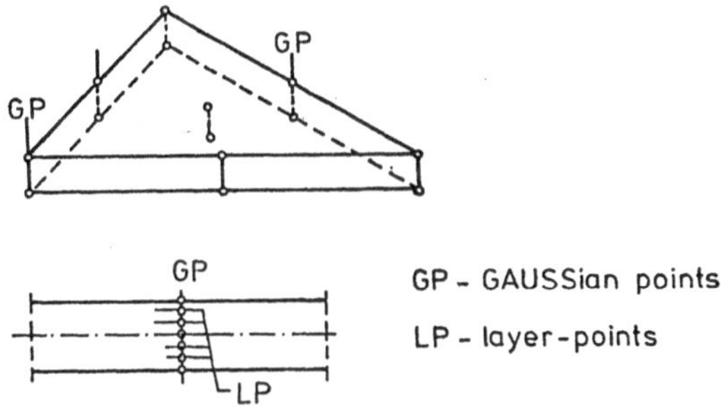

Fig. 6.3: Gaussian points and layer points

7. Flow charts for nonlinear algorithms

In fig. 7.1, a flow chart for incremental nonlinear calculations is shown. The iterations schemes for the solution of the nonlinear equations resulting from the material non-linearity are given in fig. 7.2. We use NEWTON-RAPHSON, modified NEWTON-RAPHSON and BFGS methods, respectively. The whole algorithm was implemented in our program system INA-SP which was developed within the portable program library DFGBIB, supported by the DFG, see [14].

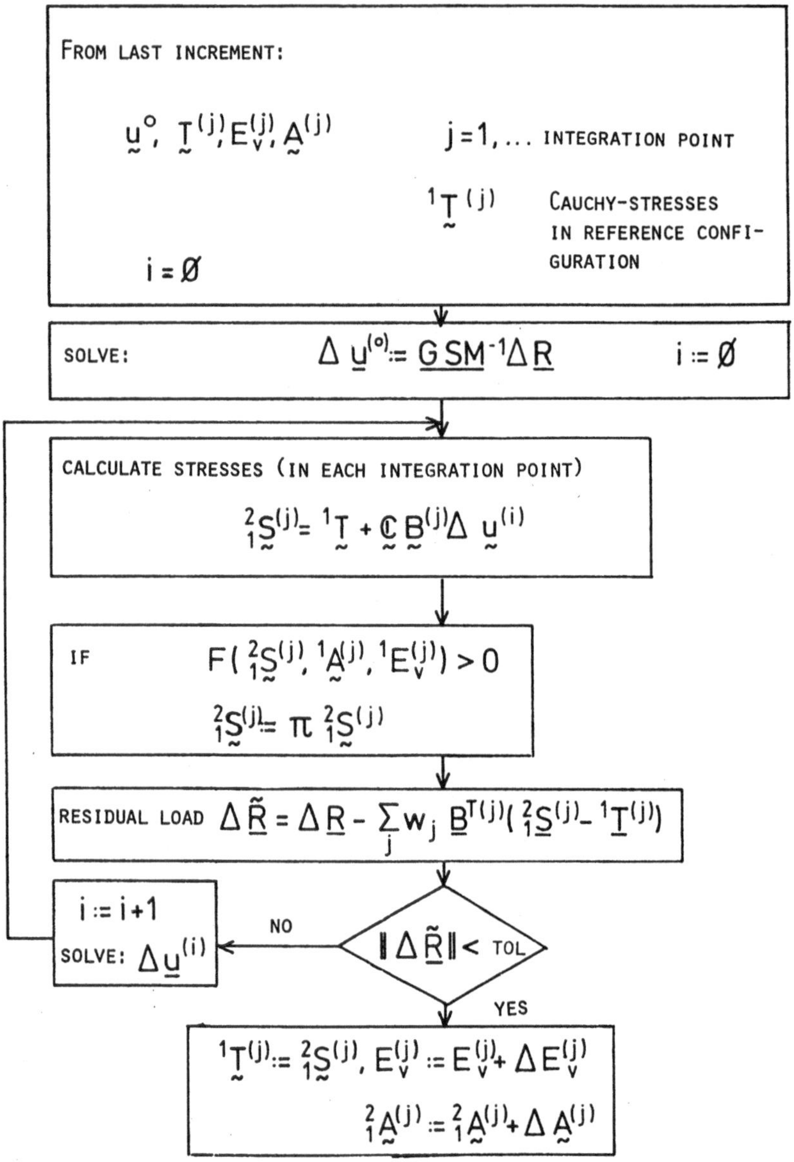

$\underset{\sim}{A}$ is the tensor of kinematic hardening

Fig. 7.1: Flow chart for incremental calculations

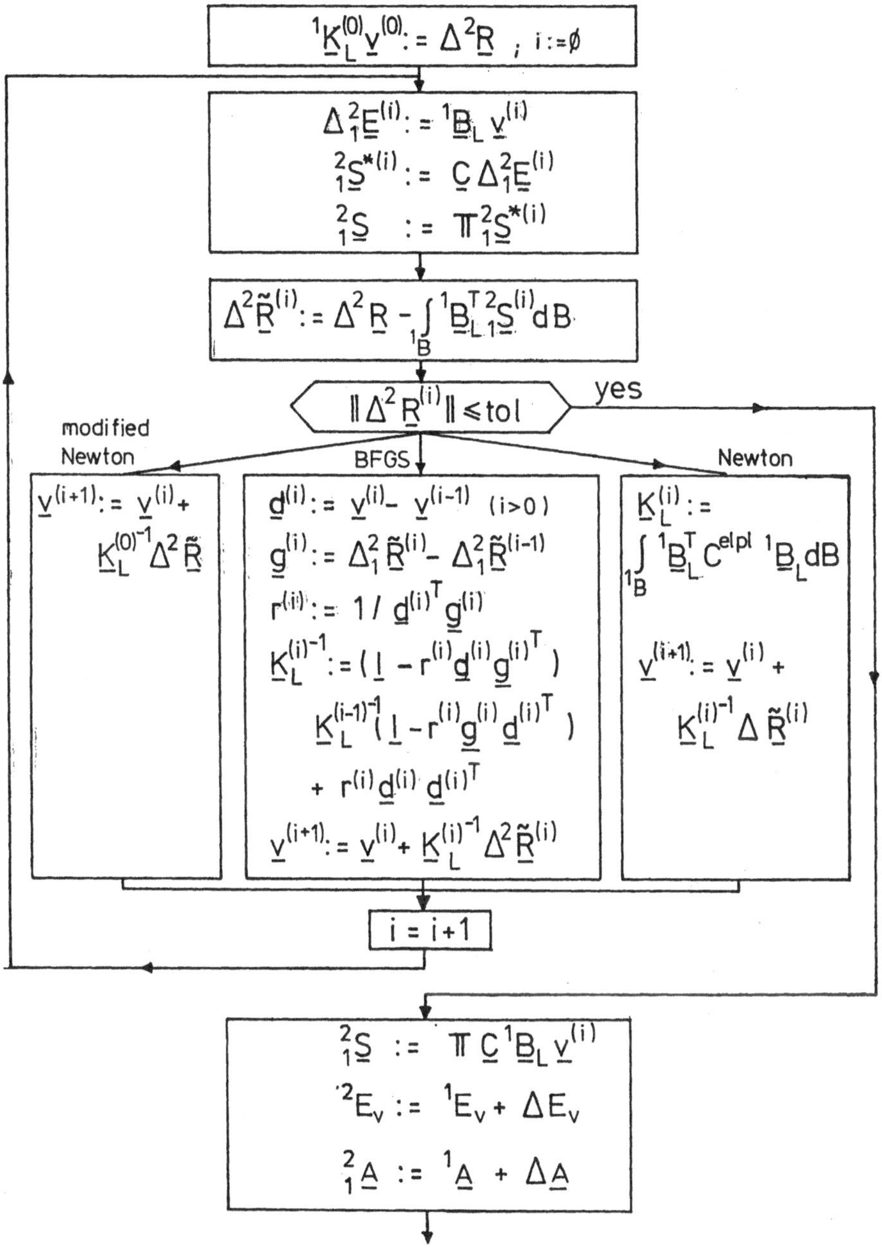

Fig. 7.2: Flow chart for elasto-plastic algorithm

8. Examples

8.1 Large deflection of a cantilever

This well known example (fig. 8.1) is commonly used in order to show the quality of elements [2] and algorithms. In this context, it is used to compare the qualities of different algorithms for the treatment of geometrically nonlinear problems. Fig. 8.2 and 8.3 show the relative errors of the displacements u and w compared with the analytical solution for different incrementations. Obviously, the iteration scheme (algorithm 2) yields better results with less numerical effort than algorithm 1.

8.2 Beam with axial and transversal load

In fig. 8.4 this structure is shown together with its deformed configurations. For F = 10 kN, the loaded end point is rotated by an angle larger than 90°. The displacement u and the rotation φ of the endpoint and the displacement w of the midpoint are displayed in fig. 8.5 and 8.6. The size of the load increments was adapted to the non-linear behaviour of the structure (the result for each increment is marked). The relative error increases with the curvature of the load-displacement curves and is reduced by smaller load increments.

8.3 Plate girder subjected to patch loading

Our special interest is the calculation of ultimate loads of thin walled structures and its confirmation with experiments [14]. The following example refers to experiments performed at the Technical University Braunschweig [15]. The measured initial imperfections were considered in the FE-analysis.

Table 8.1: Calculated ultimate loads

number of elements girder A-27		lower bound	upper bound
elastic plastic	96	220	244
	128	284	294
	257	244	254
	322	244	254
	362	244	249
elastic 322		970	980
experimental		248	

Using symmetry-conditions, one half of the girder was analyzed with up to 362 DKT-elements, see fig. 8.7. The experimental ultimate load is 248 kN. Table 8.1 gives our results obtained with different FE-meshes [14]. A purely elastic calculation results in a critical load of about 975 kN. Therefore, the influence of plastic deformations of the web (fig. 8.9) on the ultimate load is evident. In the case of this very thin web, there are no plastic deformations in the flanges. The calculated deformations (fig. 8.8) agree well with the experimental results.

9. Conclusions

In this paper, two algorithms are described for the static calculation of thin-walled structures with large rotations and displacements but small elastic-plastic strains. It is shown how the material non-linearity can be simply and effectively treated using projection methods and consistent tangents in the iteration process.

Examples for elastic and elastic-plastic material show the behaviour and the applicability of the algorithms even for difficult ultimate load calculations.

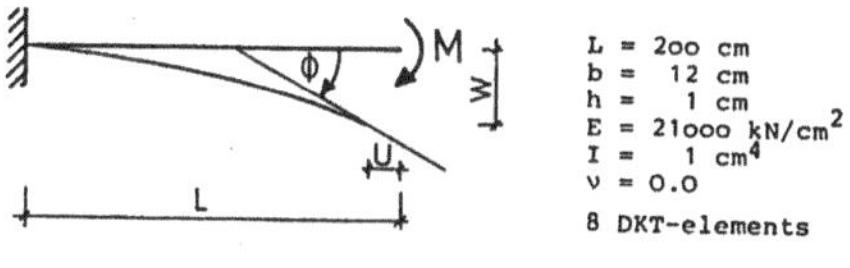

Fig. 8.1: Cantilever beam

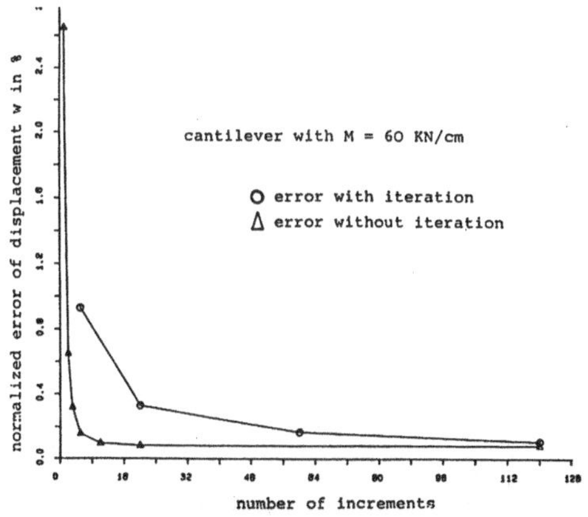

Fig. 8.2: Error of displacement w

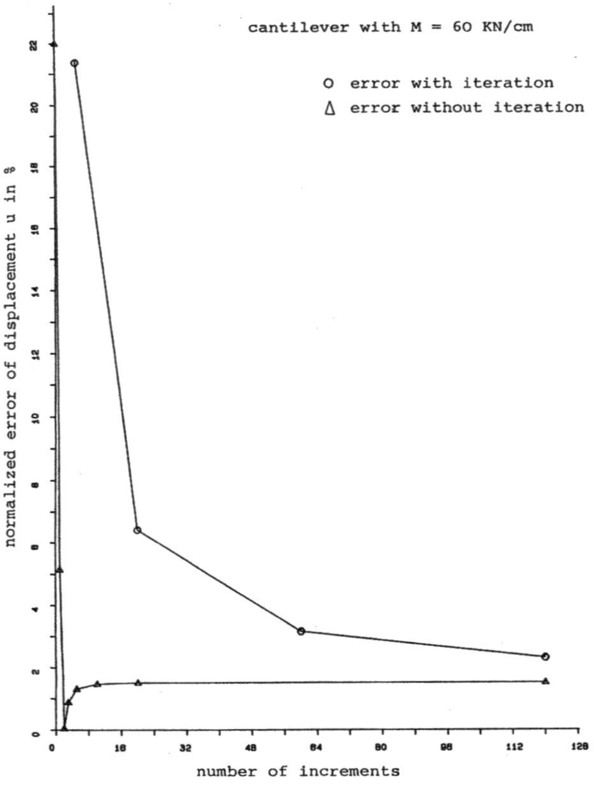

Fig. 8.3: Error of displacement u

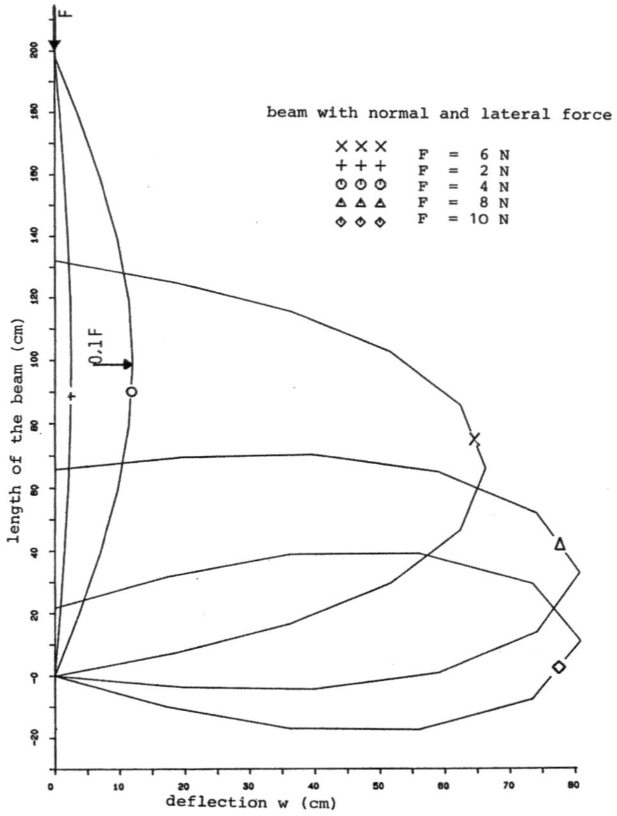

Fig. 8.4: Beam with normal and lateral force

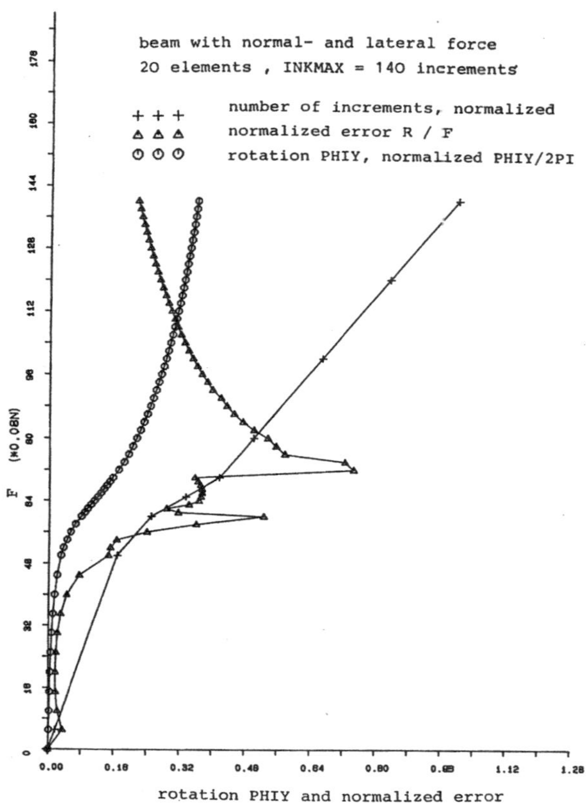

Fig. 8.5: Rotation and normalized residual load

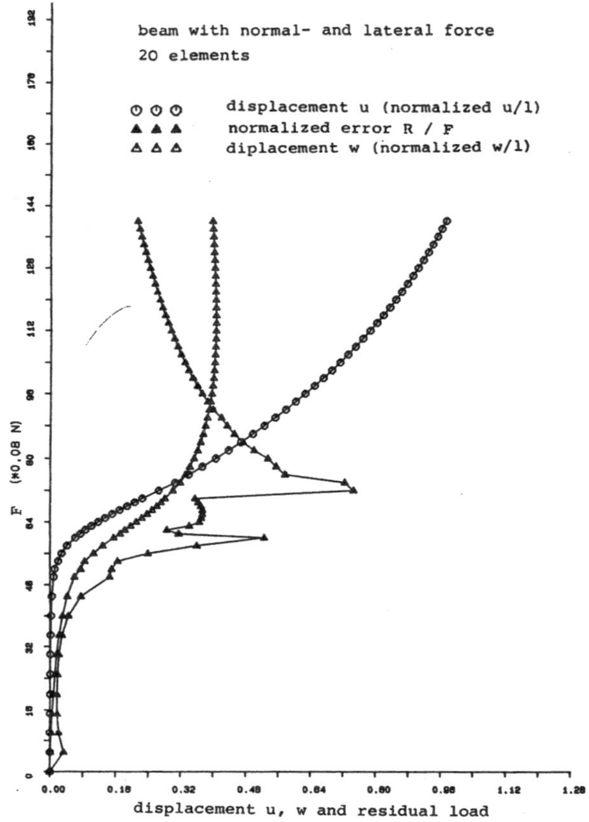

Fig. 8.6: Displacement u and normalized residual load

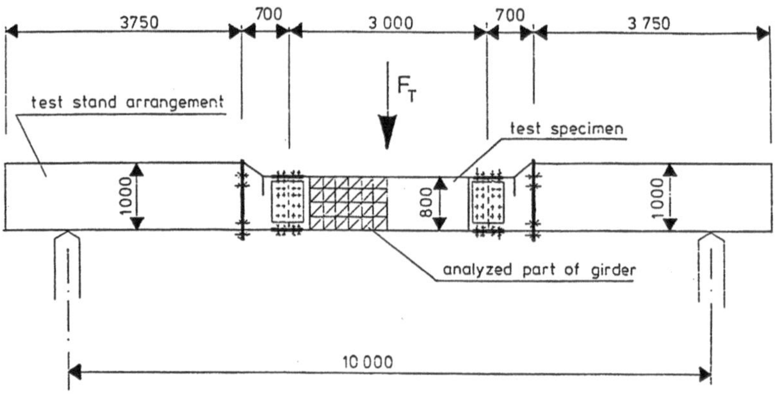

Fig. 8.7: Test stand arrangement of girder

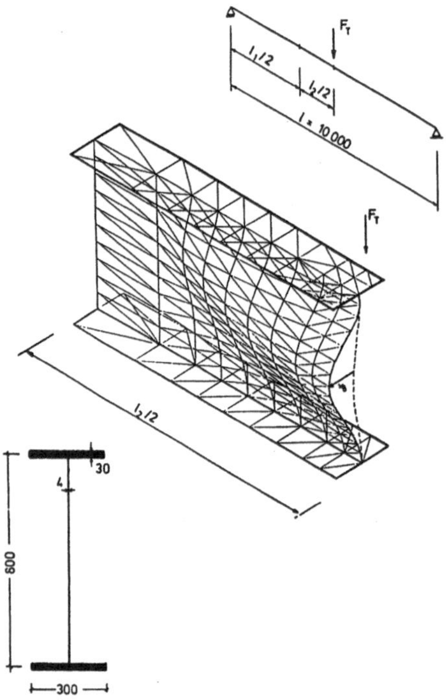

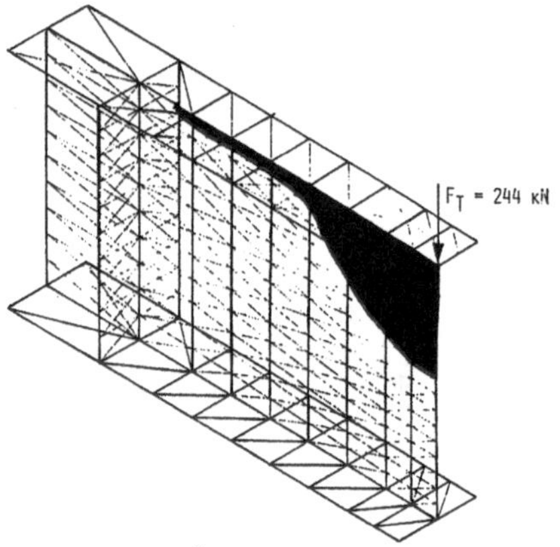

Fig. 8.9: Yield zone of girder at limit load

Fig. 8.8: Deformation of girder at limit load

References

1. BATHE K.J., RAMM E., WILSON E.L., Finite element formulations for large deformation dynamic analysis, Int. J. Num. Meth. Eng. 9 (1975), 353-386.

2. STEIN E., WAGNER W., LAMBERTZ K.H., Geometrically nonlinear theory and incremental analysis of thin shells, in: Flexible Shells, ed. by E.L. Axelrad, F.A. Emmerling, Springer Verlag, Berlin, Heidelberg, New York (1984).

3. STEIN E., LAMBERTZ K.H., PLANK L., Ultimate load analysis of folded plate structures with large elasto-plastic deformations - theoretical and practical comparisons of different FE-algorithms - NUMETA 85, Swansea, U.K. (1985), Verlag A.A. Balkema, Rotterdam, Boston.

4. BATHE K.J., HO L.W., A simple and effective element for analysis of general shell structures, Comp. & Struct. 13 (1981), 673-681.

5. JOHNSON C., Existence theorems for plasticity problems, J. Math. pures et appl. 55 (1976), 431-444.

6. HALPHEN B., NGUYEN QUOC SON, Sur les materiaux standards generalises, Journal de Mecanique 14 (1975), 39-63.

7. KRIEG R.D., KRIEG D.B., Accuracies of numerical solution methods for the elastic-perfectly plastic model, Trans. ASME, Journ. PVP, 99 (1975), 510-515.

8. MATTHIES H., STRANG G., TEMAM R., Mathematical and computational methods in plasticity, in: Variational methods in the mechanics of solids, ed. S. Nemat-Nasser, Pergamon Press, Oxford, 1980, 20-28.

9. SCHREYER H.L., KULAK R.F., KRAMER J.M., Accurate numerical solutions for elastic-plastic models, Trans. ASME, Journ. PVP, 101 (1979).

10. MARGUES J.M.M.C., Stress computation in elastoplasticity, Eng. Comp. 1 (1984), 42-51.

11. GRUTTMANN F., Konsistente Steifigkeitsmatrizen in der Elasto-Plastizitätstheorie, Workshop "Diskretisierungen in der Kontinuumsmechanik, Finite Elemente und Randelemente", Bad Honnef, 1985.

12. SIMO J., TAYLOR R.L., Consistent tangent operators for rate-independent elasto-plasticity, Comp. Meth. Appl. Mech. Eng. 48 (1985), 101-118.

13. BATHE K.J., BATOZ J.L., HO L.W., A study of three-node triangular plate bending elements, Int. J. Num. Meth. Eng., 15 (1980), 1771-1812.

14. STEIN E., LAMBERTZ K.H., PLANK L., Traglastberechnung dünnwandiger Strukturen bei großen elastoplastischen Deformationen, Stahlbau 1985, H. 1.

15. SCHEER J., Dehnungsmessungen an Versuchsträgern.., Bericht Nr. 6o92, Inst. f. Stahlbau, TU Braunschweig, 1983.

COMPATIBILITY OF ROTATIONS WITH THE CHANGE-OF-METRIC MEASURES IN A DEFORMATION OF A MATERIAL SURFACE

M. L. SZWABOWICZ
Institute of Fluid-Flow Machinery
of the Polish Academy of Sciences
ul.J.Fiszera 14, 80-952 Gdańsk,Poland

1. Introduction

Considering a deformation of a thin shell, modelled by a flexible material surface, one is confronted with diversity of shapes that can be attained by stretching and bending. Yet, not all conceivable shapes are accessible in reality. There is a reasonable limit to this richness, be it due to the fact that any material can withstand stretching only up to a certain point and then fails. Once the stretches are restricted, some deeper reaching consequences of this may be detected. For instance, a flat plate cannot be deformed into a strongly curved cap without generating large stretches. Apparently, the nature of this restraint involves something more than just mere restriction on the elongation of the material fibres of the shell. Curvature must be liable to some restrictions of a similar type and, consequently, so must be the rotations.

This aspect of the nonlinear shell theory has not, of course, been overlooked by the researchers. The lack of place prevents us from citing the multitude of papers in which it has been exploited in variety of ways in order to introduce significant simplifications to the equations of equilibrium. Starting with the Marguerre's shallow shell theory in the thirties this trend is continued up to the present moment. Let us only mention here the work by Koiter [1], Pietraszkiewicz's classification of the nonlinear theories with respect to rotations [2], Axelrad's series of publications on flexible shells (for references see [3]) and the monograph by Libai and Simmonds [4] as the latest entries on this long list. However, the argumentation used seems to be founded to a certain extent rather on the intuitive and experimental than rational grounds. This paper is intended to make a step towards a strict geometric interpretation that would allow for more solid arguments.

The essence of the matter lies undoutedly in the compatibility equations. They mark a clear division between the properties of a deforming surface belonging and not belonging to its intrinsic geometry [5]. Therefore they indicate the kinematic quantities depending directly on the change-of-metric measures of deformation (be it stretches or strains) and the others independent of them. Since the Gaussian curvature is, according to the Gauss theorema egregium [5], an intrinsic property of a surface it must be determined completely by the stretches. This idea and its implications are the pivot of the present paper.

The compatibility equations, ever since Goldenweiser appreciated their role in the theory of shells, were a matter of interest to many authors - see [1-4] and further references also there. Yet, the present author is not aware of any paper that would assume the approach followed here or that would explore and reveal explicitly the facts stated below.

2. Preliminary remarks

For the reader's convenience we shall recall some well-known facts from tensor algebra and differential geometry. For an extensive treatment see for example [5].

Let M denote a regular surface imbedded in the three-dimensional Euclidean space R^3. Let $\underset{\sim}{r} = \underset{\sim}{r}(\theta^\alpha)$, $\alpha = 1,2$ be its position vector and $\underset{\sim}{a}_\alpha = \underset{\sim}{r},_\alpha$, $\underset{\sim}{a}^\alpha$ the co- and contravariant base vectors, respectively. The orientation of M is determined by the unit normal vector $\underset{\sim}{n} = \underset{\sim}{n}(\underset{\sim}{r})$.

In the sequel we shall employ the following conventions concerning the algebra of the second-order tensors $\underset{\sim}{A} = A^{ij}\, \underset{\sim}{a}_i \otimes \underset{\sim}{a}_j$ $(i,j = 1,2,3$ $\underset{\sim}{a}_3 = \underset{\sim}{n})$ and $\underset{\sim}{B} = B^{kl}\, \underset{\sim}{a}_k \otimes \underset{\sim}{a}_l$: the composition of $\underset{\sim}{A}$ and $\underset{\sim}{B}$ is the second-order tensor $\underset{\sim}{A}\underset{\sim}{B} = A_{ij}^{\cdot\cdot}B^{jk}\, \underset{\sim}{a}^i \otimes \underset{\sim}{a}_k$, whereas the full contraction is denoted by $\underset{\sim}{A}\cdot\underset{\sim}{B} = A^{ij}B_{ij}$. A cross product of the vector $\underset{\sim}{v}$ and the tensor $\underset{\sim}{A}$ is the second-order tensor $\underset{\sim}{v} \times \underset{\sim}{A} = A^{ij}\,(\underset{\sim}{v} \times \underset{\sim}{a}_i) \otimes \underset{\sim}{a}_j$. Besides, $\underset{\sim}{A}^t$ denotes the transpose of $\underset{\sim}{A}$, whereas $\underset{\sim}{A}^s$ and $\underset{\sim}{A}^a$ its symmetric and skew-symmetric parts, respectively.

The tangent plane at a point $M \in M$ will be denoted by T_M. Let $\underset{\sim}{F} = \underset{\sim}{F}(\underset{\sim}{r})$ be any k'th order $(k = 0,1,2\ldots)$ tensor field defined on M. The differential of $\underset{\sim}{F}$ at the point M on the vector $d\underset{\sim}{r} \in T_M$ is given by the formula $d\underset{\sim}{F} = \underset{\sim}{F} \otimes \underset{\sim}{\nabla}\, d\underset{\sim}{r}$, where $(\,)\underset{\sim}{\nabla} = (\,),_\alpha \underset{\sim}{a}^\alpha$ is the derivative operator. The tensor $\underset{\sim}{F}\otimes\underset{\sim}{\nabla} = \underset{\sim}{F},_\alpha \otimes \underset{\sim}{a}^\alpha$ of the order $k+1$ is the gradient of the field $\underset{\sim}{F}$. Whenever $k > 0$ we may define the divergence of the field $\underset{\sim}{F}$ as $\operatorname{div} \underset{\sim}{F} = \underset{\sim}{F}\underset{\sim}{\nabla} = \underset{\sim}{F},_\alpha \underset{\sim}{a}^\alpha$, $\underset{\sim}{F}\underset{\sim}{\nabla}$ being itself a tensor field of the order $k-1$.

Let us now introduce the following second-order tensors on M: $\underset{\sim}{a} = \underset{\sim}{r}\otimes\underset{\sim}{\nabla}$ the metric tensor, $\underset{\sim}{\varepsilon} = \varepsilon^{\alpha\beta}\, \underset{\sim}{a}_\alpha \otimes \underset{\sim}{a}_\beta = \varepsilon_{\alpha\beta}\, \underset{\sim}{a}^\alpha \otimes \underset{\sim}{a}^\beta$ the surface permutation tensor and $\underset{\sim}{b} = -\underset{\sim}{n}\otimes\underset{\sim}{\nabla}$ the curvature tensor. In addition, we remind of the following useful identities holding on the surface M:

$$\begin{aligned} &\underset{\sim}{a}\cdot\underset{\sim}{a} = 2 \\ &\underset{\sim}{\varepsilon} = -\underset{\sim}{\varepsilon}^t = -\underset{\sim}{n}\times\underset{\sim}{a} = -\underset{\sim}{a}\times\underset{\sim}{n} \\ &\underset{\sim}{\varepsilon}\underset{\sim}{\varepsilon} = -\underset{\sim}{a} \quad , \quad \underset{\sim}{\varepsilon}\cdot\underset{\sim}{\varepsilon} = 2 \end{aligned} \tag{2.1}$$

Besides, any second-order surface tensor $\underset{\sim}{\tau} = \tau^{\alpha\beta}\, \underset{\sim}{a}_\alpha \otimes \underset{\sim}{a}_\beta$ may be represented in the form

$$\begin{aligned} &\underset{\sim}{\tau} = \underset{\sim}{\tau}^s + \underset{\sim}{\tau}^a \\ &\underset{\sim}{\tau}^s = \tau_1\, \underset{\sim}{e}_1\otimes\underset{\sim}{e}_1 + \tau_2\, \underset{\sim}{e}_2\otimes\underset{\sim}{e}_2 \quad , \quad \underset{\sim}{e}_2 = \underset{\sim}{\varepsilon}\underset{\sim}{e}_1 \\ &\underset{\sim}{\tau}^a = \tfrac{1}{2}\,(\underset{\sim}{\tau}\cdot\underset{\sim}{\varepsilon})\,\underset{\sim}{\varepsilon} \quad , \end{aligned} \tag{2.2}$$

where τ_α are the principal values and $\underset{\sim}{e}_\alpha$ the principal directions of its symmetric part. According to the Cayley-Hamilton theorem any such tensor must satisfy the equation

$$\underset{\sim}{\tau}^2 - \mathrm{tr}(\underset{\sim}{\tau})\,\underset{\sim}{\tau} + \det(\underset{\sim}{\tau})\,\underset{\sim}{a} = \underset{\sim}{0} \quad , \quad \mathrm{tr}(\underset{\sim}{\tau}) = \underset{\sim}{\tau}\cdot\underset{\sim}{a} \quad . \tag{2.3}$$

Hence, applying (2.3) to the tensor $\underset{\sim}{\tau}\underset{\sim}{\varepsilon}$ we obtain

$$\underset{\sim}{\tau}\underset{\sim}{\varepsilon}\underset{\sim}{\tau}\underset{\sim}{\varepsilon} + (\underset{\sim}{\tau}^a\cdot\underset{\sim}{\varepsilon})\,\underset{\sim}{\tau} + \det(\underset{\sim}{\tau})\,\underset{\sim}{a} = \underset{\sim}{0} \quad . \tag{2.4}$$

In particular, for the curvature tensor $\underset{\sim}{b}$ we have

$$\underset{\sim}{b}^2 - 2H\,\underset{\sim}{b} + K\,\underset{\sim}{a} = \underset{\sim}{0} \quad , \quad \underset{\sim}{b}\underset{\sim}{\varepsilon}\underset{\sim}{b}\underset{\sim}{\varepsilon} = -K\,\underset{\sim}{a} \quad , \tag{2.5}$$

where H is the mean curvature and K the Gaussian curvature.

Let C be any regular closed curve on M, parametrized by its arc length s. Then $\underset{\sim}{\nu}, \underset{\sim}{t}, \underset{\sim}{n}$ is its Darboux orthonormal triad and $d\underset{\sim}{r} = ds\, \underset{\sim}{t}$, $\underset{\sim}{\nu} = \underset{\sim}{t}\times\underset{\sim}{n}$. In the sequel we shall make use of the following integral theorems:

(a) the Stokes theorem

$$-\iint_M (\underset{\sim}{F}\underset{\sim}{\varepsilon})\underset{\sim}{\nabla}\; dA = \oint_C \underset{\sim}{F}\; d\underset{\sim}{r} \quad ; \tag{2.6}$$

(b) the Gauss-Bonnet theorem

$$\iint_M K\; dA + \oint_C \underset{\sim}{n}\cdot\underset{\sim}{t}\times d\underset{\sim}{t} + \sum_i \Theta_i = 2\pi \quad , \tag{2.7}$$

where Θ_i is the rotation angle of the tangent $\underset{\sim}{t}$ at the i-th corner of C in the parallel transport of $\underset{\sim}{t}$ along C.

3. Mappings of surfaces

Let $F: M \to \overline{M}$ be a mapping of an initially regular surface M into another regular surface $\overline{M}$. The quantities related to $\overline{M}$ will be denoted by overbars: $\bar{\mathbf{a}}$, $\bar{\boldsymbol{\varepsilon}}$, $\bar{\mathbf{b}}$ etc. The two-point second-order tensor $\mathbf{F} = \bar{\mathbf{a}}_\alpha \otimes \mathbf{a}^\alpha$ such that

$$d\bar{\mathbf{r}} = \mathbf{F}\, d\mathbf{r} \quad , \quad d\mathbf{r} \in T_M \quad , \quad d\bar{\mathbf{r}} \in T_{\overline{M}} \tag{3.1}$$

is the gradient of the mapping F.

Certainly, not every tensor field $\mathbf{F}$ is a gradient of some mapping of M into $\overline{M}$. The necessary and sufficient conditions for $\mathbf{F}$ to be a gradient, called the compatibility conditions, result from the demand that the image $\overline{C} = F(C)$ of every closed curve C on M remain a closed curve on $\overline{M}$. Since for every closed curve C in the space the equation $\oint_C d\mathbf{r} = \mathbf{0}$ holds, we can obtain this condition from (2.6) in the form

$$(\mathbf{F}\boldsymbol{\varepsilon})\boldsymbol{\nabla} = \mathbf{0} \; . \tag{3.2}$$

Thus, any $\mathbf{F}$ satisfying (3.2) is a gradient of some mapping. Especially, for the identical mapping $\mathbf{F} = \mathbf{a}$ of M into itself (3.2) yields the equation $\boldsymbol{\varepsilon}\boldsymbol{\nabla} = \mathbf{0}$.

Let us consider some particular mapping N of the surface M into the unit sphere S^2, called the Gauss (or the spherical) mapping. In this mapping to every point $M \in M$ corresponds a point $N(M) \in S^2$ determined by the direction of the unit normal $\mathbf{n}$ at the point M. This is a straightforward matter to verify that $-\mathbf{b} = \mathbf{n}\otimes\boldsymbol{\nabla}$ is its gradient [5]. Hence it follows that $\mathbf{b}$ must also satisfy the compatibility condition (3.2). The resulting vector equation

$$(\mathbf{b}\boldsymbol{\varepsilon})\boldsymbol{\nabla} = \mathbf{0} \tag{3.3}$$

is known as the Mainardi-Codazzi equations.

Note that the condition (3.2) admits also singular gradients. For instance, the Gaussian images of developable surfaces are curves on S^2 (and especially that of a plane is a point on S^2) for that matter. Yet, whenever $\mathbf{F}$ is not singular it can be represented in the form analogous to the polar decomposition of three-dimensional gradients, that is

$$\begin{aligned} &\mathbf{F} = \mathbf{R}\mathbf{U} \quad , \\ &\mathbf{U} = \lambda_1\, \mathbf{e}_1\otimes\mathbf{e}_1 + \lambda_2\, \mathbf{e}_2\otimes\mathbf{e}_2 \quad , \quad \mathbf{e}_2 = \mathbf{e}_1\boldsymbol{\varepsilon} \quad , \quad \mathbf{e}_\alpha \in T_M , \end{aligned} \tag{3.4}$$

where $\underset{\sim}{R}$ is an orthogonal tensor and $\underset{\sim}{U}$ is a symmetric surface tensor, not necessarily positive definite in general. The truthfulness of (3.4) follows from the Tissot's theorem [5] after application of the polar decomposition theorem to the tensor $\underset{\sim}{G} = \underset{\sim}{F} + \underset{\sim}{\bar{n}} \otimes \underset{\sim}{n}$.

Carrying the analysis further one may derive the following relations characterizing the mapping of M into $\bar{M}$:

$$\begin{aligned} &\underset{\sim}{\bar{\varepsilon}} = \underset{\sim}{R}\underset{\sim}{\varepsilon}\underset{\sim}{R}^t \quad , \quad \underset{\sim}{\varepsilon} = \underset{\sim}{R}^t\underset{\sim}{\bar{\varepsilon}}\underset{\sim}{R} \\ &J = \frac{d\bar{A}}{dA} = \det(U) = \lambda_1\lambda_2 \neq 0 \\ &\underset{\sim}{F}^{-1}\underset{\sim}{F} = \underset{\sim}{a} \quad , \quad \underset{\sim}{F}\underset{\sim}{F}^{-1} = \underset{\sim}{a} \quad , \quad \underset{\sim}{F}^{-1} = \underset{\sim}{U}^{-1}\underset{\sim}{R}^t \\ &(\,)\underset{\sim}{\bar{\nabla}}\underset{\sim}{F} = (\,)\underset{\sim}{\nabla} \quad . \end{aligned} \tag{3.5}$$

Remark. Note, that in the case of the Gauss mapping N the gradient $\underset{\sim}{F}_N = -\underset{\sim}{b}$ is symmetric. Hence it follows that $\underset{\sim}{R}_N = \underset{\sim}{1}$, $\underset{\sim}{U}_N = -\underset{\sim}{b}$ and, consequently, that $\underset{\sim}{a}_N = \underset{\sim}{a}$, $\underset{\sim}{\varepsilon}_N = \underset{\sim}{\varepsilon}$, $J_N = K$, $(\,)\underset{\sim}{\nabla}_N\underset{\sim}{b} = -(\,)\underset{\sim}{\nabla}$.

Let us now assume that M coincides with the middle surface of a thin shell. Evidently, not all mappings of M correspond to physical deformations. The necessary and sufficient condition for $\underset{\sim}{F}$ to be a gradient of some deformation of M may be stated as follows:

(i) $\underset{\sim}{F}$ must satisfy the compatibility condition (3.2);

(ii) $\underset{\sim}{R}$ must be proper orthogonal, that is, it has to be a rotation

$$\begin{aligned} \underset{\sim}{R} &= \underset{\sim}{i}\otimes\underset{\sim}{i} + \cos\alpha\,(\underset{\sim}{1} - \underset{\sim}{i}\otimes\underset{\sim}{i}) + \sin\alpha\,\underset{\sim}{i}\times\underset{\sim}{1} = \\ &= \underset{\sim}{\bar{e}}_1\otimes\underset{\sim}{e}_1 + \underset{\sim}{\bar{e}}_2\otimes\underset{\sim}{e}_2 + \underset{\sim}{\bar{n}}\otimes\underset{\sim}{n} \end{aligned} \tag{3.6}$$

around the axis $\underset{\sim}{i}$ by the angle α;

(iii) both principal values of the stretch tensor $\underset{\sim}{U}$ must be strictly positive $\lambda_\alpha > 0$ (consequently, $J>0$ and $\underset{\sim}{U}^{-1}$ always exists).

4. Rotation versus stretch

Henceforth we assume that $\underset{\sim}{F}$ is a gradient of some deformation, so that the conditions (i)-(iii) from the section 3 are fulfilled. Clearly, regarding stretches as free variables, the condition (i) in connection with the decomposition (3.4) imposes some sort of constraints on the rotations. After introducing (3.4) into (3.2) and some additonal transformations it can be shown to yield the vector equation

$$\underset{\sim}{R}^t\underset{\sim}{R}_{,\alpha}\underset{\sim}{U}\underset{\sim}{\varepsilon}\underset{\sim}{a}^\alpha = -(\underset{\sim}{U}\underset{\sim}{\varepsilon})\underset{\sim}{\nabla} \quad . \tag{4.1}$$

Since the tensor $\underset{\sim}{R}^t\underset{\sim}{R}_{,\alpha}$ is skew-symmetric [4] its surface part

$\underset{\sim\sim}{aR}^t\underset{\sim}{R}_{,\alpha}\underset{\sim}{a}^\alpha$, according to $(2.2)_3$, may be represented in the form

$$\underset{\sim\sim}{aR}^t\underset{\sim}{R}_{,\alpha}\underset{\sim}{a} = \tfrac{1}{2}(\underset{\sim}{\varepsilon}\cdot\underset{\sim}{R}^t\underset{\sim}{R}_{,\alpha})\,\underset{\sim}{\varepsilon} \quad . \tag{4.2}$$

Introducing this to (4.1), then multiplying both sides by $\underset{\sim}{U}$ and applying (2.4) we obtain the vector equation

$$(\underset{\sim}{\varepsilon}\cdot\underset{\sim}{R}^t\underset{\sim}{R}_{,\alpha})\underset{\sim}{a}^\alpha = 2\ J^{-1}\underset{\sim}{U}(\underset{\sim\sim}{U\varepsilon})\underset{\sim}{\nabla} \quad . \tag{4.3}$$

After substitution of (3.6) to (4.3) and further transformations we may express the compatibility condition (3.2) directly in the angle and the axis of rotation

$$(\underset{\sim}{i}\cdot\underset{\sim}{n})(\alpha\underset{\sim}{\nabla}) + (1-\cos\alpha)\ \underset{\sim\sim}{i\varepsilon}(\underset{\sim}{i}\otimes\underset{\sim}{\nabla}) + \sin\alpha\ \underset{\sim}{n}(\underset{\sim}{i}\otimes\underset{\sim}{\nabla}) = \\ = J^{-1}\ \underset{\sim}{U}(\underset{\sim\sim}{U\varepsilon})\underset{\sim}{\nabla} \quad . \tag{4.4}$$

Hence, rotations corresponding to a given stretch field $\underset{\sim}{U}$ must satisfy (4.3) or (4.4). It is difficult to state something more about the class of all admissible rotation fields basing only on the equation (4.4), for treating $\underset{\sim}{U}$ as a parameter it becomes a system of two complex nonlinear partial differential equations. However, this problem becomes more clear in the context of the Gauss mapping.

Let us consider a subregion $M'\subset M$ with $\underset{\sim}{b}$ nowhere singular, i.e. such that its Gaussian curvature $K \neq 0$ for all $M \in M'$. The underlying idea is illustrated by the following diagram

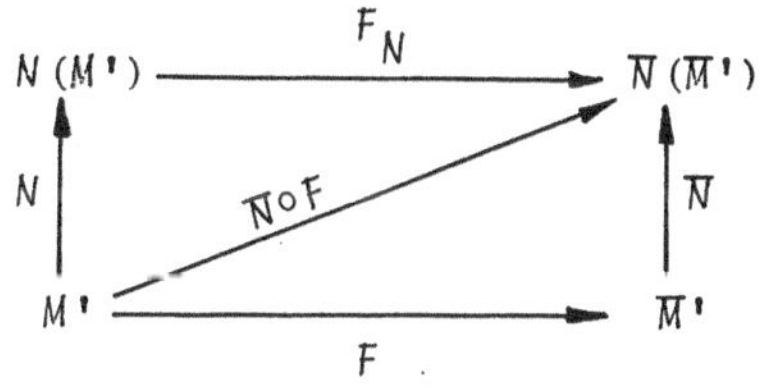

(4.5)

To every deformation of M' into $\bar{M}' = F(M')$ there corresponds a unique mapping of the Gaussian images $F_N\colon N(M') \to \bar{N}(\bar{M}')$. Note, that the mapping F_N depicts directly the rotation field $\underset{\sim}{R}$ generated by the deformation F, for the Gaussian images are formed by the ends of the unit normal vectors to the surface M before and after the deformation. Thus, examination of F_N or $\bar{N}\circ F$ yields direct information about the rotations.

Since the diagram (4.5) is commutative the gradient of F_N can be found as $\underset{\sim}{F}_N = \underset{\sim\sim\sim}{\bar{b}Fb}^{-1}$, and that of the composition $N\circ F$ as $\underset{\sim}{F}_{\bar{N}\,F} = -\underset{\sim\sim}{\bar{b}F}$. According to (3.2) and the remark after (3.5), they must satify the two, equvalent to each other, compatibility conditions

$$(\underset{\sim\sim\sim}{bFb\varepsilon})\underset{\sim}{\nabla}_N = \underset{\sim}{0} \qquad \text{or} \qquad (\underset{\sim\sim}{\overline{b}F\varepsilon})\underset{\sim}{\nabla} = \underset{\sim}{0} \tag{4.6}$$

(However, note that $(4.6)_2$ must hold even in the case of singular N). These are the Mainardi-Codazzi equations for the deformation of surfaces.

Let us now examine how the area of the Gaussian image $N(M')$ changes in the deformation of M' into $\overline{M}'$. It follows from (3.5) and the Gauss-Bonnet theorem (2.7) that for a regular closed curve C the following equation holds true

$$\begin{aligned}\iint_{\overline{N}(\overline{M}')} d\overline{A}_{\overline{N}} - \iint_{N(M')} dA_N &= \iint_{M'} (J\overline{K} - K)\, dA = \\ &= -\oint_{\overline{C}} \underset{\sim}{n}\cdot\underset{\sim}{t}\times d\underset{\sim}{t} + \oint_C \underset{\sim}{n}\cdot\underset{\sim}{t}\times d\underset{\sim}{t} \quad .\end{aligned} \tag{4.7}$$

Let φ denote the angle measured from the direction of the principal stretch $\underset{\sim}{e}_1$ to the tangent $\underset{\sim}{t}$ at a point $M \in C$. Then, according to the lemma 2 from [5] (p.251) and due to the continuity of φ along C we have

$$\oint_C \underset{\sim}{n}\cdot(\underset{\sim}{t}\times d\underset{\sim}{t} - \underset{\sim}{e}_1\times d\underset{\sim}{e}_1) = \oint_C d\varphi = 0 \quad . \tag{4.8}$$

Since the same must hold true along the deformed curve $\overline{C}$ and due to the fact that the directions of the principal stretches are material, that is $\overline{\underset{\sim}{e}}_\alpha = \underset{\sim\sim}{Re}_\alpha$, we may transform the right-hand side of (4.7) to the form

$$\begin{aligned}-\oint_{\overline{C}} \overline{\underset{\sim}{n}}\cdot\overline{\underset{\sim}{t}}\times d\overline{\underset{\sim}{t}} + \oint_C \underset{\sim}{n}\cdot\underset{\sim}{t}\times d\underset{\sim}{t} &= -\oint_{\overline{C}} \overline{\underset{\sim}{e}}_2\cdot d\overline{\underset{\sim}{e}}_1 + \oint_C \underset{\sim}{e}_2\cdot d\underset{\sim}{e}_1 = \\ &= -\oint_C \underset{\sim}{e}_2(\underset{\sim}{R}^t d\underset{\sim}{R})\underset{\sim}{e}_1 = \tfrac{1}{2}\oint_C (\underset{\sim}{\varepsilon}\cdot\underset{\sim}{R}^t\underset{\sim}{R}_{,\alpha})\underset{\sim}{a}^\alpha\cdot d\underset{\sim}{r} \quad .\end{aligned} \tag{4.9}$$

Further, substituting the condition (4.3) and applying the Stokes theorem (2.6) we obtain eventually the result in the global form

$$\iint_{\overline{N}(\overline{M}')} d\overline{A}_{\overline{N}} - \iint_{N(M')} dA_N = \iint_{M'} [J^{-1}\underset{\sim\sim}{\varepsilon U}(\underset{\sim\sim}{U\varepsilon})\underset{\sim}{\nabla}]\underset{\sim}{\nabla}\, dA \tag{4.10}$$

or in the local form

$$J\overline{K} - K = [J^{-1}\underset{\sim\sim}{\varepsilon U}(\underset{\sim\sim}{U\varepsilon})\underset{\sim}{\nabla}]\underset{\sim}{\nabla} \quad . \tag{4.11}$$

This result is the extention of the Gauss theorema egregium to the deformation of surfaces. In particular, if M is a plane in R^3 the equation (4.11) yields the Gauss formula for the surface M [5]. Note that (4.11) also holds in the case of singular Gaussian images $N(M)$ and $\overline{N}(\overline{M})$.

The scalar equation (4.11) and the vector equation (or two scalar equations) (4.6) form the complete set of the compatibility equations for an arbitrary deformation of the surface M. They acquire a meaningful interpretation in terms of the diagram (4.5), for they state, in fact, something about the mapping F_N of the Gaussian image $N(M)$ into $\overline{N}(\overline{M})$. For a given stretch field $\underset{\sim}{U}$ the Gauss formula (4.11) restricts the local change of area in the mapping F_N, making it a mapping with an internal constraint. In consequence this imposes a restriction on the class of the admissible patterns of the resultant Gaussian images $\overline{N}(\overline{M})$ and thus also on the admissible rotations. On the other hand the Mainardi-Codazzi equations (4.6) prevent from occurrance of irregularities such as creases after the deformation F of the originally regular surface M.

5. Consequences for small strains

In this section we shall try to settle what conclusions concerning the rotations can be drawn from the Gauss formula (4.11) for the deformations admitting small changes of the metric properties of the surface M. To put things into a more precise setting we shall supplement the conditions (i)-(iii) from the section 3 by the following additional demands:

(iv) the principal values of the strain tensor $\underset{\sim}{\gamma} = \gamma_1 \underset{\sim}{e}_1 \otimes \underset{\sim}{e}_1 + \gamma_2 \underset{\sim}{e}_2 \otimes \underset{\sim}{e}_2$ are small in comparison with unity everywhere on the surface M, $\max(|\gamma_\alpha|) = \eta \ll 1$;

(v) the magnitudes of the gradients $\gamma_\alpha \underset{\sim}{\nabla}$ are of the order not lower than $|\gamma_\alpha \underset{\sim}{\nabla}| = 0\left(\frac{\eta}{L}\right)$, where L is the smallest wavelength of the deformation pattern [1,2,4], comparable with the dimensions of M, and higher gradients are of higher orders;

(vi) the geodesic curvatures $k_{g\alpha} = \underset{\sim}{n} \cdot \underset{\sim}{e}_\alpha \times \frac{d\underset{\sim}{e}_\alpha}{ds_\alpha}$ of the lines of principal strains or stretches are of the order not lower than $0\left(\frac{1}{L}\right)$.

The conditions (v) and (vi) are equivalent to the demand that the deformation pattern of M have fluent character.

Let us estimate the quantities on the right-hand sides of the equations (4.3) and (4.11). Since the strain and stretch tensors are linked by the algebraic equation

$$\underset{\sim}{U}^2 = 2\underset{\sim}{\gamma} + \underset{\sim}{a} \quad , \quad \lambda_\alpha^2 = 2\gamma_\alpha + 1 \quad , \tag{5.1}$$

it can be shown by means of (2.2-4) that

$$\begin{aligned} \lambda_\alpha &= 1 + \gamma_\alpha + O(\eta^2) = 1 + O(\eta) \\ J &= 1 + \gamma_1 + \gamma_2 + O(\eta^2) = 1 + O(\eta) \\ J^{-1} &= 1 - \gamma_1 - \gamma_2 + O(\eta^2) = 1 + O(\eta) \\ |J^{-1}\, \underset{\sim}{U}(\underset{\sim}{U}\underset{\sim}{\varepsilon})\underset{\sim}{\nabla}| &= O\left(\frac{\eta}{L}\right) \\ [J^{-1}\, \underset{\sim}{\varepsilon}\underset{\sim}{U}(\underset{\sim}{U}\underset{\sim}{\varepsilon})\underset{\sim}{\nabla}]\underset{\sim}{\nabla} &= O\left(\frac{\eta^2}{L^2}\right) \quad . \end{aligned} \tag{5.2}$$

On the other hand, due to the fact that the rotation tensor $\underset{\sim}{R}$ can be represented in the form $\underset{\sim}{R} = \bar{\underset{\sim}{e}}_1 \otimes \underset{\sim}{e}_1 + \bar{\underset{\sim}{e}}_2 \otimes \underset{\sim}{e}_2 + \bar{\underset{\sim}{n}} \otimes \underset{\sim}{n}$, where $\bar{\underset{\sim}{e}}_\alpha$ denote the principal directions of stretch on the deformed surface $\bar{M}$, the left-hand side of (4.3) can be interpreted in terms of the geodesic curvatures $\bar{k}_{g\alpha}$ and $k_{g\alpha}$ of the lines of principal stretches on $\bar{M}$ and M, respectively, according to the equation

$$\tfrac{1}{2}\,(\underset{\sim}{\varepsilon} \cdot \underset{\sim}{R}^t \underset{\sim}{R},_\beta)\,\underset{\sim}{e}_\alpha \cdot \underset{\sim}{a}^\beta = -\,\bar{k}_{g\alpha}\lambda_\alpha + k_{g\alpha} \quad . \tag{5.3}$$

Thus, the equation (4.3) sets bounds to the admissible values of the geodesic curvatures of the lines of principal stretches on the deformed surface $\bar{M}$. By virtue of $(5.4)_4$ we have the estimation

$$\bar{k}_{g\alpha}\lambda_\alpha - k_{g\alpha} = O\left(\frac{\eta}{L}\right) \quad . \tag{5.4}$$

The estimation for the bounds to the Gaussian curvature of the deformed surface $\bar{M}$, set by the equation (4.11), follows from $(5.2)_5$ in the form

$$J\bar{K} - K = O\left(\frac{\eta^2}{L^2}\right) \quad . \tag{5.5}$$

The estimations (5.4) and (5.5) have important implications concerning the receptivity of thin shells to the rotations. The difference $\bar{k}_{g\alpha}\lambda_\alpha - k_g$ describes the increment of the in-surface component of the rotation, whereas the difference $J\bar{K} - K$ the change of area of an elementary spherical triangle contained on the Gaussian image of the middle surface $N(M)$. Moreover, all middle surfaces of slowly varying Gaussian curvatures may be divided into two classes from the point of view of the estimation (5.5).

(a) Strongly curved middle surfaces.

These are surfaces with the Gaussian curvatures of the order $|K| = O\left(\frac{\eta}{L^2}\right)$ or lower. In this case the compatibility condition (4.11) may be approximated by the particularily simple algebraic equation

$$J\overline{K} - K \approx 0 \quad , \qquad (5.6)$$

which interpreted in terms of the mapping between the Gaussian images F_N: $N(M) \to \overline{N}(\overline{M})$ means that F_N is an equiareal (or an area-preserving) mapping. These surfaces will, therefore, resist strongly to the rotations up to the critical point, at which large rotations occur rapidly through a snap-through process. This phenomenon is connected indispensably with the violation of the Mainardi-Codazzi equations (4.6) along a curve encircling the buckled region of the middle surface. In reality this manifests itself frequently in the form of a plastic crease along such curve.

(b) The surfaces with the Gaussian curvatures of the order $|K| = 0\left(\frac{n^2}{L^2}\right)$ or higher.

Since the discriminant $|K| = 0\left(\frac{n^2}{L^2}\right)$ of this class can be replaced by the equivalent one $|K|L^2 = 0(n^2) \ll 1$, this is immediately noticable that it coincides fully with what is known in the literature as the quasi-shallow shells [1,4]. Therefore, we may call them quasi-shallow middle surfaces. Note that the developable surfaces belong to this class. Here the full compatibility condition (4.11) must be taken into account, because the admissible local changes of the area of the Gaussian image $N(M)$ are of the order of the initial area itself. These surfaces can submit to even unrestricted rotations still preserving their regularity, though buckling is also possible.

It is not possible to carry out more detailed analysis within the limited space of this paper. Let us, however, point out the following conclusions. All thin shells of regular uncomplicated shapes fall into one of the two classes described above. The division between these classes depends on the magnitudes of the allowed strains and their derivatives. The difference between them consists mainly in their dissimilar receptivity to the rotations. Due to the stiffness of the strongly curved shells, for which any drastic change of shape may be attained only through the loss-of-stability processes, significant simplifications in the governing equationsshould be available. One may expect that suitably simplified equilibrium equations, referred to the undeformed configuration before the loss of stability and to the buckled configuration after, should completely suffice in this case. The level of the nonlinearity taken into account would depend solely on the allowed magnitude of strain. Much less can be said about the quasi-shallow shells. Their flexibility enables them to assume variety of shapes, though even here the available deformation patterns are limited. This case requires further investigation.

References

1. KOITER W.T., On the nonlinear theory of thin elastic shells, Proc. Kon.Ned.Ak.Wet., Ser.B, vol.69, No.1 (1966), 1-54.
2. PIETRASZKIEWICZ W., Finite Rotations and Lagrangean Description in the Non-Linear Theory of Shells, Polish Sci. Publ., Warszawa 1979.
3. AXELRAD E.L., EMMERLING F.A., Intrinsic shell-theory formulation effective for large rotations and an application, Proof of the paper presented at the Euromech-Colloquium Nr.197, Jabłonna 17-20 Sept. 1985.
4. LIBAI A., SIMMONDS J.G., Nonlinear Elastic Shell Theory, in: Advances in Applied Mechanics, Ed. by Hutchinson & Wu, Vol.23, 271-371, Academic Press, 1983.
5. do CARMO M.P., Differential Geometry of Curves and Surfaces, Prentice-Hall, Inc., Enlewood Cliffs, New Jersey 1976.

FINITE ROTATIONS, VARIATIONAL PRINCIPLES AND BUCKLING IN SHELL THEORY

R. VALID
Office National d'Etudes et de Recherches Aérospatiales
29, Avenue de la Division Leclerc - 92320 Châtillon Cedex, France

SUMMARY

A non linear principle including stresses and finite rotations using a direct surface polar decomposition, is presented with some variants, following the three-dimensional approach of FRAEIJS DE VEUBEKE.

A mixed stability criterion which has to be applied for mixed functionals is also derived. Its applications holds for the principle variants.

I. Introduction

Since several years the non linear analysis of structures has made remarkable progress thanks to the extension to non linear computations by the finite element method of large computer programs and to the utilization of powerful computers. In regard to this general and difficult problem, the search of variational principles submitted to finite element discretization has registered significant advances in particular for thin shell problems.

Generally speaking, thin structures like shells or slender ones like beams are subject to large rotations, their deformation remaining small. These considerations have oriented the researchers to introducing large rotations in their theories through exact or more or less approximated formulae and associated variational principles [1-4].

In this aspect important studies, due in particular to PIETRASZKIEWICZ [5,7], have provided, through rigorous formulae of approximation, a classification of different approximations according to the magnitude of rotations. These discussions lie on a polar decomposition of the transformation gradient. One finds in [5,7] an extended bibliography on the question.

Besides, the need of accurate stresses in shell computation, in order to introduce in rupture, damage, buckling, ... criteria, trustworthy values, has prompted the use of mixed principles where stresses like displacements are taken as independent variables. Various principles are described in particular in

[6-10]. As a limit case pure complementary principles were investigated.

We are indebted to KOITER [11] the demonstration of a three-dimensional complementary principle which utilizes the Piola-Lagrange (or Boussinesq) stress tensor, the gradient polar decomposition, and a local inversion of the constitutive equation which admits the non uniqueness of the gradient with respect to the stress, the stability of the equilibrium being a necessary condition for this inversion. Still in three-dimension, FRAEIJS DE VEUBEKE [12] proposed a complementary energy principle. These two studies had the result to take away some difficulties [13, 14]. This way opened by FRAEIJS DE VEUBEKE found applications in the case of membranes or plates [15, 16].

In the case of shells, SIMMONDS and DANIELSON [17] proposed a mixed variational principle lying on the introduction of rotations before deformations and where only a membrane stress function and the finite rotation were finally present as independent unknowns. These authors used to that aim a compatibility condition of the transformation gradient of the midsurface. SCHMIDT [18], starting from the principle of virtual stresses, provided a non linear mixed complementary principle, and STUMPF [19] a complementary principle with moderate rotations. In the treatment of buckling and post-buckling shell problems with moderate rotations, STEIN [20] and STUMPF [21] used the primal principle.

Thus in the case of shells, the search of mixed complementary principles with finite rotations might answer three criteria : a non linear shell variational principle to be discretized, stresses and finite rotations as independent variables, utilization for analyses and stability problems.

Using a purely surface polar decomposition of the transformation gradient of the middle surface, which was some variant compared to the preceding methods, we proposed a mixed complementary principle with finite rotations based on the PIOLA-LAGRANGE membrane stress [22].

In what follows, after having recalled, to be self-contained, this theory, we shall indicate some derived formulations, which would be useful and efficient for numerical analyses, and some variants. Then, from a generalization to mixed functionals of the primal classical stability criterion, we provide the stability criterion associated with the preceding mixed principles. Due to the non trivial form of this criterion, we propose a theoretical verification.

2. Notations [23]

Let us consider a shell whose middle surface Σ_m, of generic point m in the deformed state $\mathcal{E}$, is a differentiable compact manifold of dimension 2, embedded in the three-dimensional Euclidian space E_3. We shall call $\overrightarrow{E_2}$ the tangent plane and N the unit normal at m. The middle surface is supposed to be Riemannian, i.e.

- $\exists$ a scalar product in $\overrightarrow{E_2}$ induced from $\overrightarrow{E_3}$;

- $\exists$ an orthogonal projector field π which projects $\overrightarrow{E_3}$ onto $\overrightarrow{E_2}$ and which defines the Riemannian connection. Noting the transposition operator by a superbar, we have :

$$\Pi = \Pi^2 = \overline{\Pi} \in \mathcal{L}(\overrightarrow{E_3}, \overrightarrow{E_2}).$$

Let us call $\underset{\sim}{m}$ a map such that (s.t.) :

$$m = \underset{\sim}{m}(X), \quad \forall X \in O \subset \mathbb{R}^2, \quad X = \begin{bmatrix} {}^1X \\ {}^2X \end{bmatrix},$$

where O is an open set of $\mathbb{R}^2$.

(In the sequel the underlined tilde will mean "function of ").

This map gives at point m the natural basis

$$\frac{\partial m}{\partial X} = \left[\frac{\partial m}{\partial {}^1X} \; \frac{\partial m}{\partial {}^2X}\right] = [S_1 \; S_2] = S,$$

s.t.

$$dm = \frac{\partial m}{\partial X} dX, \quad \forall dX \in \mathbb{R}^2.$$

Then $\forall$ M $\in$ three-dimensional (3-d) shell body :

$$M = m + Nz, \tag{2.1}$$

where z is the normal coordinate.

An arbitrary virtual variation of M will be :

$$\delta M = \delta m + \delta N z + N \delta z \tag{2.2}$$

We shall assume the classical Kirchhoff-Love (K-L) assumptions :

(i) The geometric normal remains identical to the material one ;

(ii) The normal strain energy is negligible compared with the others.

From hypothesis (i), it comes for any kinematically admissible (K.A.) displacement $\delta m = \delta V$ of m :

$$\delta N = -\overline{\frac{\partial \delta m}{\partial m}} N \tag{2.3}$$

From the preceding assumptions and starting from the 3-d virtual strain energy referred to the deformed state ζ, it can be easily shown that the virtual strain energy of the shell referred to any reference state ζ_0 is written [23, 24] :

$$\delta w = \int_{\Sigma_{m_0}} T_r(\mathbb{n}\,\delta\Gamma + \mathbb{m}\,\delta K) \quad \forall \delta V \text{ K.A.} \tag{2.4}$$

where $\mathbb{n} = \overline{\mathbb{n}}, \; \mathbb{m} = \overline{\mathbb{m}} \in (\overrightarrow{E_{20}}, \overrightarrow{E_{20}})$

are the Piola-Kirchhoff membrane and bending stresses respectively, the dual quantities of $\delta\Gamma$ and δK, the virtual variations of the membrane deformation and change of curvature of Σ_m, s.t.

$$\left[\begin{array}{l} \overline{dm_0}\,\Gamma\,dm_0 \overset{\text{def}}{=} \frac{1}{2}\left[\overline{dm}\,dm - \overline{dm_0}\,dm_0\right] \\ \overline{dm_0}\,K\,dm_0 \overset{\text{def}}{=} \overline{dm}\,\frac{\partial N}{\partial m}\,dm - \overline{dm_0}\,\frac{\partial N_0}{\partial m_0}\,dm_0 \end{array}\right\} \forall dm_0 \in \overrightarrow{E_{20}} \tag{2.5}$$

$\overrightarrow{E_{20}}$ is the tangent plane at m_0 in the reference state ζ_0. In $(2.5)_2$, $\frac{\partial N}{\partial m}$ is the symmetric curvature mapping at $m \in \Sigma_m$ and $\frac{\partial N_0}{\partial m_0}$ at $m_0 \in \Sigma_{m_0}$.

It comes from (2.5) :

$$\left[\begin{array}{l} \Gamma = \frac{1}{2}\left[\overline{\frac{\partial m}{\partial m_0}}\frac{\partial m}{\partial m_0} - 1_{E_{20}}\right] = \overline{\Gamma} \\ K = \overline{\frac{\partial m}{\partial m_0}}\frac{\partial N}{\partial m}\frac{\partial m}{\partial m_0} - \frac{\partial N_0}{\partial m_0} = \overline{K} \end{array}\right\} \in \mathcal{L}(\overrightarrow{E_{20}}, \overrightarrow{E_{20}}) \tag{2.6}$$

$$\left[\begin{array}{l} \delta\Gamma = \frac{1}{2}\left[\overline{\frac{\partial m}{\partial m_0}}\frac{\partial \delta m}{\partial m_0} - \overline{\frac{\partial \delta m}{\partial m_0}}\frac{\partial m}{\partial m_0}\right] = \overline{\delta\Gamma} \\ \delta K = \overline{\frac{\partial \delta m}{\partial m_0}}\frac{\partial N}{\partial m_0} + \overline{\frac{\partial m}{\partial m_0}}\frac{\partial \delta N}{\partial m_0} = \overline{\delta K} \\ \delta N = -\overline{\frac{\partial \delta m}{\partial m}}\, N \end{array}\right. \quad \in \mathcal{L}(\overrightarrow{E_{20}}, \overrightarrow{E_{20}}) \tag{2.7}$$

It is worth noting that Γ and K give the variations of the first and second fundamental forms of Σ_{m_0} (Bonnet's theorem) to within a rigid body displacement.

In the case of hyperelastic shells, one supposes the existence of a surface strain density α function of Γ and K, s.t.

$$\delta w = \int_{\Sigma_{m_0}} \delta\alpha \tag{2.8}$$

Thus

$$\delta\alpha = \frac{\partial\alpha}{\partial\Gamma}\delta\Gamma + \frac{\partial\alpha}{\partial K}\delta K = T_r(\mathrm{n}\,\delta\Gamma + \mathrm{m}\,\delta K) = \mathrm{n}^{\alpha\beta}\delta\Gamma_{\alpha\beta} + \mathrm{m}^{\alpha\beta}\delta K_{\alpha\beta} \tag{2.9}$$
$$\alpha, \beta = 1, 2,$$

in conventional notations.

3. Complementary energy principle with rotation

3.1. Surface polar decomposition

The derivative $\frac{\partial m}{\partial m_0} \in \mathcal{L}(\overrightarrow{E_{20}}, \overrightarrow{E_2})$ can be decomposed in a strictly surface polar decomposition through the following arguments just recalled. Let us consider the positive definite quadratic form

$$\overline{dm}\, dm = \overline{dm_0}\, \overline{\frac{\partial m}{\partial m_0}}\frac{\partial m}{\partial m_0}\, dm_0, \ \forall dm_0 \neq 0 \in \overrightarrow{E_{20}},$$

Sylvester's theorem gives that ∃ a unitary basis S_0 of $\overrightarrow{E_{20}}$ s.t.

$$\overline{\frac{\partial m}{\partial m_0}}\frac{\partial m}{\partial m_0} = S_0 \Delta \overline{S_0}, \quad \overline{S_0} S_0 = 1_{\mathbb{R}^2},$$

where Λ is a diagonal positive matrix. Then setting down $\sqrt{\Delta}$ the diagonal matrix of

Δ square roots components

$$\overline{\frac{\partial m}{\partial m_o}}\frac{\partial m}{\partial m_o} = S_o\sqrt{\Delta}\,\bar{S}_o\sqrt{\Delta}\,\bar{S}_o = \bar{B}B \qquad \text{with } B = \bar{B} = S_o\sqrt{\Delta}\,\bar{S}_o \in \mathcal{L}(\vec{E_{20}}, \vec{E_{20}}).$$

Now for any unitary and "normal" rotation mapping R s.t.

$$\bar{R}R = R\bar{R} = 1_{E_3}$$

$$\overline{\frac{\partial m}{\partial m_o}}\frac{\partial m}{\partial m_o} = \bar{B}\,\overline{R\pi_o}\,R\pi_o B, \tag{3.1}$$

where the range of B is extended on $\vec{E_3}$ by the connection projector π_o. As B admits an inverse, given $\frac{\partial m}{\partial m_o}$ and setting down

$$\frac{\partial m}{\partial m_o} = R\pi_o B \quad , \; B = 1_{E_{20}} + h, \tag{3.2}$$

it may be seen that B is unique, and that $h \in \mathcal{L}(\vec{E_{20}}, \vec{E_{20}})$ is the stretch of Σ_{m_o}. Finally by (3.1)-(3.3) and $(2.6)_1$:

$$\left[\begin{array}{l} \frac{\partial m}{\partial m_o} = R\pi_o\left[1_{E_{20}} + h\right] \; , \; h = \bar{h} \in \mathcal{L}(\vec{E}_{20}, \vec{E}_{20}) \\ \overline{\frac{\partial m}{\partial m_o}}\frac{\partial m}{\partial m_o} = \left[1_{E_{20}} + h\right]^2 = 2\Gamma + 1_{E_{20}} \end{array}\right. \tag{3.3}$$

$(3.3)_1$ is called the surface polar decomposition of $\frac{\partial m}{\partial m_o}$.

Remarks :

1) $R\pi_o$ presents a matrix representation s.t. ${}^i\mathcal{R}_\alpha$, i=1,2,3 ; α=1,2.
2) $\overline{R\pi_o}\, R\pi_o = 1_{E_{20}}$, so that $\overline{R\pi_o}$ is a left inverse of $R\pi_o$.
3) Decomposition $(3.3)_1$ where the rotation follows the stretch, gives a constitutive equation independent of R as it must be.
4) This decomposition of the midsurface gradient is quite sufficient instead of the 3-d gradient one, due to the fact that the tangent plane defines the unit normal. This is a consequence of the K-L assumptions as regard to the shell structure.

3.2. Mixed principle with finite rotation

The primal principle for hyperelastic shells, from the natural reference state $\mathcal{C}_{oo}$ reads

$$\left[\begin{array}{l} \delta\left[\int_{\Sigma_{m_{oo}}} \alpha \, d\Sigma_{m_{oo}} - \mathcal{L}\right] = 0 \quad \forall\; V \text{ K.A.} \\ \text{with} \quad \alpha = \underset{\sim}{\alpha}(\Gamma, K) \; , \; \Gamma = \underset{\sim}{\Gamma}(V) \; , \; K = \underset{\sim}{K}(V), \; m = m_o + V, \; N = \underset{\sim}{N}(V) \end{array}\right. \tag{3.4}$$

where $\mathcal{L}$ is the work of given external loads.

Now (2.9) and $(2.6)_1$ give :

$$
\left[
\begin{array}{l}
\delta\alpha = T_r(\mathbb{n}\,\delta\Gamma + \mathbb{m}\,\delta K) = T_r\left(\mathbb{n}\frac{\partial m}{\partial m_{oo}}\frac{\partial \delta m}{\partial m_{oo}} + \mathbb{m}\,\delta K\right) \\
\delta\alpha = T_r\left(\mathbb{t}\frac{\partial\delta m}{\partial m_{oo}} + \mathbb{m}\,\delta K\right) \\
\text{with} \quad \mathbb{t} = \mathbb{n}\frac{\partial m}{\partial m_{oo}} \in \mathcal{L}(\vec{E}_2, \vec{E}_{2oo}) \; .
\end{array}
\right.
\tag{3.5}
$$

The stress $\mathbb{t} \neq \bar{\mathbb{t}}$, dual of $\frac{\partial\delta m}{\partial m_{oo}}$, will be called the membrane Boussinesq (or Piola) stress and is represented by a matrix s.t. ${}^{\alpha}\mathbb{t}_i$, $\alpha=1,2$; $i=1,2,3$.

Now quantities h, R, K may be taken as independent variables in (3.4) provided that the constraints

$$
\frac{\partial m}{\partial m_{oo}} - R\Pi_{oo}\left[1_{E_{2oo}} + h\right] = 0 \;, \quad \underset{\sim}{K}(V) - K = 0
$$

be introduced in $(3.4)_1$ through Lagrange multipliers (L.M.) t and s, which gives the principle :

$$
\left[
\begin{array}{l}
\delta\underset{\sim}{\mathcal{F}}(V, h, R, K, t, s) = \\
\delta\left[\int_{\Sigma_{m_{oo}}}\left[\alpha + T_r\left(\mathbb{t}\left[\frac{\partial m}{\partial m_{oo}} - R\Pi_{oo}\left[1_{E_{2oo}} + h\right]\right] + s\left[\frac{\partial m}{\partial m_{oo}}\frac{\partial N}{\partial m_{oo}} - \frac{\partial N_{oo}}{\partial m_{oo}} - K\right]\right)\right] - \mathcal{L}\right] = 0 \\
\forall\, V\ K.A. ;\ h = \bar{h},\ K = \bar{K},\ s = \bar{s} \in \mathcal{L}(\vec{E}_{2oo}, \vec{E}_{2oo}) ;\ \mathbb{t} \in \mathcal{L}(\vec{E}_2, \vec{E}_{2oo}) ; \\
R\,|\,\bar{R}R = 1_{E_3} \;;\ \alpha = \underset{\sim}{\alpha}(h, K) .
\end{array}
\right.
\tag{3.6}
$$

Obviously,the L.M. t, which is dual of the derivative $\frac{\partial\delta m}{\partial m_{oo}}$, may be identified with the unsymmetric stress $\mathbb{t}$. Here, due to $(3.3)_2$, α is a function of h and K.

3.3. Euler's equations of principle (3.6)

A virtual variation of α gives now :

$$
\delta\alpha = \frac{\partial\alpha}{\partial h}\delta h + \frac{\partial\alpha}{\partial K}\delta K = T_r(\mathbb{r}\,\delta h + \mathbb{m}\,\delta K) . \tag{3.7}
$$

The stress $\mathbb{r}$, dual of the stretch δh is a symmetric stress called the Jauman membrane stress.

Varying h, R, K in (3.6), the following Euler equations are found successively :

$$
\left[
\begin{array}{l}
2\,\mathbb{r} = \mathbb{t}R\Pi_{oo} + \overline{\mathbb{t}R\Pi_{oo}} \\
\frac{\partial m}{\partial m_{oo}}\mathbb{t} = \overline{\frac{\partial m}{\partial m_{oo}}\mathbb{t}} \iff \mathbb{n} = \bar{\mathbb{n}} \\
\mathbb{m} = s .
\end{array}
\right.
\tag{3.8}
$$

Let us point out that the skew symmetric component of $\mathbb{t}R\Pi_{oo}$ will give no strain energy and may be ignored.

Let us now suppose that $(3.8)_1$ and $(3.8)_3$ are strictly satisfied, the result is :

$$\frac{\partial \mathcal{F}}{\partial h} = \frac{\partial \mathcal{F}}{\partial K} = 0 \ ,$$

which means that h and K must be eliminated, in fact through the constitutive equation. A Legendre transformation gives the complementary energy density

$$\left[\begin{array}{l} \beta = \underset{\sim}{\beta}(\mathbb{r}, \mathbb{m}) = T_r(\mathbb{r} h + \mathbb{m} K) - \alpha \text{, with } \alpha = \underset{\sim}{\alpha}(h, K), \\ h = \underset{\sim}{h}(\mathbb{r}, \mathbb{m}) \ , \ K = \underset{\sim}{K}(\mathbb{r}, \mathbb{m}). \end{array} \right. \tag{3.9}$$

So that with (3.7) :

$$\delta\beta = T_r(\delta\mathbb{r}\, h + \delta\mathbb{m}\, K).$$

Then principle (3.6) becomes :

$$\left| \begin{array}{l} \delta \underset{\sim}{\mathcal{F}}(\mathbb{t}, \mathbb{m}, V, R) = \delta \left[\int_{\Sigma_{m_{oo}}} \left[T_r\left(\mathbb{t} \frac{\partial m}{\partial m_{oo}} + \mathbb{m}\, \underset{\sim}{K}(V) - \mathbb{t} R \Pi_{oo}\right) - \beta \right] - \mathcal{E} \right] = 0 \\ \text{with} \quad \beta = \underset{\sim}{\beta}(\mathbb{r}, \mathbb{m}) \ , \ \mathbb{r} = \overline{\mathbb{r}} = \mathbb{t} R \Pi_{oo} \ , \ R \,|\, \overline{R} R = 1_{E_3} \ , \end{array} \right. \tag{3.10}$$

which is the mixed complementary energy principle with finite rotation.

Remarks

1) (3.10) is a total Lagrangian (T.L.) formulation.
2) In linear elasticity $\underset{\sim}{\beta}(\mathbb{r},\mathbb{m})$ is a quadratic positive definite form, the introduction of which by (3.9) avoids difficulties evoqued in § 1.
3) The unknowns $\mathbb{t}$, $\mathbb{m}$, V, R represent 15 independent scalar unknowns.
4) The constraint $\overline{R}R = 1_{E_3}$ may be taken into account easer directly (see [17, 5]) for large or moderate rotations, or through a L.M. whose value will be given in § 5.

4. Modified Principles

4.1. Linearized principle from a prestressed state

In view of numerical applications (for instance by the Newton-Raphson procedure) a T.L. linearized principle will be used. Noting by a subscript 0 the prestressed state and by a subscript 1 the increments of the unknowns $\mathbb{t}$, $\mathbb{m}$, V, R, linearization of (3.10) is straightforward from a prestressed state $\mathcal{C}_0$. Hence the T.L. principle :

$$\left[\begin{array}{l} \delta \underset{\sim}{\mathcal{F}_L}(\mathbb{t}_1, \mathbb{m}_1, V_1, R_1) = \\ \delta \left[\int_{\Sigma_{m_{oo}}} \left[T_r\left(\mathbb{t}_1 \frac{\partial V_1}{\partial m_{oo}} + \mathbb{m}_1 \underset{\sim}{K}'(V_0)(V_1) + \frac{\mathbb{m}_0}{2} \underset{\sim}{K}''(V_0)(V_1)(V_1) - \mathbb{t}_1 R_1 \Pi_{oo}\right) - \right. \right. \\ \vdots \end{array} \right. \tag{4.1}$$

$$- \frac{1}{2}\underset{\sim}{\beta}''(C_0)(C_1)(C_1)] - \frac{1}{2}\underset{\sim}{\delta}''(V_0)(V_1)(V_1)\Big] = 0 \text{ , where } C = \begin{bmatrix} \mathbb{r} \\ \mathbb{m} \end{bmatrix} \quad (4.1) \text{ cont}^{d}$$

Remarks

1) If the bending prestress is negligible, (4.1) is simplified.

2) Note that $\mathbb{r} = \mathbb{t} R \pi_{00}$ with $\mathbb{t} R \pi_{00} - \overline{\mathbb{t} R \pi_{00}} = 0$, and that $\mathbb{t}$, R being the independent variables :

$$\mathbb{r}_1 = \mathbb{t}_1 R_0 \pi_{00} + \mathbb{t}_0 R_1 \pi_{00}$$

Moreover (2.3) gives

$$N_1 = -\overline{\frac{\partial V_1}{\partial m_0}} N_0$$

3) In (4.1) it is supposed that the shell is hyperelastic from the reference state $\mathcal{C}_{00}$, so that the prestress appears only in term β''.

4) Term $\mathbb{t}_1$ may be written $\mathbb{t}_1 \pi_0$ because it multiplies linearized quantities.

4.2. Linearized principle for a shell linear from the prestressed state $\mathcal{C}_0$ only

If the material is inelastic from the reference state $\mathcal{C}_{00}$, the computation will be carried out through an incremental step by step iterative process. In that conditions it may be assumed that the material is linear elastic from the prestress only for the incremental solution. The corresponding T.L. principle then becomes :

$$\begin{aligned} &\delta \underset{\sim}{\mathcal{F}}_L(\mathbb{t}_1, \mathbb{m}_1, R_1, V_1) = \\ &\delta\Big[\int_{\Sigma_{m_{00}}}\Big[T_r\Big(\mathbb{t}_1 \frac{\partial V_1}{\partial m_{00}} + \mathbb{m}_1 \underset{\sim}{K}'(V_0)(V_1) + \frac{\mathbb{m}_0}{2} \underset{\sim}{K}''(V_0)(V_1)(V_1) - \mathbb{t}_1 R_1 \pi_{00}\big[1_{E_{300}} + h\big]\Big) \\ &\qquad - \beta_1\Big] - \frac{1}{2}\underset{\sim}{\delta}''(V_0)(V_1)(V_1)\Big] = 0 \end{aligned} \quad (4.2)$$

Remarks

1) It is worth noting that in (4.2) $T_r(h_0 \delta \mathbb{r}) \neq \frac{\partial \beta}{\partial \mathbb{r}} \delta \mathbb{r}$ in general contrarily as in (4.1).

2) In an updated formulation (U.D.L.), where the prestressed state $\mathcal{C}_0$ is the reference state, (4.2) becomes (with $\mathbb{m}_0 = 0$) :

$$\begin{aligned} &\delta\Big[\int_{\Sigma_{m_0}}\Big[T_r\Big(\mathbb{t}_1 \pi_0 \frac{\partial V_1}{\partial m_0} + \mathbb{m}_1\Big[\frac{\partial N_1}{\partial m_0} + \overline{\frac{\partial V_1}{\partial m_0}} \frac{\partial N_0}{\partial m_0}\Big] - \mathbb{t}_1 R_1 \pi_0\Big) - \frac{1}{2}\underset{\sim}{\beta}''(C_0)(C_1)(C_1)\Big] \\ &\qquad - \frac{1}{2}\underset{\sim}{\delta}''(V_0)(V_1)(V_1)\Big] = 0. \end{aligned} \quad (4.3)$$

4.3. Modified linearized principles

Let us consider for instance principle (4.3). We may integrate by parts the terms in V_1. To that purpose we use the following formulae, which may be considered as a definition of the Riemannian divergence of a tangent vector W, or of a mapping A, together with the Stokes formula :

$$\left[\begin{array}{l} \widehat{\mathrm{div}}.W \overset{\mathrm{def.}}{=} T_r\left(\Pi_0 \frac{\partial W}{\partial m_0}\right), \ \forall W \in \vec{E}_{20} \\ \widehat{\mathrm{div}}.A.V \overset{\mathrm{def.}}{=} \widehat{\mathrm{div}}.A.V, \ \forall A \in \mathcal{L}(\vec{E}_3, \vec{E}_{20}), \ \forall V = C^{te} \in \vec{E}_3 \\ \int_{\Sigma_{m_0}} \widehat{\mathrm{div}}.W = \int_{\partial \Sigma_{m_0}} \bar{\nu}_0 W \, ds \quad (\mathrm{Stokes}), \end{array}\right. \tag{4.4.}$$

where s is the linear coordinate of the boundary $\partial\Sigma_{m_0}$ and ν_0 its unit external normal in $\vec{E}_{20}$.

With (4.4), (4.3) now becomes, still in U.D.L. formulation :

$$\left[\begin{array}{l} \delta \mathcal{F}_{Lm}(\mathbb{t}_1, \mathbb{m}_1, V_1, R_1) = \\ \delta\Big[\tilde{\int}_{\Sigma_{m_0}} \Big\{ -\Big[\widehat{\mathrm{div}}.\big[\mathbb{t}_1 \Pi_0 + \mathbb{m}_1 \frac{\partial N_0}{\partial m_0} + \overline{N_0 \widehat{\mathrm{div}}\, \mathbb{m}}\big] + \bar{f}_1\Big] V_1 + T_r(\mathbb{t}_1 R_1 \Pi_0) - \\ \quad - \frac{1}{2} \underset{\sim}{\beta}''(C_0)(C_1)(C_1)\Big\} + \mathrm{B.T.} - \frac{1}{2} \underset{\sim}{\ell}''(V_0)(V_1)(V_1)\Big] = 0 \end{array}\right. \tag{4.5}$$

In (4.5), B.T. represents the boundary terms coming from the integration by parts. Details will be given in [25].

<u>Remark</u>

In formulation (4.5) the term in $\widehat{\mathrm{div}}$, as well as V_1 and the surface load density f_1 given on Σ_{m_0}, must be square summable on Σ_{m_0}. We pose

$$\mathbb{t}_1, \mathbb{m}_1 \in H(\widehat{\mathrm{div}}, \Sigma_{m_0}).$$

But this condition implies that the stress fluxes must be continuous at element interfaces (which simplifies the B.T.), but not the displacement V_1, and the assembling of degrees of freedom results.

4.4. Fraeijs de Veubeke-Pian linearized principle for shells

If in (4.5) we take statically admissible (S.A.) stresses s.t.

$$\widehat{\mathrm{div}}\Big[\mathbb{t}_1 \Pi_0 + \mathbb{m}_1 \frac{\partial N_0}{\partial m_0} + \overline{N_0 \widehat{\mathrm{div}}\, \mathbb{m}_1}\Big] + \bar{f}_1 = 0, \text{ piecewise,} \tag{4.6}$$

principle (4.5) becomes the so-called Fraeijs de Veubeke-Pian shell principle :

$$\delta\Big[\int_{\Sigma_{m_0}}\Big[T_r(\mathbb{t}_1 \Pi_0 R_1 \Pi_0) - \frac{1}{2}\underset{\sim}{\beta}''(C_0)(C_1)(C_1)\Big] + \mathrm{B.T.} - \frac{1}{2}\underset{\sim}{\ell}''(V_0)(V_1)(V_1)\Big] = 0 \tag{4.7}$$

<u>Remarks</u>

1) It must be emphasized that in (4.7) the stress fluxes do not need to be continuous anymore, because (4.6) is supposed to hold piecewise, but still $\mathbb{t}_1$, $\mathbb{m}_1 \in H(\widehat{\mathrm{div}}, \Sigma_{m_0})$ because the fluxes must be square summable on the boundary of each element e in order that their work be bounded. (The preceding condition

results from an integration by parts of that work).

2) This functional (4.7) would give good numerical results due to (4.6), because, contrarily to other mixed principles in linearized incremental form, (4.6) is a complete equilibrium equation without any prestress term. Incidentally (4.6) could be taken into account eather in a weak form, which would give again (4.5), or through stress-functions.

3) Stress - functions can be exhibited resulting from the following closure conditions :

$$\widehat{\mathrm{div}}\left[\mathbb{t}_1 \Pi_0 + \mathbb{m}_1 \frac{\partial N_0}{\partial m_0} + \overline{N_0\, \widehat{\mathrm{div}}\, \mathbb{m}_1} \right] = 0$$

It is found [22] :

$$\left[\begin{array}{l} \exists\ V_1^*, \Omega_1^* \in \vec{E}_3 \text{ s.t.} \\ \mathbb{t}_1 \Pi_0 = i_2 \dfrac{\partial \Omega_1^*}{\partial m_0} - \dfrac{1}{2} i_2 \left[\dfrac{\partial V_1^*}{\partial m_0} + \overline{\dfrac{\partial V_1^*}{\partial m_0}} \right] i_2 \dfrac{\partial N_0}{\partial m_0} \\ \mathbb{m}_1 = \dfrac{1}{2} i_2 \left[\dfrac{\partial V_1^*}{\partial m_0} + \overline{\dfrac{\partial V_1^*}{\partial m_0}} \right] i_2 \\ \forall\ V_1^*, \Omega_1^* \in \vec{E}_3, \text{ s.t. } \Pi_0 N_0 \Omega_1^* = -\dfrac{1}{2} i_2 \left[\dfrac{\partial \Pi_0 V_1^*}{\partial m_0} - \overline{\dfrac{\partial \Pi_0 V_1^*}{\partial m_0}} \right]. \\ i_2 \Rightarrow \text{rotation of } +90° \text{ in } \vec{E}_{20}\,. \end{array}\right. \qquad (4.8)$$

Due to $(4.8)_3$, this gives 5 independent scalar stress functions. This form is different from that associated with the conventional symmetric stresses $\mathbb{n}, \mathbb{m}$ [23] due to the non symmetry of $\mathbb{t}_1$.

4.5. Modified non linear principles

In the linearized principles, the normal increment $N_1 = -\overline{\dfrac{\partial V_1}{\partial m_0}}\, N_0$ may be replaced by $R_1 N_0$, s.t. R_1 will be also the independent variable in the curvature term $\underset{\sim}{K}(V_1)$. In fact it is easily shown and well known that the introduction of new unknowns through L.M. may be done in all or parts of the terms of the functional.

But then principle (4.5) cannot be written with complete S.A. stresses. This way may be used also in the non linear principle, putting

$$N = R\, N_{00}$$

in T.L. for instance.

Another variant consists in taking directly the Jauman membrane stress $\mathbb{r}$ as independent unknown instead of $\mathbb{t}$, by

$$\mathbb{r} = \overline{\mathbb{r}} = \mathbb{t} R \Pi_{00} \Rightarrow \mathbb{t} = \mathbb{r} \overline{R \Pi_{00}} \qquad (4.9)$$

where one takes account of the constraint $\mathbb{r} = \overline{\mathbb{r}}$:

$$\mathbb{t} R \Pi_{00} = \overline{\mathbb{t} R \Pi_{00}}$$

This way spares directly 3 unknowns and is also interesting in the buckling criterion.

4. Stability criterion

If we intend to use proper mixed functionals in stability problems, the primal criterion which states that a stable equilibrium meets a minimum of the total potential energy [26], cannot be applied anymore directly. Now all the mixed functionals were written in bringing constraints into the primal functional, so that the classical necessary conditions of equilibrium and stability are mofified as follows.

Let us consider a functional y of the state variable x which is supposed to belong to a linear normed space X and is submitted to a constraint $z = \underset{\sim}{z}(x) = 0$ where z belongs to a linear vector space $\vec{E}$. It is easily shown [27] that the necessary conditions of equilibrium and stability are respectively with a L.M. Λ :

$$\left[\begin{array}{l} \delta[y-\Lambda z] = 0 \,, \forall\, \delta x \in \vec{X} \,,\, \delta\Lambda \in \vec{E}' \\ \delta^2[y-\Lambda z] \geqslant 0 \,, \forall\, \delta x \in \vec{X} \text{ s.t. } \delta z = 0 \end{array}\right. \tag{5.1}$$

Remarks

1) Equations $(5.1)_1$ gives the equilibrium solution (x_o, Λ_o) whose stability is tested by $(5.1)_2$ which must be calculated at this point solution.
2) The second variation $(5.1)_2$ must be calculated with $\delta^2 x = \delta^2\Lambda = 0$, or constant field δx and $\delta\Lambda$, due du $(5.1)_1$.
3) Calculation of $(5.1)_2$ must obviously take account of all the terms of the left side member of $(5.1)_1$ and with all the constraints, even if some Euler's equations are admitted from (5.1), this equation being applied afterwards.
4) In $(5.1)_2$, the constraints appear in a linearized form, due to the local sense of the criterion in the state space.

Due to preceding remarks, if one wants to apply this criterion to one of the preceding mixed functionals, it is necessary to go back to the primary one (3.6). But observing that the constraint $\bar{R}\,R = 1_{E_3}$ was taken into account "outside of the principle", this constraint must imperatively be brought into (3.6) with a L.M. ζ s.t. (3.6) becomes (in U.D.L. for instance) :

$$\left[\begin{array}{l} \delta\Big[\int_{\Sigma_{m_o}}\Big[\underset{\sim}{d}(h,\kappa) + T_r\Big(\mathbb{t}\Big[\frac{\partial m}{\partial m_o} - R\Pi_o[1_{E_{2o}} + h]\Big] + \mathcal{A}\Big[\overline{\frac{\partial m}{\partial m_o}}\frac{\partial N}{\partial m_o} - \frac{\partial N_o}{\partial m_o} - \kappa\Big] + \\ \qquad + \qquad + \zeta\big[\overline{R\Pi_o}\,R\Pi_o - 1_{E_{2o}}\big]\Big] - \mathcal{G}\Big] = 0 \\ \forall\, V\,\text{K.A.}, \mathbb{t} \in \mathcal{L}(\vec{E}_2, \vec{E}_{2o}) \,,\, R\Pi_o \in \mathcal{L}(\vec{E}_{2o}, \vec{E}_3) \,,\, \mathcal{A}, \zeta, h \text{ sym.} \in \mathcal{L}(\vec{E}_{2o}, \vec{E}_{2o}) \end{array}\right. \tag{5.2}$$

Besides the preceding Euler equations, one finds for δR :

$$\zeta + \bar{\zeta} = [1_{E_{2o}} + h]\,\mathbb{t}\,R\Pi_o = \overline{[1_{E_{2o}} + h]\,\mathbb{t}\,R\Pi_o} \tag{5.3}$$

Interpretation of (5.3) is again incidentally that the medium is a non-polarized one, or the symmetry of the membrane stress.

Then, by a strict application of (5.1), it is easily found the stability criterion :

$$\left[\begin{array}{l} \int_{\Sigma_{m_0}} \Big[\underset{\sim}{\alpha}''(h,\kappa)\binom{\delta h}{\delta\kappa}\binom{\delta h}{\delta\kappa} + T_r\Big(-2\mathbb{t}_0\,\delta R\,\pi_0\,\delta h + \mathbb{m}_0\Big[2\frac{\overline{\partial\delta V}}{\partial m_0}\frac{\partial\delta \underset{\sim}{N}(V)}{\partial m_0} + \frac{\partial\delta^2\underset{\sim}{N}(V)}{\partial m_0}\Big] + \\ \qquad\qquad + \big[1_{E_{20}} + h_0\big]\,\mathbb{r}_0\,\overline{\delta R\pi_0}\,\delta R\pi_0\Big)\Big] - \underset{\sim}{\ell}''(V_0)(\delta V)(\delta V) \geq 0 \\ \forall \delta V\ \text{K.A.};\ \delta h, \delta\kappa\ \text{sym.} \in \mathcal{L}(\vec{E}_{20}, \vec{E}_{20})\ ;\ \delta R \,|\, \overline{R}R = 1_{E_3}\ ;\ \text{s.t.}: \\ \left[\begin{array}{l} \int_{\Sigma_{m_0}} T_r\Big(\underline{\delta\mathbb{t}}\Big\{\frac{\partial\delta V}{\partial m_0} - \delta\big[R\pi_0[1_{E_{20}} + h]\big]\Big\} + \underline{\delta\mathbb{m}}\,\delta\Big[\frac{\overline{\partial m}}{\partial m_0}\frac{\partial \underset{\sim}{N}(V)}{\partial m_0} - \kappa\Big]\Big) = 0 \\ \forall\ \underline{\delta\mathbb{t}} \in \mathcal{L}(\vec{E}_2, \vec{E}_{20}),\ \underline{\delta\mathbb{m}} = \overline{\underline{\delta\mathbb{m}}} \in \mathcal{L}(\vec{E}_{20}, \vec{E}_{20}). \end{array}\right. \end{array}\right. \tag{5.4}$$

Going back now to (3.9), if we put

$$D = \binom{h}{\kappa}\ ,\ C = \binom{\mathbb{r}}{\mathbb{m}}\ ,\ T_r(\mathbb{r}h + \mathbb{m}\kappa) = \tilde{\mathbb{r}}h + \tilde{\mathbb{m}}\kappa = \tilde{C}D\ ,\ \text{(scalar product)}$$

we see that :

$$\begin{array}{l} \tilde{C} = \underset{\sim}{\alpha}'(D)\ ,\ \tilde{D} = \underset{\sim}{\beta}'(C)\ ,\ \widetilde{\delta C} = \underset{\sim}{\alpha}''(D)(\delta D)\ ,\ \widetilde{\delta D} = \underset{\sim}{\beta}''(C)(\delta C) \\ \widetilde{\delta C}.\delta D = \widetilde{\delta D}.\delta C = \underset{\sim}{\alpha}''(D)(\delta D)(\delta D) = \underset{\sim}{\beta}''(C)(\delta C)(\delta C). \end{array} \tag{5.5}$$

Taking into account (5.5), (5.4) reads finally :

$$\left[\begin{array}{l} \int_{\Sigma_{m_0}} \Big[\underset{\sim}{\beta}''(C_0)(\delta C)(\delta C) + T_r\Big(-2\mathbb{t}_0\,\delta R\,\pi_0\,\delta h + \mathbb{m}_0\Big[2\frac{\overline{\partial\delta V}}{\partial m_0}\frac{\partial\delta \underset{\sim}{N}(V)}{\partial m_0} + \frac{\partial\delta^2\underset{\sim}{N}(V)}{\partial m_0}\Big] + \\ \qquad\qquad + \big[1_{E_{20}} + h_0\big]\,\mathbb{r}_0\,\overline{\delta R\pi_0}\,\delta R\pi_0\Big)\Big] - \underset{\sim}{\ell}''(V_0)(\delta V)(\delta V) \geq 0 \\ \left[\begin{array}{l} \forall \delta V\ \text{K.A.};\ \delta\mathbb{t} \in \mathcal{L}(\vec{E}_2, \vec{E}_{20})\ ;\ \delta R \,|\, \overline{R}R = 1_{E_3}\ ;\ \delta h = \frac{\widetilde{\partial\beta}}{\partial\mathbb{r}}\ ,\ \delta\kappa = \frac{\widetilde{\partial\beta}}{\partial\mathbb{m}}\ ; \\ \text{s.t.} \int_{\Sigma_{m_0}} T_r\Big(\underline{\delta\mathbb{t}}.\Big\{\frac{\partial\delta V}{\partial m_0} - \delta\big[R\pi_0[1_{E_{20}} + h]\big] + \underline{\delta\mathbb{m}}.\delta\Big[\frac{\overline{\partial m}}{\partial m_0}\frac{\partial \underset{\sim}{N}(V)}{\partial m_0} - \kappa\Big]\Big) = 0 \\ \qquad \forall\ \underline{\delta\mathbb{t}}\ ,\ \underline{\delta\mathbb{m}} \end{array}\right. \end{array}\right. \tag{5.6}$$

<u>Remarks</u>

1) In $(5.6)_2$, one has $T_r(\underline{\delta\mathbb{t}}\,R\,\pi_0\,\delta h + \underline{\delta\mathbb{m}}\,\delta\kappa) = \underset{\sim}{\beta}''(C_0)(\delta C)(\delta C)$.

2) The criterion is the same for the so-called modified principles, but the functional spaces are conformly changed. In particular if $\mathbb{r}$ is taken as independent variable instead of $\mathbb{t}$, because equation (4.9) is an Euler equation, due to a preceding remark.

3) The variation $\delta \underset{\sim}{N}(V)$ and $\delta^2\underset{\sim}{N}(V)$ of the non rational term $\underset{\sim}{N}(V)$ can be easily calculated [25] using expansions of the two equations $\bar{N}N = 1$, $\bar{N}dm = 0$.

<u>Verification of the criterion</u>

Criterion (5.6), which is not trivial indeed, can be verified in the following way. The second variation of the primal strain energy is, with

$$D = \begin{pmatrix} h \\ \kappa \end{pmatrix} = \underset{\sim}{D}(V)$$

$$\delta^2\alpha = \frac{\partial^2\alpha}{\partial D\partial D}\left(\delta \underset{\sim}{D}(V)\right)\left(\delta \underset{\sim}{D}(V)\right) + \widetilde{\mathbb{r}}\, \delta^2 \underset{\sim}{h}(V) + \widetilde{\mathbb{m}}\, \delta^2 \underset{\sim}{\kappa}(V) \tag{5.7}$$

But $(3.4)_2$ gives

$$\delta^2 h \left[1_{E_{20}} + h\right] = \delta^2\Gamma - [\delta h]^2 \tag{5.8}$$

Now $\delta^2\Gamma$ comes from $(2.6)_1$:

$$\delta^2 \underset{\sim}{\Gamma}(V) = \overline{\frac{\partial \delta V}{\partial m_0}}\, \frac{\partial \delta V}{\partial m_0} \tag{5.9}$$

and with $(3.4)_1$

$$\frac{\partial \delta V}{\partial m_0} = \delta R\pi_0 \left[1_{E_{20}} + h\right] + R\pi_0\, \delta h . \tag{5.10}$$

From (5.8)-(5.10), and neglecting h compared with identity $1_{E_{20}}$, for simplification's sake, it becomes :

$$\delta^2 h = \overline{\delta R\pi_0}\, \delta R\pi_0 + \overline{\delta R\pi_0}\, R\pi_0 \delta h + \delta h\, \overline{R\pi_0}\, \delta R\pi_0$$

so that with $\mathbb{r} = \overline{\mathbb{r}} = \mathbb{t} R\pi_0$, (5.7) reads :

$$\underset{\sim}{\alpha}''(D)(\delta D)(\delta D) + T_r\left(\mathbb{r}\, \overline{\delta R\pi_0}\, \delta R\pi_0 - 2\mathbb{t}\overline{R\pi_0}\, \delta R\pi_0 \delta h + \mathbb{m}\delta^2 \underset{\sim}{\kappa}(V)\right)$$

which is exactly the integrand of $(5.4)_1$ Q.E.D.

4. Conclusions

From a direct surface theory of Kirchhoff-Love shells using finite rotations and a complementary energy density, various mixed variational principles have been derived for numerical analyses, in particular linearized ones for step by step iterative procedures.

The best will be those with statically admissible stresses to improve the stress accuracy. This would be realized eather with complete stress functions, which have been recalled, or in a weak way by a Lagrange-multiplier and projection on its discretized space.

It should be observed that statically admissible stresses, strictly speaking, cannot be used in other mixed variational principles due to the non linear prestress term appearing in the local equilibrium equations.

We did not try to derive a fully complementary principle as Koiter did in three-dimensional theory, because this would need the elimination of intermediate variables, what is practically achieved through the discretized process.

Discretization of some spaces like $H(\widehat{\text{div}}, \Sigma_{m_0})$ will nevertheless present some difficulties which could be overcome in relaxing the stress symmetry.

Extension to shells which undergo transverse shear strains do not present any difficulty if we add the shear energy to the preceding ones, due to the fact that the polar decomposition affects only membrane quantities.

Stability criteria have been presented for those mixed principles from a general method, which differ only by the functional spaces involved.

Post-buckling analysis following Koiter's primal method has been extended to the case of mixed functionals for shells without or with imperfections and also in the elasto-plastic case [28].

Let us mention that an interesting utilization of the preceding non linear principle is to allow for the writing of coupled primal-dual principles in non linear elasto-dynamics in a substructuring method as was done in the linear case [29, 30].

References

1. SANDERS J.L., Non linear theories of thin shells. Q. Applied. Math. 21, 21-36, (1963).

2. KOITER W.T. On the non linear theory of thin elastic shells. Proc. Phys. Sc. Mech. Series B., Vol. 69, n° 1 (1965).

3. BUDIANKSKY B., Notes on the non linear shell theory. Trans. ASME, Series E, J. Appl. Mech. 35, 393-401 (1968).

4. SIMMONDS J.G., DANIELSON D.A., Non linear shell theory with a finite rotation vector. Proc. Kon. Ned. Ak. Wet., Series B, vol. 73, (1970) p. 460-478.

5. PIETRASZKIEWICZ W., Finite rotations and Lagrangian description in the non linear theory of shells. Polish Sc. Pub. Warszawa-Poznan, 1979.

6. PIETRASZKIEWICZ W. in Theory of Shells, p. 445-471. North-Holland (1980).

7. STUMPF H., Generating functionals and extremum principles in non linear elasticity with application to non linear plates and shallow shells theory. In Lecture Notes in Math. 503. Appl. of Math. of Functional Analysis to Problems in Mech. Joint. Symp. IUTAM/IMU. Marseille 1975, p. 500-510, Springer, (1976).

8. SCHMIDT R., PIETRASZKIEWICZ W., Variational principles in the geometrically non linear theory of shells undergoing moderate rotations. Ing. Arch. 50 (1981), p. 187-201.

9. SCHMIDT R., A current trend in shell theory : Constrained geometrically non linear Kirchhoff-Love type theories based on polar decomposition of strains and rotations. Proc. Symp. on Advances and Trends in Structures and Dynamics. Washington D.C. 1984. A.K. Noor and R.J. Hayduk Ed. Pergamon 1985.

10. VALID R., Principes mixtes et critères de flambage. Conf. CEA-INRIA-EDF 1983, T.P. ONERA 1983-149.

11. KOITER W.T., On the complementary energy theorem in non linear elasticity theory. Trends in Appl. of Pure Math. to Mech. G. Fichera ed. Pitman Pub. Conf. at the Univ. of Leece (Italy) 1975.

12. FRAEIJS de VEUBEKE B., A new variational principle for finite elastic displacements. Int. J. Engng. Sc. Vol. 10 (1972) p. 745-763.

13. ZUBOV L.M., The stationary principle of complementary work in non linear theory of elasticity. J. Appl. Math. Mech., 34 (1970) p. 228-232.

14. KOITER W.T., On the principle of stationary complementary energy in the non linear theory of elasticity. SIAM J. Appl. Math. Vol. 25, n° 3 (1973).

15. ATLURI S.N. and MURAKAWA H., On hybrid finite element models in non linear solid mechanics. Int. Conf. on Finite Elements in Non linear Solid and Structural Mechanics. Gello (Nordway), Aug. 29-Sept. 1, 1977.

16. SANDER C., CARNOY E., Hybrid stress finite element models for elasto-plastic analysis. Int. Conf. on Finite Elements in Non linear Solid and Structural Mechanics. Gello (Nordway) Aug. 29-Sept. 1, 1977.

17. SIMMONDS J.G., DANIELSON D.A., Non linear shell theory with finite rotation and stress function vectors. J. of Appl. Mech. Dec. 1972, Trans. ASME, p. 1085-1090.

18. SCHMIDT R., Thin elastic shell undergoing small strains and large rotations. A simple consistent theory and variational principles. In Numerical Meth. for Non linear Probl. Proc. Int. Conf. Univ. Politec. de Barcelona, (Spain), p. 9-13 Apr. 1984.

19. STUMPF H., The derivation of dual extremum and complementary stationary principles in geometrically non linear shell theory. Ing-Arch. 48 (1979), p. 221-237.

20. STEIN E., Variational functionals in the geometrical non linear theory of thin shells and finite element discretization with application to stability problems. In Theory of Shells. W.T. Koiter and G.K. Mikhailov ed. North-Holland (1980) p. 509-535.

21. STUMPF H., Unified operator description, non linear bukcling and post-buckling analysis of thin elastic shells. In Proc. Euromech Coll. 165 : Flexible Shells, Theory and Applications. Munich (Germany) May 17-20, 1983. Springer (1984) p. 91-105.

22. VALID R., The principle of complementary energy in non linear shell theory. 15th IUTAM Congress, Aug. 17-23, 1980, Toronto (Canada). La Rech. Aerosp. 1981-1. Engl. Transl.

23. VALID R., Mechanics of Continuous Media and Analysis of Structures. North-Holland 1981.

24. VALID R., On intrinsic formulation for non linear theory of shells and some approximations. Trends in Computerized Structural Analysis and Synthesis. Oct. 30-Nov. 1, 1978, Washington D.C. Comp. and Struct. Vol. 10, 1979, p. 143-194. T.P. ONERA 1979-2.

25. VALID R., Some Aspects of the Non linear Theory of Shells (To appear).

26. KOITER W.T., On the stability of elastic equilibrium. Thesis, Delft Univ., H.J. Paris Amsterdam 1945 ; Engl. Transl. (a) NASA TT-10883 (1967), (b) AFF DL-TR-70-25 (1970).

27. VALID R., The structural stability criterion for mixed principles. Int. Symp. on Hybrid and Mixed Finite Elements Methods. Atlanta (Georgia), 8-10 April 1981. In Hybrid and Mixed Finite Element Methods, S.N. Atluri, R.H. Gallagher, O.C. Zienkiewicz ed. J. Wiley 1983.

28. VALID R., On buckling and post-buckling of elastic and inelastic thin shells in primal and mixed formulation. IUTAM Symp. on Inelastic Behavior of Plates and Shells, Rio de Janeiro (Bresil) 5-9 Aug. 1985. T.P. ONERA n° 1985-87 (Proc. to appear).

29. OHAYON R., VALID R., Principes variationnels, symétriques couplés de type primal-dual en élastodynamique linéaire. C.R. Ac. Sc. Paris t. 297, série II. 1983.

30. OHAYON, R., VALID R., Principes variationnels couplés primal-dual en élastodynamique linéaire : cas des coques minces. C.R. Ac. Sc. Paris t. 298, série II, n° 9, 1984.

NUMERICAL ANALYSIS OF THIN-WALLED STRUCTURE FINITE DISPLACEMENTS

N.V.VALISHVILI and A.K.TVALCHRELIDZE
Kutaisi Polytechnical Institute
pr.Molodezhi 62,384014 Kutaisi, USSR

I. Introduction

The behavior of the structure thin-walled elements upon finite displacements is described in terms of the boundary value problem for non-linear differential equations. The dependence of the mechanical system state vector $\vec{X}$ on external effect vector $\vec{Q}$ may be very complicated. Thus Fig.I shows the relation between deflection W_0 in the vertex of the gently sloping spherical shell and load parameter q [I]. To obtain the solution of such problems numerical methods are usually used. This paper discusses numerical algorithms based on the cannon

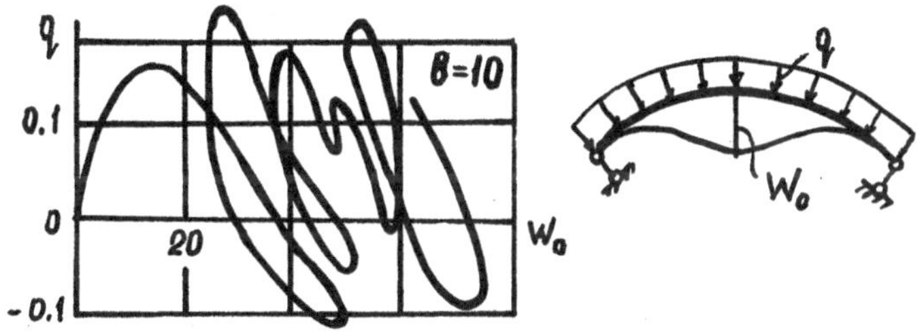

Fig.I

method [I,2]; analyses features arising when they are used;ways of improvement, as well as poses non-linear problems and gives their solutions.

2. Two-points boundary value problem and cannon method

Finite axis-symmetrical displacements of the rotation shells are described by the system of non-linear ordinary differential equations [I]

$$\vec{X}' = \vec{f}(t, q, \vec{X}), \qquad t \in [t_0, t_1] \tag{1}$$

with boundary conditions

$$\vec{\Psi}(\vec{X}(t_0)) = 0, \qquad \vec{\varphi}(\vec{X}(t_1)) = 0 \tag{2}$$

The following notations are adopted here

$$\vec{X} = (x_1, x_2, \dots, x_n), \quad \vec{f} = (f_1, f_2, \dots, f_n),$$

$$\vec{\Psi} = (\Psi_1, \Psi_2, \dots, \Psi_K), \quad \vec{\varphi} = (\varphi_1, \varphi_2, \dots, \varphi_j), \quad K + j = n$$

q is scalar parameter whose numerical value influences the problem solution.

To solve boundary value problems (I,2) different variants of the cannon method are effectively used. The core of the method is self-explanatory. With the help of boundary conditions (2) primary value of vector $\vec{X}(t_0, q)$ can be expressed through j of cannon parameters $(A_1, \dots A_j) = \vec{A}$:

$$\vec{X}(t_0, q) = \vec{X}_0(q, \vec{A}) \tag{3}$$

Then we numerically solve the Cauchy problem for system (I) with primary conditions (3). The obtained solution is at the same time the solution of the boundary value problem (I,2) if the value of vector $\vec{X}(t_1, q) = \vec{X}_1(q, \vec{A})$ at the end of the integration segment meets boundary conditions (2)

$$\vec{\varphi}(\vec{X}_1(q, \vec{A})) = 0 \tag{4}$$

The iteration process to define cannon parameters providing the solution of the given boundary value problem [I] can be developed on the basis of the Newton's method for the non-linear equation system

$$\vec{A}^{i+1} = \vec{A}^{i} - W^{-1}(q, \vec{A}^{i}) \cdot \vec{\varphi}(q, \vec{A}^{i}) \tag{5}$$

Jacobi matrix elements

$$W = \begin{pmatrix} \frac{\partial \varphi_1}{\partial A_1} & \dots & \frac{\partial \varphi_1}{\partial A_j} \\ \cdot & \dots & \cdot \\ \frac{\partial \varphi_j}{\partial A_1} & \dots & \frac{\partial \varphi_j}{\partial A_j} \end{pmatrix} \tag{6}$$

are defined using difference formulas by integrating the initial system (I) with small increments of cannon parameters ΔA_i.

3. Primary approximation and parameter continuation method

The application of the parameter continuation method in combination with the cannon method is most effective when we obtain the problem solution successively with the values of parameter q_o, $q_1, \ldots,$ q_k, q_k ,... . The solution of boundary value problem (I,2) on step (k+1) is derived from the Cauchy problem solutions with cannon parameters $\vec{A}=\vec{A}_k+\Delta\vec{A}$. Vectors $\vec{A}_k$ and $\Delta\vec{A}$ are also used for the numerical determination of elements of matrix (6). The iteration process can be initiated from the approximation (Fig.2):

$$A_i^{o}(q_k+\Delta q) = A_i(q_k) + \frac{dA_i}{dq}\Delta q + \frac{d^2A_i}{2\,dq^2}\Delta q^2$$

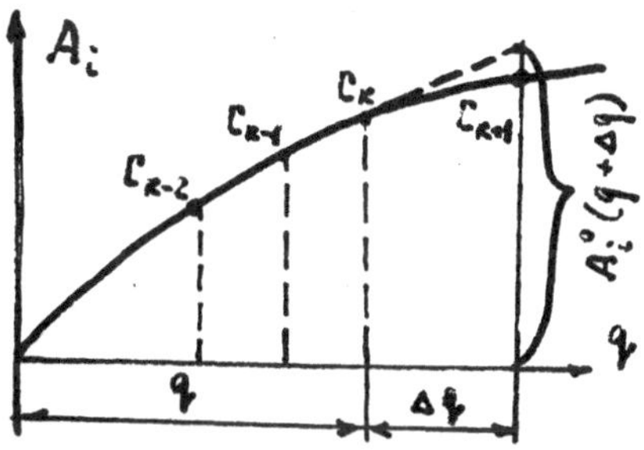

Fig.2

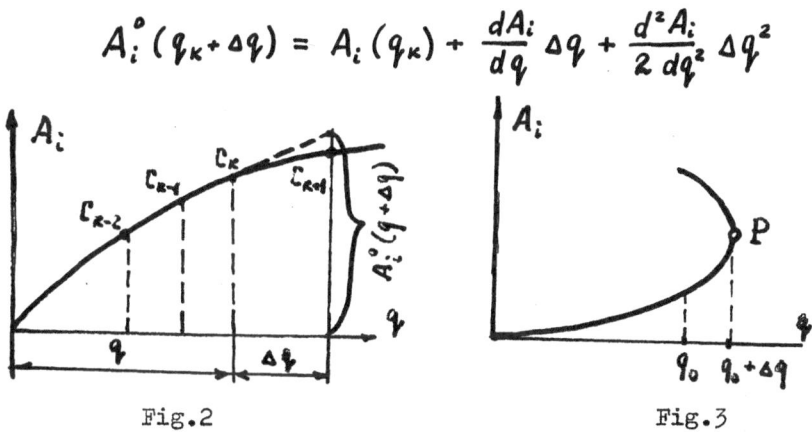

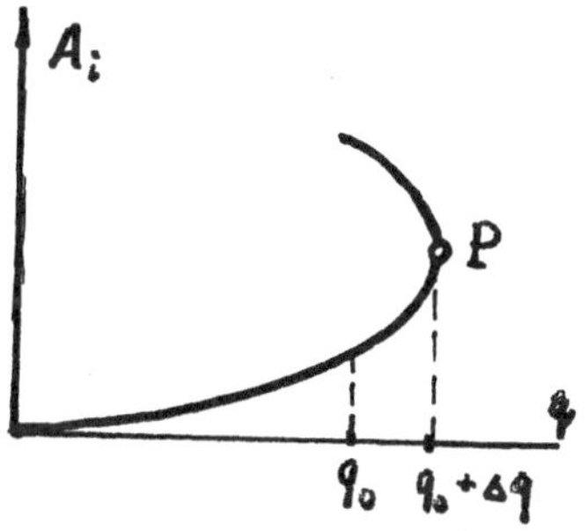

Fig.3

Approximate difference formulas are used to define $\frac{dA_i}{dq}$ and $\frac{d^2A_i}{dq^2}$.

In the segments with parameter q monotonous changing (Fig.3) the solution is obtained by the above method. The monotonous condition is violated in point P and matrix (6) is degenerated. Let us introduce notation $q=A_{j+1}$ and consider the augmented matrix

$$W^* = \begin{pmatrix} \frac{\partial\varphi_1}{\partial A_1} & \cdots & \frac{\partial\varphi_1}{\partial A_j} & \frac{\partial\varphi_1}{\partial A_{j+1}} \\ \vdots & \cdots & \vdots & \\ \frac{\partial\varphi_j}{\partial A_1} & \cdots & \frac{\partial\varphi_j}{\partial A_j} & \frac{\partial\varphi_j}{\partial A_{j+1}} \end{pmatrix} \qquad (7)$$

The point where the rank of matrix (6) is smaller than j and rank (7) equals j is the ultimate point (point P). The solution in the ultimate point neighbourhood can be obtained in the following way. In the course of solving the problem we monitor increments of all cannon parameters A_i, i=1,...,j+1 and the parameter with a maximum increment on the last step is taken as independent. We forcely change its value then the other parameters are defined by the iteration process.

In the ultimate point neighbourhood the solution can be obtained without parameter changing if an artificial parameter monotonously changing along the deformation curve is used [3,4,5]. For example Valishvili [6] proposes to define this parameter s from the additional relation

$$\varphi_{j+1}(s, \vec{A}) \equiv L(\vec{A}-\vec{A}^{\circ}) - (s-s_{\circ}) = 0$$

Auxiliary function $L(\vec{A}-\vec{A}^{\circ})$ should provide the monotone of s changing in the course of solution, e.g.

$$L(\vec{A}-\vec{A}^{\circ}) = \left(\sum_{i=1}^{j+1} |A_i - A_i^{\circ}|^{\ell} \right)^{1/\ell}, \quad \ell = 0,1,2,\dots$$

4. On accuracy setting of solution

Let us consider the problems of the required accuracy for the Cauchy problem solution and the boundary conditions to be met. Let the exact solution at some value q be represented by point T (Fig.4). The approximate solution is considered to be obtained if the appropriate point is located in the ε-neighbourhood of T. The possibility to obtain a solution is defined by $\vec{A}^{\circ}$ approximation. If $\|\vec{A}^{\circ}-\vec{A}_T\| \leq \rho_{\circ}$ then the iteration process converges; otherwise it diverges. The solution search process will be stable if $\rho_{\circ} \geq \rho_{\varepsilon}$ is valid for all q values.

In the parameter q value region where $\rho_{\circ} < \rho_{\varepsilon}$ (Fig.4) the calculation process may stop when the solution curve reaches the boundary of permissible primary approximations. The deformation curve seems to be

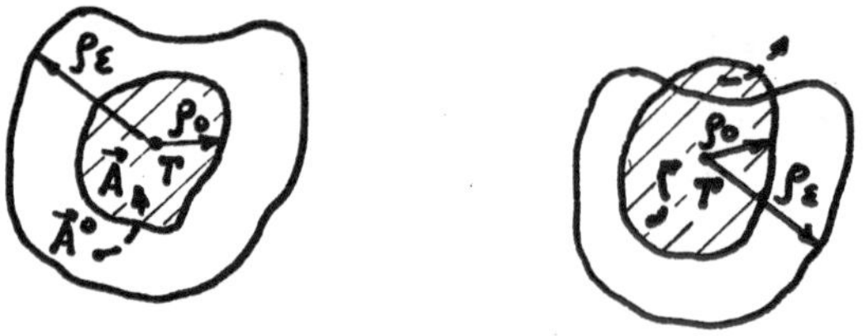

Fig.4

discontinued. As a matter of fact the applied solution accuracy is insufficient and we may move further along q if the accuracy is increased.

The choice of accuracy influences the computer time τ necessary to obtain the solution. With the increase in accuracy two tendences are observed: τ increases due to the iteration number increase to achieve the required accuracy and τ decreases due to a more accurate forecast of primary approximation. This may result in the optimum accuracy providing minimum computer time loss.

5. Parallel cannon and the choice of the segment breakdown

A specific feature of the shell and plate theory equations is great dependence of the Cauchy problem solution on primary conditions and, hence, on cannon parameters at the beginning of the integration segment. This may result in the corresponding perturbed solution being sharply increased or sharply decreased if the cannon parameters are perturbed even slightly relative to the support parameters. Such solution cannot be obtained on a computer as the overflow occurs and the calculation process stops. To expand the algorithm application the parallel cannon method [I,2] is used.

Consider the choice of the integration segment breakdown. When solving a specific problem it is difficult to determine beforehand how to carry out the breakdown. Moreover, the solution often happens to be greatly dependent on the parameter q value. This necessitates the integration segment breakdown to be changed during the solution of the problem by moving along q. It is undesirable to use a great number of intermediate nodes. This brings about a corresponding growth in the cannon parameter number and accumulation of errors when defined. The condition of solution continuity in the intermediate nodes is met approximately, the same is true of any other condition when the numerical method is realized on the computer.

Let us present a variant of the parallel cannon method with intermediate node choice regularization which provides breakdown with a minimum possible number of nodes. The variant is based on the idea of breakdown adaptation to the solution being obtained. The condition of the intermediate node appearence in point $t \in (t_0, t_1)$ is introduced:

$$\| \vec{X}(t, \vec{A} + \Delta\vec{A}) \| \geq M \tag{8}$$

If in the course of numerical solution (8) is fulfilled in some point t^* then an intermediate node appears there and the perturbed solution returns to the neighbourhood of the support solution (Fig.5).

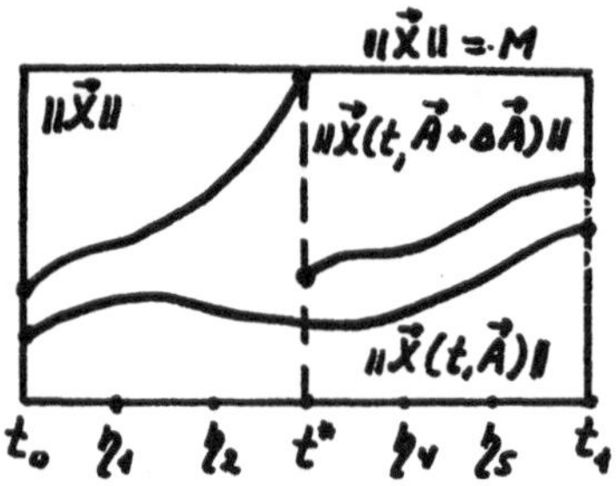

Fig.5

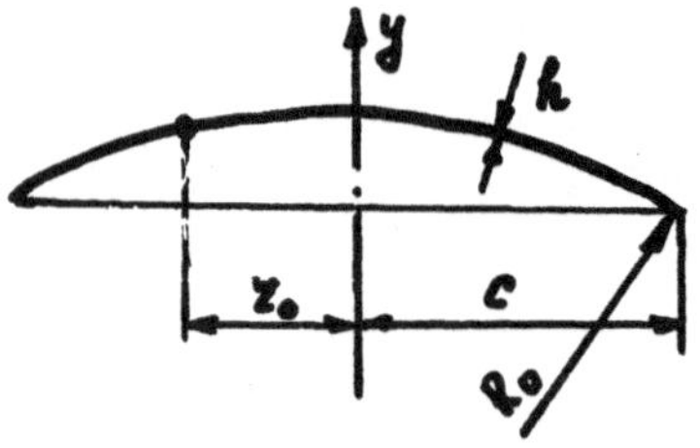

Fig.6

Thus the number of intermediate nodes and their location are unknown beforehand but are defined in the course of solution. When realizing the algorithm, a uniform network of auxiliary nodes $\{\eta_i\}$ is introduced into the integration segment. The support solution values are memorized only in these nodes and it is checked whether condition (8) is met.

6. Boundary value problem for the sloping rotation shell

Examine a family of sloping rotation shells of the non-negative Gaussian curvature loaded by external pressure. Their geometry is defined by the following formulas (Fig.6):

$$\frac{1}{\rho_2} = \frac{1}{R_0}\left(\frac{c}{r_0}\right)^m, \qquad y = \frac{c^2}{(2-m)R_0}\left[1-\left(\frac{r_0}{c}\right)^{2-m}\right] \tag{9}$$

where ρ_2 is the curvature radius of the midsurface in the circular direction;

m is the parameter which determines the shell shape.

Equations for determination of the stressed and strained state of the shell under consideration can be written as [7]:

$$\nabla^2\nabla^2 W = -\lambda_2 \frac{\partial^2 F}{\partial r^2} - \lambda_1\left(\frac{\partial F}{r\,\partial\varphi} + \frac{\partial^2 F}{r^2\,\partial\varphi^2}\right) + N(W,F) + 4q$$
$$\nabla^2\nabla^2 F = \lambda_2 \frac{\partial^2 W}{\partial r^2} + \lambda_1\left(\frac{\partial W}{r\,\partial\varphi} + \frac{\partial^2 W}{r^2\,\partial\varphi^2}\right) - \frac{1}{2}N(W,W) \tag{10}$$

where W, F are parameters of normal displacement W^* and force function F^*, respectively, defined by formulas

$$W = \sqrt{\eta}\,\frac{W^*}{h}, \qquad F = \eta\,\frac{F^*}{E h^2}, \qquad \eta = 12\,(1-\nu^2) \tag{11}$$

The parameters of curvatures and load are equal:

$$\lambda_1 = (1-m)\,\lambda_2, \quad \lambda_2 = \left(\frac{b}{r}\right)^m, \quad q = \frac{\sqrt{\eta}}{4}\,\frac{q^*}{E}\left(\frac{R_0}{h}\right)^2 \tag{12}$$

We obtain (10) from

$$W = \sum_{k=0}^{\infty} W_k \cos k\varphi , \qquad F = \sum_{k=0}^{\infty} F_k \cos k\varphi \tag{13}$$

Assuming the smallness of non-axis-symmetrical components of (13) we obtain the following two equation systems:

$$\begin{aligned} \theta'' &= -\frac{\theta'}{r} + \frac{\theta}{r^2} + \phi\left(\lambda_2 + \frac{\theta}{r}\right) + 2qr , \quad W_0' = -\theta \\ \phi'' &= -\frac{\phi'}{r} + \frac{\phi}{r^2} - \theta\left(\lambda_2 + \frac{\theta}{2r}\right), \quad F_0' = \phi \end{aligned} \tag{14}$$

$$\begin{aligned} W_k^{IV} &= -\frac{2}{r} W_k''' + \frac{1+2k^2}{r^2} W_k'' - \frac{1+2k^2}{r^3} W_k' + \frac{k^2(k^2-4)}{r^4} W_k - \\ &\quad - F_k''\left(\lambda_2 + \frac{\theta}{r}\right) + W_k'' \frac{\phi}{r} - (\lambda_1 + \theta')\left(\frac{F_k'}{r} - k^2 \frac{F_k}{r^2}\right) + \\ &\quad + \phi'\left(\frac{W_k'}{r} - k^2 \frac{W_k}{r^2}\right) \\ F_k^{IV} &= -\frac{2}{r} F_k''' + \frac{1+2k^2}{r^2} F_k'' - \frac{1+2k^2}{r^3} F_k' - \frac{k^2(k^2-4)}{r^4} F_k + \\ &\quad + W_k''\left(\lambda_2 + \frac{\theta}{r}\right) + (\lambda_1 + \theta')\left(\frac{W_k'}{r} - k^2 \frac{W_k}{r^2}\right) \end{aligned} \tag{15}$$

Thus a non-linear boundary value problem is solved independently for the principal axis-symmetrical state described by (14). System (15) is linear relative to unknowns W_k, F_k and their derivatives. It should be solved simultaneously with the integration of system(14).

7. Sloping spherical shell

For the closed spherical shell (m=0) integration of Equations (13) is carried out in segment $[\Delta, b]$. The parallel cannon method is used for large b ($b > 20$) due to existence of quickly increasing solutions. If the integration segment is divided into d parts then the number of cannon parameters is p=4d-2.

Table I shows the number of cannon parameters p used to solve boundary value problem (14) and the values of maximum loads obtained for a sloping spherical shell depending on the parameter. The problems are numbered according to the mount. of the contour of support: non-fixed-hinge (problem I), fixed-hinge (problem 2), non-fixed-constraint (problem 3), fixed-constraint (problem 4).

Table I

p	2			14			38		58	
b	6	10	16	30	50	60	80	100	120	150
problem N 1	0.166	0.179	0.186	0.192	0.194	0.196	0.196	0.196	0.196	0.197
2	0.743	0.752	0.844	0.940	0.965	0.971	0.979	0.984	0.987	0.990
3	0.311	0.386	0.409	0.425	0.432	0.434	0.436	0.437	0.438	0.439
4	0.972	0.811	0.900	0.951	0.998	0.983	0.988	0.999	0.999	0.997

8. Round plate

$$R_0 = \sqrt{p}\,\frac{c^2}{h}, \quad \lambda_1 = \lambda_2 = 0, \quad \beta = 1,$$

$$q = \frac{p^{3/2}}{4}\,\frac{q^* c^4}{E h^4}, \quad K = \frac{q^* c^4}{E h^4}$$

should be assumed for a round plate in (11-15).

It is of interest to analyse the stability loss of a plate with non-fixed-constraint contour (problem 3) under pressure. All above difficulties of the calculation process organization are observed in such an analysis.

Fig.7 shows the relationship between computer time losses due to the movement along the plate deformation curve and the accuracy of meeting boundary conditions (for computer BESM-4M, speed 20,000 operations per second). It is seen that with increasing load parameter K values it is necessary to increase the solution accuracy. For example, value K=400 cannot be achieved with accuracy $\varepsilon_2=10^{-2}$ due to the calculation divergence. Optimum calculation mode corresponds to the increase in the problem solution accuracy with growing K.

Our investigation demonstrated that the plate stability loss occurs when load parameter K values exeed 1800, while in the case of k 12 the equilibrium form ramification does not happen. Critical load values are given in Table 2. Fig.8 shows plots of dimensionless deflections of the plate central point $V_0 = \frac{W_0}{h}$.

For a constraint plate maximum stress intensity $\sigma_0 = 220\ E\ (h/c)^2$ corresponds to the load parameter K=800. Our calculations show that the analysis of such plates at such loads should usually take into account the plastic strain.

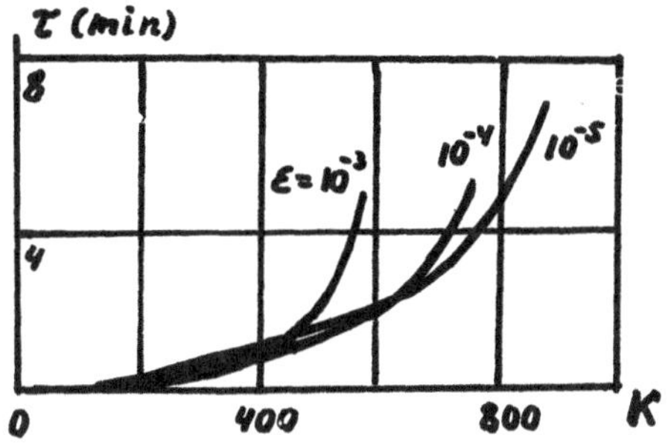

Fig.7

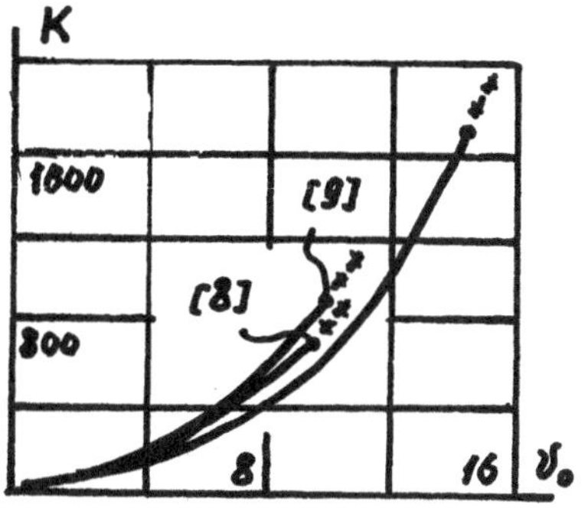

Fig.8

Table 2

k	38	89	12	13	14	15	16	17
K	715	880	1900	1820	1800	1840	1900	1980

9. Compound rotation shell

The algorithm considered allows one to carry out numerical analysis of the stressed and strained state of compound and simple multiply connected rotation shells.

Calculation results of a compound shell based on equations from [I] are quoted as an example. The meridian shape of the midsurface is set by relations

$$\psi_o' = \begin{cases} 0 & 0 \le X < X_c \\ \dfrac{X_1^2 \cos\varphi_2 \,(2X^2 - X_o^2)}{X_2^2 (X_2^2 - X_o^2)\sqrt{X^2 - X_o^2}} & X_c \le X < X_1 \\ \dfrac{X^2 \cos\varphi_2 \,(2X^2 - 3X_o^2)}{X_2^2 (X^2 - X_o^2)^{3/2}} & X_1 \le X < X_2 \end{cases}$$

where X_o, X_1, X_2, φ_2 are parameters defining the geometry of the special shell [10] which comes into contact with spherical shell in the zenith region (Fig.9).

Fig.10 shows the dependence of the shell central point W_o on the load parameter q for different conditions of mounting of contour of support.

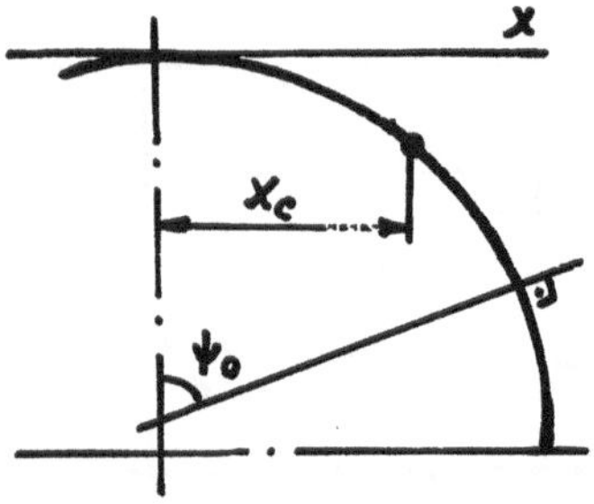

Fig. 9

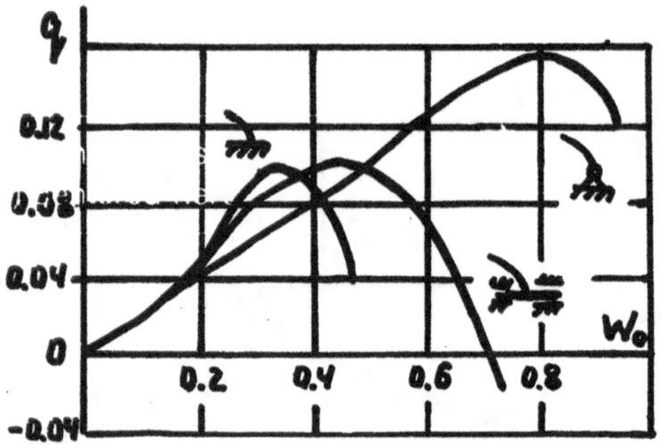

Fig. 10

References

1. VALISHVILI N.V., Calculation methods for rotation shells on a digital computer (in Russian), Moscow, Mashinostroentie, 1976.
2. KELLER H.B., Numerical methods for the two-points boundary value problems. Waltham, Mass. Toronto, London, Blaisdell Publ.Comp., 1968.
3. VOROVICH I.I.,Zipalova V.F., On the solution of the non-linear boundary value problems of the elasticity theory by transition to the Cauchy problem (in Russian), Prikl. Mat. Mekh. 29(1965),5
4. GRIGOLYUK E.I.,SHALASHILIN V.I., Parameter continuation method in problems of non-linear deformation of rods, plates and shells (in Russian), Uspekhi mehaniki, 4 (1981), 2,
5. RICS E., The application of Newton's method to problem of elastic stability. J. Appl. Mech., (1972), 39.
6. VALISHVILI N.V., On flexible structure element analysis by numrical methods (in Russian), Izv. Vuzov, Mashinostroenie,1984,2.
7. VLASOV V.Z., General shell theory and its application in technology (in Russian), Moscow, Gostekhizdat, 1949.
8. SHIL'KRUT D.I.,GERLAKU I.D., Determination of critical loads of axis-symmetrically loaded sloping, geometrically non-linear rotation shells with account taken of non-axis-symmetrical equilibrium forms (in Russian), In: Problems of shell and rod non-linear theory, Kishinev, 1969.
9. PANOV D.Yu., FEODOS'EV V.I., On equilibrium and stability loss of sloping shells under large deflections (in Russian), Prikl. Mat. Mekh. 12 (1948), 4, 389-406.
10. ELPAT'EVSKIĬ A.N., VASIL'EV V.V., Strength of cylindrical shells from reinforced materials (in Russian), Moscow, Mashinostroenie, 1972.

ELASTIC-PLASTIC STRUCTURES UNDER VARIABLE LOADS AT SMALL STRAINS AND MODERATE ROTATIONS

D. Weichert

Institut für Mechanik, Ruhr-Universität Bochum, West-Germany.

ABSTRACT

The long-time behaviour of elastic-plastic structures under variable loads is investigated. In particular, shakedown-conditions are derived for shell-like structures undergoing moderate rotations at small strains.

INTRODUCTION

In many practical engineering problems the question if a structure under loads collapses or not due to a large number of loading cycles is more important than to determine the precise shape and the state of stress in the deformed configuration of the considered structure. In particular for elastic-plastic material behaviour, failure of the structure may occur due to the unlimited accumulation of plastic deformations during the loading program. Whilst the application of conventional step-by-step methods in the numerical analysis of boundary-value problems is limited to a finite number of loading cycles, for this kind of problems MELAN [1] has developed a powerful tool by deriving through a statical approach a sufficient condition under which the plastic deformation in an elastic plastic body remains finite. If it can be shown that for a particular body this happens, one says that "Shakedown" occurs. Complementary to this, KOITER [2] gave a sufficient condition for failure due to "Non-Shakedown" via a kinematical approach. Extensions of these basic concepts to other classes of problems including the change of temperature, hardening, creep and dynamical effects have been made and the numerical application basing on the discrete formulation due to MAIER [3,4] is widely used in the analysis of structures. Reviews of recent developments can be found in KÖNIG and MAIER [5] and MAIER and MUNRO [6].

The great majority of these papers is based on the small displacement theory neglecting the influences of geometrical changes on the equilibrium equations, although in many practical problems the unservability of the structure may occur only when the displacements are considerable and beyond the range of the applicability of the geometrically linear theories. MAIER [3,4] took for discrete systems "second order effects" into account using linearized yield surfaces and a linear strain measure. KÖNIG [7,8] treated in several papers the problem of stabilizing and destabilizing effects of geometrical changes on the incremental collapse process and obtained a criterion for stable and unstable processes, provided that either the collapse mode remains invariant or only the beginning of the incremental process is considered.

In this paper the effect of geometrical changes on the shakedown behaviour is studied in the framework of continuum mechanics (see also WEICHERT [13, 14]). For particular situations, namely if informations about the deformation pattern of the considered structure are available, practicable methods are derived for the prediction of the shakedown behaviour.

The requirement of the additive decomposition of the strain tensor into a purely elastic and a purely plastic part restricts the range of validity of the proposed theory. Following the classification given by CASEY [10], this decomposition is justified for small total strains accompanied by moderate rotations, so that a large class of problems of shell-like structures can be dealt with by the presented methods. In particular, thin shells undergoing moderate rotations about tangents and small rotations about normals to the midsurface are considered.

I. THE THREE-DIMENSIONAL PROBLEM

I.1 Definitions

An elastic-plastic body **B** occupies at time $\tau_o=0$ in the undeformed state the volume V. The vectors $\mathbf{X} = (X^1,X^2,X^3)$ and $\mathbf{x} = (x^1,x^2,x^3)$, referred to a Cartesian coordinate system describe the positions of the particles of **B** during their motion in the initial state at $\tau_o=0$ and in an arbitrary state, respectively. The displacement-function $\mathbf{u}$, the deformation-gradient $\mathbf{F}$ and GREEN's strain tensor $\mathbf{E}$ are defined in the usual way :

$$\mathbf{u} = \mathbf{x} - \mathbf{X} \qquad \text{in } V \qquad (1.1)$$

$$\mathbf{F} = \mathbf{1} + \nabla\mathbf{u} \qquad \text{in } V \qquad (1.2)$$

$$\mathbf{E} = (1/2)\,(\mathbf{F}^T.\mathbf{F} - \mathbf{1}) \qquad \text{in } V \qquad (1.3)$$

Here, $\mathbf{1}$ denotes the unit tensor, ∇ stands for differentiation with respect to the coordinates of the undeformed configuration and the upper index "T" indicates transposed tensors. The motion of $\mathbf{B}$ is assumed to be quasistatical under the action of external agencies $\mathbf{a}(\mathbf{X},\tau)$ which consist of the surface tractions $\mathbf{f}^*$, surface displacements $\mathbf{u}^*$, acting on the disjoint parts S_F and S_K of the surface S of $\mathbf{B}$, respectively, and the volume forces $\mathbf{p}^*$. The equilibrium conditions and the boundary conditions are then defined by

$$\text{Div}\ \mathbf{t} = \mathbf{p}^* \qquad \text{in } V \qquad (1.4)$$

$$\mathbf{n}\cdot\mathbf{t} = \mathbf{f}^* \qquad \text{on } S_F \qquad (1.5)$$

$$\mathbf{u} = \mathbf{u}^* \qquad \text{on } S_K \qquad (1.6)$$

with $\mathbf{t}$ as first Piola-Kirchhoff stress tensor, $\mathbf{n}$ as outer normal vector on S and "Div" as divergence-operator. Point "•" denotes the contraction of two tensors over one index (in index-notation: $\mathbf{n}\cdot\mathbf{t} \mathrel{\hat=} n_A\ t^{Ai}$; $i,A = 1,2,3$).

I.2 Constitutive relations

It is assumed that the total strains $\mathbf{E}$ can be decomposed additively into a purely elastic part $\mathbf{E}^e$ and a purely plastic part $\mathbf{E}^p$:

$$\mathbf{E} = \mathbf{E}^e + \mathbf{E}^p \quad . \qquad (1.7)$$

For the elastic part $\mathbf{E}^e$ HOOKE's law is assumed to hold with

$$\mathbf{E}^e = \mathbf{L}:\mathbf{s} \qquad (1.8)$$

relating $\mathbf{E}^e$ to the second Piola-Kirchhoff stress tensor $\mathbf{s}$, where $\mathbf{L}$ is the constant and positive definit tensor of elastic coefficients with the known symmetries. Double point ":" stands for the contractive product of two tensors over two indices (in index-notation: $\mathbf{a}:\mathbf{b} \mathrel{\hat=} a^{AB}\ b_{AB}$; $A,B = 1,2,3$).
For the plastic part of the material behaviour we assume the existence

of a convex elastic domain C defined by the yield-condition in terms of **s** and the validity of the normality rule for the plastic strain rate $\dot{\mathbf{E}}^p$.

In the light of the 'multiplicative decomposition of the deformation-gradient **F** into a purely elastic part and a purely plastic part (see e.g. LEE [9]), the decomposition (1.7) is suitable only for particular cases of deformations within certain defined measures of approximation. A classification of such cases for the general three-dimensional theory has been given by CASEY [10]. Introducing measures of smallness either for rotations and strains he proves that for defined orders of approximation the following practically important situations fit within the concept of the additive decomposition (1.7):

(i) Small plastic deformations/Moderate elastic deformations
(ii) Small elastic deformations/Moderate plastic deformations
(iii) Small strains/Moderate rotations.

For the definitions of measures of smallness and orders of approximation used in this approach we refer to the paper of CASEY.

I.3 Extension of MELAN`S theorem

It is assumed, that the solution of a comparison problem is given differing from the original one only by the fact that the considered comparison body $\mathbf{B}^o$ behaves purely elastically. The time-dependent difference-fields between those fields describing the state of the real body **B** and those describing the state of $\mathbf{B}^o$ are denoted by $\Delta\mathbf{u}$, $\Delta\mathbf{F}$, $\Delta\mathbf{E}$ and $\Delta\mathbf{s}$ with the definitions

$$\mathbf{u} = \mathbf{u}^o + \Delta\mathbf{u} \quad , \quad \mathbf{F} = \mathbf{F}^o + \Delta\mathbf{F}$$

$$\mathbf{E} = \mathbf{E}^o + \Delta\mathbf{E} \quad , \quad \mathbf{s} = \mathbf{s}^o + \Delta\mathbf{s} \quad , \qquad (1.9)$$

where upper index "o" refers to the comparison body $\mathbf{B}^o$. Let $\mathbf{s}^+$ be a stress field that satisfies for all times τ the yield condition, so that $\mathbf{s}^+$ lies in the interior of the elastic domain C

$$\mathbf{s}^+ \in C(\mathbf{s}) \ . \qquad (1.10)$$

Here, $\mathbf{s}^+$ is defined by

$$\mathbf{s}^+ = \mathbf{s}^o + \Delta\mathbf{s}^+ \ , \qquad (1.11)$$

where $\Delta\mathbf{s}^+$ is a time-independent stress field. It will be proved under

which conditions shakedown occurs in **B** in the case that such a field Δs^+ exists.
Assuming the existence of Δs^+, the quadratic form

$$W = (1/2) < \Delta s-\Delta s^+, \Delta s-\Delta s^+ >_L = (1/2) \int_{(V)} [(\Delta s-\Delta s^+):L:(\Delta s-\Delta s^+)]dV \tag{1.12}$$

is positive from definition, as **L** is assumed to be positive definit. The rate $\dot{W}$ of W is then given by

$$\dot{W} = < \Delta s - \Delta s^+, \dot{\varepsilon} - \dot{\varepsilon}^+ > , \tag{1.13}$$

where the superposed dot denotes derivation with respect to time τ and ε and ε^+ are defined by

$$\varepsilon = L:\Delta s \quad , \varepsilon^+ = L:\Delta s^+ . \tag{1.14}$$

The strain difference ΔE, defined by the equation (1.9c) is partly due to plastic deformations causing the residual stresses Δs, which have their elastic strain-counterpart defined by equation (1.14), so that one arrives through the equations (1.6,1.8,1.14,1.15) at

$$\Delta E = E^p + \varepsilon \tag{1.15}$$

Substituting equation (1.15) into (1.9c) gives

$$E = E^o + \varepsilon + E^p . \tag{1.16}$$

From the assumption, that Δs^+ is time independent follows with equation (1.14) that $\dot{\varepsilon}^+$ is equal to zero. Substituting equation (1.16) into (1.12) delivers

$$\dot{W} = < \Delta s - \Delta s^+, \dot{E} - \dot{E}^o - \dot{E}^p > . \tag{1.17}$$

By the use of the definition (1.3), GREEN's strain tensor of the comparison body B^o is given by

$$E^o = (1/2) [(F^o)^T \cdot F^o - 1] \tag{1.18}$$

with

$$F^o = 1 + \nabla u^o . \tag{1.19}$$

Substituting (1.9c) and (1.18) into (1.17) delivers immediately

$$\dot{W} = (1/2) < \Delta s-\Delta s^{+},\dot{F}^{T}\cdot F+F^{T}\cdot\dot{F}-(\dot{F}^{o})^{T}\cdot F^{o}-(F^{o})^{T}\cdot\dot{F}^{o}-2\dot{E}^{p} >. \quad (1.20)$$

Using the symmetry of Δs and Δs^{+}, equation (1.20) can be replaced by

$$\dot{W} = < \Delta s-\Delta s^{+},F^{T}\cdot\dot{F}-(F^{o})^{T}\cdot\dot{F}^{o}- \dot{E}^{p} >. \quad (1.21)$$

With the definitions of the deformation-gradient F and the difference-field ΔF by (1.2) and (1.9c), respectively, we get

$$F^{o} = F - \nabla(\Delta u). \quad (1.22)$$

Substitution into (1.21) then delivers

$$\dot{W} = <\Delta s-\Delta s^{+}, F^{T}\cdot(\dot{F}-\dot{F}^{o})+\nabla(\Delta u)^{T}\cdot\dot{F}^{o} > - < \Delta s-\Delta s^{+}, \dot{E}^{p} >. \quad (1.23)$$

With (1.9d) and (1.11) this becomes

$$\dot{W} = < s-s^{+},F^{T}\cdot(\dot{F}-\dot{F}^{o})+\Delta(\nabla u\cdot\dot{F}^{o}) > - < s-s^{+}, \dot{E}^{p} >. \quad (1.24)$$

Relating first and second Piola-Kirchhoff stress tensors by the definitions

$$t = F \cdot s \quad , \quad t^{+} = F \cdot s^{+} \quad (1.25)$$

and restricting to only those tensor fields t^{+} which satisfy the equilibrium conditions

$$\mathrm{Div}\ t^{+} = -p^{*} \qquad \text{in } V \quad (1.26)$$

$$n \cdot t^{+} = f^{*} \qquad \text{on } S_{F} , \quad (1.27)$$

the first summand in the equation (1.24) takes the form $<t-t^{+},\dot{F}-\dot{F}^{o}>$ and vanishes as the differences of the weak forms of the equilibrium conditions (1.26,1.27) through the application of GAUSS' theorem. The remaining part of (1.24) is

$$\dot{W} = < s-s^{+}, \Delta(\nabla u)^{T}\cdot \dot{F}^{o} > - < s-s^{+}, \dot{E}^{p} > , \quad (1.28)$$

which can be split into the integrals $\dot{W}_{D}$ and $\dot{W}_{G}$:

$$\dot{W}_{G} = < s - s^{+}, \Delta(\nabla u)^{T}\cdot \dot{F}^{o} > \quad (1.29)$$

$$\dot{W}_D = -\langle s - s^+, \dot{E}^p \rangle . \tag{1.30}$$

From the convexity of C and the validity of the normality rule follows the non-positiveness of $\dot{W}_D$. As from definition W is non-negative, a time will be reached for which the dissipation process ceases and the body shakes down, provided that the integral W_G

$$W_G = \int_0^\infty \dot{W}_G(\tau)\, d\tau \tag{1.31}$$

is bounded from above. This follows immediately if time-integration is performed

$$W - W_G = -\int_0^\infty \langle s - s^+, \dot{E}^p \rangle\, d\tau \quad . \tag{1.32}$$

If the l.h.s. of (1.32) is bounded from below, then the r.h.s. cannot decrease indefinitely. This means however that plastic dissipation and by that plastic strains remains finite, what meets the used definition of shakedown.

I.4 Two particular cases allowing for practical methods

Two particular problems are investigated here:

(A) The given data of the problem allow to predict the deformation pattern of **B** within a certain range of unprecision. Namely, we assume that a range $\mathbf{U}^a$ of displacement-fields $\mathbf{u}^a$ is known, defining the range of unprecision of the prediction of the behaviour of **B** (see figure 1). Under which conditions does **B** shake down?

(B) The external agencies **a** are of special type: Up to an instant τ^R the body **B** undergoes finite and given displacements $\mathbf{u}^R$ under the loads $\mathbf{a}^R$ in such a way that **B** is in equilibrium in the known configuration V^R at the time τ^R. For times τ greater than τ^R the body is submitted to oscillating loads

$$\mathbf{a}^*(\mathbf{X},\tau) = \mathbf{a}^R(\mathbf{X}) + \mathbf{r}(\mathbf{X},\tau) \tag{1.33}$$

such that a comparison body $\mathbf{B}^C$ occupying also V^R but reacting purely elastical to the additional loads **r** would perform quasistatical oscillations with amplitudes $\mathbf{u}^r$ small in comparison with $\mathbf{u}^R$ (see figure 2). Under which conditions will **B** shake down now?

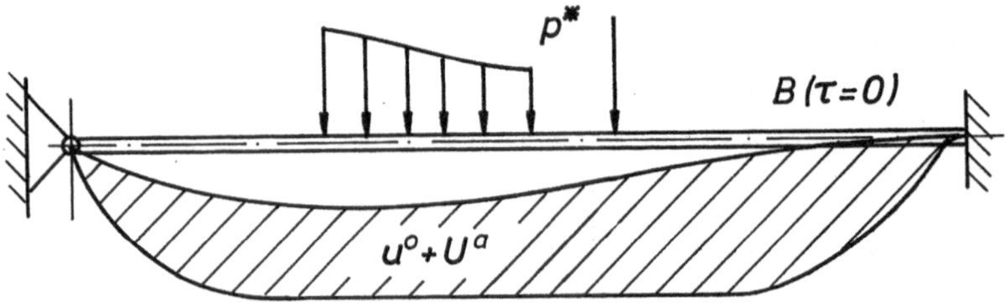

Figure 1: Statical admissibility with respect to a domain.

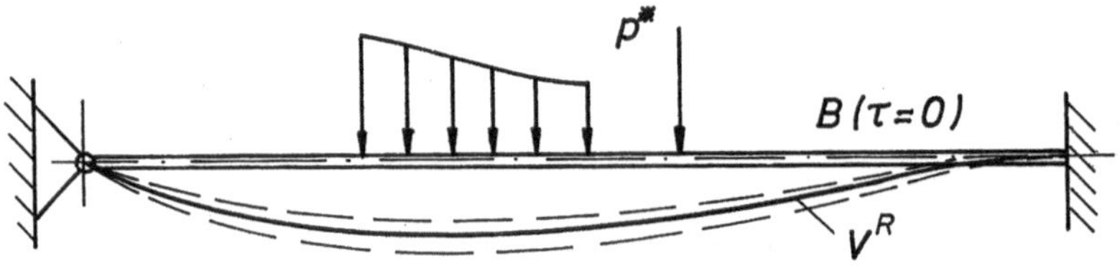

Figure 2: Statical admissibility with respect to a reference-configuration V^R

SOLUTION OF PROBLEM (A)

A class U^a of sufficiently smooth and time-independent displacement fields $\mathbf{u}^a$, dense in U^a, is assumed to be given, satisfying the homogeneous boundary conditions. For convenience it is assumed that U^a is bounded by functions $\mathbf{u}^+$ and $\mathbf{u}^-$ which are chosen according to a known or estimated deformation pattern. U^a be convex so that $\mathbf{u}^a=\mathbf{u}^-+\alpha(\mathbf{u}^+-\mathbf{u}^-)$, with $\alpha \in [0,1]$.

Then, the equilibrium conditions for **B** are given by:

$$\mathrm{Div}[(\mathbf{1}+\nabla\mathbf{u}^o+\nabla\mathbf{u}^a)\cdot\mathbf{s}] = -\mathbf{p}^* \qquad (1.34)$$

$$\mathbf{n}\cdot[(1+\nabla\mathbf{u}^{o}+\nabla\mathbf{u}^{a})\cdot\mathbf{s}] = \mathbf{f}^{*} \tag{1.35}$$

for $\mathbf{u}^a \in U^a$. The according set of deformation gradients $\mathbf{F}^a(\mathbf{u}^a)$ related to the deformation pattern described by U^a is given by:

$$\mathbf{F}^{a} = 1 + \nabla\mathbf{u}^{o} + \nabla\mathbf{u}^{a} \tag{1.36}$$

Then, stress fields $\mathbf{s}^{as}$ and $\mathbf{t}^{as}$ can be defined which are statically admissible with respect to all deformation gradients $\mathbf{F}^a(U^a)$ as those fields $\mathbf{s}^+$ which satisfy:

$$\text{Div}\,(\mathbf{F}^{a}\cdot\mathbf{s}^{+}) = -\mathbf{p}^{*} \tag{1.37}$$

$$\mathbf{n}\cdot(\mathbf{F}^{a}\cdot\mathbf{s}^{+}) = \mathbf{f}^{*}\ , \tag{1.38}$$

where the definition $\mathbf{t}^a = \mathbf{F}^a.\mathbf{s}^+$ has been used. From the formulation of the problem (A) follows that at least in one instant τ of the deformation process one of the displacement fields $\mathbf{u}^a$ will describe the actual configuration of the body **B**. As a consequence, at this instant τ the equilibrium conditions are fulfilled by the actual stress field $\mathbf{s}$ and the equations (1.34,1.35) hold. Inserting $\mathbf{t}^{as}$ and $\mathbf{t}^a$, defined by

$$\mathbf{t}^{a} = \mathbf{F}^{a}\cdot\mathbf{s} \tag{1.39}$$

into the r.h.s. of (1.29), it follows by the same argument as used in equations (1.26-1.28) that

$$<\mathbf{t}^{a} - \mathbf{t}^{as},\ \dot{\mathbf{F}} - \dot{\mathbf{F}}^{o}> = 0 \tag{1.40}$$

and the equation (1.29) becomes

$$\dot{W}^{a} = <\mathbf{s} - \mathbf{s}^{as},\ (\nabla\mathbf{u}^{a})^{T}.\nabla\dot{\mathbf{u}}^{o}> - <\mathbf{s} - \mathbf{s}^{as},\ \dot{\mathbf{E}}^{p}> \tag{1.41}$$

Analogously to the equations (1.28-1.32) it follows that a time will be reached for which the dissipation process ceases if one finds for every field $\mathbf{u}^a$ from the set U^a a stress field $\mathbf{s}^+ \in C$, provided that the integral

$$W_{G}^{a} = \int_{o}^{\infty} <\mathbf{s} - \mathbf{s}^{as},(\nabla\mathbf{u}^{a})^{T}\cdot\nabla\dot{\mathbf{u}}^{o}> d\tau \tag{1.42}$$

is bounded from above. We call the construction of statically admissible functions $\mathbf{s}^+$ with respect to a set U^a of displacement fields $\mathbf{u}^a$ "LINEARIZATION WITH RESPECT TO A DOMAIN OF DEFORMATIONS".

SOLUTION OF PROBLEM (B)

This problem is a particular case of problem (A): The set $U^a+\mathbf{u}^o$ reduces here to one single displacement field $\mathbf{u}^R$ describing the reference configuration V^R. Equation (1.24) then becomes:

$$\dot{W} = < \mathbf{s} - \mathbf{s}^+, (\mathbf{F}^R + \partial\mathbf{F})^T \cdot \mathbf{F} - (\mathbf{F}^R + \partial\mathbf{F}^C)^T \cdot \dot{\mathbf{F}}^C - \dot{\mathbf{E}}^p >. \quad (1.43)$$

Here, $\partial\mathbf{F}$ denotes the change of the true deformation gradient and $\partial\mathbf{F}^C$ denotes the change of the deformation gradient of a comparison body $\mathbf{B}^C$ reacting purely elastically to the additional superimposed load $\mathbf{r}$. The influence of $\partial\mathbf{F}$ and $\partial\mathbf{F}^C$ is assumed to be neglectible in the equilibrium equations. Whereas for $\partial\mathbf{F}^C$ this is justified by the formulation of the problem (B), the neglection of $\partial\mathbf{F}$ should be interpreted as a linearization of the equilibrium equations for the body $\mathbf{B}$ in the geometrical neighbourhood of the reference configurations V^R. By this follows:

$$\dot{W} = < \mathbf{F}^R \cdot (\mathbf{s} - \mathbf{s}^+), \partial\dot{\mathbf{F}} - \partial\dot{\mathbf{F}}^o > - < \mathbf{s} - \mathbf{s}^+, \dot{\mathbf{E}}^p > . \quad (1.44)$$

The actual linearized stress $\mathbf{t}^R$, defined by

$$\mathbf{t}^R = \mathbf{F}^R \cdot \mathbf{s} \quad (1.45)$$

fulfills the linearized equilibrium conditions

$$\mathrm{Div}\, \mathbf{t}^R = -\mathbf{p}^R - \mathbf{p}^r \quad (1.46)$$

$$\mathbf{n} \cdot \mathbf{t}^R = \mathbf{f}^R + \mathbf{f}^r \quad (1.47)$$

where the index "r" refers to the superimposed load $\mathbf{r}(\mathbf{X},\tau)$. If only those fields $\mathbf{t}^{Rs}$, defined by

$$\mathbf{t}^{Rs} = \mathbf{F}^R \cdot \mathbf{s}^s \quad (1.48)$$

are admitted which satisfy the equilibrium equations, then the first summand in (1.28) is equal to zero and MELAN's argument holds again: As W is non-negative from definition, a time will be reached when plastic flow ceases, as the term $- < s-s^S, \dot{E}^p >$ is always non-positive. We identify the herein used construction of statically admissible functions s^{Rs} as a "LINEARIZATION WITH RESPECT TO A GIVEN DISPLACEMENT FIELD".

II. APPLICATION TO SHELL-LIKE STRUCTURES WITH MODERATE ROTATIONS

The case of moderate rotations and small strains fits within the concept of the presented theory (CASEY [10]). It is of particular interest for the prediction of the behaviour of plates and shells and by that will serve as an illustration of the presented method. As example thin shells with moderate rotations about tangents and small rotations about normals to the midsurface Σ are considered. Following [12], the equilibrium equations of an appropriate shell theory are

$$T^{\alpha\beta}|_\beta - b^\alpha_\beta T^\beta + p^\alpha = 0 \qquad \text{in } \Sigma$$

$$T^\beta|_\beta + b_{\alpha\beta} T^{\alpha\beta} + p_3 = 0 \qquad \text{in } \Sigma \tag{2.1}$$

$$\left.\begin{aligned} T^{\alpha\beta}\nu_\alpha\nu_\beta + \tau_t M_{t\nu} - T^*_{\nu\nu} - \tau_t H^*_{t\nu} &= 0 \\ T^{a\beta} t_\alpha \nu_\beta - \sigma_t M_{t\nu} - T^*_{t\nu} + \sigma_t H^*_{t\nu} &= 0 \\ T^\beta \nu_\beta + M_{t\nu,s} - T^*_{n\nu} - H^*_{t\nu} &= 0 \\ M_{\nu\nu} - H^*_{\nu\nu} &= 0 \end{aligned}\right\} \text{ on } \partial\Sigma_F \tag{2.2}$$

with

$$T^{\alpha\beta} = N^{\alpha\beta} - (1/2)(b^\alpha_\lambda M^{\lambda\beta} + b^\beta_\lambda M^{\lambda\alpha}) - (1/2)(b^\alpha_\lambda M^{\lambda\beta} - b^\beta_\lambda M^{\alpha\lambda})$$

$$T^\beta = M^{\alpha\beta}|_\alpha + \varphi_\alpha N^{\alpha\beta} \quad . \tag{2.3}$$

Greek indices run here from one to two, **N**, **M**, **b** and $\boldsymbol{\varphi}$ denote membrane forces, bending moments, curvature and rotation about tangents to Σ, respectively. At the boundary $\partial\Sigma_F$, where static quatities are prescribed, σ_t, τ_t denote normal curvature and geodetic torsion; lower indices ν, t, s are related to the outer normal unit vectors $\boldsymbol{\nu}$ and $\mathbf{t}$, respectively, and boundary coordinate s. Upper index "*" denotes prescribed quantities at $\partial\Sigma_F$; in particular $H^*_{\nu\nu}$, $H^*_{t\nu}$ and $T^*_{\nu\nu}$ denote prescribed bending moment and prescribed boundary forces, respectively.

For convenience, the sandwich model of the shell is chosen (DUSZEK

[11]) to describe the material behaviour and the shell is treated as locally flat. Then, HUBER-MISES yield condition is formulated in the upper and lower layer of the shell by

$$G^{+} = (\bar{n}_1 + \bar{m}_1)^2 + (\bar{n}_2 + \bar{m}_2)^2 - (\bar{n}_1 + \bar{m}_1)(\bar{n}_2 + \bar{m}_2) \leq 1$$

$$G^{-} = (\bar{n}_1 - \bar{m}_1)^2 + (\bar{n}_2 - \bar{m}_2)^2 - (\bar{n}_1 - \bar{m}_1)(\bar{n}_2 - \bar{m}_2) \leq 1, \quad (2.4)$$

where n_α and m_α are dimensionless quantities defined by

$$\bar{n}_\alpha = (N_\alpha / 2 s_o T) \; , \; \bar{m}_\alpha = (M_\alpha / 2 s_o H_o T) \; .$$

Here, N_α and M_α are the stress resultants of the principal stresses with coinciding principal directions, s_o, T and $2H_o$ are the uniaxial yield stress, thickness of the stress-bearing layers and the thickness of the sandwich-shell, respectively. Upper index "R" refers in the following to the given reference state according to problem (B). Then the set $[\mathbf{N}^R, \mathbf{M}^R, \boldsymbol{\varphi}^R]$ equilibrates the external agencies $\mathbf{a}^R$ according to equations (2.1-2.3). The additional superimposed loads $\mathbf{r}(\tau)$ cause additional membrane forces $\mathbf{N}^r$, bending moments $\mathbf{M}^r$ and the rotation $\boldsymbol{\varphi}^r$ much smaller than $\boldsymbol{\varphi}^R$ in a comparison body $\mathbf{B}^C$ reacting purely elastically on the loads $\mathbf{r}$ in the geometrical vicinity of the reference state. The equilibrium equations are then similar to (2.1-2.3) if it is taken into account that now

$$T^{\alpha\beta} = (\overset{R}{N}{}^{\alpha\beta} + \overset{r}{N}{}^{\alpha\beta}) - (1/2)[b^\alpha_\lambda(\overset{R}{M}{}^{\lambda\beta} + \overset{r}{M}{}^{\lambda\beta}) + b^\beta_\lambda(\overset{R}{M}{}^{\lambda\alpha} + \overset{r}{M}{}^{\lambda\alpha})]$$
$$- (1/2)\,[b^\alpha_\lambda(\overset{R}{M}{}^{\lambda\beta} + \overset{r}{M}{}^{\lambda\beta}) - b^\beta_\lambda(\overset{R}{M}{}^{\lambda\alpha} - \overset{r}{M}{}^{\lambda\alpha})]$$

$$T^{\beta} = (\overset{R}{M}{}^{\alpha\beta} + \overset{r}{M}{}^{\alpha\beta})|_\alpha + \overset{R}{\varphi}_\alpha (\overset{R}{N}{}^{\alpha\beta} + \overset{r}{N}{}^{\alpha\beta}) \quad (2.5)$$

$$M^{\alpha\beta} = \overset{R}{M}{}^{\alpha\beta} + \overset{r}{M}{}^{\alpha\beta} \; , \; N^{\alpha\beta} = \overset{R}{N}{}^{\alpha\beta} + \overset{r}{N}{}^{\alpha\beta} \; . \quad (2.6)$$

In these equations the additional rotation $\boldsymbol{\varphi}^r$ is dropped according to the linearization procedure described in the solution of the problem (B). In following (1.46-1.48) now a set $[\Delta\mathbf{N}, \Delta\mathbf{M}]$ has to be constructed so that the homogeneous equilibrium conditions

$$\Delta T^{\alpha\beta}|_\beta - b^\alpha_\beta \Delta T^\beta = 0 \; , \; \Delta T^\beta|_\beta + b_{\alpha\beta}\Delta T^{\alpha\beta} = 0 \quad \text{in } \Sigma \quad (2.7)$$

$$\left.\begin{aligned} \Delta T^{\alpha\beta}\nu_\alpha\nu_\beta + \tau_t \Delta M_{t\nu} &= 0 \\ \Delta T^{\alpha\beta}T_\alpha\nu_\beta - \sigma_t \Delta M_{t\nu} &= 0 \end{aligned}\right\} \text{ on } \partial\Sigma_F \quad (2.8)$$

$$\Delta T^{\beta}\nu_{\beta} + \Delta M_{t\nu,s} = 0$$

$$\Delta M_{\nu\nu} = 0 \qquad \text{on } \partial\Sigma_F$$

with

$$\Delta T^{\alpha\beta} = \Delta N^{\alpha\beta} - (1/2)(b^{\alpha}_{\lambda}\ \Delta M^{\lambda\beta} + b^{\beta}_{\lambda}\ \Delta M^{\lambda\alpha}) - (1/2)(b^{\alpha}_{\lambda}\ \Delta M^{\lambda\beta} - b^{\beta}_{\lambda}\ \Delta M^{\lambda\alpha})$$

$$\Delta T^{\beta} = \Delta M^{\alpha\beta}|_{\alpha} + \overset{R}{\varphi}\Delta N^{\alpha\beta} \qquad (2.9)$$

are fulfilled together with the condition of plastic admissibility, requiring that

$$N^{\alpha\beta} = (\overset{R}{N}{}^{\alpha\beta} + \overset{r}{N}{}^{\alpha\beta})\ k_1 + \Delta N^{\alpha\beta}$$

$$M^{\alpha\beta} = (\overset{R}{M}{}^{\alpha\beta} + \overset{r}{M}{}^{\alpha\beta})\ k_2 + \Delta M^{\alpha\beta} \qquad (2.10)$$

satisfy the yield criterion (2.4), where k_1 and k_2 denote safety factors greater than one assuring that the safe state of stress lies in the strict interior of the elastic domain.

LITERATURE

[1] MELAN, E. Theorie statisch unbestimmter Tragwerke aus ideal-plastischem Baustoff, Sitzungsbericht der Akademie der Wissenschaften, Wien, Abt. IIa, p.195 (1938).

[2] KOITER, W.T. Progress in solid mechanics (Ed. Sneddon, I.N. & Hill, R.), North-Holland, p. 167 (1960).

[3] MAIER, G A shakedown matrix theory allowing for workhardening and second-order geometric effects, Foundations in Plasticity (Ed. Cohn, M.Z. & Maier, G.) North-Holland, p.417 (1973).

[4] MAIER, G. Shakedown analysis, Structural Plasticity and Mathematical Programming (Ed. Sawczuk, A.) Pergamon Press, p.107 (1979).

[5] KÖNIG, J.A. & MAIER, G. Shakedown analysis of elastic plastic structures: A review of recent developements, Nucl. Engng. & Design, **6**, p.81 (1981).

[6] MAIER, G. & MUNRO, J. Mathematical programming application to engi neering plastic analysis, Appl. Mec. Rev., **35**,12, p.1631 (1982).

[7] KÖNIG, J.A. On stability of the incremental collapse process, Arch. Inz. Lad., XXVI,z.1, p.219 (1980).

[8] KÖNIG, J.A. On some recent developments in the shakedown theory, Advances in Mechanics **5**, 1/2, p.237 (1982).

[9] LEE, E.H. Elastic-plastic deformations at finite strains, J. Appl. Mech. **36**, p.6 (1965).

[10] CASEY, J. Approximate kinematic relations in plasticity, Int. J. Sol. Struct., **21**, 7, p.671 (1985).

[11] DUSZEK, M.K. Foundations of the non-linear plastic shell theory, Rep. **31**, Inst. of Mech., Ruhr-Univers. Bochum, (1982).

[12] SCHMIDT, R. On geometrically non-linear theories for thin shells, Flexible Shells, Theory and Applications (Ed. Axelrad, E.L. and Emmerling, F.A.)Springer-Verlag, p.76 (1984).

[13] WEICHERT, D. Shakedown at finite displacements, a note on MELAN's theorem, Mech. Res. Com., **11** (2/3), p. 127 (1983).

[14] WEICHERT, D. On the influence of geometrical nonlinearities on the shakedown of elastic-plastic structures, Int. J. of Plasticity, in print.

FINITE ROTATIONS IN THE APPROXIMATION OF SHELLS

GERALD WEMPNER
Georgia Institute of Technology
Atlanta, GA 30332

1. Introduction

In the spirit of this colloquium, our attention is focused upon finite rotations in structures and, specifically, their role in the approximation of thin shells. Here, the theory of shells is recast; the motion is decomposed into strains and rotations with no restrictions on their magnitudes. With a view toward the further approximation via finite elements, the general theory is couched in alternative forms: A potential admits variations of displacements whereas a complementary functional admits variations of displacements, stresses and strains. To admit very simple approximations, transverse-shear deformations are included. Interelement continuity is then preserved even when kinks occur in the surface.

Although the underlying theory encompasses finite strains, our interest focuses on the more prevalent circumstance wherein strains remain small. Then the geometrical nonlinearities are attributed to rotations and the consequent changes of curvature. Some explicit geometrical results are given in an earlier article [1] on the deformation of thin shells. Small strains also imply that relative rotations are small within small regions. The latter observation enables us to exploit the linear theory in the formulation of nonlinear approximations via finite elements [2,3] and even admits continuous solutions of simple nonlinear problems [4]. Specifically, small hookean elements are adequately approximated according to linear theory while the nonlinearities are incorporated in the assembly. Some illustrations are given in earlier writings [2,5].

2. Variational Theory in Three-Dimensions

A primitive functional (P) of the stress vector ($\underset{\sim}{T}^i$) and position ($\underset{\sim}{R}$) was given in an earlier article [6] for the three-dimensional body:

$$P \equiv \int_V [\underset{\sim}{T}^i \cdot \underset{\sim}{R},_i - \underset{\sim}{f} \cdot \underset{\sim}{R}]\, dv - \int_a \underset{\sim}{t} \cdot \underset{\sim}{R}\, da - \int_{a_V} \underset{\sim}{t} \cdot (\underset{\sim}{R} - \tilde{\underset{\sim}{R}})\, da \qquad (1)$$

Here $\underset{\sim}{f}$ is the body force (per unit volume $\underset{\sim}{v}$), $\underset{\sim}{t}$ is the traction on the bounding surface (per unit of initial area a) and $\underset{\sim}{R}$ is the prescribed position on a portion a_V of the boundary. Variations of position ($\underset{\sim}{R}$) are subject to continuity of the vector and derivative, and to the constraint ($\underset{\sim}{R} = \tilde{\underset{\sim}{R}}$) on a_V. Variations of the stress ($\underset{\sim}{T}^i$) are subject to the equilibrium conditions and to the prescribed values on $a_t = a - a_V$. The functional (P) is stationary for all admissible variations of the stress and displacement. If $P_V(\underset{\sim}{R})$ and $P_c(\underset{\sim}{T}^i)$ denote the potential and complementary potential [6], then

$$P = P_V + P_c \qquad (2)$$

The functional P_c is that given by Fraeijs de Veubeke [7] and Koiter [8]. It is based upon the decomposition of strain and rotation. The stress ($\underset{\sim}{T}_i$) is referred to a rigidly rotated triad of vectors. The rotation is that which carries the principal lines of strain to their orientation in the deformed state.

A forthcoming article [9] gives two-dimensional counterparts of the functionals P, P_V and P_c. The underlying concepts are the same, but the formulation differs in certain essentials: The rotation of principal lines is the natural choice in three dimensions, both mathematically and physically. However, when transverse-shear strains occur in a shell, the same rotation (of principal lines) carries the tangent vectors out of the reference surface. Consequently, a slightly different, more natural, rotation is employed to describe the motion of the shell:

The basic triad ($\underset{\sim}{a}_\alpha$, $\underset{\sim}{n}$) of the reference state consists of the tangents ($\underset{\sim}{a}_\alpha \equiv \underset{\sim}{r}^o,_\alpha$) and normal ($\underset{\sim}{n}$) to the surface. The basic triad ($\underset{\sim}{b}_\alpha$, $\underset{\sim}{N}$) of the deformed state is similar, but rotated; the vectors remain tangent ($\underset{\sim}{b}_\alpha$) and normal ($\underset{\sim}{N}$) to the deformed surface. The reference triad ($\underset{\sim}{a}_\alpha$, $\underset{\sim}{n}$) and the similar, but rotated, triad ($\underset{\sim}{b}_\alpha$, $\underset{\sim}{N}$) are depicted in Fig. 1.

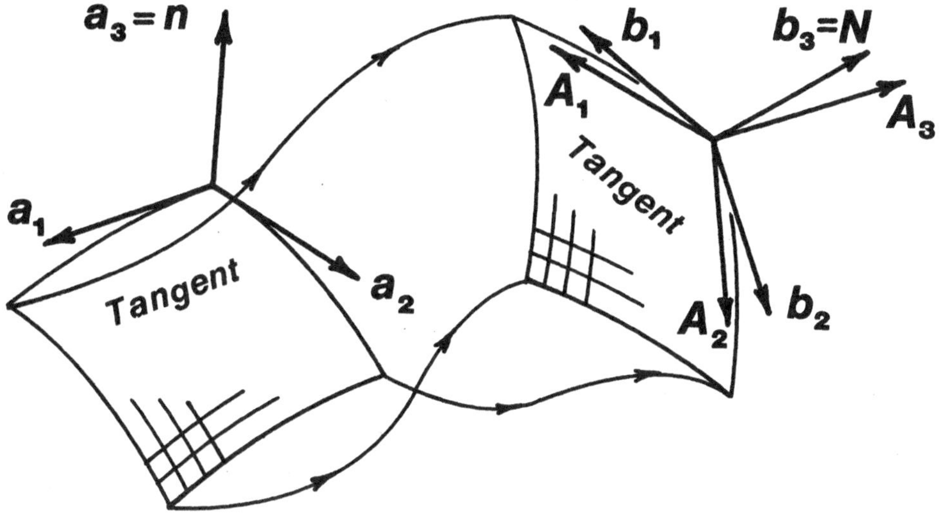

Figure 1

3. Basic Equations in Two-Dimensions

Our theory of shells is founded upon the one underlying assumption: The displacement is a linear function of distance (θ^3) from the reference surface. In the reference state and in the deformed state, the position is given by vectors $\underset{\sim}{r}$ and $\underset{\sim}{R}$, respectively;

$$\underset{\sim}{r} = \underset{\sim}{r}^o + \theta^3 \underset{\sim}{n} \quad , \quad \underset{\sim}{R} = \underset{\sim}{R}^o + \theta^3 \underset{\sim}{A}_3 \qquad (3,a,b)$$

We follow the usual conventions: θ^α denotes an arbitrary coordinate of the reference surface ($\alpha = 1,2$). $\underset{\sim}{n}$ and $\underset{\sim}{N}$ ($\neq A_3$) denote the unit normal vectors. Where possible, minuscules and majuscules signify the reference state and deformed state, respectively. $\underset{\sim}{A}_\alpha \equiv R^o,_\alpha$ is tangent to the coordinate line on the reference surface. Top and bottom surfaces lie at $\theta^3 = h_+, h_-$; s denotes the reference surface; c denotes the bounding edge. If h and k denote the mean and gaussian curvatures of the reference surface, then the differentials of volume (v) and surface (s) follow:

$$dv = \mu\ (\theta^3)\ ds \qquad (4a)$$

$$\mu \equiv 1 - h\ \theta^3 + k\ (\theta^3)^2 \qquad (4b)$$

The two-dimensional counterpart of (1) follows by integration with respect to the coordinate θ^3 through the thickness ($-h_-$ to $+h_+$):

$$P = \int_s [\underset{\sim}{N}^\alpha \cdot \underset{\sim}{R}^o_{,\alpha} - \underset{\sim}{F} \cdot \underset{\sim}{R}^o + \underset{\sim}{M}^\alpha \cdot \underset{\sim}{A}_{3,\alpha} + \underset{\sim}{T} \cdot \underset{\sim}{A}_3 - \underset{\sim}{C} \cdot \underset{\sim}{A}_3]\, ds \qquad (5)$$

$$- \int_c \underset{\sim}{N} \cdot \underset{\sim}{R}^o\, dc - \int_c \underset{\sim}{M} \cdot \underset{\sim}{A}_3\, dc$$

$$- \int_{c_v} \underset{\sim}{N} \cdot (\underset{\sim}{R}^o - \tilde{\underset{\sim}{R}})\, dc - \int_{c_v} \underset{\sim}{M} \cdot (\underset{\sim}{A}_3 - \tilde{\underset{\sim}{A}}_3)\, dc$$

As usual, the two-dimensional version is more complicated than the three-dimensional version: One displacement ($\underset{\sim}{R}$) is replaced by two ($\underset{\sim}{R}^o$ and $\underset{\sim}{A}_3$). Three stresses ($\underset{\sim}{T}^i$) are supplanted by five, which follow:

$$\underset{\sim}{N}^\alpha \equiv \int_{-h_-}^{h_+} \underset{\sim}{T}^\alpha\, \mu d\theta^3 \qquad (6a)$$

$$\underset{\sim}{M}^\alpha \equiv \int_{-h_-}^{h_+} \underset{\sim}{T}^\alpha\, \theta^3 \mu d\theta^3 \qquad (6b)$$

$$\underset{\sim}{T} \equiv \int_{-h_-}^{h_+} \underset{\sim}{T}^3\, \mu d\theta^3 \qquad (6c)$$

The external force ($\underset{\sim}{f}$) is replaced by external force ($\underset{\sim}{F}$) and couple ($\underset{\sim}{C}$); these include tractions upon the upper and lower surfaces. Force ($\underset{\sim}{N}$) and couple ($\underset{\sim}{M}$) act upon the edge (c) and are prescribed upon a portion (c_t). Displacements ($\underset{\sim}{R}^o$ and $\underset{\sim}{A}_3$) are prescribed on a portion ($c_v = c - c_t$).

Variations of the displacements ($\underset{\sim}{R}^o$ and $\underset{\sim}{A}_3$) in the functional P produce the virtual work. The equations of equilibrium follow after the usual integrations; with the notation $ds = \eta\, d\theta^1 d\theta^2$:

$$(\eta\, \underset{\sim}{N}^\alpha)_{,\alpha} + \underset{\sim}{F} = 0 \qquad \text{in } s \qquad (7a)$$

$$(\eta\, \underset{\sim}{M}^\alpha)_{,\alpha} - \underset{\sim}{T} + \underset{\sim}{C} = 0 \qquad \text{in } s \qquad (7b)$$

$$\underset{\sim}{N}^\alpha n_\alpha = \underset{\sim}{N}\ , \quad \underset{\sim}{M}^\alpha n_\alpha = \underset{\sim}{M} \qquad \text{on } c_t \qquad (8a,b)$$

Components of the stresses ($\underset{\sim}{N}^{\alpha}$, $\underset{\sim}{M}^{\alpha}$, $\underset{\sim}{T}$) are expressed in terms of the convected triad ($\underset{\sim}{b}_{\alpha}$, $\underset{\sim}{N}$):

$$N^{\alpha i} \equiv \underset{\sim}{b}^{i} \cdot \underset{\sim}{N}^{\alpha}, \quad M^{\alpha i} = \underset{\sim}{b}^{i} \cdot \underset{\sim}{M}^{\alpha} \qquad (9a,b)$$

$$T^{i} \equiv \underset{\sim}{b}^{i} \cdot \underset{\sim}{T} \qquad (9c)$$

Here, the reciprocal vectors are defined in the customary way: $\underset{\sim}{b}^{i} \cdot \underset{\sim}{b}_{j} = \delta^{i}_{j}$ and $\underset{\sim}{b}_{3} \equiv \underset{\sim}{b}^{3} \equiv \underset{\sim}{N}$.

As previously noted, our rigid rotation carries the reference system ($\underset{\sim}{a}_{\alpha}$, $\underset{\sim}{n}$) to a convected system ($\underset{\sim}{b}_{\alpha}$, $\underset{\sim}{N}$) in which the vectors are tangent ($\underset{\sim}{b}_{\alpha}$) and normal ($\underset{\sim}{N}$) to the surface. The rotation is such that the principal lines of the surface are carried to their deformed orientations; the stretch of the surface is given by a symmetric tensor:

$$C_{\alpha\beta} = \underset{\sim}{b}_{\alpha} \cdot \underset{\sim}{A}_{\beta} = \underset{\sim}{b}_{\beta} \cdot \underset{\sim}{A}_{\alpha} \quad , \qquad \underset{\sim}{A}_{\alpha} \equiv \underset{\sim}{R}^{o}_{,\alpha} \qquad (10a,b)$$

The deformation then carries the normal $\underset{\sim}{N}$ to the vector $\underset{\sim}{A}_3 \equiv \underset{\sim}{R}_{,3}$. Additional components of stretch are measures of transverse shear and extension:

$$C_{3i} = \underset{\sim}{b}_{i} \cdot \underset{\sim}{A}_{3} \qquad (10c)$$

In a primitive form, the power of the stresses is

$$\dot{W} = \underset{\sim}{N}^{\alpha} \cdot \dot{\underset{\sim}{R}}^{o}_{,\alpha} + \underset{\sim}{M}^{\alpha} \cdot \dot{\underset{\sim}{A}}_{3,\alpha} + \underset{\sim}{T} \cdot \dot{\underset{\sim}{A}}_{3}$$

An examination reveals that work is expended upon stretches C_{ij}, as defined by (10a,b,c), but also upon flexures D_{ij}, as follow

$$D_{\alpha\beta} = \underset{\sim}{b}_{\beta} \cdot \underset{\sim}{A}_{3,\alpha} \qquad (11a)$$

$$D_{\alpha 3} \equiv \underset{\sim}{N} \cdot \underset{\sim}{A}_{3,\alpha} \qquad (11b)$$

Components of the spin $\dot{\underset{\sim}{\Omega}}$ [of the triad ($\underset{\sim}{b}_{\alpha}$, $\underset{\sim}{N}$)] has components ($\dot{\Omega}^{i}$), which are also expressed by a skew-symmetric tensor ($\dot{\omega}_{ij}$):

$$\dot{\Omega}^{i} \equiv \underset{\sim}{b}^{i} \cdot \dot{\underset{\sim}{\Omega}} = \frac{1}{2} \varepsilon^{ijk} \dot{\omega}_{kj}$$

The power expended upon the spin must vanish, since inertial effects are not presently included. Three equations of equilibrium follow:

$$(N^{\alpha 3} - K^{\alpha}_{\eta} M^{\eta 3})\, C^{\mu}_{\alpha} - T^{\mu} C_{33} + T^{3} C^{\mu}_{3} - M^{\alpha\mu} D_{\alpha 3} = 0 \tag{12a}$$

$$(N^{\alpha\gamma} - K^{\alpha}_{\beta} M^{\beta\gamma})\, C^{\mu}_{\alpha} + T^{\gamma} C^{\mu}_{3} = (N^{\alpha\mu} - K^{\alpha}_{\beta} M^{\beta\mu})\, C^{\gamma}_{\alpha} + T^{\mu} C^{\gamma}_{3} \tag{12b}$$

where

$$K^{\alpha}_{\beta} \equiv - \underset{\sim}{A}^{\alpha} \cdot \underset{\sim}{A}_{3,\beta} \tag{13}$$

Extensional and flexural strains are defined to vanish in the reference state:

$$h_{\alpha\beta} \equiv C_{\alpha\beta} - a_{\alpha\beta}, \qquad h_{3\alpha} \equiv C_{3\alpha} \tag{14a,b}$$

$$k_{\alpha\beta} \equiv D_{\alpha\beta} + b_{\alpha\beta} \tag{14c}$$

$$h_{33} \equiv C_{33} - 1, \qquad k_{\alpha 3} \equiv D_{\alpha 3} \tag{14d,e}$$

Here, $a_{\alpha\beta}$ and $b_{\alpha\beta}$ are components of the metric and curvature tensors of the surface in the reference state. In terms of the components of stress (9) and strain (14) and in view of equilibrium (12), the power assumes the form:

$$\dot{W} = N^{\alpha\beta}\dot{h}_{\alpha\beta} + M^{\alpha\beta}\dot{k}_{\alpha\beta} + T^{\alpha}\dot{h}_{3\alpha} + T^{3}\dot{h}_{33} + M^{\alpha 3}\dot{k}_{\alpha 3} \tag{15}$$

<u>All</u> of the foregoing equations apply to any thin shell, elastic or inelastic, <u>if</u> the underlying approximation (3b) is adequate.

4. <u>Complementary Potentials of the Elastic Shell</u>

With the definitions of the stresses (9a, b, c) and strains (14a, b, c, d, e) and the form of the power (15), the strain energy W_V (per unit initial area s) and complementary energy W_c follow:

$$W_V = W_V(h_{\alpha\beta}, k_{\alpha\beta}, h_{3\alpha}, h_{33}, k_{\alpha 3}) \tag{16a}$$

$$W_c = W_c\,(N^{\alpha\beta}, M^{\alpha\beta}, T^{\alpha}, T^{3}, M^{\alpha 3}) \tag{16b}$$

$$W_c = (N^{\alpha\beta} h_{\alpha\beta} + M^{\alpha\beta} k_{\alpha\beta} + T^{i} h_{3i} - M^{\alpha 3} k_{\alpha 3}) - W_V \,. \tag{17}$$

The potential of the shell follows:

$$P_v = \int_S W_v \, ds - \int_S (\underset{\sim}{F} \cdot \underset{\sim}{R}^o + \underset{\sim}{C} \cdot \underset{\sim}{A}_3) \, ds \qquad (18)$$

$$- \int_{C_t} (\underset{\sim}{N} \cdot \underset{\sim}{R}^o + \underset{\sim}{M} \cdot \underset{\sim}{A}_3) \, dc$$

Since the strains are expressed in terms of displacements by (14), (10) and (11), the potential is a functional of the displacements ($\underset{\sim}{R}^o$ and $\underset{\sim}{A}_3$). The stationary conditions are the differential equations of equilibrium, versions of (7) and (8).

A complementary potential $P_c = P - P_v$ can be expressed in terms of the stress components and the rotation [9]. Another functional is perhaps more useful. It is a modification of the potential (18) analogous to the functional of Hu-Washizu [10,11]:

$$\bar{P}_v = \int_S \{W_v - N^{\alpha\beta}[h_{\alpha\beta} - \underset{\sim}{b}_\beta \cdot (\underset{\sim}{R}^o_{,\alpha} - \underset{\sim}{b}_\alpha)] \qquad (19)$$

$$- M^{\alpha\beta}[k_{\alpha\beta} - \underset{\sim}{b}_\beta \cdot (\underset{\sim}{A}_{3,\alpha} + b^\mu_\alpha \underset{\sim}{b}_\mu)]$$

$$+ N^{\alpha 3}[\underset{\sim}{N} \cdot \underset{\sim}{R}^o_{,\alpha}] - M^{\alpha 3}[k_{\alpha 3} - \underset{\sim}{N} \cdot \underset{\sim}{A}_{3,\alpha}] - T^\alpha[h_{3\alpha} - \underset{\sim}{b}_\alpha \cdot \underset{\sim}{A}_3]$$

$$- T^3[h_{33} - \underset{\sim}{N} \cdot (\underset{\sim}{A}_3 - \underset{\sim}{N})] - \underset{\sim}{F} \cdot \underset{\sim}{R}^o - \underset{\sim}{C} \cdot \underset{\sim}{A}_3\} \, ds$$

$$+ \int_{C_v} [\underset{\sim}{N} \cdot (\underset{\sim}{R}^o - \bar{\underset{\sim}{R}}^o) + \underset{\sim}{M} \cdot (\underset{\sim}{A}_3 - \bar{\underset{\sim}{A}}_3)] \, dc$$

$$- \int_{C_t} [\underset{\sim}{N} \cdot \underset{\sim}{R}^o + \underset{\sim}{M} \cdot \underset{\sim}{A}_3] \, dc$$

The latter is dependent on all variables, the displacements ($\underset{\sim}{R}^o$, $\underset{\sim}{A}_3$), strains (h_{ij}, $k_{\alpha i}$), stresses ($N^{\alpha i}$, $M^{\alpha i}$, T^i) and the rotation of the triad ($\underset{\sim}{b}_i$). The functional (19) is stationary with respect to the rotation if the stresses satisfy the equations (12 a,b). It is stationary with respect to the stresses, if the strain - displacement equations are satisfied, and with respect to the strains, if the stress-strain relations are satisfied, e.g., $h_{\alpha\beta} = \partial W_v / \partial N^{\alpha\beta}$.

5. Theory of Small Strain

Usually strains remain small in structures, though rotations become large. This circumstance has far-reaching implications. In some instances the solutions of linear problems lead readily to a description of large rotations. An example is the torsion of a long thin hookean rod [4, p. 234]; locally the deformations are described by the linear solution of St. Venant. More generally, if strains are small, the flexure and torsion of thin rods is accompanied by small relative rotations [4, pp. 355-362]; finite rotation is the accumulative effect. Similarly, the rotations and then the displacements can be determined from the strains and the curvatures of a thin shell [1].

Small strains have immediate practical consequences in the approximation of shells via finite elements. If the elements are small enough to provide a general tool for the description of the unknown configurations of the shell, then the individual element can be described by small strain and small relative rotation. In other words, the description of the element is geometrically linear. The essential nonlinearities arise in the assembly. Since our present theory is based upon the decomposition and offers the alternative variational formulations, it is particularly suited to the approximation of the small strains and finite rotations via finite elements. To that end, let us consider the simplifications which accompany the small strains:

When transverse shear and extension are small, equation (13) gives the approximation: $K^{\alpha}_{\beta} \doteq B^{\alpha}_{\beta}$, the curvature of the deformed surface.

The work expended by the transverse stress, T^3 and $M^{\alpha 3}$, upon the strains, h_{33} and $k_{\alpha 3}$, is usually neglected in thin shells. Accordingly, these components are suppressed throughout; a state of stress, and strain, is described by the remaining components $(N^{\alpha\beta}, M^{\alpha\beta}, T^{\alpha})$ and $(h_{\alpha\beta}, k_{\alpha\beta}, h_{3\alpha})$.

Small strains imply that products can be neglected in the equations (12a,b); then,

$$N^{\mu 3} \doteq T^{\mu} \tag{20a}$$

$$N^{\mu\gamma} - B^{\mu}_{\beta} M^{\beta\gamma} = N^{\gamma\mu} - B^{\gamma}_{\beta} M^{\beta\mu} \tag{20b}$$

Finally, to correlate with the classical theory [12, 13, 14, 15], the flexural strain (14c) can be replaced by another measure:

Then

$$\bar{k}^{\alpha}_{\beta} \equiv - K^{\alpha}_{\beta} + b^{\alpha}_{\beta}$$

$$\dot{k}_{\alpha\beta} = - \dot{\bar{k}}^{\mu}_{\alpha} C_{\mu\beta} - k^{\mu}_{\alpha} \dot{h}_{\mu\beta}$$

Then the power (15) assumes the form:

$$\dot{W} = (N^{\alpha\beta} - K^{\alpha}_{\gamma} M^{\gamma\beta}) \dot{h}_{\alpha\beta} - M^{\alpha\beta} C_{\mu\beta} \dot{\bar{k}}^{\mu}_{\alpha} + T^{\alpha} \dot{h}_{3\alpha} \tag{21}$$

In the case of small strain, $C_{\mu\beta} \doteq a_{\mu\beta}$, a component of the initial metric tensor. Also, according to the Kirchhoff-Love hypothesis, $K^{\alpha}_{\beta} = B^{\alpha}_{\beta}$, a component of the curvature tensor; B^{α}_{β} and k^{α}_{β} are therefore symmetrical. Then, the stresses in (21) are the symmetrical components of the classical theory:

$$n^{\alpha\beta} = N^{\alpha\beta} - B^{\alpha}_{\mu} M^{\mu\beta}, \qquad m^{\alpha\beta} = M^{\alpha\beta}$$

5. On Approximation via Finite Elements

In the approximation of the shell by finite elements, each function is approximated by discrete values. To insure convergence of an approximation founded on the potential, the displacements ($\underset{\sim}{R}^{o}$ and $\underset{\sim}{A}_3$) must be continuous [16]. In the approximations of strains, or stresses, in an alternative functional, or complementary potential, it is necessary to insure that all deformational modes are restrained [17]. Stated otherwise, each deformational mode must produce strain energy (W_V) or complementary energy (W_C).

To appreciate such approximation and correlate it to the continuous theory, consider the shell subdivided into quadrilaterals by coordinate lines. Then the simplest approximation of $\underset{\sim}{R}^{o}$ (also $\underset{\sim}{r}^{o}$, $\underset{\sim}{n}$ and $\underset{\sim}{A}_3$) is the bilinear form within each element:

$$\underset{\sim}{R}^{o} = \bar{\underset{\sim}{R}} + \underset{\sim}{R}_1 \xi^1 + \underset{\sim}{R}_2 \xi^2 + \underset{\sim}{R}_{12} \xi^1 \xi^2 \tag{22}$$

Here ξ^{α} is a local coordinate which originates at the midpoint of the element; the coordinate is normalized, such that the corners have coordinates $\xi^{\alpha} = \pm 1$.

$$\xi^\alpha = \frac{\theta^\alpha}{\Delta\theta^\alpha}$$

The opposing edges are separated by amount $(2\Delta\theta^\alpha)$. The displacement is approximated by four constants ($\bar{\underset{\sim}{R}}$, $\underset{\sim}{R}_1$, $\underset{\sim}{R}_2$ and $\underset{\sim}{R}_{12} = \underset{\sim}{R}_{21}$) within each element. The assembly is continuous, if the elements are joined at the corners (nodes). The approximation of all coordinate lines is piecewise linear; these lines remain straight and the interfaces conform.

Mathematically, the constants ($\bar{\underset{\sim}{R}}$, $\underset{\sim}{R}_1$, $\underset{\sim}{R}_2$, $\underset{\sim}{R}_{12}$) are linearly dependent on the nodal values $\underset{\sim}{R}_N$.

$$\underset{\sim}{R}^o = \sum_N f_N\,(\theta^1,\theta^2)\,\underset{\sim}{R}_N \tag{23}$$

With the approximation (22), the "shape function" f_N is nonvanishing only in the patch of four elements which are joined at node N [18]. The variation of $\underset{\sim}{R}_N$ implies a nonzero variation $f_N\dot{\underset{\sim}{R}}_N$ in that patch and the stationary equation of equilibrium follows from the functional (5):

$$\int_{sp} [\underset{\sim}{N}^\alpha f_{N,\alpha} - \underset{\sim}{F}\, f_N]\ \eta\ d\xi^1\, d\xi^2 = 0 \tag{24a}$$

Here s_p signifies the integral over the surface of the patch, the four contiguous elements. The specific form of the equation (24a) depends on the approximation of the stress $(\underset{\sim}{N}^\alpha)$ and external force $(\underset{\sim}{F})$. If the model is to be based upon the potential (P_V), then the form of the stress is implicit in the stress-strain-displacement equations. If the model is to be founded upon the modified functional $(\bar{P}_V)$, then the stress is also independently variable; however, the approximation must possess a mean value and must inhibit each of the deformational modes implicit in the approximation (22), and (23).

Now, equation (24a) is the sum of four integrals, taken over each of the four adjoining elements. Terms of each integral contain powers of the coordinates (ξ^α); such terms vanish in the limit, if only the stresses $(\underset{\sim}{N}^\alpha)$ and loads $(\underset{\sim}{F})$ are bounded. Let us identify the four quadrants (elements) of our patch by numerals, (I, II, III, IV) in the customary order and denote mean values in each quadrant by the numerical subscript. Then, the dominant terms of the equilibrium equation (24a) follow:

$$(\frac{1}{\Delta\theta^1}\left[-\frac{1}{4}(\underset{\sim}{N}^1 n)_I + \frac{1}{4}(\underset{\sim}{N}^1 n)_{II} + \frac{1}{4}(\underset{\sim}{N}^1 n)_{III} - \frac{1}{4}(\underset{\sim}{N}^1 n)_{IV}\right]$$

$$+ \frac{1}{\Delta\theta^2}\left[-\frac{1}{4}(\underset{\sim}{N}^2 n)_I - \frac{1}{4}(\underset{\sim}{N}^2 n)_{II} + \frac{1}{4}(\underset{\sim}{N}^2 n)_{III} + \frac{1}{4}(\underset{\sim}{N}^2 n)_{IV}\right]$$

$$+ \left[-\frac{1}{4}\underset{\sim}{F}_I - \frac{1}{4}\underset{\sim}{F}_{II} - \frac{1}{4}\underset{\sim}{F}_{III} - \frac{1}{4}\underset{\sim}{F}_{IV}\right]) = 0 \qquad (24b)$$

The bracketed sum must vanish. These are the dominant terms in this algebraic equation of equilibrium. In the limit, the equation approaches the differential equation (7a).

Now, the equilibrium equation (24b) involves the differences of the stress $\underset{\sim}{N}^\alpha$, as the equation (7a) involves the derivatives. In particular,

$$\frac{\Delta\underset{\sim}{N}^\alpha}{\Delta\theta^\alpha} \doteq \frac{\Delta N^{\alpha i}}{\Delta\theta^\alpha}\underset{\sim}{b}_i + N^{\alpha i}\frac{\Delta\underset{\sim}{b}_i}{\Delta\theta^\alpha}$$

Clearly, the differences in the convected vectors are essential, just as the curvatures of the deformed surface are needed in a geometrically nonlinear theory of the continuum. Note that the finite rotation can only appear in the difference or differential equation, if it enters into the description of the load $\underset{\sim}{F}$. Note also that the essential nonlinear product ($N^{\alpha\beta}\,\Delta\underset{\sim}{b}_\beta \cdot \underset{\sim}{N}$) appears. In the usual procedures of formulation, such terms are produced in the assembly, but the assembly must account for the relative rotations of the adjacent elements.

7. Summary

The foregoing theory incorporates the decomposition of strains and rotations; it is cast in alternative variational forms. Consequently, the formulation is particularly suited to the treatment of finite rotation.

The theory is founded upon the one assumption (3b): Displacement is a linear function of the distance along the normal. The inclusion of transverse shear and the alternative formulations make the theory particularly useful for approximations via finite elements.

A linear formulation of the individual elements can be adapted for the approximation of the geometrically nonlinear shell. Indeed, most practical methods of analysis and computation are based upon incremental procedures, wherein successive increments are governed by linear equations [19]. The present formulation is "tailor-made" for such procedures: The kinematical equations (10), (11) and (14) express the strains as linear functions of the displacements ($\underset{\sim}{R}^o$ and $\underset{\sim}{A}_3$) relative to the rigidly convected triad ($\underset{\sim}{b}_\alpha, \underset{\sim}{N}$).

References

1. Wempner, G., The Deformations of Thin Shells, in Developments in Theoretical Mechanics, 3, Pergamon Press, 1967; Proc. Third SECTAM Conf., pp. 245-254, 1966.

2. Wempner, G., New Concepts for Finite Elements of Shells, Z. Angew. Math. Mech., T-174-176, 1968.

3. Wempner, G., Finite Elements, Finite Rotations and Small Strains of Flexible Shells, Int. J. Solids Structures, 5, pp. 117-153, 1969.

4. Wempner, G., Mechanics of Solids, McGraw-Hill, 1978.

5. Wempner, G., and Patrick, G., Finite Deflections, Buckling and Postbuckling of an Arch, Proc.Midwest.Con., pp. 439-450 1969.

6. Wempner, G., Complementary Theorems of Solid Mechanics, in Variational Methods in the Mechanics of Solids (ed. S. Nemat-Nasser) Pergamon Press, 1980; Proc. I.U.T.A.M. Symp. N. W. Univ., pp. 127-135, 1978.

7. Fraeijs de Veubeke, B., A New Variational Principle for Finite Elastic Displacements, Int. J. Eng. Sci., 10, pp. 745-763, 1972.

8. Koiter, W. T., On the Principle of Stationary Complementary Energy in the Nonlinear Theory of Elasticity, SIAM J. Appl. Math., 25, pp. 424-434, 1973.

9. Wempner, G., A General Theory of Shells and the Complementary Functionals, (to appear), Journal of Applied Mechanics.

10. Hu, H. C., On Some Variational Principles in the Theory of Elasticity and Plasticity, Scientia Sinica, 4, 1955.

11. Washizu, K., On the Variational Principles of Elasticity and Plasticity, Tech. Report 25-18, Mass. Inst. Tech., 1955.

12. Koiter, W. T., A Consistent First Approximation in the General Theory of Thin Elastic Shells, Proc. IUTAM Symp., Delft, pp. 12-33, North-Holland, 1960.

13. Leonard, R. W., Nonlinear First-Approximation Thin Shell and Membrane Theory, Thesis, Virginia Poly. Inst., 1961.

14. Sanders, J. L., Jr., Nonlinear Theories for Thin Shells, Q. Appl. Math., 21, pp. 21-36, 1963.

15. Naghdi, P. M., The Theory of Plates and Shells, in Encyclopedia of Physics, VI a/2 (ed. by S. Flügge), Springer Verlag, pp. 425-640, 1972.

16. Johnson, M. W., and McLay, R. W., Convergence of the Finite Element Method in the Theory of Elasticity, J. Appl. Mech., 35, pp. 274-278, 1968.

17. Wempner, G., Talaslidis, D., and Hwang, C.-M., A Simple and Efficient Approximation of Shells by Quadrilateral Elements, J. Appl. Mech., 49, pp. 115-120, 1982.

18. Zienkiewicz, O. C., The Finite Element Method, McGraw-Hill, 1971.

19. Wempner, G., Discrete Approximations Related to Nonlinear Theories of Solids, Int. J. Solids Structures, 7, pp. 1581-1599, 1971.

FINITE ROTATIONS OF LINEAR ELASTIC BODIES

Cz. WOŹNIAK
University of Warsaw, Dept. of Mathematics,
Computer Science and Mechanics, Institute
of Mechanics, Pałac Kultury i Nauki,
00-901 Warszawa, Poland

1. Introduction

The purpose of the paper is to derive from the nonlinear elastodynamics the governing relations for problems in which small strains and finite rotations and/or finite displacements are involved. Such problems arise, for example, when a motion of a linear elastic body is restricted by a system of obstacles. The main result is a proof of the two following facts:

1° The dynamics of a linear elastic body subject to finite motion can be described by the equations of the rigid body dynamics coupled with the equations of the linear elastodynamics,

2° The aforementioned coupling is due to the interaction between the body and the obstacles.

Although there is no formal difficulty in introducing more general interactions between the body and the obstacles, we shall neglect for simplicity the effect of friction and we shall assume that the motion of the obstacles is known a priori.

2. Preliminaries

Let $(x,t) \in R^3 \times R$ be the inertial coordinates in the Gallilean space-time, B be the known regular region in R^3 occupied by the body in the reference configuration and $y(\cdot,t)\colon B \to R^3$ be a deformation of the body at a time instant $t \in R$. Moreover, let $T_R(X,t)$, $X \in B$, be the first Piola-Kirchhoff stress tensor, $n_R(X)$ be the unit normal vector to ∂B at X, $X \in \partial B$, let $b_R(X,t)$, $t_R(X,t)$ be densities of external forces acting at the body (defined for a.e. $X \in B$ and a.e. $X \in \partial B$, respectively) and $\varrho_R(X)$, $X \in B$, be the mass density in the reference configuration. The laws of motion then yield

$$\mathrm{Div}T_R(X,t) + b_R(X,t) = \varrho_R(X)\ddot{y}(X,t) ,$$

$$T_R(X,t)\,\nabla y^T(X,t) = \nabla y(X,t)T_R^T(X,t) \quad \text{for a.e. } X\in B, \tag{2.1}$$

$$T_R(X,t)\,n_R(X) = t_R(X,t) \quad \text{for a.e.} \quad X\in \partial B .$$

The stress relation will be given by

$$T_R(X,t) = J(X,t)\,\nabla y(X,t)g_R(E(X,t),X) \quad \text{for a.e. } X\in B , \tag{2.2}$$

where $J(X,t) = \det \nabla y(X,t)$ and

$$E(X,t) = \frac{1}{2}[\,\nabla y^T(X,t)\,\nabla y(X,t) - 1\,] . \tag{2.3}$$

The interaction between the body and its exterior will be assumed in the form

$$b_R(X,t) = \varrho_R(X)b(y(X,t),t) \quad \text{for} \quad \text{a.e.} \quad X\in B ,$$
$$\tag{2.4}$$
$$t_R(X,t) = l_R(y(X,t), \nabla y(X,t),X,t) + s_R(X,t) \text{ for a.e. } X\in\partial B,$$

where $b(x,t)$, $(x,t)\in R^3\times R$, are body forces, $l_R(y(X,t), \nabla y(X,t),X,t)$ are boundary tractions due to the loadings acting at the body (functions $b(\cdot,t)$, $l_R(\cdot,\cdot,X,t)$ are assumed to be known and sufficiently smooth) and $s_R(X,t)$ are reactions due the possible contact between the body and the obstacles. The system of obstacles restricts the motion of the body, so that its configurations are subjected to certain unilateral constraints. To describe this fact we assume that for every $(x,t)\in R^3\times R$ there is know a set $A(x,t)$ in R^3 (possible empty) of all permissible velocities of a material point which at time instant $t\in R$ occupies place $x\in R^3$. Hence the part D_t of the physical space R^3, which at time instant $t\in R$ is occupied by the system of obstacles, is given by

$$D_t := \{\, x\in R^3 \mid A(x,t) \neq \emptyset \,\} ,$$

and hence the constraints for deformations are determined by $\overline{y(B,t)}\subset R^3\setminus D_t$, $t\in R$. We shall assume that every D_t is a set of regular regions in R^3 which are bounded by the moving smooth surfaces.

Under assumptions of the general theory of constraints, [1], we have

$$s_R(X,t) \in -N_{A(y(X,t),t)}(\dot{y}(X,t)) \quad \text{for a.e.} \quad X \in \partial B, \qquad (2.5)$$

where $N_{A(x,t)}(v)$ is the normal cone to set $A(x,t)$ in R^3 at the point $v \in R^3$ (1). Formulas (2.1) - (2.5) constitute foundations of the nonlinear dynamics for problems of bodies motion of which is subjected to the unilateral constraints.

Now assume that the body under consideration can suffer small strains only (but is subjected to finite motions) and that its material properties are determined exclusively by the mass density distribution and by the tensor field of elastic moduli, both related to the known reference configuration (to region B in R^3). Then the crucial point of the problem stated in Introduction lies in the passage from formulas (2.1) - (2.5) to the relations describing the elastodynamics of a linear elastic body. One could suppose that mapping $B \times R \ni (X,t) \mapsto y(X,t) \in R^3$ in this case has to represent a certain rigid motion on which there is superimposed a time dependent field of "small" displacements $\varepsilon w(x,t)$, $x \in y(B,t)$, with $\varepsilon \to 0$. It can be seen, however, that on this way we introduce into elastodynamics the extra assumption which, for finite motions and hence for finite external forces, is not compatible with Eqs. $(2.1)_1$ and $(2.1)_3$ (2). Thus we see, that the known theory of small elastic

(1) If $A(x,t)$ is a closed and convex in R^3, then

$$N_{A(x,t)}(v) = \begin{cases} \{s \in R^3 \mid (w-v)\cdot s \leqslant 0 \text{ for every } w \in A(x,t)\} & \text{if } v \in A(x,t), \\ \emptyset \quad \text{if} \quad v \in R^3 \setminus A(x,t) . \end{cases}$$

(2) Mind, that from $\varepsilon \to 0$ we would obtain (under the known smoothness conditions concerning functions $g_R(\cdot, X)$) that also $T_R(X,t) \to 0$. Hence Eqs. $(2.1)_{1,3}$ would imply conditions $[b_R(X,t) - \varrho_R(X)\ddot{y}(X,t)] \to 0$, $t_R(X,t) \to 0$, which for finite motions (and hence for finite $b_R(X,t)$, $t_R(X,t)$) cannot be satisfied.

deformations superimposed on finite (here: rigid) motions cannot be taken as a basis for a class of problems specified in Introduction.

In order to formulate the elastodynamics of small strains and finite displacements (and/or finite rotations) let us observe that for finite motions (and hence for finite external forces $b(x,t)$, $l_R(y, \nabla y, X, t)$) also stresses have to be finite. It follows that we have to deal with such elastic materials in which "small" strains produce finite stresses. Such class of materials will constitue the object of our analysis and will be defined in the next Section.

3. Elastodynamics of nearly rigid bodies

Let δ be an arbitrary but fixed positive number and let us take into account the class of elastic materials depending on the parameter $\varepsilon \in (0, \delta)$, substituting into Eq. (2.2)

$$g_R(E,X) = \frac{1}{\varepsilon}\, g(E,X) \;, \qquad \varepsilon \in (0, \delta) \;,$$

where $g(\cdot,X)$ is, for a.e. $X \in B$, an arbitrary but fixed material function, such that

$$g(E,X) = L(X)[E] + O(\|E\|^2) \;, \quad L(X) \equiv \frac{\partial g}{\partial E}\Big|_{E=0} \;,$$

and where $L(X)$ can be interpreted as a tensor of elastic moduli. It means that we are to deal with the class of elastic materials given by

$$T_R(X,t) = \frac{1}{\varepsilon} J(X,t)\, \nabla y(X,t) \{L(X)[E(X,t)] + O(\|E(X,t)\|^2) \;, \qquad \varepsilon \in (0, \delta) \;. \tag{3.1}$$

Under assumption that $\varrho_R(\cdot)$, $g(\cdot)$, $b(\cdot)$, $l_R(\cdot)$ and $A(\cdot)$ are known and independent of ε, we shall discuss the class of problems given by formulas (2.1), (2.3), (2.4), (2.5) and (3.1), for the case in which $\varepsilon \to 0$. For any positive integer k and any continuous function $f : (0, \delta) \ni \varepsilon \mapsto f(\varepsilon) \in R^k$, such that $f(\varepsilon) \to 1$ for $\varepsilon \to 0$, we shall write $f \in \theta(\varepsilon^n)$, $n = 0,1,2,\ldots$, if and only if $f(\varepsilon)\varepsilon^{-n} \to m$ for $\varepsilon \to 0$, where: $m \neq 0; 1, m \in R^k$. We shall also write $f_1 \simeq f_2$ if and only if $f_1 - f_2 \in \theta(\varepsilon^2)$. If $f \in \theta(\varepsilon^0)$ then function f will be called finite. It has to be emphasized that from now on, all unknown functions in formulas (2.1), (2.3) - (2.5),

(3.1) (i.e. all functions apart from the mentioned above) depend also on parameter $\varepsilon \in (0,\delta)$. Since $y(X,t)$, $\ddot{y}(X,t)$, $\nabla y(X,t)$ are finite (we deal with the finite motions) then $b_R(X,t)$, $t_R(X,t)$ have to be finite. Taking into account Eqs $(2.1)_{1,3}$ we shall assume that also $T_R(X,t)$ has to be finite. Hence from Eq. (3.1) we obtain

Fact 1. For the class of materials under consideration, condition $E(X,t) \in \Theta(\varepsilon)$ has to hold for every $(X,t) \in B \times R$.

Thus Eq. (3.1) determines the class of materials in which "small" strains produce finite stresses (cf. Sec.2). In what follows bodies made of such materials will be referred to as "nearly rigid bodies".

Now define $F \equiv \nabla y(X,t)$ for an arbitrary $(X,t) \in B \times R$ and take into account the polar decomposition $F = RU$.

Proposition 1. For the nearly rigid bodies condition $\nabla R(X,t) \in \Theta(\varepsilon)$ holds for every $(X,t) \in B \times R$.

In order to prove the proposition let us take into account the known formulas (3):

$$y^i_{,\alpha\beta} = \Gamma^\gamma_{\alpha\beta}\, y^i_{,\gamma} \;, \quad \Gamma^\gamma_{\alpha\beta} = \frac{1}{2} C^{\gamma\delta}(C_{\alpha\gamma,\beta} + C_{\gamma\beta,\alpha} - C_{\alpha\beta,\gamma}), \quad C_{\alpha\beta} = \delta_{\alpha\beta} + 2E_{\alpha\beta} \;. \tag{3.2}$$

Hence $\Gamma^\gamma_{\alpha\beta} = C^{\gamma\delta}(E_{\alpha\gamma,\delta} + E_{\gamma\beta,\alpha} - E_{\alpha\beta,\gamma})$ and by virtue of $E(X,t) \in \Theta(\varepsilon)$ we have $\nabla E(X,t) \in \Theta(\varepsilon)$ and then $\Gamma(X,t) \in \Theta(\varepsilon)$. Since $\nabla y(X,t)$ are finite, then from $(3.2)_1$ it follows that $\nabla\nabla y(X,t) \in \Theta(\varepsilon)$, i.e. $\nabla F(X,t) \in \Theta(\varepsilon)$. But from $\nabla F = \nabla R U + R \nabla U$ and $U^2 = C = 1 + 2E$ we obtain $U \simeq 1+E$ and hence $\nabla F \simeq \nabla R(1+E) + R\nabla E$. Now by means of $\nabla F \in \Theta(\varepsilon)$ and $E \in \Theta(\varepsilon)$, we arrive at $\nabla R(X,t) \in \Theta(\varepsilon)$ which ends the proof. Thus we have arrived at the conclusion that for the nearly rigid bodies rotations are finite but their material gradients are "small".

From Proposition 1 we obtain the following

Corollary 1. For every $X_1, X_2 \in B$ and every $t \in R$ condition $R(X_1,t) - R(X_2,t) \in \Theta(\varepsilon)$ holds.

Corollary 2. For every nearly rigid body there exist function $R \ni t \mapsto Q(t)$, where $Q(t)$ is the rotation matrix and $\dot{Q}(t)$, $\ddot{Q}(t)$ exist for a.e. $t \in R$, such that $Q(t) - R(X,t) \in \Theta(\varepsilon)$ for every $(X,t) \in B \times R$.

---

(3) Indices $\alpha, \beta, \gamma, \delta$ are related to the material coordinates $X = (X^\alpha) \in B$ and index "i" is related here to the spatial coordinates $x = (x^i) \in R^3$; all indices run over 1,2,3 ; summation convention holds.

Mind that function $t \mapsto Q(t)$ is not uniquely determined and that rotation matrices $Q(t)$ are finite. From now on let $t \mapsto Q(t)$ be arbitrary but fixed. Now define

$$H(X,t) \equiv F(X,t) - Q(t) , \quad (X,t) \in B \times R . \tag{3.3}$$

From $F = RU \simeq R(1+E)$, $H = F - Q \simeq R + RE - Q$ and by virtue of $R - Q \in \Theta(\varepsilon)$, $RE \in \Theta(\varepsilon)$ we obtain

Fact 2. For the nearly rigid bodies condition $H(X,t) \in \Theta(\varepsilon)$ holds for every $(X,t) \in B \times R$.

Taking into account conditions $y^i{}_{,[\alpha\beta]} = 0$ (4) we obtain from (3.3) that $H^k{}_{[\alpha,\beta]}(X,t) = 0$ and hence we can assume that there exist smooth functions $u(\cdot,t) : B \to R^3$, $t \in R$, such that $H(X,t) = \nabla u(X,t)$, $(X,t) \in B \times R$. Thus Eq. (3.3) can be rewritten to the form

$$\nabla y(X,t) = Q(t) + \nabla u(X,t) , \tag{3.4}$$

and finally we obtain the following

Proposition 2. For the nearly rigid bodies the deformation function $B \times R \ni (X,t) \mapsto y(X,t) \in R^3$ has the form

$$y(X,t) = p(t) + Q(t)X + u(X,t) , \tag{3.5}$$

where $\nabla u(X,t) \in \Theta(\varepsilon)$, $Q(t)$ are finite rotation matrices and $p(t)$ are arbitrary vectors, $p(t) \in R^3$.

It has to be emphasized that all terms in formula (3.5) have to satisfy the well known regularity conditions. We shall refer $u(\cdot,t)$ to as incremental displacement fields.

Remark. In general, incremental displacement fields may be not "small", i.e. conditions $u(X,t) \in \Theta(\varepsilon)$ may not hold.

It can be easily observed that from $\nabla u(X,t) \in \Theta(\varepsilon)$ and from Eq. (3.4) we obtain

Fact 3. For the nearly rigid bodies the Lagrangian strain tensor $E(X,t)$ for every $(X,t) \in B \times R$ satisfies condition

$$E(X,t) \simeq \frac{1}{2}[Q^T(t) \nabla u(X,t) + \nabla u^T(X,t)Q(t)] . \tag{3.6}$$

Analogously, taking into account that $R = FU^{-1} \simeq F(1+E)^{-1} \simeq (Q + \nabla u)(1-E)$ and that the aforementioned relation hold, we arrive

(4) We define $y^i{}_{[\alpha,\beta]} \equiv \frac{1}{2}(y^i{}_{,\alpha\beta} - y^i{}_{,\beta\alpha})$.

at

Fact 4. For the nearly rigid bodies the rotation matrix $R(X,t)$ for every $(X,t) \in B \times R$ satisfies condition

$$R(X,t) \simeq Q(t)\{1 + \frac{1}{2}[Q^T(t)\nabla u(X,t) - \nabla u^T(X,t)Q(t)]\} . \quad (3.7)$$

Let us define $T_\varepsilon(X,t) \equiv \varepsilon\, T_R(X,t)$. Formula (3.1) now yields

$$T_\varepsilon \simeq J\, \nabla y L(X)[E] \simeq (\det(1+E))R(1+E)L(X)[E]$$

and hence from (3.6), (3.7) we obtain the following

Fact 5. For the nearly rigid bodies relation

$$\varepsilon\, T_R(X,t) \simeq Q(t)L(X)[E(X,t)] \quad (3.8)$$

holds for every $t \in R$ and every $X \in B$, for which $L(X)$ is defined.

Let us also observe that from $T_\varepsilon F^T \simeq T_\varepsilon[R(1+E)]^T \simeq T_\varepsilon R^T \simeq T_\varepsilon Q^T$ and from Eq. $(2.1)_2$ we conclude

Fact 6. For the nearly rigid bodies relation

$$\varepsilon\, T_R(X,t)Q^T(t) \simeq \varepsilon\, Q(t)T_R^T(X,t) \quad (3.9)$$

holds for every $t \in R$ and a.e. $X \in B$.

Formula (3.5) (in which $\nabla u(X,t) \in \Theta(\varepsilon)$) together with formulas (3.6), (3.8) and (3.9) constitute the basis for the formulation of elastodynamics of nearly rigid bodies. To this aid we shall replace in Eqs. (3.6), (3.8), (3.9), under assumption that ε is sufficiently small with respect to the unity, the "approximate" equalities $\simeq$ by the strict equalities. After that, putting $L_R(X) \equiv \frac{1}{\varepsilon}L(X)$ and taking into account Eqs. $(2.1)_1$, we arrive at relations

$$\begin{aligned}
&\mathrm{Div}T_R(X,t) + b_R(X,t) - \varrho_R(X)[\ddot{p}(t) + \ddot{Q}(t)X] = \varrho_R(X)\ddot{u}(X,t), \\
&T_R(X,t)Q^T(t) = Q(t)T_R^T(X,t)\,, \qquad (3.10) \\
&T_R(X,t) = Q(t)L_R(X)[E(X,t)]\,, \\
&E(X,t) = \frac{1}{2}[Q^T(t)\nabla u(X,t) + \nabla u^T(X,t)Q(t)]\,,
\end{aligned}$$

which have to hold for a.e. $X \in B$ and $t \in R$. At the same time for a.e. $X \in \partial B$ and $t \in R$ we assume that Eq. $(2.1)_3$ holds in the unchanged form

$$T_R(X,t)n_R(X) = t_R(X,t) . \qquad (3.11)$$

Let us observe that up to now functions $p(\cdot)$, $Q(\cdot)$ in the decomposition formula (3.5) are not uniquely determined. Now assume that

$$\int_B \varrho_R(X)X dV_R = 0 , \qquad dV_R \equiv dX^1dX^2dX^3 ,$$

and define

$$m \equiv \int_B \varrho_R(X)dV_R , \qquad I \equiv \int_B \varrho_R(X) X \otimes X \, dV_R .$$

To conceal the ambiguity in the determination of functions $p(\cdot)$, $Q(\cdot)$, we shall introduce the extra postulate (5)

$$\begin{aligned} m\ddot{p}(t) &= \int_B \varrho_R(X)\ddot{y}(X,t)dV_R, \\ \operatorname{ant}[\ddot{Q}(t)IQ(t)] &= \operatorname{ant}\Big[\int_B \varrho_R(X)\ddot{y}(X,t) \otimes X dV_R Q^T(t)\Big], \end{aligned} \qquad (3.12)$$

which, by means of formula (3.5), is equivalent to

$$\int_B \varrho_R(X)\ddot{u}(X,t)dV_R = 0, \quad \operatorname{ant}\Big[\int_B \varrho_R(X)\ddot{u}(X,t) \otimes X dV_R Q^T(t)\Big] = 0. \qquad (3.13)$$

The physical sense of Eqs. (3.12) and (3.13) is evident. Taking into account (3.12), from $(3.10)_{1,2}$ and (3.11), after simple calculations (6) we obtain the formulas

$$\begin{aligned} m\ddot{p}(t) &= \oint_{\partial B} t_R(X,t)dA_R + \int_B b_R(X,t)dV_R , \\ \operatorname{ant}[\ddot{Q}(t)IQ(t)] &= \operatorname{ant}\Big\{\Big[\oint_{\partial B} t_R(X,t) \otimes X dA_R + \int_B b_R(X,t) \otimes X dV_R\Big]Q^T(t)\Big\} \end{aligned} \qquad (3.14)$$

which have to hold for every $t \in R$ such that $\ddot{p}(t)$, $\ddot{Q}(t)$ exist.

---

(5) Symbol $\operatorname{ant}C$, where C is an arbitrary square matrix, stands for the antysymmetric part of C, i.e., $\operatorname{ant}C \equiv \frac{1}{2}(C - C^T)$.

(6) Eq. $(3.14)_1$ results directly from Eqs. $(3.10)_1$, (3.11) and $(3.13)_1$. In order to obtain $(3.14)_2$ we have to multiply $(3.10)_1$ by $Q(t)X$ and take into account $(3.10)_2$, (3.11) and $(3.13)_2$.

Eqs. (3.10), (3.11) and Eqs. (3.14) will be taken as the governing relations of the elastodynamics of the nearly rigid bodies. It can be seen that the obtained relations are represented by the equations of the linear elastodynamics coupled with the equations of the rigid body dynamics. It can be also easily proved that Eqs. (3.10), (3.11), (3.14) imply conditions (3.13) and hence conditions (3.12), where $y(X,t)$ is given by means of formula (3.5). The elastodynamics of the nearly rigid bodies constitute the theoretical foundations for problems in which linear elastic bodies are subjected to finite displacements and rotations. A certain class of such problems will be discussed in the next Section.

4. Problems with unilateral constraints

In order to pass to the problems in which a motion of the linear elastic body is restricted by the known system of obstacles (cf. Sec.2), we have to combine Eqs. (3.10), (3.11), (3.14) with conditions (2.4), (2.5). Moreover, instead of $(3.12)_1$ we shall introduce stronger condition

$$p(t) = \frac{1}{m} \int_B \rho_R(X) y(X,t) dV_R \tag{4.1}$$

which will be assumed to hold for every $t \in R$. Hence we conclude, by means of (3.5), that also

$$\int_B \rho_R(X) u(X,t) dV_R = 0 \tag{4.2}$$

has to hold for every $t \in R$. Since $\nabla u(X,t) \in \Theta(\varepsilon)$ (cf. Proposition 2) then formula (4.2) implies that $u(X,t) \in \Theta(\varepsilon)$ holds for every $(X,t) \in B \times R$, i.e., under condition (4.1) also incremental displacements $u(X,t)$ are "small". Setting $f_1 \sim f_2$ if and only if $f_1 - f_2 \in \Theta(\varepsilon)$ and assuming that functions $b(\cdot,t)$, $l_R(\cdot,\cdot,X,t)$ in Eqs. (2.4) are differentiable, we obtain from (3.5) and from $u(X,t) \in \Theta(\varepsilon)$ the following "approximate" equalities:

$$\begin{aligned} b_R(X,t) &\sim \rho_R(X) b(p(t) + Q(t)X,t) \quad \text{for a.e. } X \in B , \\ t_R(X,t) &\sim l_R(p(t) + Q(t)X, Q(t), X, t) + s_R(X,t) \text{ for a.e. } x \in \partial B. \end{aligned} \tag{4.3}$$

Under assumption that ε is sufficiently small with respect to the unity we shall replace the "approximate" equalities (4.3) by the strict equalities. Putting for simplicity $h_R(p(t),Q(t),X,t) \equiv l_R(p(t) + Q(t)X,Q(t),X,t)$ and taking into account formulas (2.5) and (3.5), we arrive at the following relations describing, for every $t \in R$, the interaction between the nearly rigid body and its exterior:

$$\begin{aligned} b_R(X,t) &= \varrho_R b(p(t) + Q(t)X,t) \quad \text{for a.e. } X \in B , \\ t_R(X,t) &= h_R(p(t),Q(t),X,t) + s_R(X,t), \ \text{for a.e. } X \in \partial B , \end{aligned} \tag{4.4}$$

where

$$s_R(X,t) \in -N_{A(p(t)+Q(t)X+u(X,t),t)}(\dot{p}(t)+\dot{Q}(t)X+\dot{u}(X,t)) \tag{4.5}$$

holds for a.e. $X \in \partial B$.

Summing up we see that the problems in which a motion of the linear elastic body is subjected to the unilateral constraints are governed by Eqs. (3.10), (3.11), (3.14) of the elastodynamics of the nearly rigid bodies and by Eqs. (4.4), (4.5) which specify the interaction between the body and its exterior. Substituting the RHS of Eqs. (4.4) into Eqs. (3.14) we arrive finally at the following system of governing relations:

1° Eqs. (3.14) with $t_R(X,t)$, $b_R(X,t)$ given by formulas (4.4), which are equations of the rigid body dynamics and have to be treated together with initial conditions for $p(\cdot)$, $Q(\cdot)$,

2° Eqs. (3.10), (3.11) which, for every rigid motion $t \mapsto p(t) + Q(t)X$, are equations of the linear elastodynamics (initial conditions for $u(X,\cdot)$, $X \in B$, have to be known),

3° Formula (4.5) which describes the interaction between the body and the system of obstacles.

Thus we have arrived at the equations of the rigid body dynamics coupled with the equations of the linear elastodynamics. Mind, that if for some $t \in R$ there is no contact between the body and the system of obstacles, then $A(p(t) + Q(t)X + u(X,t),t) = R^3$ for every $X \in \partial B$ and hence formula (4.5) yields $s_R(X,t) = 0$. It means that the coupling between Eqs. (3.10), (3.11) and Eqs. (3.14) combined with Eqs. (4.4), is due exclusively to the interaction with obstacles. If $\overline{y(B,t)} \subset R^3 \setminus \bar{D}_t$ for $t \in (t_o,t_1)$ (cf. Sec.2), then in the time interval (t_o,t_1) we deal with the "free" motion of the linear elastic body, the problem is uncpupled, and we can firstly determine functions $p(\cdot)$, $Q(\cdot)$ from Eqs. (3.14) combined with (4.4) (where now $s_R(X,t) = 0$) and then to solve the problem of linear elastodynamics, given by Eqs. (3.10), (3.11), provided that all initial conditions for $t = t_o$ are known.

5. Final remarks

The resulting relations (3.10), (3.11), (3.14) and (4.4), (4.5), from a physical point of view, have rather evident sense and the form which could be expected. Thus, one may suppose that the forementioned relations could be simply postulated as a certain generalization of the linear elastodynamics. However, the concept of the nearly rigid body and the line of approach applied in the note leads to the elastodynamics of small strains and finite motions directly from the general elastodynamics. Using this approach we can also obtain more general results then those obtained in the note, for example by replacing stress relation (2.2) by a certain more general or alternative constitutive relation, which holds under assumption that the strains are small. On this way we can analyse problems of different nearly rigid bodies, i.e., the bodies which can suffer exclusively small strains but finite rotations and finite displacements.

References

1. WOŹNIAK Cz., Constraints in constitutive relations of mechanics, Mech. Teor. i Stos. 22 (1984), 3-4, 323-341.

Lecture Notes in Engineering

Edited by C.A. Brebbia and S.A. Orszag

Vol. 1: J. C. F. Telles,
The Boundary Element Method
Applied to Inelastic Problems
IX, 243 pages. 1983.

Vol. 2: Bernard Amadei,
Rock Anisotropy and
the Theory of Stress Measurements
XVIII, 479 pages. 1983.

Vol. 3: Computational Aspects of
Penetration Mechanics
Proceedings of the Army Research
Office Workshop on Computational
Aspects of Penetration Mechanics
held at the Ballistic Research Laboratory
at Aberdeen Proving Ground, Maryland,
27–29 April, 1982
Edited by J. Chandra and J.E. Flaherty
VII, 221 pages. 1983.

Vol. 4: W.S. Venturini
Boundary Element Method in Geomechanics
VIII, 246 pages. 1983.

Vol. 5: Madassar Manzoor
Heat Flow Through Extended
Surface Heat Exchangers
VII, 286 pages. 1984.

Vol. 6: Myron B. Allen III
Collocation Techniques for Modeling
Compositional Flows in Oil Reservoirs
VI, 210 pages. 1984.

Vol. 7: Derek B. Ingham,
Mark A. Kelmanson
Boundary Integral Equation
Analyses of Singular, Potential,
and Biharmonic Problems
IV, 173 pages. 1984.

Vol. 8: Linda M. Abriola
Multiphase Migration of Organic
Compounds in a Porous Medium
A Mathematical Model
VIII, 232 pages. 1984.

Vol. 9: Theodore V. Hromadka II
The Complex Variable Boundary
Element Method
XI, 243 pages. 1984.

Vol. 10: C. A. Brebbia, H. Tottenham,
G. B. Warburton, J. M. Wilson, R. R. Wilson
Vibrations of Engineering Structures
VI, 300 pages. 1985.

Vol. 11: M. B. Beck
Water Quality Management:
A Review of the Development and
Application of Mathematical Models
VIII, 108 pages. 1985.

Vol. 12: G. Walker, J. R. Senft
Free Piston Stirling Engines
XIV, 286 pages. 1985.

Vol. 13: Nonlinear Dynamics
of Transcritical Flows
Proceedings of a DFVLR International
Colloquium, Bonn, Germany, March 26, 1984
VI, 203 pages. 1985.

Vol. 14: A. A. Bakr
The Boundary Integral
Equation Method in Axisymmetric
Stress Analysis Problems
XI, 213 pages. 1986.

Vol. 15: I. Kinnmark
The Shallow Water Wave
Equation: Formulation,
Analysis and Application
XXIII, 187 pages, 1986.

Vol. 16: G. J. Creus
Viscoelasticity – Basic
Theory and Applications
to Concrete Structures
VII, 161 pages. 1986.

Vol. 17: S. M. Baxter
C. L. Morfey
Angular Distribution
Analysis in Acoustics
VII, 202 pages. 1986.

Vol. 18: N. C. Markatos,
D. G. Tatchell, M. Cross, N. Rhodes
Numerical Simulation of Fluid Flow
and Heat/Mass Tranfer Processes
VIII, 482 pages. 1986.

Vol. 19: Finite Rotations
in Structural Mechanics
Proceedings of the Euromech
Colloquium 197, Jabłonna, Poland, 1985
XII, 379 pages. 1986.